CCNP and CCIE Security Core

SCOR 350-701

Official Cert Guide

OMAR SANTOS

Cisco Press
221 River St.
Hoboken, NJ 07030 USA

CCNP and CCIE Security Core SCOR 350-701 Official Cert Guide

Omar Santos

Copyright © 2020 Cisco Systems, Inc.

Published by:
Cisco Press
221 River St.
Hoboken, NJ 07030 USA

1 2020

Library of Congress Control Number: 2020901233

ISBN-10: 0-13-597197-7

ISBN-13: 978-0-13-597197-0

Warning and Disclaimer

This book is designed to provide information about the Implementing and Operating Cisco Security Core Technologies (SCOR 350-701) exam. Every effort has been made to make this book as complete and accurate as possible, but no warranty or fitness is implied. The information provided is on an "as is" basis. The author and the publisher shall have neither liability nor responsibility to any person or entity with respect to any loss or damages arising from the information contained in this book or from the use of the supplemental online content or programs accompanying it.

Special Sales

For information about buying this title in bulk quantities, or for special sales opportunities (which may include electronic versions; custom cover designs; and content particular to your business, training goals, marketing focus, or branding interests), please contact our corporate sales department at corpsales@pearsoned.com or (800) 382-3419.

For government sales inquiries, please contact governmentsales@pearsoned.com.

For questions about sales outside the U.S., please contact intlcs@pearson.com.

Trademark Acknowledgments

All terms mentioned in this book that are known to be trademarks or service marks have been appropriately capitalized. Cisco Press cannot attest to the accuracy of this information. Use of a term in this book should not be regarded as affecting the validity of any trademark or service mark.

Feedback Information

At Cisco Press, our goal is to create in-depth technical books of the highest quality and value. Each book is crafted with care and precision, undergoing rigorous development that involves the unique expertise of members from the professional technical community.

Readers' feedback is a natural continuation of this process. If you have any comments regarding how we could improve the quality of this book, or otherwise alter it to better suit your needs, you can contact us through email at feedback@ciscopress.com. Please make sure to include the book title and ISBN in your message.

We greatly appreciate your assistance.

Editor-in-Chief: Mark Taub

Alliances Manager, Cisco Press: Arezou Gol

Director, Product Manager: Brett Bartow

Managing Editor: Sandra Schroeder

Development Editor: Christopher A. Cleveland

Project Editor: Mandie Frank

Copy Editor: Bart Reed

Technical Editor: John Stuppi

Editorial Assistant: Cindy Teeters

Designer: Chuti Prasertsith

Composition: codeMantra

Indexer: Ken Johnson

Proofreader: Abigail Manheim

··|···|··
CISCO.

Americas Headquarters
Cisco Systems, Inc.
San Jose, CA

Asia Pacific Headquarters
Cisco Systems (USA) Pte. Ltd.
Singapore

Europe Headquarters
Cisco Systems International BV Amsterdam,
The Netherlands

Cisco has more than 200 offices worldwide. Addresses, phone numbers, and fax numbers are listed on the Cisco Website at **www.cisco.com/go/offices.**

Cisco and the Cisco logo are trademarks or registered trademarks of Cisco and/or its affiliates in the U.S. and other countries. To view a list of Cisco trademarks, go to this URL: www.cisco.com/go/trademarks. Third party trademarks mentioned are the property of their respective owners. The use of the word partner does not imply a partnership relationship between Cisco and any other company. (1110R)

Credits

Figure 1-1	Screenshot of The Exploit Database (Exploit-DB) © OffSec Services Limited 2020
Figure 1-2	Screenshot of Using searchsploit © OffSec Services Limited 2020
Figure 1-4	Screenshot of Ghidra Software Reverse Engineering Framework, ghidra
Figure 1-6	Screenshot of SQL injection vulnerability © Webgoat SQL Injection
Figure 3-27	Screenshot of Installing the Python requests package using pip © Python Software Foundation
Figure 3-28	Screenshot of Using the Python requests package © Python Software Foundation
Figure 3-29	Screenshot of Using curl to obtain information from an API © GitHub, Inc.
Figure 3-30	Screenshot of Using curl to obtain additional information from the Deck of Cards API © GitHub, Inc.
Figure 9-11	Screenshot of AWS Lamda © 2020, Amazon Web Services, Inc
Figure 9-14	Screenshot of Docker © 2020 Docker Inc.
Figure 9-15	Screenshot of Docker © 2020 Docker Inc.
Figure 9-16	Screenshot of Docker © 2020 Docker Inc.
Figure 9-17	Deploying your first app on Kubernetes, Google Inc.
Figure 9-19	Screenshot of The Kubernetes Authors © Google Inc.
Figure 9-20	Screenshot of The Kubernetes Authors © Google Inc.
Figure 9-21	Screenshot of The Kubernetes Authors © Google Inc.
Figure 10-2	Screenshot of macOS © Apple 2019

The International Organization for Standardization (ISO), ISO/IEC 27001:2005(en)

The International Organization for Standardization (ISO)

Malware Tunneling in IPv6, June 22, 2012. United States Department of Homeland Security

The International Organization for Standardization (ISO)

NIST Special Publication 800-61

NIST Special Publication 800-61

NIST Special Publication 800-61

NIST Special Publication 800-61

US-CERT Description Document - RFC 2350

Cybersecurity and Infrastructure Security Agency (CISA), U.S. Department of Homeland Security

NIST Special Publication 800-63B

Contents at a Glance

Contents

About the Author

Omar Santos is an active member of the security community, where he leads several industry-wide initiatives and standard bodies. His active role helps businesses, academic institutions, state and local law enforcement agencies, and other participants dedicated to increasing the security of the critical infrastructure.

Omar is the author of more than 20 books and video courses as well as numerous white papers, articles, and security configuration guidelines and best practices. Omar is a Principal Engineer of the Cisco Product Security Incident Response Team (PSIRT), where he mentors and leads engineers and incident managers during the investigation and resolution of security vulnerabilities.

Omar has been quoted by numerous media outlets, such as TheRegister, Wired, ZDNet, ThreatPost, CyberScoop, TechCrunch, Fortune Magazine, Ars Technica, and more. You can follow Omar on Twitter @santosomar.

About the Technical Reviewer

John Stuppi, CCIE No. 11154, is a Technical Leader in the Customer Experience Security Programs (CXSP) organization at Cisco where he consults with Cisco customers on protecting their networks against existing and emerging cyber security threats, risks, and vulnerabilities. Current projects include working with newly acquired entities to integrate them into the Cisco PSIRT Vulnerability Management processes. John has presented multiple times on various network security topics at Cisco Live, Black Hat, as well as other customer-facing cyber security conferences. John is also the co-author of the *Official Certification Guide for CCNA Security 210-260* published by Cisco Press. Additionally, John has contributed to the Cisco Security Portal through the publication of white papers, Security Blog posts, and Cyber Risk Report articles. Prior to joining Cisco, John worked as a network engineer for JPMorgan, and then as a network security engineer at Time, Inc., with both positions based in New York City. John is also a CISSP (No. 25525) and holds AWS Cloud Practitioner and Information Systems Security (INFOSEC) Professional Certifications. In addition, John has a BSEE from Lehigh University and an MBA from Rutgers University. John lives in Ocean Township, New Jersey (down on the "Jersey Shore") with his wife, two kids, and his dog.

Dedication

I would like to dedicate this book to my lovely wife, Jeannette, and my two beautiful children, Hannah and Derek, who have inspired and supported me throughout the development of this book.

Acknowledgments

I would like to thank the technical editor and my good friend, John Stuppi, for his time and technical expertise.

I would like to thank the Cisco Press team, especially James Manly and Christopher Cleveland, for their patience, guidance, and consideration.

Finally, I would like to thank Cisco and the Cisco Product Security Incident Response Team (PSIRT), Security Research, and Operations for enabling me to constantly learn and achieve many goals throughout all these years.

Introduction

The Implementing and Operating Cisco Security Core Technologies (SCOR 350-701) exam is the required "core" exam for the CCNP Security and CCIE Security certifications. If you pass the SCOR 350-701 exam, you also obtain the Cisco Certified Specialist – Security Core Certification. This exam covers core security technologies, including cybersecurity fundamentals, network security, cloud security, identity management, secure network access, endpoint protection and detection, and visibility and enforcement.

The Implementing and Operating Cisco Security Core Technologies (SCOR 350-701) is a 120-minute exam.

> **TIP**   You can review the exam blueprint from Cisco's website at https://learningnetwork. cisco.com/community/certifications/ccnp-security/scor/exam-topics.

This book gives you the foundation and covers the topics necessary to start your CCNP Security or CCIE Security journey.

The CCNP Security Certification

The CCNP Security certification is one of the industry's most respected certifications. In order for you to earn the CCNP Security certification, you must pass two exams: the SCOR exam covered in this book (which covers core security technologies) and one security concentration exam of your choice, so you can customize your certification to your technical area of focus.

> **TIP**   The SCOR core exam is also the qualifying exam for the CCIE Security certification. Passing this exam is the first step toward earning both of these certifications.

The following are the CCNP Security concentration exams:

- Securing Networks with Cisco Firepower (SNCF 300-710)
- Implementing and Configuring Cisco Identity Services Engine (SISE 300-715)
- Securing Email with Cisco Email Security Appliance (SESA 300-720)
- Securing the Web with Cisco Web Security Appliance (SWSA 300-725)
- Implementing Secure Solutions with Virtual Private Networks (SVPN 300-730)
- Automating Cisco Security Solutions (SAUTO 300-735)

TIP   CCNP Security now includes automation and programmability to help you scale your security infrastructure. If you pass the Developing Applications Using Cisco Core Platforms and APIs v1.0 (DEVCOR 350-901) exam, the SCOR exam, and the Automating Cisco Security Solutions (SAUTO 300-735) exam, you will achieve the CCNP Security and DevNet Professional certifications with only three exams. Every exam earns an individual Specialist certification, allowing you to get recognized for each of your accomplishments, instead of waiting until you pass all the exams.

There are no formal prerequisites for CCNP Security. In other words, you do not have to pass the CCNA Security or any other certifications in order to take CCNP-level exams. The same goes for the CCIE exams. On the other hand, CCNP candidates often have three to five years of experience in IT and cybersecurity.

Cisco considers ideal candidates to be those that possess the following:

■ Knowledge of implementing and operating core security technologies

■ Understanding of cloud security

■ Hands-on experience with next-generation firewalls, intrusion prevention systems (IPSs), and other network infrastructure devices

■ Understanding of content security, endpoint protection and detection, and secure network access, visibility, and enforcement

■ Understanding of cybersecurity concepts with hands-on experience in implementing security controls

The CCIE Security Certification

The CCIE Security certification is one of the most admired and elite certifications in the industry. The CCIE Security program prepares you to be a recognized technical leader. In order to earn the CCIE Security certification, you must pass the SCOR 350-701 exam and an 8-hour, hands-on lab exam. The lab exam covers very complex network security scenarios. These scenarios range from designing through deploying, operating, and optimizing security solutions.

Cisco considers ideal candidates to be those who possess the following:

■ Extensive hands-on experience with Cisco's security portfolio

■ Experience deploying Cisco's next-generation firewalls and next-generation IPS devices

■ Deep understanding of secure connectivity and segmentation solutions

■ Hands-on experience with infrastructure device hardening and infrastructure security

■ Configuring and troubleshooting identity management, information exchange, and access control

■ Deep understanding of advanced threat protection and content security

The Exam Objectives (Domains)

The Implementing and Operating Cisco Security Core Technologies (SCOR 350-701) exam is broken down into six major domains. The contents of this book cover each of the domains and the subtopics included in them, as illustrated in the following descriptions.

The following table breaks down each of the domains represented in the exam.

Domain	Percentage of Representation in Exam
1: Security Concepts	25%
2: Network Security	20%
3: Securing the Cloud	15%
4: Content Security	15%
5: Endpoint Protection and Detection	10%
6: Secure Network Access, Visibility, and Enforcement	15%
	Total 100%

Here are the details of each domain:

Domain 1: Monitoring and Reporting: This domain is covered in Chapters 1, 2, 3, and 8.

- 1.1 Explain common threats against on-premises and cloud environments

    - 1.1.a On-premises: viruses, trojans, DoS/DDoS attacks, phishing, rootkits, man-in-the-middle attacks, SQL injection, cross-site scripting, malware

    - 1.1.b Cloud: data breaches, insecure APIs, DoS/DDoS, compromised credentials

- 1.2 Compare common security vulnerabilities such as software bugs, weak and/or hardcoded passwords, SQL injection, missing encryption, buffer overflow, path traversal, cross-site scripting/forgery

- 1.3 Describe functions of the cryptography components such as hashing, encryption, PKI, SSL, IPsec, NAT-T IPv4 for IPsec, pre-shared key, and certificate-based authorization

- 1.4 Compare site-to-site VPN and remote access VPN deployment types such as sVTI, IPsec, Cryptomap, DMVPN, FLEXVPN, including high availability considerations, and AnyConnect

- 1.5 Describe security intelligence authoring, sharing, and consumption

- 1.6 Explain the role of the endpoint in protecting humans from phishing and social engineering attacks

- 1.7 Explain northbound and southbound APIs in the SDN architecture

- 1.8 Explain DNAC APIs for network provisioning, optimization, monitoring, and troubleshooting

- 1.9 Interpret basic Python scripts used to call Cisco Security appliances APIs

Domain 2: Network Security: This domain is covered primarily in Chapters 5, 6, and 7.

2.1   Compare network security solutions that provide intrusion prevention and firewall capabilities

2.2   Describe deployment models of network security solutions and architectures that provide intrusion prevention and firewall capabilities

2.3   Describe the components, capabilities, and benefits of NetFlow and Flexible NetFlow records

2.4   Configure and verify network infrastructure security methods (router, switch, wireless)

2.4.a   Layer 2 methods (network segmentation using VLANs and VRF-lite; Layer 2 and port security; DHCP snooping; dynamic ARP inspection; storm control; PVLANs to segregate network traffic; and defenses against MAC, ARP, VLAN hopping, STP, and DHCP rogue attacks)

2.4.b   Device hardening of network infrastructure security devices (control plane, data plane, management plane, and routing protocol security)

2.5   Implement segmentation, access control policies, AVC, URL filtering, and malware protection

2.6   Implement management options for network security solutions such as intrusion prevention and perimeter security (single vs. multidevice manager, in-band vs. out-of-band, CDP, DNS, SCP, SFTP, and DHCP security and risks)

2.7   Configure AAA for device and network access (authentication and authorization, TACACS+, RADIUS and RADIUS flows, accounting, and dACL)

2.8   Configure secure network management of perimeter security and infrastructure devices (secure device management, SNMPv3, views, groups, users, authentication, encryption, secure logging, and NTP with authentication)

2.9   Configure and verify site-to-site VPN and remote access VPN

2.9.a   Site-to-site VPN utilizing Cisco routers and IOS

2.9.b   Remote access VPN using Cisco AnyConnect Secure Mobility client

2.9.c   Debug commands to view IPsec tunnel establishment and troubleshooting

Domain 3: Securing the Cloud: This domain is covered primarily in Chapter 9.

3.1   Identify security solutions for cloud environments

3.1.a   Public, private, hybrid, and community clouds

3.1.b   Cloud service models: SaaS, PaaS, and IaaS (NIST 800-145)

3.2   Compare the customer vs. provider security responsibility for the different cloud service models

3.2.a   Patch management in the cloud

3.2.b   Security assessment in the cloud

3.2.c Cloud-delivered security solutions such as firewall, management, proxy, security intelligence, and CASB

3.3 Describe the concept of DevSecOps (CI/CD pipeline, container orchestration, and security)

3.4 Implement application and data security in cloud environments

3.5 Identify security capabilities, deployment models, and policy management to secure the cloud

3.6 Configure cloud logging and monitoring methodologies

3.7 Describe application and workload security concepts

Domain 4: Content Security: This domain is covered primarily in Chapter 10.

4.1 Implement traffic redirection and capture methods

4.2 Describe web proxy identity and authentication, including transparent user identification

4.3 Compare the components, capabilities, and benefits of local and cloud-based email and web solutions (ESA, CES, WSA)

4.4 Configure and verify web and email security deployment methods to protect on-premises and remote users (inbound and outbound controls and policy management)

4.5 Configure and verify email security features such as SPAM filtering, antimalware filtering, DLP, blacklisting, and email encryption

4.6 Configure and verify secure Internet gateway and web security features such as blacklisting, URL filtering, malware scanning, URL categorization, web application filtering, and TLS decryption

4.7 Describe the components, capabilities, and benefits of Cisco Umbrella

4.8 Configure and verify web security controls on Cisco Umbrella (identities, URL content settings, destination lists, and reporting)

Domain 5: Endpoint Protection and Detection: This domain is covered primarily in Chapter 11.

5.1 Compare Endpoint Protection Platforms (EPPs) and Endpoint Detection & Response (EDR) solutions

5.2 Explain antimalware, retrospective security, Indication of Compromise (IOC), antivirus, dynamic file analysis, and endpoint-sourced telemetry

5.3 Configure and verify outbreak control and quarantines to limit infection

5.4 Describe justifications for endpoint-based security

5.5 Describe the value of endpoint device management and asset inventory such as MDM

5.6 Describe the uses and importance of a multifactor authentication (MFA) strategy

5.7 Describe endpoint posture assessment solutions to ensure endpoint security

5.8 Explain the importance of an endpoint patching strategy

Domain 6: Secure Network Access, Visibility, and Enforcement: This domain is covered primarily in Chapters 4 and 5.

6.1 Describe identity management and secure network access concepts such as guest services, profiling, posture assessment, and BYOD

6.2 Configure and verify network access device functionality such as 802.1X, MAB, and WebAuth

6.3 Describe network access with CoA

6.4 Describe the benefits of device compliance and application control

6.5 Explain exfiltration techniques (DNS tunneling, HTTPS, email, FTP/SSH/SCP/ SFTP, ICMP, Messenger, IRC, and NTP)

6.6 Describe the benefits of network telemetry

6.7 Describe the components, capabilities, and benefits of these security products and solutions:

   6.7.a Cisco Stealthwatch

   6.7.b Cisco Stealthwatch Cloud

   6.7.c Cisco pxGrid

   6.7.d Cisco Umbrella Investigate

   6.7.e Cisco Cognitive Threat Analytics

   6.7.f Cisco Encrypted Traffic Analytics

   6.7.g Cisco AnyConnect Network Visibility Module (NVM)

Steps to Pass the SCOR Exam

There are no prerequisites for the SCOR exam. However, students must have an understanding of networking and cybersecurity concepts.

Signing Up for the Exam

The steps required to sign up for the SCOR exam as follows:

1. Create an account at https://home.pearsonvue.com/cisco.

2. Complete the Examination Agreement, attesting to the truth of your assertions regarding professional experience and legally committing to the adherence of the testing policies.

3. Submit the examination fee.

Facts About the Exam

The exam is a computer-based test. The exam consists of multiple-choice questions only. You must bring a government-issued identification card. No other forms of ID will be accepted.

TIP   Refer to the Cisco Certification site at https://cisco.com/go/certifications for more information regarding this, and other, Cisco certifications.

About the CCNP and CCIE Security Core SCOR 350-701 Official Cert Guide

This book maps directly to the topic areas of the SCOR exam and uses a number of features to help you understand the topics and prepare for the exam.

Objectives and Methods

This book uses several key methodologies to help you discover the exam topics that need more review, to help you fully understand and remember those details, and to help you prove to yourself that you have retained your knowledge of those topics. This book does not try to help you pass the exam only by memorization; it seeks to help you to truly learn and understand the topics. This book is designed to help you pass the Implementing and Operating Cisco Security Core Technologies (SCOR 350-701) exam by using the following methods:

- Helping you discover which exam topics you have not mastered

- Providing explanations and information to fill in your knowledge gaps

- Supplying exercises that enhance your ability to recall and deduce the answers to test questions

- Providing practice exercises on the topics and the testing process via test questions on the companion website

Book Features

To help you customize your study time using this book, the core chapters have several features that help you make the best use of your time:

- **Foundation Topics:** These are the core sections of each chapter. They explain the concepts for the topics in that chapter.

- **Exam Preparation Tasks:** After the "Foundation Topics" section of each chapter, the "Exam Preparation Tasks" section lists a series of study activities that you should do at the end of the chapter:

  - **Review All Key Topics:** The Key Topic icon appears next to the most important items in the "Foundation Topics" section of the chapter. The Review All Key Topics activity lists the key topics from the chapter, along with their page numbers. Although the contents of the entire chapter could be on the exam, you should definitely know the information listed in each key topic, so you should review these.

- **Define Key Terms:** Although the Implementing and Operating Cisco Security Core Technologies (SCOR 350-701) exam may be unlikely to ask a question such as "Define this term," the exam does require that you learn and know a lot of cybersecurity terminology. This section lists the most important terms from the chapter, asking you to write a short definition and compare your answer to the glossary at the end of the book.

- **Review Questions:** Confirm that you understand the content you just covered by answering these questions and reading the answer explanations.

- **Web-based practice exam:** The companion website includes the Pearson Cert Practice Test engine, which allows you to take practice exam questions. Use it to prepare with a sample exam and to pinpoint topics where you need more study.

How This Book Is Organized

This book contains 11 core chapters—Chapters 1 through 11. Chapter 12 includes preparation tips and suggestions for how to approach the exam. Each core chapter covers a subset of the topics on the Implementing and Operating Cisco Security Core Technologies (SCOR 350-701) exam. The core chapters map to the SCOR topic areas and cover the concepts and technologies you will encounter on the exam.

The Companion Website for Online Content Review

All the electronic review elements, as well as other electronic components of the book, exist on this book's companion website.

To access the companion website, which gives you access to the electronic content with this book, start by establishing a login at www.ciscopress.com and registering your book.

To do so, simply go to www.ciscopress.com/register and enter the ISBN of the print book: 9780135971970. After you have registered your book, go to your account page and click the **Registered Products** tab. From there, click the **Access Bonus Content** link to get access to the book's companion website.

Note that if you buy the *Premium Edition eBook and Practice Test* version of this book from Cisco Press, your book will automatically be registered on your account page. Simply go to your account page, click the **Registered Products** tab, and select **Access Bonus Content** to access the book's companion website.

Please note that many of our companion content files can be very large, especially image and video files.

If you are unable to locate the files for this title by following the steps at left, please visit www.pearsonITcertification.com/contact and select the **Site Problems/Comments** option. Our customer service representatives will assist you.

How to Access the Pearson Test Prep (PTP) App

You have two options for installing and using the Pearson Test Prep application: a web app and a desktop app. To use the Pearson Test Prep application, start by finding the registration code that comes with the book. You can find the code in these ways:

- Print book: Look in the cardboard sleeve in the back of the book for a piece of paper with your book's unique PTP code.

- Premium Edition: If you purchased the Premium Edition eBook and Practice Test directly from the Cisco Press website, the code will be populated on your account page after purchase. Just log in at www.ciscopress.com, click **account** to see details of your account, and click the **digital purchases** tab.

- Amazon Kindle: For those who purchased a Kindle edition from Amazon, the access code will be supplied directly from Amazon.

- Other bookseller e-books: Note that if you purchase an e-book version from any other source, the practice test is not included because other vendors to date have not chosen to provide the required unique access code.

NOTE   Do not lose the activation code because it is the only means with which you can access the QA content for the book.

Once you have the access code, to find instructions about both the PTP web app and the desktop app, follow these steps:

Step 1.   Open this book's companion website, as was shown earlier in this Introduction under the heading "How to Access the Companion Website."

Step 2.   Click the **Practice Exams** button.

Step 3.   Follow the instructions listed there both for installing the desktop app and for using the web app.

Note that if you want to use the web app only at this point, just navigate to www.pearsontestprep.com, establish a free login if you do not already have one, and register this book's practice tests using the registration code you just found. The process should take only a couple of minutes.

NOTE   Amazon eBook (Kindle) customers: It is easy to miss Amazon's email that lists your PTP access code. Soon after you purchase the Kindle eBook, Amazon should send an email. However, the email uses very generic text and makes no specific mention of PTP or practice exams. To find your code, read every email from Amazon after you purchase the book. Also, do the usual checks for ensuring your email arrives, such as checking your spam folder.

NOTE Other eBook customers: As of the time of publication, only the publisher and Amazon supply PTP access codes when you purchase their eBook editions of this book.

Customizing Your Exams

Once you are in the exam settings screen, you can choose to take exams in one of three modes:

- **Study mode:** Allows you to fully customize your exams and review answers as you are taking the exam. This is typically the mode you would use first to assess your knowledge and identify information gaps.

- **Practice Exam mode:** Locks certain customization options, as it is presenting a realistic exam experience. Use this mode when you are preparing to test your exam readiness.

- **Flash Card mode:** Strips out the answers and presents you with only the question stem. This mode is great for late-stage preparation when you really want to challenge yourself to provide answers without the benefit of seeing multiple-choice options. This mode does not provide the detailed score reports that the other two modes do, so you should not use it if you are trying to identify knowledge gaps.

In addition to these three modes, you will be able to select the source of your questions. You can choose to take exams that cover all of the chapters or you can narrow your selection to just a single chapter or the chapters that make up specific parts in the book. All chapters are selected by default. If you want to narrow your focus to individual chapters, simply deselect all the chapters and then select only those on which you wish to focus in the Objectives area.

You can also select the exam banks on which to focus. Each exam bank comes complete with a full exam of questions that cover topics in every chapter. The two exams printed in the book are available to you as well as two additional exams of unique questions. You can have the test engine serve up exams from all four banks or just from one individual bank by selecting the desired banks in the exam bank area.

There are several other customizations you can make to your exam from the exam settings screen, such as the time of the exam, the number of questions served up, whether to randomize questions and answers, whether to show the number of correct answers for multiple-answer questions, and whether to serve up only specific types of questions. You can also create custom test banks by selecting only questions that you have marked or questions on which you have added notes.

Updating Your Exams

If you are using the online version of the Pearson Test Prep software, you should always have access to the latest version of the software as well as the exam data. If you are using the Windows desktop version, every time you launch the software while connected to the Internet, it checks if there are any updates to your exam data and automatically downloads any changes that were made since the last time you used the software.

Sometimes, due to many factors, the exam data may not fully download when you activate your exam. If you find that figures or exhibits are missing, you may need to manually update your exams. To update a particular exam you already activated and downloaded, simply click the **Tools** tab and click the **Update Products** button. Again, this is only an issue with the desktop Windows application.

If you wish to check for updates to the Pearson Test Prep exam engine software, Windows desktop version, simply click the **Tools** tab and click the **Update Application** button. This ensures that you are running the latest version of the software engine.

CHAPTER 1

Cybersecurity Fundamentals

This chapter covers the following topics:

Introduction to Cybersecurity: Cybersecurity programs recognize that organizations must be vigilant, resilient, and ready to protect and defend every ingress and egress connection as well as organizational data wherever it is stored, transmitted, or processed. In this chapter, you will learn concepts of cybersecurity and information security.

Defining What Are Threats, Vulnerabilities, and Exploits: Describe the difference between cybersecurity threats, vulnerabilities, and exploits.

Exploring Common Threats: Describe and understand the most common cybersecurity threats.

Common Software and Hardware Vulnerabilities: Describe and understand the most common software and hardware vulnerabilities.

Confidentiality, Integrity, and Availability: The CIA triad is a concept that was created to define security policies to protect assets. The idea is that confidentiality, integrity and availability should be guaranteed in any system that is considered secured.

Cloud Security Threats: Learn about different cloud security threats and how cloud computing has changed traditional IT and is introducing several security challenges and benefits at the same time.

IoT Security Threats: The proliferation of connected devices is introducing major cybersecurity risks in today's environment.

An Introduction to Digital Forensics and Incident Response: You will learn the concepts of digital forensics and incident response (DFIR) and cybersecurity operations.

This chapter starts by introducing you to different cybersecurity concepts that are foundational for any individual starting a career in cybersecurity or network security. You will learn the difference between cybersecurity threats, vulnerabilities, and exploits. You will also explore the most common cybersecurity threats, as well as common software and hardware vulnerabilities. You will learn the details about the CIA triad—confidentiality, integrity, and availability. In this chapter, you will learn about different cloud security and IoT security threats. This chapter concludes with an introduction to DFIR and security operations.

The following SCOR 350-701 exam objectives are covered in this chapter:

- 1.1 Explain common threats against on-premises and cloud environments

  - 1.1.a On-premises: viruses, trojans, DoS/DDoS attacks, phishing, rootkits, man-in-the-middle attacks, SQL injection, cross-site scripting, malware

  - 1.1.b Cloud: data breaches, insecure APIs, DoS/DDoS, compromised credentials

- 1.2 Compare common security vulnerabilities such as software bugs, weak and/or hardcoded passwords, SQL injection, missing encryption, buffer overflow, path traversal, cross-site scripting/forgery

- 1.5 Describe security intelligence authoring, sharing, and consumption

- 1.6 Explain the role of the endpoint in protecting humans from phishing and social engineering attacks

"Do I Know This Already?" Quiz

The "Do I Know This Already?" quiz allows you to assess whether you should read this entire chapter thoroughly or jump to the "Exam Preparation Tasks" section. If you are in doubt about your answers to these questions or your own assessment of your knowledge of the topics, read the entire chapter. Table 1-1 lists the major headings in this chapter and their corresponding "Do I Know This Already?" quiz questions. You can find the answers in Appendix A, "Answers to the 'Do I Know This Already?' Quizzes and Q&A Sections."

Table 1-1 "Do I Know This Already?" Section-to-Question Mapping

Foundation Topics Section	Questions
Introduction to Cybersecurity	1
Defining What Are Threats, Vulnerabilities, and Exploits	2–6
Common Software and Hardware Vulnerabilities	7–10
Confidentiality, Integrity, and Availability	11–13
Cloud Security Threats	14–15
IoT Security Threats	16–17
An Introduction to Digital Forensics and Incident Response	18

CAUTION The goal of self-assessment is to gauge your mastery of the topics in this chapter. If you do not know the answer to a question or are only partially sure of the answer, you should mark that question as wrong for purposes of the self-assessment. Giving yourself credit for an answer you incorrectly guess skews your self-assessment results and might provide you with a false sense of security.

1. Which of the following is a collection of industry standards and best practices to help organizations manage cybersecurity risks?

   a. MITRE

   b. NIST Cybersecurity Framework

   c. ISO Cybersecurity Framework

   d. CERT/cc

2. _____ is any potential danger to an asset.

   a. Vulnerability

   b. Threat

   c. Exploit

   d. None of these answers is correct.

3. A _____ is a weakness in the system design, implementation, software, or code, or the lack of a mechanism.

   a. Vulnerability

   b. Threat

   c. Exploit

   d. None of these answers are correct.

4. Which of the following is a piece of software, a tool, a technique, or a process that takes advantage of a vulnerability that leads to access, privilege escalation, loss of integrity, or denial of service on a computer system?

   a. Exploit

   b. Reverse shell

   c. Searchsploit

   d. None of these answers is correct.

5. Which of the following is referred to as the knowledge about an existing or emerging threat to assets, including networks and systems?

   a. Exploits

   b. Vulnerabilities

   c. Threat assessment

   d. Threat intelligence

6. Which of the following are examples of malware attack and propagation mechanisms?

   a. Master boot record infection

   b. File infector

   c. Macro infector

   d. All of these answers are correct.

7. Vulnerabilities are typically identified by a _____.?

   a. CVE

   b. CVSS

   c. PSIRT

   d. None of these answers is correct.

8. SQL injection attacks can be divided into which of the following categories?
   a. Blind SQL injection
   b. Out-of-band SQL injection
   c. In-band SQL injection
   d. None of these answers is correct.
   e. All of these answers are correct.

9. Which of the following is a type of vulnerability where the flaw is in a web application but the attack is against an end user (client)?
   a. XXE
   b. HTML injection
   c. SQL injection
   d. XSS

10. Which of the following is a way for an attacker to perform a session hijack attack?
    a. Predicting session tokens
    b. Session sniffing
    c. Man-in-the-middle attack
    d. Man-in-the-browser attack
    e. All of these answers are correct.

11. A denial-of-service attack impacts which of the following?
    a. Integrity
    b. Availability
    c. Confidentiality
    d. None of these answers is correct.

12. Which of the following are examples of security mechanisms designed to preserve confidentiality?
    a. Logical and physical access controls
    b. Encryption
    c. Controlled traffic routing
    d. All of these answers are correct.

13. An attacker is able to manipulate the configuration of a router by stealing the administrator credential. This attack impacts which of the following?
    a. Integrity
    b. Session keys
    c. Encryption
    d. None of these answers is correct.

14. Which of the following is a cloud deployment model?
    a. Public cloud
    b. Community cloud
    c. Private cloud
    d. All of these answers are correct.

15. Which of the following cloud models include all phases of the system development life cycle (SDLC) and can use application programming interfaces (APIs), website portals, or gateway software?

    a. SaaS

    b. PaaS

    c. SDLC containers

    d. None of these answers is correct.

16. Which of the following is not a communications protocol used in IoT environments?

    a. Zigbee

    b. INSTEON

    c. LoRaWAN

    d. 802.1X

17. Which of the following is an example of tools and methods to hack IoT devices?

    a. UART debuggers

    b. JTAG analyzers

    c. IDA

    d. Ghidra

    e. All of these answers are correct.

18. Which of the following is an adverse event that threatens business security and/or disrupts service?

    a. An incident

    b. An IPS alert

    c. A DLP alert

    d. A SIEM alert

Foundation Topics

Introduction to Cybersecurity

We live in an interconnected world where both individual and collective actions have the potential to result in inspiring goodness or tragic harm. The objective of cybersecurity is to protect each of us, our economy, our critical infrastructure, and our country from the harm that can result from inadvertent or intentional misuse, compromise, or destruction of information and information systems.

Cybersecurity risk includes not only the risk of a data breach but also the risk of the entire organization being undermined via business activities that rely on digitization and accessibility. As a result, learning how to develop an adequate cybersecurity program is crucial for any organization. Cybersecurity can no longer be something that you delegate to the information technology (IT) team. Everyone needs to be involved, including the board of directors.

Cybersecurity vs. Information Security (InfoSec)

Many individuals confuse traditional information security with cybersecurity. In the past, information security programs and policies were designed to protect the confidentiality, integrity, and availability of data within the confines of an organization. Unfortunately, this is no longer sufficient. Organizations are rarely self-contained, and the price of interconnectivity is exposure to attack. Every organization, regardless of size or geographic location, is a potential target. Cybersecurity is the process of protecting information by preventing, detecting, and responding to attacks.

Cybersecurity programs recognize that organizations must be vigilant, resilient, and ready to protect and defend every ingress and egress connection as well as organizational data wherever it is stored, transmitted, or processed. Cybersecurity programs and policies expand and build upon traditional information security programs, but also include the following:

- Cyber risk management and oversight

- Threat intelligence and information sharing

- Third-party organization, software, and hardware dependency management

- Incident response and resiliency

The NIST Cybersecurity Framework

The National Institute of Standards and Technology (NIST) is a well-known organization that is part of the U.S. Department of Commerce. NIST is a nonregulatory federal agency within the U.S. Commerce Department's Technology Administration. NIST's mission is to develop and promote measurement, standards, and technology to enhance productivity, facilitate trade, and improve quality of life. The Computer Security Division (CSD) is one of seven divisions within NIST's Information Technology Laboratory. NIST's Cybersecurity Framework is a collection of industry standards and best practices to help organizations manage cybersecurity risks. This framework is created in collaboration among the United States government, corporations, and individuals. The NIST Cybersecurity Framework can be accessed at https://www.nist.gov/cyberframework.

The NIST Cybersecurity Framework is developed with a common taxonomy, and one of the main goals is to address and manage cybersecurity risk in a cost-effective way to protect critical infrastructure. Although designed for a specific constituency, the requirements can serve as a security blueprint for any organization.

Additional NIST Guidance and Documents

Currently, there are more than 500 NIST information security–related documents. This number includes FIPS, the SP 800 series, information, Information Technology Laboratory (ITL) bulletins, and NIST interagency reports (NIST IR):

- **Federal Information Processing Standards (FIPS):** This is the official publication series for standards and guidelines.

- **Special Publication (SP) 800 series:** This series reports on ITL research, guidelines, and outreach efforts in information system security and its collaborative activities with industry, government, and academic organizations. SP 800 series documents can be downloaded from https://csrc.nist.gov/publications/sp800.

- **Special Publication (SP) 1800 series:** This series focuses on cybersecurity practices and guidelines. SP 1800 series document can be downloaded from https://csrc.nist.gov/publications/sp1800.

- **NIST Internal or Interagency Reports (NISTIR):** These reports focus on research findings, including background information for FIPS and SPs.

- **ITL bulletins:** Each bulletin presents an in-depth discussion of a single topic of significant interest to the information systems community. Bulletins are issued on an as-needed basis.

From access controls to wireless security, the NIST publications are truly a treasure trove of valuable and practical guidance.

The International Organization for Standardization (ISO)

ISO is a network of the national standards institutes of more than 160 countries. ISO has developed more than 13,000 international standards on a variety of subjects, ranging from country codes to passenger safety.

The ISO/IEC 27000 series (also known as the ISMS Family of Standards, or ISO27k for short) comprises information security standards published jointly by the ISO and the International Electrotechnical Commission (IEC).

The first six documents in the ISO/IEC 27000 series provide recommendations for "establishing, implementing, operating, monitoring, reviewing, maintaining, and improving an Information Security Management System":

- ISO 27001 is the specification for an Information Security Management System (ISMS).

- ISO 27002 describes the Code of Practice for information security management.

- ISO 27003 provides detailed implementation guidance.

- ISO 27004 outlines how an organization can monitor and measure security using metrics.

- ISO 27005 defines the high-level risk management approach recommended by ISO.

- ISO 27006 outlines the requirements for organizations that will measure ISO 27000 compliance for certification.

In all, there are more than 20 documents in the series, and several more are still under development. The framework is applicable to public and private organizations of all sizes. According to the ISO website, "the ISO standard gives recommendations for information security management for use by those who are responsible for initiating, implementing or maintaining security in their organization. It is intended to provide a common basis for developing organizational security standards and effective security management practice and to provide confidence in inter-organizational dealings."

Defining What Are Threats, Vulnerabilities, and Exploits

In the following sections you will learn about the characteristics of threats, vulnerabilities, and exploits.

What Is a Threat?

A threat is any potential danger to an asset. If a vulnerability exists but has not yet been exploited—or, more importantly, it is not yet publicly known—the threat is latent and not yet realized. If someone is actively launching an attack against your system and successfully accesses something or compromises your security against an asset, the threat is realized. The entity that takes advantage of the vulnerability is known as the malicious actor, and the path used by this actor to perform the attack is known as the threat agent or threat vector.

What Is a Vulnerability?

A vulnerability is a weakness in the system design, implementation, software, or code, or the lack of a mechanism. A specific vulnerability might manifest as anything from a weakness in system design to the implementation of an operational procedure. The correct implementation of safeguards and security countermeasures could mitigate a vulnerability and reduce the risk of exploitation.

Vulnerabilities and weaknesses are common, mainly because there isn't any perfect software or code in existence. Some vulnerabilities have limited impact and are easily mitigated; however, many have broader implications.

Vulnerabilities can be found in each of the following:

- **Applications:** Software and applications come with tons of functionality. Applications might be configured for usability rather than for security. Applications might be in need of a patch or update that may or may not be available. Attackers targeting applications have a target-rich environment to examine. Just think of all the applications running on your home or work computer.

- **Operating systems:** Operating system software is loaded on workstations and servers. Attackers can search for vulnerabilities in operating systems that have not been patched or updated.

- **Hardware:** Vulnerabilities can also be found in hardware. Mitigation of a hardware vulnerability might require patches to microcode (firmware) as well as the operating system or other system software. Good examples of well-known hardware-based vulnerabilities are Spectre and Meltdown. These vulnerabilities take advantage of a feature called "speculative execution" common to most modern processor architectures.

- **Misconfiguration:** The configuration file and configuration setup for the device or software may be misconfigured or may be deployed in an unsecure state. This might be open ports, vulnerable services, or misconfigured network devices. Just consider wireless networking. Can you detect any wireless devices in your neighborhood that have encryption turned off?

- **Shrinkwrap software:** This is the application or executable file that is run on a workstation or server. When installed on a device, it can have tons of functionality or sample scripts or code available.

Vendors, security researchers, and vulnerability coordination centers typically assign vulnerabilities an identifier that's disclosed to the public. This identifier is known as the Common Vulnerabilities and Exposures (CVE). CVE is an industry-wide standard. CVE is sponsored by US-CERT, the office of Cybersecurity and Communications at the U.S. Department of Homeland Security. Operating as DHS's Federally Funded Research and Development Center (FFRDC), MITRE has copyrighted the CVE list for the benefit of the community in order to ensure it remains a free and open standard, as well as to legally protect the ongoing use of it and any resulting content by government, vendors, and/or users. MITRE maintains the CVE list and its public website, manages the CVE Compatibility Program, oversees the CVE Naming Authorities (CNAs), and provides impartial technical guidance to the CVE Editorial Board throughout the process to ensure CVE serves the public interest.

The goal of CVE is to make it easier to share data across tools, vulnerability repositories, and security services.

More information about CVE is available at http://cve.mitre.org.

What Is an Exploit?

An *exploit* refers to a piece of software, a tool, a technique, or a process that takes advantage of a vulnerability that leads to access, privilege escalation, loss of integrity, or denial of service on a computer system. Exploits are dangerous because all software has vulnerabilities; hackers and perpetrators know that there are vulnerabilities and seek to take advantage of them. Although most organizations attempt to find and fix vulnerabilities, some organizations lack sufficient funds for securing their networks. Sometimes no one may even know the vulnerability exists, and it is exploited. That is known as a *zero-day exploit*. Even when you do know there is a problem, you are burdened with the fact that a window exists between when a vulnerability is disclosed and when a patch is available to prevent the exploit. The more critical the server, the slower it is usually patched. Management might be afraid of interrupting the server or afraid that the patch might affect stability or performance. Finally, the time required to deploy and install the software patch on production servers and workstations exposes an organization's IT infrastructure to an additional period of risk.

There are several places where people trade exploits for malicious intent. The most prevalent is the "dark web." The dark web (or darknet) is an overlay of networks and systems that use the Internet but require specific software and configurations to access it. The dark web is just a small part of the "deep web." The deep web is a collection of information and systems on the Internet that is not indexed by web search engines. Often people incorrectly confuse the term deep web with dark web.

Not all exploits are shared for malicious intent. For example, many security researchers share proof-of-concept (POC) exploits in public sites such as The Exploit Database (or Exploit-DB) and GitHub. The Exploit Database is a site maintained by Offensive Security where security researchers and other individuals post exploits for known vulnerabilities. The Exploit Database can be accessed at https://www.exploit-db.com. Figure 1-1 shows different publicly available exploits in the Exploit Database.

There is a command-line tool called searchsploit that allows you to download a copy of the Exploit Database so that you can use it on the go. Figure 1-2 shows an example of how you can use searchsploit to search for specific exploits. In the example illustrated in Figure 1-2, searchsploit is used to search for exploits related to SMB vulnerabilities.

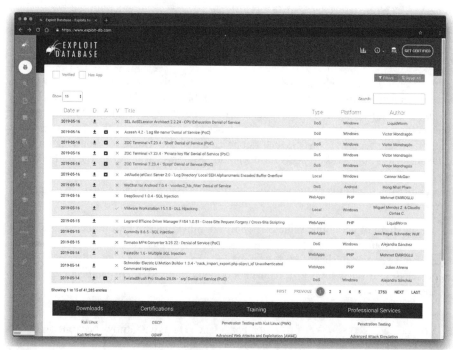

Figure 1-1   *The Exploit Database (Exploit-DB)*

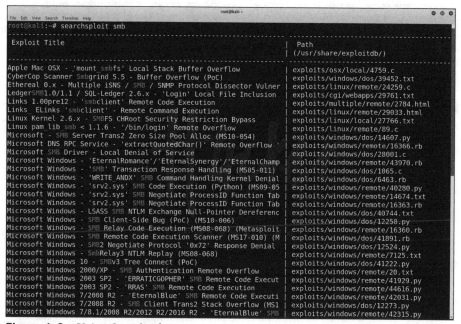

Figure 1-2   *Using Searchsploit*

Risk, Assets, Threats, and Vulnerabilities

As with any new technology topic, to better understand the security field, you must learn the terminology that is used. To be a security professional, you need to understand the relationship between risk, threats, assets, and vulnerabilities.

Risk is the probability or likelihood of the occurrence or realization of a threat. There are three basic elements of risk: assets, threats, and vulnerabilities. To deal with risk, the U.S. federal government has adopted a risk management framework (RMF). The RMF process is based on the key concepts of mission- and risk-based, cost-effective, and enterprise information system security. NIST Special Publication 800-37, "Guide for Applying the Risk Management Framework to Federal Information Systems," transforms the traditional Certification and Accreditation (C&A) process into the six-step Risk Management Framework (RMF). Let's look at the various components associated with risk, which include assets, threats, and vulnerabilities.

An asset is any item of economic value owned by an individual or corporation. Assets can be real—such as routers, servers, hard drives, and laptops—or assets can be virtual, such as formulas, databases, spreadsheets, trade secrets, and processing time. Regardless of the type of asset discussed, if the asset is lost, damaged, or compromised, there can be an economic cost to the organization.

NOTE   No organization can ever be 100 percent secure. There will always be some risk left over. This is known as residual risk, which is the amount of risk left after safeguards and controls have been put in place to protect the asset.

A threat sets the stage for risk and is any agent, condition, or circumstance that could potentially cause harm, loss, or damage, or compromise an IT asset or data asset. From a security professional's perspective, threats can be categorized as events that can affect the confidentiality, integrity, or availability of the organization's assets. These threats can result in destruction, disclosure, modification, corruption of data, or denial of service. Examples of the types of threats an organization can face include the following:

- **Natural disasters, weather, and catastrophic damage:** Hurricanes, storms, weather outages, fire, flood, earthquakes, and other natural events compose an ongoing threat.

- **Hacker attacks:** An insider or outsider who is unauthorized and purposely attacks an organization's infrastructure, components, systems, or data.

- **Cyberattack:** Attacks that target critical national infrastructures such as water plants, electric plants, gas plants, oil refineries, gasoline refineries, nuclear power plants, waste management plants, and so on. Stuxnet is an example of one such tool designed for just such a purpose.

- **Viruses and malware:** An entire category of software tools that are malicious and are designed to damage or destroy a system or data.

- **Disclosure of confidential information:** Anytime a disclosure of confidential information occurs, it can be a critical threat to an organization if such disclosure causes loss of revenue, causes potential liabilities, or provides a competitive advantage to an adversary. For instance, if your organization experiences a breach and detailed customer information is exposed (for example, personally identifiable information [PII]), such a breach could have potential liabilities and loss of trust from your customers. Another example is when a threat actor steals source code or design documents and sells them to your competitors.

■ **Denial of service (DoS) or distributed DoS (DDoS) attacks:** An attack against availability that is designed to bring the network, or access to a particular TCP/IP host/ server, to its knees by flooding it with useless traffic. Today, most DoS attacks are launched via botnets, whereas in the past tools such as the Ping of Death or Teardrop may have been used. Like malware, hackers constantly develop new tools so that Storm and Mariposa, for example, are replaced with other, more current threats.

NOTE   If the organization is vulnerable to any of these threats, there is an increased risk of a successful attack.

Defining Threat Actors

Threat actors are the individuals (or group of individuals) who perform an attack or are responsible for a security incident that impacts or has the potential of impacting an organization or individual. There are several types of threat actors:

■ **Script kiddies:** People who use existing "scripts" or tools to hack into computers and networks. They lack the expertise to write their own scripts.

■ **Organized crime groups:** Their main purpose is to steal information, scam people, and make money.

■ **State sponsors and governments:** These agents are interested in stealing data, including intellectual property and research-and-development data from major manufacturers, government agencies, and defense contractors.

■ **Hacktivists:** People who carry out cybersecurity attacks aimed at promoting a social or political cause.

■ **Terrorist groups:** These groups are motivated by political or religious beliefs.

Originally, the term *hacker* was used for a computer enthusiast. A hacker was a person who enjoyed understanding the internal workings of a system, computer, and computer network and who would continue to hack until he understood everything about the system. Over time, the popular press began to describe hackers as individuals who broke into computers with malicious intent. The industry responded by developing the word *cracker*, which is short for a criminal hacker. The term cracker was developed to describe individuals who seek to compromise the security of a system without permission from an authorized party. With all this confusion over how to distinguish the good guys from the bad guys, the term *ethical hacker* was coined. An ethical hacker is an individual who performs security tests and other vulnerability-assessment activities to help organizations secure their infrastructures. Sometimes ethical hackers are referred to as white hat hackers.

Hacker motives and intentions vary. Some hackers are strictly legitimate, whereas others routinely break the law. Let's look at some common categories:

- **White hat hackers:** These individuals perform ethical hacking to help secure companies and organizations. Their belief is that you must examine your network in the same manner as a criminal hacker to better understand its vulnerabilities.

- **Black hat hackers:** These individuals perform illegal activities, such as organized crime.

- **Gray hat hackers:** These individuals usually follow the law but sometimes venture over to the darker side of black hat hacking. It would be unethical to employ these individuals to perform security duties for your organization because you are never quite clear where they stand.

Understanding What Threat Intelligence Is

Threat intelligence is referred to as the knowledge about an existing or emerging threat to assets, including networks and systems. Threat intelligence includes context, mechanisms, indicators of compromise (IoCs), implications, and actionable advice. Threat intelligence is referred to as the information about the observables, indicators of compromise (IoCs) intent, and capabilities of internal and external threat actors and their attacks. Threat intelligence includes specifics on the tactics, techniques, and procedures of these adversaries. Threat intelligence's primary purpose is to inform business decisions regarding the risks and implications associated with threats.

Converting these definitions into common language could translate to threat intelligence being evidence-based knowledge of the capabilities of internal and external threat actors. This type of data can be beneficial for the security operations center (SOC) of any organization. Threat intelligence extends cybersecurity awareness beyond the internal network by consuming intelligence from other sources Internet-wide related to possible threats to you or your organization. For instance, you can learn about threats that have impacted different external organizations. Subsequently, you can proactively prepare rather than react once the threat is seen against your network. Providing an enrichment data feed is one service that threat intelligence platforms would typically provide.

Figure 1-3 shows a five-step threat intelligence process for evaluating threat intelligence sources and information.

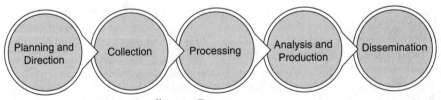

Figure 1-3 *The Threat Intelligence Process*

Many different threat intelligence platforms and services are available in the market nowadays. Cyber threat intelligence focuses on providing actionable information on adversaries, including IoCs. Threat intelligence feeds help you prioritize signals from internal systems against unknown threats. Cyber threat intelligence allows you to bring more focus to cybersecurity investigation because instead of blindly looking for "new" and "abnormal" events, you can search for specific IoCs, IP addresses, URLs, or exploit patterns.

A number of standards are being developed for disseminating threat intelligence information. The following are a few examples:

- **Structured Threat Information eXpression (STIX):** An express language designed for sharing of cyber-attack information. STIX details can contain data such as the IP addresses or domain names of command-and-control servers (often referred to C2 or CnC), malware hashes, and so on. STIX was originally developed by MITRE and is now maintained by OASIS. You can obtain more information at http://stixproject.github.io.

- **Trusted Automated eXchange of Indicator Information (TAXII):** An open transport mechanism that standardizes the automated exchange of cyber-threat information. TAXII was originally developed by MITRE and is now maintained by OASIS. You can obtain more information at http://taxiiproject.github.io.

- **Cyber Observable eXpression (CybOX):** A free standardized schema for specification, capture, characterization, and communication of events of stateful properties that are observable in the operational domain. CybOX was originally developed by MITRE and is now maintained by OASIS. You can obtain more information at https://cyboxproject.github.io.

- **Open Indicators of Compromise (OpenIOC):** An open framework for sharing threat intelligence in a machine-digestible format. Learn more at http://www.openioc.org.

- **Open Command and Control (OpenC2):** A language for the command and control of cyber-defense technologies. OpenC2 Forum was a community of cybersecurity stakeholders that was facilitated by the U.S. National Security Agency. OpenC2 is now an OASIS technical committee (TC) and specification. You can obtain more information at https://www.oasis-open.org/committees/tc_home.php?wg_abbrev=openc2.

It should be noted that many open source and non-security-focused sources can be leveraged for threat intelligence as well. Some examples of these sources are social media, forums, blogs, and vendor websites.

TIP   The following GitHub repository includes thousands of references and resources related to threat intelligence, threat hunting, ethical hacking, penetration testing, digital forensics, incident response, vulnerability research, exploit development, reverse engineering, and more:

https://github.com/The-Art-of-Hacking/h4cker

You will learn more about these resources throughout this book.

Viruses and Worms

One thing that makes viruses unique is that a virus typically needs a host program or file to infect. Viruses require some type of human interaction. A worm can travel from system to system without human interaction. When a worm executes, it can replicate again and infect even more systems. For example, a worm can email itself to everyone in your address book and then repeat this process again and again from each user's computer it infects. That massive amount of traffic can lead to a denial of service very quickly.

Spyware is closely related to viruses and worms. Spyware is considered another type of malicious software. In many ways, spyware is similar to a Trojan because most users don't know that the program has been installed, and the program hides itself in an obscure location. Spyware steals information from the user and also eats up bandwidth. If that's not enough, spyware can also redirect your web traffic and flood you with annoying pop-ups. Many users view spyware as another type of virus.

This section covers a brief history of computer viruses, common types of viruses, and some of the most well-known virus attacks. Also, some tools used to create viruses and the best methods of prevention are discussed.

Types and Transmission Methods

Although viruses have a history that dates back to the 1980s, their means of infection has changed over the years. Viruses depend on people to spread them. Viruses require human activity, such as booting a computer, executing an autorun on digital media (for example, CD, DVD, USB sticks, external hard drives, and so on), or opening an email attachment. Malware propagates through the computer world in several basic ways:

- **Master boot record infection:** This is the original method of attack. It works by attacking the master boot record of the hard drive.

- **BIOS infection:** This could completely make the system inoperable or the device could hang before passing Power On Self-Test (POST).

- **File infection:** This includes malware that relies on the user to execute the file. Extensions such as .com and .exe are usually used. Some form of social engineering is normally used to get the user to execute the program. Techniques include renaming the program or trying to mask the .exe extension and make it appear as a graphic (.jpg, .bmp, .png, .svg, and the like).

- **Macro infection:** Macro viruses exploit scripting services installed on your computer. Manipulating and using macros in Microsoft Excel, Microsoft Word, and Microsoft PowerPoint documents have been very popular in the past.

- **Cluster:** This type of virus can modify directory table entries so that it points a user or system process to the malware and not the actual program.

- **Multipartite:** This style of virus can use more than one propagation method and targets both the boot sector and program files. One example is the NATAS (Satan spelled backward) virus.

> **NOTE**   Know the primary types of malware attack mechanisms: master boot record, file infector, macro infector, and others listed previously.

After your computer is infected, the malware can do any number of things. Some spread quickly. This type of virus is known as a *fast infection*. Fast-infection viruses infect any file that they are capable of infecting. Others limit the rate of infection. This type of activity is known as *sparse infection*. Sparse infection means that the virus takes its time in infecting other files or spreading its damage. This technique is used to try to help the virus avoid infection. Some viruses forgo a life of living exclusively in files and load themselves into RAM, which is the only way that boot sector viruses can spread.

As the antivirus and security companies have developed better ways to detect malware, malware authors have fought back by trying to develop malware that is harder to detect. For example, in 2012, Flame was believed to be the most sophisticated malware to date. Flame has the ability to spread to other systems over a local network. It can record audio, screenshots, and keyboard activity, and it can turn infected computers into Bluetooth beacons that attempt to download contact information from nearby Bluetooth-enabled devices. Another technique that malware developers have attempted is polymorphism. A polymorphic virus can change its signature every time it replicates and infects a new file. This technique makes it much harder for the antivirus program to detect it. One of the biggest changes is that malware creators don't massively spread viruses and other malware the way they used to. Much of the malware today is written for a specific target. By limiting the spread of the malware and targeting only a few victims, malware developers make finding out about the malware and creating a signature to detect it much harder for antivirus companies.

When is a virus not a virus? When is the virus just a hoax? A virus hoax is nothing more than a chain letter, meme, or email that encourages you to forward it to your friends to warn them of impending doom or some other notable event. To convince readers to forward the hoax, the email will contain some official-sounding information that could be mistaken as valid.

Malware Payloads

Malware must place their payload somewhere. They can always overwrite a portion of the infected file, but to do so would destroy it. Most malware writers want to avoid detection for as long as possible and might not have written the program to immediately destroy files. One way the malware writer can accomplish this is to place the malware code either at the beginning or the end of the infected file. Malware known as a prepender infects programs by placing its viral code at the beginning of the infected file, whereas an appender places its code at the end of the infected file. Both techniques leave the file intact, with the malicious code added to the beginning or the end of the file.

No matter the infection technique, all viruses have some basic common components, as detailed in the following list. For example, all viruses have a search routine and an infection routine.

- **Search routine:** The *search routine* is responsible for locating new files, disk space, or RAM to infect. The search routine could include "profiling." Profiling could be used to identify the environment and morph the malware to be more effective and potentially bypass detection.

- **Infection routine:** The search routine is useless if the virus doesn't have a way to take advantage of these findings. Therefore, the second component of a virus is an *infection routine*. This portion of the virus is responsible for copying the virus and attaching it to a suitable host. Malware could also use a re-infect/restart routine to further compromise the affected system.

- **Payload:** Most viruses don't stop here and also contain a *payload*. The purpose of the payload routine might be to erase the hard drive, display a message to the monitor, or possibly send the virus to 50 people in your address book. Payloads are not required, and without one, many people might never know that the virus even existed.

- **Antidetection routine:** Many viruses might also have an *antidetection routine*. Its goal is to help make the virus more stealth-like and avoid detection.

- **Trigger routine:** The goal of the trigger routine is to launch the payload at a given date and time. The trigger can be set to perform a given action at a given time.

Trojans

Trojans are programs that pretend to do one thing but, when loaded, actually perform another, more malicious act. Trojans gain their name from Homer's epic tale *The Iliad*. To defeat their enemy, the Greeks built a giant wooden horse with a trapdoor in its belly. The Greeks tricked the Trojans into bringing the large wooden horse into the fortified city of Troy. However, unknown to the Trojans and under cover of darkness, the Greeks crawled out of the wooden horse, opened the city's gate, and allowed the waiting soldiers into the city.

A software Trojan horse is based on this same concept. A user might think that a file looks harmless and is safe to run, but after the file is executed, it delivers a malicious payload. Trojans work because they typically present themselves as something you want, such as an email with a PDF, a Word document, or an Excel spreadsheet. Trojans work hard to hide their true purposes. The spoofed email might look like it's from HR, and the attached file might purport to be a list of pending layoffs. The payload is executed if the attacker can get the victim to open the file or click the attachment. That payload might allow a hacker remote access to your system, start a keystroke logger to record your every keystroke, plant a backdoor on your system, cause a denial of service (DoS), or even disable your antivirus protection or software firewall.

Unlike a virus or worm, Trojans cannot spread themselves. They rely on the uninformed user.

Trojan Types

A few Trojan categories are command-shell Trojans, graphical user interface (GUI) Trojans, HTTP/HTTPS Trojans, document Trojans, defacement Trojans, botnet Trojans, Virtual Network Computing (VNC) Trojans, remote-access Trojans, data-hiding Trojans, banking Trojans, DoS Trojans, FTP Trojans, software-disabling Trojans, and covert-channel Trojans. In reality, it's hard to place some Trojans into a single type because many have more than one function. To better understand what Trojans can do, refer to the following list, which outlines a few of these types:

- **Remote access:** Remote-access Trojans (RATs) allow the attacker full control over the system. Poison Ivy is an example of this type of Trojan. Remote-access Trojans are usually set up as client/server programs so that the attacker can connect to the infected system and control it remotely.

- **Data hiding:** The idea behind this type of Trojan is to hide a user's data. This type of malware is also sometimes known as ransomware. This type of Trojan restricts access to the computer system that it infects, and it demands a ransom paid to the creator of the malware for the restriction to be removed.

- **E-banking:** These Trojans (Zeus is one such example) intercept and use a victim's banking information for financial gain. Usually, they function as a transaction authorization number (TAN) grabber, use HTML injection, or act as a form grabber. The sole purpose of these types of programs is financial gain.

- **Denial of service (DoS):** These Trojans are designed to cause a DoS. They can be designed to knock out a specific service or to bring an entire system offline.

- **Proxy:** These Trojans are designed to work as proxy programs that help a hacker hide and allow him to perform activities from the victim's computer, not his own. After all, the farther away the hacker is from the crime, the harder it becomes to trace him.

- **FTP:** These Trojans are specifically designed to work on port 21. They allow the hacker or others to upload, download, or move files at will on the victim's machine.

- **Security-software disablers:** These Trojans are designed to attack and kill antivirus or software firewalls. The goal of disabling these programs is to make it easier for the hacker to control the system.

Trojan Ports and Communication Methods

Trojans can communicate in several ways. Some use overt communications. These programs make no attempt to hide the transmission of data as it is moved on to or off of the victim's computer. Most use covert communication channels. This means that the hacker goes to lengths to hide the transmission of data to and from the victim. Many Trojans that open covert channels also function as backdoors. A *backdoor* is any type of program that will allow a hacker to connect to a computer without going through the normal authentication process. If a hacker can get a backdoor program loaded on an internal device, the hacker can then come and go at will. Some of the programs spawn a connection on the victim's computer connecting out to the hacker. The danger of this type of attack is the traffic moving from the inside out, which means from inside the organization to the outside Internet. This is usually the least restrictive because companies are usually more concerned about what comes in the network than they are about what leaves the network.

TIP   One way an attacker can spread a Trojan is through a *poison apple attack* or *USB key drop*. Using this technique, the attacker leaves a thumb drive (USB stick) in the desk drawer of the victim or maybe in the cafeteria of the targeted company, perhaps in a key chain along with some keys and a photo of a cat to introduce a personal touch. The attacker then waits for someone to find it, insert it in the computer, and start clicking on files to see what's there. Instead of just one bite of the apple, it's just one click, and the damage is done!

Trojan Goals

Not all Trojans were designed for the same purpose. Some are destructive and can destroy computer systems, whereas others seek only to steal specific pieces of information. Although not all of them make their presence known, Trojans are still dangerous because they represent a loss of confidentiality, integrity, and availability. Common targets of Trojans include the following:

- **Credit card data:** Credit card data and banking information have become huge targets. After the hacker has this information, he can go on an online shopping spree or use the card to purchase services, such as domain name registration.

- **Electronic or digital wallets:** Individuals can use an electronic device or online service that allows them to make electronic transactions. This includes buying goods online or using a smartphone to purchase something at a store. A digital wallet can also be a cryptocurrency wallet (such as Bitcoin, Ethereum, Litecoin, Ripple, and so on).

- **Passwords:** Passwords are always a big target. Many of us are guilty of password reuse. Even if we are not, there is always the danger that a hacker can extract email passwords or other online account passwords.

- **Insider information:** We have all had those moments in which we have said, "If only I had known this beforehand." That's what insider information is about. It can give the hacker critical information before it is made public or released.

- **Data storage:** The goal of the Trojan might be nothing more than to use your system for storage space. That data could be movies, music, illegal software (warez), or even pornography.

- **Advanced persistent threat (APT):** It could be that the hacker has targeted you as part of a nation-state attack or your company has been targeted because of its sensitive data. Two examples include Stuxnet and the APT attack against RSA in 2011. These attackers might spend significant time and expense to gain access to critical and sensitive resources.

Trojan Infection Mechanisms

After a hacker has written a Trojan, he will still need to spread it. The Internet has made this much easier than it used to be. There are a variety of ways to spread malware, including the following:

- **Peer-to-peer networks (P2P):** Although users might think that they are getting the latest copy of a computer game or the Microsoft Office package, in reality, they might be getting much more. P2P networks and file-sharing sites such as The Pirate Bay are generally unmonitored and allow anyone to spread any programs they want, legitimate or not.

- **Instant messaging (IM):** IM was not built with security controls. So, you never know the real contents of a file or program that someone has sent you. IM users are at great risk of becoming targets for Trojans and other types of malware.

- **Internet Relay Chat (IRC):** IRC is full of individuals ready to attack the newbies who are enticed into downloading a free program or application.

- **Email attachments:** Attachments are another common way to spread a Trojan. To get you to open them, these hackers might disguise the message to appear to be from a legitimate organization. The message might also offer you a valuable prize, a desired piece of software, or similar enticement to pique your interest. If you feel that you must investigate these attachments, save them first and then run an antivirus on them. Email attachments are the number-one means of malware propagation. You might investigate them as part of your information security job to protect network users.

- **Physical access:** If a hacker has physical access to a victim's system, he can just copy the Trojan horse to the hard drive (via a thumb drive). The hacker can even take the attack to the next level by creating a Trojan that is unique to the system or network. It might be a fake login screen that looks like the real one or even a fake database.

- **Browser and browser extension vulnerabilities:** Many users don't update their browsers as soon as updates are released. Web browsers often treat the content they receive as trusted. The truth is that nothing in a web page can be trusted to follow any guidelines. A website can send to your browser data that exploits a bug in a browser, violates computer security, and might load a Trojan.

- **SMS messages:** SMS messages have been used by attackers to propagate malware to mobile devices and to perform other scams.

- **Impersonated mobile apps:** Attackers can impersonate apps in mobile stores (for example, Google Play or Apple Store) to infect users. Attackers can perform visual impersonation to intentionally misrepresents apps in the eyes of the user. Attackers can do this to repackage the application and republish the app to the marketplace under a different author. This tactic has been used by attackers to take a paid app and republish it to the marketplace for less than its original price. However, in the context of mobile malware, the attacker uses similar tactics to distribute a malicious app to a wide user audience while minimizing the invested effort. If the attacker repackages a popular app and appends malware to it, the attacker can leverage the user's trust of their favorite apps and successfully compromise the mobile device.

- **Watering hole:** The idea is to infect a website the attacker knows the victim will visit. Then the attacker simply waits for the victim to visit the watering hole site so the system can become infected.

- **Freeware:** Nothing in life is free, and that includes most software. Users are taking a big risk when they download freeware from an unknown source. Not only might the freeware contain a Trojan, but freeware also has become a favorite target for adware and spyware.

TIP Be sure that you understand that email is one of the most widely used forms of malware propagation.

Effects of Trojans

The effects of Trojans can range from the benign to the extreme. Individuals whose systems become infected might never even know; most of the creators of this category of malware don't want to be detected, so they go to great lengths to hide their activity and keep their actions hidden. After all, their goal is typically to "own the box." If the victim becomes aware of the Trojan's presence, the victim will take countermeasures that threaten the attacker's ability to keep control of the computer. In some cases, programs seemingly open by themselves or the web browser opens pages the user didn't request. However, because the hacker is in control of the computer, he can change its background, reboot the systems, or capture everything the victim types on the keyboard.

Distributing Malware

Technology changes, and that includes malware distribution. The fact is that malware detection is much more difficult today than in the past. Today, it is not uncommon for attackers to use multiple layers of techniques to obfuscate code, make malicious code undetectable from antivirus, and employ encryption to prevent others from examining malware. The result is that modern malware improves the attackers' chances of compromising a computer without being detected. These techniques include wrappers, packers, droppers, and crypters.

Wrappers offer hackers a method to slip past a user's normal defenses. A *wrapper* is a program used to combine two or more executables into a single packaged program. Wrappers are also referred to as binders, packagers, and EXE binders because they are the functional equivalent of binders for Windows Portable Executable files. Some wrappers only allow programs to be joined; others allow the binding together of three, four, five, or more programs. Basically, these programs perform like installation builders and setup programs. Besides allowing you to bind a program, wrappers add additional layers of obfuscation and encryption around the target file, essentially creating a new executable file.

Packers are similar to programs such as WinZip, Rar, and Tar because they compress files. However, whereas compression programs compress files to save space, packers do this to obfuscate the activity of the malware. The idea is to prevent anyone from viewing the malware's code until it is placed in memory. Packers serve a second valuable goal to the attacker in that they work to bypass network security protection mechanisms, such as host- and network-based intrusion detection systems. The malware packer will decompress the program only when in memory, revealing the program's original code only when executed. This is yet another attempt to bypass antimalware detection.

Droppers are software designed to install malware payloads on the victim's system. Droppers try to avoid detection and evade security controls by using several methods to spread and install the malware payload.

Crypters function to encrypt or obscure the code. Some crypters obscure the contents of the Trojan by applying an encryption algorithm. Crypters can use anything from AES, RSA, to even Blowfish, or might use more basic obfuscation techniques such as XOR, Base64 encoding, or even ROT13. Again, these techniques are used to conceal the contents of the executable program, making it undetectable by antivirus and resistant to reverse-engineering efforts.

Ransomware

Over the past few years, ransomware has been used by criminals making money out of their victims and by hacktivists and nation-state attackers causing disruption. Ransomware can propagate like a worm or a virus but is designed to encrypt personal files on the victim's hard drive until a ransom is paid to the attacker. Ransomware has been around for many years but made a comeback in recent years. The following are several examples of popular ransomware:

- WannaCry

- Pyeta

- Nyeta

- Sodinokibi

- Bad Rabbit

- Grandcrab

- SamSam

- CryptoLocker

- CryptoDefense

- CryptoWall

- Spora

Ransomware can encrypt specific files in your system or all your files, in some cases including the master boot record of your hard disk drive.

Covert Communication

Distributing malware is just half the battle for the attacker. The attacker will need to have some way to exfiltrate data and to do so in a way that is not detected. If you look at the history of covert communications, you will see that the Trusted Computer System Evaluation Criteria (TCSEC) was one of the first documents to fully examine the concept of covert communications and attacks. TCSEC divides covert channel attacks into two broad categories:

- **Covert timing channel attacks:** Timing attacks are difficult to detect because they are based on system times and function by altering a component or by modifying resource timing.

- **Covert storage channel attacks:** Use one process to write data to a storage area and another process to read the data.

It is important to examine covert communication on a more focused scale because it will be examined here as a means of secretly passing information or data. For example, most everyone has seen a movie in which an informant signals the police that it's time to bust the criminals. It could be that the informant lights a cigarette or simply tilts his hat. These small signals are meaningless to the average person who might be nearby, but for those who know what to look for, they are recognized as a legitimate signal.

In the world of hacking, covert communication is accomplished through a covert channel. A *covert channel* is a way of moving information through a communication channel or protocol in a manner in which it was not intended to be used. Covert channels are important for security professionals to understand. For the ethical hacker who performs attack and penetration assessments, such tools are important because hackers can use them to obtain an initial foothold into an otherwise secure network. For the network administrator, understanding how these tools work and their fingerprints can help her recognize potential entry points into the network. For the hacker, these are powerful tools that can potentially allow him control and access.

How do covert communications work? Well, the design of TCP/IP offers many opportunities for misuse. The primary protocols for covert communications include Internet Protocol (IP), Transmission Control Protocol (TCP), User Datagram Protocol (UDP), Internet Control Message Protocol (ICMP), and Domain Name Service (DNS).

The Internet layer offers several opportunities for hackers to tunnel traffic. Two commonly tunneled protocols are IPv6 and ICMP.

IPv6 is like all protocols in that it can be abused or manipulated to act as a covert channel. This is primarily possible because edge devices might not be configured to recognize IPv6 traffic even though most operating systems have support for IPv6 turned on. According to US-CERT, Windows misuse relies on several factors:

- Incomplete or inconsistent support for IPv6

- The IPv6 autoconfiguration capability

- Malware designed to enable IPv6 support on susceptible hosts

- Malicious application of traffic "tunneling," a method of Internet data transmission in which the public Internet is used to relay private network data

There are plenty of tools to tunnel over IPv6, including 6tunnel, socat, nt6tunnel, and relay6. The best way to maintain security with IPv6 is to recognize that even devices supporting IPv6 may not be able to correctly analyze the IPv6 encapsulation of IPv4 packets.

The second protocol that might be tunneled at the Internet layer is Internet Control Message Protocol (ICMP). ICMP is specified by RFC 792 and is designed to provide error messaging, best path information, and diagnostic messages. One example of this is the **ping** command. It uses ICMP to test an Internet connection.

The transport layer offers attackers two protocols to use: TCP and UDP. TCP offers several fields that can be manipulated by an attacker, including the TCP Options field in the TCP header and the TCP Flag field. By design, TCP is a connection-oriented protocol that provides robust communication. The following steps outline the normal TCP process:

1. **A three-step handshake:** This ensures that both systems are ready to communicate.

2. **Exchange of control information:** During the setup, information is exchanged that specifies maximum segment size.

3. **Sequence numbers:** This indicates the amount and position of data being sent.

4. **Acknowledgments:** This indicates the next byte of data that is expected.

5. **Four-step shutdown:** This is a formal process of ending the session that allows for an orderly shutdown.

Although SYN packets occur only at the beginning of the session, ACKs may occur thousands of times. They confirm that data was received. That is why packet-filtering devices build their rules on SYN segments. It is an assumption on the firewall administrator's part that ACKs occur only as part of an established session. It is much easier to configure, and it reduces workload. To bypass the SYN blocking rule, a hacker may attempt to use TCP ACK packets as a covert communication channel. Tools such as AckCmd serve this exact purpose.

UDP is stateless and, as such, may not be logged in firewall connections; some UDP-based applications such as DNS are typically allowed through the firewall and might not be watched closely by network and firewall administrators. UDP tunneling applications typically act in a client/server configuration. Also, some ports like UDP 53 are most likely open. This means it's also open for attackers to use as a potential means to exfiltrate data. There are several UDP tunnel tools that you should check out, including the following:

- **UDP Tunnel:** Also designed to tunnel TCP traffic over a UDP connection. You can find UDP Tunnel at https://code.google.com/p/udptunnel/.

- **dnscat:** Another option for tunneling data over an open DNS connection. You can download the current version, dnscat2, at https://github.com/iagox86/dnscat2.

Application layer tunneling uses common applications that send data on allowed ports. For example, a hacker might tunnel a web session, port 80, through SSH port 22 or even through port 443. Because ports 22 and 443 both use encryption, it can be difficult to monitor the difference between a legitimate session and a covert channel.

HTTP might also be used. Netcat is one tool that can be used to set up a tunnel to exfiltrate data over HTTP. If HTTPS is the transport, it is difficult for the network administrator to inspect the outbound data. Cryptcat (http://cryptcat.sourceforge.net) can be used to send data over HTTPS.

Finally, even Domain Name System (DNS) can be used for application layer tunneling. DNS is a request/reply protocol. Its queries consist of a 12-byte fixed-size header followed by one or more questions. A DNS response is formatted in much the same way in that it has a header, followed by the original question, and then typically a single-answer resource record. The most straightforward way to manipulate DNS is by means of these request/replies. You can easily detect a spike in DNS traffic; however, many times attackers move data using DNS without being detected for days, weeks, or months. They schedule the DNS exfiltration packets in a way that makes it harder for a security analyst or automated tools to detect.

 Keyloggers

Keystroke loggers (keyloggers) are software or hardware devices used to record everything a person types. Some of these programs can record every time a mouse is clicked, a website is visited, and a program is opened. Although not truly a covert communication tool, these devices do enable a hacker to covertly monitor everything a user does. Some of these devices secretly email all the amassed information to a predefined email address set up by the hacker.

The software version of this device is basically a shim, as it sits between the operating system and the keyboard. The hacker might send a victim a keystroke-logging program wrapped up in much the same way as a Trojan would be delivered. Once installed, the logger can operate in stealth mode, which means that it is hard to detect unless you know what you are looking for.

There are ways to make keyloggers completely invisible to the OS and to those examining the file system. To accomplish this, all the hacker has to do is use a hardware keylogger. These devices are usually installed while the user is away from his desk. Hardware keyloggers are completely undetectable except for their physical presence. Even then, they might be overlooked because they resemble an extension. Not many people pay close attention to the plugs on the back of their computer.

To stay on the right side of the law, employers who plan to use keyloggers should make sure that company policy outlines their use and how employees are to be informed. The CERT Division of the Software Engineering Institute (SEI) recommends a warning banner similar to the following: "This system is for the use of authorized personnel only. If you continue to access this system, you are explicitly consenting to monitoring."

Keystroke recorders have been around for years. Hardware keyloggers can be wireless or wired. Wireless keyloggers can communicate via 802.11 or Bluetooth, and wired keyloggers must be retrieved to access the stored data. One such example of a wired keylogger is KeyGhost, a commercial device that is openly available worldwide from a New Zealand firm that goes by the name of KeyGhost Ltd (http://www.keyghost.com). The device looks like a small adapter on the cable connecting one's keyboard to the computer. This device requires no external power, lasts indefinitely, and cannot be detected by any software.

Numerous software products that record all keystrokes are openly available on the Internet. You have to pay for some products, but others are free.

Spyware

Spyware is another form of malicious code that is similar to a Trojan. It is installed without your consent or knowledge, hidden from view, monitors your computer and Internet usage, and is configured to run in the background each time the computer starts. Spyware has grown to be a big problem. It is usually used for one of two purposes:

- **Surveillance:** Used to determine your buying habits, discover your likes and dislikes, and report this demographic information to paying marketers.

- **Advertising:** You're targeted for advertising that the spyware vendor has been paid to deliver. For example, the maker of a rhinestone cell phone case might have paid the spyware vendor for 100,000 pop-up ads. If you have been infected, expect to receive more than your share of these unwanted pop-up ads.

Many times, spyware sites and vendors use droppers to covertly drop their spyware components to the victim's computer. Basically, a *dropper* is just another name for a wrapper, because a dropper is a standalone program that drops different types of standalone malware to a system.

Spyware programs are similar to Trojans in that there are many ways to become infected. To force the spyware to restart each time the system boots, code is usually hidden in the

Registry run keys, the Windows Startup folder, the Windows **load=** or **run=** lines found in the Win.ini file, or the **Shell=** line found in the Windows System.ini file. Spyware, like all malware, may also make changes to the hosts file. This is done to block the traffic to all the download or update servers of the well-known security vendors or to redirect traffic to servers of their choice by redirecting traffic to advertisement servers and replacing the advertisements with their own.

If you are dealing with systems that have had spyware installed, start by looking at the hosts file and the other locations discussed previously or use a spyware removal program. It's good practice to use more than one antispyware program to find and remove as much spyware as possible.

Analyzing Malware

Malware analysis can be extremely complex. Although an in-depth look at this area of cybersecurity is beyond this book, you should have a basic understanding of how analysis is performed. There are two basic methods to analyze viruses and other malware:

- Static analysis

- Dynamic analysis

Static Analysis

Static analysis is concerned with the decompiling, reverse engineering, and analysis of malicious software. The field is an outgrowth of the field of computer virus research and malware intent determination. Consider examples such as Conficker, Stuxnet, Aurora, and the Black Hole Exploit Kit. Static analysis makes use of disassemblers and decompilers to format the data into a human-readable format. Several useful tools are listed here:

- **IDA Pro:** An interactive disassembler that you can use for decompiling code. It's particularly useful in situations in which the source code is not available, such as with malware. IDA Pro allows the user to see the source code and review the instructions that are being executed by the processor. IDA Pro uses advanced techniques to make that code more readable. You can download and obtain additional information about IDA Pro at https://www.hex-rays.com/products/ida/.

- **Evan's Debugger (edb):** A Linux cross-platform AArch32/x86/x86-64 debugger. You can download and obtain additional information about Evan's Debugger at https://github.com/eteran/edb-debugger.

- **BinText:** Another tool that is useful to the malware analyst. BinText is a text extractor that will be of particular interest to programmers. It can extract text from any kind of file and includes the ability to find plain ASCII text, Unicode (double-byte ANSI) text, and resource strings, providing useful information for each item in the optional "advanced" view mode. You can download and obtain additional information about BinText from the following URL: https://www.aldeid.com/wiki/BinText.

- **UPX:** A packer, compression, and decompression tool. You can download and obtain additional information about UPX at https://upx.github.io.

- **OllyDbg:** A debugger that allows for the analysis of binary code where source is unavailable. You can download and obtain additional information about OllyDbg at http://www.ollydbg.de.

- **Ghidra:** A software reverse engineering tool developed by the U.S. National Security Agency (NSA) Research Directorate. Figure 1-4 shows an example of a file being reversed engineered using Ghidra. You can download and obtain additional information about Ghidra at https://ghidra-sre.org.

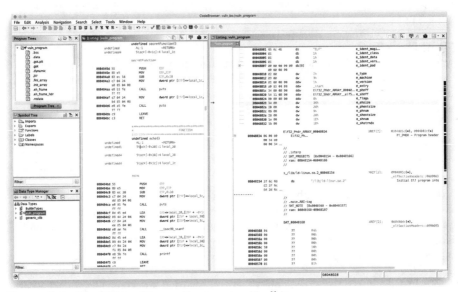

Figure 1-4   *The Ghidra Reverse Engineering Toolkit*

Dynamic Analysis

Dynamic analysis of malware and viruses is the second method that may be used. Dynamic analysis relates to the monitoring and analysis of computer activity and network traffic. This requires the ability to configure the network device for monitoring, look for unusual or suspicious activity, and try not to alert attackers. This approach requires the preparation of a testbed. Before you begin setting up a dynamic analysis lab, remember that the number-one goal is to keep the malware contained. If you allow the host system to become compromised, you have defeated the entire purpose of the exercise. Virtual systems share many resources with the host system and can quickly become compromised if the configuration is not handled correctly. Here are a few pointers for preventing malware from escaping the isolated environment to which it should be confined:

1. Install a virtual machine (VM).

2. Install a guest operating system on the VM.

3. Isolate the system from the guest VM.

4. Verify that all sharing and transfer of data is blocked between the host operating system and the virtual system.

5. Copy the malware over to the guest operating system and prepare for analysis.

Malware authors sometimes use anti-VM techniques to thwart attempts at analysis. If you try to run the malware in a VM, it might be designed not to execute. For example, one simple way is to get the MAC address; if the Organizationally Unique Identifier (OUI) matches a VM vendor, the malware will not execute.

The malware may also look to see whether there is an active network connection. If not, it may refuse to run. One tool to help overcome this barrier is FakeNet. FakeNet simulates a network connection so that malware interacting with a remote host continues to run. If you are forced to detect the malware by discovering where it has installed itself on the local system, there are some known areas to review:

- Running processes
- Device drivers
- Windows services
- Startup programs
- Operating system files

Malware has to install itself somewhere, and by a careful analysis of the system, files, memory, and folders, you should be able to find it.

Several sites are available that can help analyze suspect malware. These online tools can provide a quick and easy analysis of files when reverse engineering and decompiling is not possible. Most of these sites are easy to use and offer a straightforward point-and-click interface. These sites generally operate as a sandbox. A *sandbox* is simply a standalone environment that allows you to safely view or execute the program while keeping it contained. A good example of sandbox services is the Cisco ThreatGrid. This great tool and service tracks changes made to the file system, Registry, memory, and network.

During a network security assessment, you may discover malware or other suspected code. You should have an incident response plan that addresses how to handle these situations. If you're using only one antivirus product to scan for malware, you may be missing a lot. As you learned in the previous section, websites such as the Cisco Talos File Reputation Lookup site (https://www.talosintelligence.com/reputation) and VirusTotal (https://virustotal.com) allow you to upload files to verify if they may be known malware.

These tools and techniques listed offer some insight as to how static malware analysis is performed, but don't expect malware writers to make the analysis of their code easy. Many techniques can be used to make disassembly challenging:

- Encryption
- Obfuscation
- Encoding
- Anti-VM
- Anti-debugger

Common Software and Hardware Vulnerabilities

The number of disclosed vulnerabilities continues to rise. You can keep up with vulnerability disclosures by subscribing to vulnerability feeds and searching public repositories such as the National Vulnerability Database (NVD). The NVD can be accessed at https://nvd.nist.gov.

> **TIP**   Vulnerabilities are typically identified by a Common Vulnerabilities and Exposures (CVE) identifier. CVE is an identifier for publicly known security vulnerabilities. This is a standard created and maintained by MITRE and used by numerous organizations in the industry, as well as security researchers. You can find more information about the CVE specification and search the CVE list at https://cve.mitre.org.

There are many different software and hardware vulnerabilities and related categories. The sections that follow include a few examples.

Injection Vulnerabilities

The following are examples of injection-based vulnerabilities:

- SQL injection vulnerabilities
- HTML injection vulnerabilities
- Command injection vulnerabilities

Code injection vulnerabilities are exploited by forcing an application or a system to process invalid data. An attacker takes advantage of this type of vulnerability to inject code into a vulnerable system and change the course of execution. Successful exploitation can lead to the disclosure of sensitive information, manipulation of data, denial-of-service conditions, and more. Examples of code injection vulnerabilities include the following:

- SQL injection
- HTML script injection
- Dynamic code evaluation
- Object injection
- Remote file inclusion
- Uncontrolled format string
- Shell injection

SQL Injection

SQL injection (SQLi) vulnerabilities can be catastrophic because they can allow an attacker to view, insert, delete, or modify records in a database. In an SQL injection attack, the attacker inserts, or injects, partial or complete SQL queries via the web application. The attacker injects SQL commands into input fields in an application or a URL in order to execute predefined SQL commands.

Web applications construct SQL statements involving SQL syntax invoked by the application mixed with user-supplied data, as shown in Figure 1-5.

```
SELECT * FROM Users WHERE UserName LIKE        '%santos%';
```

This is sent by the application to the database behind the scenes.

This can be input from a user in a web form.

Figure 1-5 *An Explanation of an SQL Statement*

The first portion of the SQL statement shown in Figure 1-5 is not shown to the user. Typically, the application sends this portion to the database behind the scenes. The second portion of the SQL statement is typically user input in a web form.

If an application does not sanitize user input, an attacker can supply crafted input in an attempt to make the original SQL statement execute further actions in the database. SQL injections can be done using user-supplied strings or numeric input. Figure 1-6 shows an example of a basic SQL injection attack.

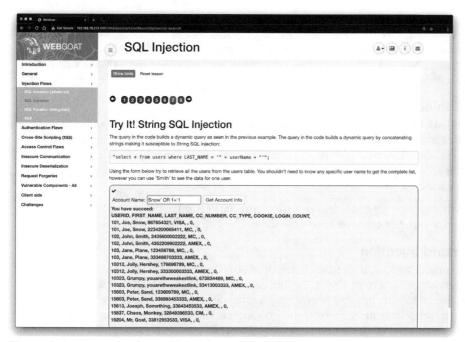

Figure 1-6 *Example of an SQL Injection Vulnerability*

Figure 1-6 shows an intentionally vulnerable application (WebGoat) being used to demonstrate the effects of an SQL injection attack. When the string Snow' OR 1='1 is entered in the web form, it causes the application to display all records in the database table to the attacker.

One of the first steps when finding SQL injection vulnerabilities is to understand when the application interacts with a database. This is typically done with web authentication forms, search engines, and interactive sites such as e-commerce sites.

SQL injection attacks can be divided into the following categories:

- **In-band SQL injection:** With this type of injection, the attacker obtains the data by using the same channel that is used to inject the SQL code. This is the most basic form of an SQL injection attack, where the data is dumped directly in a web application (or web page).

- **Out-of-band SQL injection:** With this type of injection, the attacker retrieves data using a different channel. For example, an email, a text, or an instant message could be sent to the attacker with the results of the query. Alternatively, the attacker might be able to send the compromised data to another system.

- **Blind (or inferential) SQL injection:** With this type of injection, the attacker does not make the application display or transfer any data; rather, the attacker is able to reconstruct the information by sending specific statements and discerning the behavior of the application and database.

To perform an SQL injection attack, an attacker must craft a syntactically correct SQL statement (query). The attacker may also take advantage of error messages coming back from the application and might be able to reconstruct the logic of the original query to understand how to execute the attack correctly. If the application hides the error details, the attacker might need to reverse engineer the logic of the original query.

HTML Injection

An HTML injection is a vulnerability that occurs when an unauthorized user is able to control an input point and able to inject arbitrary HTML code into a web application. Successful exploitation could lead to disclosure of a user's session cookies; an attacker might do this to impersonate a victim or to modify the web page or application content seen by the victims.

HTML injection vulnerabilities can lead to cross-site scripting (XSS). You will learn details about the different types of XSS vulnerabilities and attacks later in this chapter.

Command Injection

A command injection is an attack in which an attacker tries to execute commands that he or she is not supposed to be able to execute on a system via a vulnerable application. Command injection attacks are possible when an application does not validate data supplied by the user (for example, data entered in web forms, cookies, HTTP headers, and other elements). The vulnerable system passes that data into a system shell.

With command injection, an attacker tries to send operating system commands so that the application can execute them with the privileges of the vulnerable application. Command injection is not the same as code execution and code injection, which involve exploiting a buffer overflow or similar vulnerability.

Authentication-based Vulnerabilities

An attacker can bypass authentication in vulnerable systems by using several methods.

The following are the most common ways to take advantage of authentication-based vulnerabilities in an affected system:

- Credential brute forcing

- Session hijacking

- Redirecting

- Exploiting default credentials

- Exploiting weak credentials

- Exploiting Kerberos vulnerabilities

Credential Brute Force Attacks and Password Cracking

In a credential brute-force attack, the attacker attempts to log in to an application or a system by trying different usernames and passwords. There are two major categories of brute-force attacks:

- **Online brute-force attacks:** In this type of attack, the attacker actively tries to log in to the application directly by using many different combinations of credentials. Online brute-force attacks are easy to detect because you can easily inspect for large numbers of attempts by an attacker.

- **Offline brute-force attacks:** In this type of attack, the attacker can gain access to encrypted data or hashed passwords. These attacks are more difficult to prevent and detect than online attacks. However, offline attacks require significantly more computation effort and resources from the attacker.

The strength of user and application credentials has a direct effect on the success of brute-force attacks. Weak credentials are one of the major causes of credential compromise. The more complex and the longer a password (credential), the better. An even better approach is to use multifactor authentication (MFA). The use of MFA significantly reduces the probability of success for these types of attacks.

An attacker may feed to an attacking system a word list containing thousands of words in order to crack passwords or associated credentials. The following site provides links to millions of real-world passwords: http://wordlists.h4cker.org.

Weak cryptographic algorithms (such as RC4, MD5, and DES) allow attackers to easily crack passwords.

TIP The following site lists the cryptographic algorithms that should be avoided and the ones that are recommended, as well as several other recommendations: https://www.cisco.com/c/en/us/about/security-center/next-generation-cryptography.html.

Attackers can also use statistical analysis and rainbow tables against systems that improperly protect passwords with a one-way hashing function. A rainbow table is a precomputed table for reversing cryptographic hash functions and for cracking password hashes. Such tables can be used to accelerate the process of cracking password hashes.

For a list of publicly available rainbow tables, see http://project-rainbowcrack.com/table.htm.

In addition to weak encryption or hashing algorithms, poorly designed security protocols such as Wired Equivalent Privacy (WEP) introduce avenues of attack to compromise user and application credentials. Also, if hashed values are stored without being rendered unique first (that is, without a salt), it is possible to gain access to the values and perform a rainbow table attack.

An organization should implement techniques on systems and applications to throttle login attempts and prevent brute-force attacks. Those attempts should also be logged and audited.

Session Hijacking

There are several ways an attacker can perform a session hijack and several ways a session token may be compromised:

- **Predicting session tokens:** This is why it is important to use non-predictable tokens.

- **Session sniffing:** This can occur through collecting packets of unencrypted web sessions.

- **Man-in-the-middle attack:** With this type of attack, the attacker sits in the path between the client and the web server.

- **Man-in-the-browser attack:** This attack is similar in approach to a man-in-the-middle attack; however, in this case, a browser (or an extension or a plugin) is compromised and used to intercept and manipulate web sessions between the user and the web server.

If web applications do not validate and filter out invalid session ID values, they can potentially be used to exploit other web vulnerabilities, such as SQL injection (if the session IDs are stored on a relational database) or persistent XSS (if the session IDs are stored and reflected back afterward by the web application).

Default Credentials

A common adage in the security industry is, "Why do you need hackers if you have default passwords?" Many organizations and individuals leave infrastructure devices such as routers, switches, wireless access points, and even firewalls configured with default passwords.

Attackers can easily identify and access systems that use shared default passwords. It is extremely important to always change default manufacturer passwords and restrict network access to critical systems. A lot of manufacturers now require users to change the default passwords during initial setup, but some don't.

Attackers can easily obtain default passwords and identify Internet-connected target systems. Passwords can be found in product documentation and compiled lists available on the Internet. An example is http://www.defaultpassword.com, but there are dozens of other sites that contain default passwords and configurations on the Internet. It is easy to identify devices that have default passwords and that are exposed to the Internet by using search engines such as Shodan (https://www.shodan.io).

Insecure Direct Object Reference Vulnerabilities

Insecure Direct Object Reference vulnerabilities can be exploited when web applications allow direct access to objects based on user input. Successful exploitation could allow attackers to bypass authorization and access resources that should be protected by the system (for example, database records and system files). This vulnerability occurs when an application does not sanitize user input and does not perform appropriate authorization checks.

An attacker can take advantage of Insecure Direct Object References vulnerabilities by modifying the value of a parameter used to directly point to an object. In order to exploit this type of vulnerability, an attacker needs to map out all locations in the application where user input is used to reference objects directly. Example 1-1 shows how the value of a parameter can be used directly to retrieve a database record.

Example 1-1   *A URL Parameter Used Directly to Retrieve a Database Record*

```
https://store.h4cker.org/buy?customerID=1245
```

In this example, the value of the **customerID** parameter is used as an index in a table of a database holding customer contacts. The application takes the value and queries the database to obtain the specific customer record. An attacker may be able to change the value 1245 to another value and retrieve another customer record.

In Example 1-2, the value of a parameter is used directly to execute an operation in the system.

Example 1-2   *Direct Object Reference Example*

```
https://store.h4cker.org/changepassd?user=omar
```

In Example 1-2, the value of the **user** parameter (omar) is used to have the system change the user's password. An attacker can try other usernames and see if it is possible to modify the password of another user.

Mitigations for this type of vulnerability include input validation, the use of per-user or -session indirect object references, and access control checks to make sure the user is authorized for the requested object.

Cross-site Scripting (XSS)

Cross-site scripting (commonly known as XSS) vulnerabilities have become some of the most common web application vulnerabilities. XSS vulnerabilities are classified in three major categories:

- Reflected XSS

- Stored (persistent) XSS

- DOM-based XSS

Attackers can use obfuscation techniques in XSS attacks by encoding tags or malicious portions of the script using Unicode so that the link or HTML content is disguised to the end user browsing the site.

TIP   Dozens of examples of XSS vectors are listed at the GitHub repository https://github.com/The-Art-of-Hacking/h4cker, along with numerous other cybersecurity references.

Reflected XSS attacks (non-persistent XSS) occur when malicious code or scripts are injected by a vulnerable web application using any method that yields a response as part of a valid HTTP request. An example of a reflected XSS attack is a user being persuaded to follow a malicious link to a vulnerable server that injects (reflects) the malicious code back to the user's browser. This causes the browser to execute the code or script. In this case, the vulnerable server is usually a known or trusted site.

Examples of methods of delivery for XSS exploits are phishing emails, messaging applications, and search engines.

Stored, or persistent, XSS attacks occur when the malicious code or script is permanently stored on a vulnerable or malicious server, using a database. These attacks are typically carried out on websites hosting blog posts (comment forms), web forums, and other permanent storage methods. An example of a stored XSS attack is a user requesting the stored information from the vulnerable or malicious server, which causes the injection of the requested malicious script into the victim's browser. In this type of attack, the vulnerable server is usually a known or trusted site.

The Document Object Model (DOM) is a cross-platform and language-independent application programming interface that treats an HTML, XHTML, or XML document as a tree structure. DOM-based attacks are typically reflected XSS attacks that are triggered by sending a link with inputs that are reflected to the web browser. In DOM-based XSS attacks, the payload is never sent to the server. Instead, the payload is only processed by the web client (browser).

In a DOM-based XSS attack, the attacker sends a malicious URL to the victim, and after the victim clicks on the link, it may load a malicious website or a site that has a vulnerable DOM route handler. After the vulnerable site is rendered by the browser, the payload executes the attack in the user's context on that site.

One of the effects of any type of XSS attack is that the victim typically does not realize that an attack has taken place. DOM-based applications use global variables to manage client-side information. Often developers create unsecured applications that put sensitive information in the DOM (for example, tokens, public profile URLs, private URLs for information access, cross-domain OAuth values, and even user credentials as variables). It is a best practice to avoid storing any sensitive information in the DOM when building web applications.

Successful exploitation could result in installation or execution of malicious code, account compromise, session cookie hijacking, revelation or modification of local files, or site redirection.

The results of XSS attacks are the same regardless of the vector. Even though XSS vulnerabilities are flaws in a web application, the attack typically targets the end user. You typically find XSS vulnerabilities in the following:

- Search fields that echo a search string back to the user

- HTTP headers

- Input fields that echo user data

- Error messages that return user-supplied text

- Hidden fields that may include user input data

- Applications (or websites) that display user-supplied data

Example 1-3 demonstrates an XSS test that can be performed from a browser's address bar.

Example 1-3 *XSS Test from a Browser's Address Bar*

```
javascript:alert("Omar_s_XSS test");
javascript:alert(document.cookie);
```

Example 1-4 demonstrates an XSS test that can be performed in a user input field in a web form.

Example 1-4 *XSS Test from a Web Form*

```
<script>alert("XSS Test")</script>
```

Cross-site Request Forgery

Cross-site request forgery (CSRF or XSRF) attacks occur when unauthorized commands are transmitted from a user who is trusted by the application. CSRF attacks are different from XSS attacks because they exploit the trust that an application has in a user's browser. CSRF vulnerabilities are also referred to as "one-click attacks" or "session riding."

CSRF attacks typically affect applications (or websites) that rely on a user's identity. Attackers can trick the user's browser into sending HTTP requests to a target website. An example of a CSRF attack is a user authenticated by the application by a cookie saved in the browser unwittingly sending an HTTP request to a site that trusts the user, subsequently triggering an unwanted action.

Cookie Manipulation Attacks

Cookie manipulation attacks are often referred to as stored DOM-based attacks (or vulnerabilities). Cookie manipulation is possible when vulnerable applications store user input and then embed that input in a response within a part of the DOM. This input is later processed in an unsafe manner by a client-side script. An attacker can use a JavaScript string (or other scripts) to trigger the DOM-based vulnerability. Such scripts can write controllable data into the value of a cookie.

An attacker can take advantage of stored DOM-based vulnerabilities to create a URL that sets an arbitrary value in a user's cookie. The impact of a stored DOM-based vulnerability depends on the role that the cookie plays within the application.

Race Conditions

A race condition occurs when a system or an application attempts to perform two or more operations at the same time. However, due to the nature of such a system or application, the operations must be done in the proper sequence in order to be done correctly. When an attacker exploits such a vulnerability, he or she has a small window of time between when a security control takes effect and when the attack is performed. The attack complexity in race conditions is very high. In other words, race conditions are very difficult to exploit.

Race conditions are also referred to as *time of check to time of use (TOCTOU)* attacks. An example of a race condition is a security management system pushing a configuration to a security device (such as a firewall or an intrusion prevention system) such that the process rebuilds access control lists and rules from the system. An attacker might have a very small time window in which it could bypass those security controls until they take effect on the managed device.

Unprotected APIs

Application programming interfaces (APIs) are used everywhere today. A large number of modern applications use some type of APIs to allow other systems to interact with the application. Unfortunately, many APIs lack adequate controls and are difficult to monitor. The breadth and complexity of APIs also make it difficult to automate effective security testing. There are a few methods or technologies behind modern APIs:

- **Simple Object Access Protocol (SOAP):** This standards-based web services access protocol was originally developed by Microsoft and has been used by numerous legacy applications for many years. SOAP exclusively uses XML to provide API services. XML-based specifications are governed by XML Schema Definition (XSD) documents. SOAP was originally created to replace older solutions such as the Distributed Component Object Model (DCOM) and Common Object Request Broker Architecture (CORBA). You can find the latest SOAP specifications at https://www.w3.org/TR/soap.

- **Representational State Transfer (REST):** This API standard is easier to use than SOAP. It uses JSON instead of XML, and it uses standards such as Swagger and the OpenAPI Specification (https://www.openapis.org) for ease of documentation and to encourage adoption.

- **GraphQL:** GraphQL is a query language for APIs that provides many developer tools. GraphQL is now used for many mobile applications and online dashboards. Many different languages support GraphQL. You can learn more about GraphQL at https://graphql.org/code.

SOAP and REST use the HTTP protocol; however, SOAP limits itself to a more strict set of API messaging patterns than REST.

An API often provides a roadmap that describes the underlying implementation of an application. API documentation can provide a great level of detail that can be very valuable to a security professional, as well to attackers. API documentation can include the following:

- **Swagger (OpenAPI):** Swagger is a modern framework of API documentation and development that is the basis of the OpenAPI Specification (OAS). Additional information about Swagger can be obtained at https://swagger.io. The OAS specification is available at https://github.com/OAI/OpenAPI-Specification.

- **Web Services Description Language (WSDL) documents:** WSDL is an XML-based language that is used to document the functionality of a web service. The WSDL specification can be accessed at https://www.w3.org/TR/wsdl20-primer.

- **Web Application Description Language (WADL) documents:** WADL is an XML-based language for describing web applications. The WADL specification can be obtained from https://www.w3.org/Submission/wadl.

Return-to-LibC Attacks and Buffer Overflows

A "return-to-libc" (or ret2libc) attack typically starts with a buffer overflow. In this type of attack, a subroutine return address on a call stack is replaced by an address of a subroutine that is already present in the executable memory of the process. This is done to potentially bypass the no-execute (NX) bit feature and allow the attacker to inject his or her own code.

Operating systems that support non-executable stack help protect against code execution after a buffer overflow vulnerability is exploited. On the other hand, a non-executable stack cannot prevent a ret2libc attack because in this attack, only existing executable code is used. Another technique, called stack-smashing protection, can prevent or obstruct code execution exploitation because it can detect the corruption of the stack and can potentially "flush out" the compromised segment.

TIP The following video provides a detailed explanation of what buffer overflow attacks are: https://www.youtube.com/watch?v=1S0aBV-Waeo.

A technique called ASCII armoring can be used to mitigate ret2libc attacks. When you implement ASCII armoring, the address of every system library (such as libc) contains a NULL byte (0x00) that you insert in the first 0x01010101 bytes of memory. This is typically a few pages more than 16MB and is called the ASCII armor region because every address up to (but not including) this value contains at least one NULL byte. When this methodology is implemented, an attacker cannot place code containing those addresses using string manipulation functions such as **strcpy()**.

Of course, this technique doesn't protect the system if the attacker finds a way to overflow NULL bytes into the stack. A better approach is to use the address space layout randomization (ASLR) technique, which mitigates the attack on 64-bit systems. When you implement ASLR, the memory locations of functions are random. ASLR is not very effective in 32-bit systems, though, because only 16 bits are available for randomization, and an attacker can defeat such a system by using brute-force attacks.

OWASP Top 10

The Open Web Application Security Project (OWASP) is non-profit charitable organization that leads several industry-wide initiatives to promote the security of applications and software. They list the top 10 most common vulnerabilities against applications at their website at the following address:

https://www.owasp.org/index.php/Category:OWASP_Top_Ten_Project

> **TIP**   It is recommended that you become familiar and always keep up with the OWASP Top 10 list. OWASP not only defines each of the vulnerabilities, but they also provide a list of techniques to prevent and mitigate those vulnerabilities. OWASP also has local chapters around the world that are free and open to anyone. Many chapters also have meetings, presentations, and training that help the community. Information about the OWASP local chapters can be obtained at https://www.owasp.org/index.php/OWASP_Chapter.

Security Vulnerabilities in Open Source Software

Security vulnerability patching for commercial and open source software is one of the most important processes of any organization. An organization might use the following technologies and systems to maintain an appropriate vulnerability management program:

- Vulnerability management software and scanners, such as Qualys, Nexpose, and Nessus

- Software composition analysis tools, such as BlackDuck Hub, Synopsys Protecode (formerly known as AppCheck), FlexNet Code Insight (formerly known as Palamida), SourceClear, and WhiteSource

- Security vulnerability feeds, such as MITRE's CVE list, NIST's National Vulnerability Database (NVD), VulnDB, and Recorded Future

Confidentiality, Integrity, and Availability

The elements of confidentiality, integrity, and availability are often described as the CIA model. It is easy to guess that the first thing that popped into your mind when you read those three letters was the United States Central Intelligence Agency. In the world of cybersecurity, these three letters represent something we strive to attain and protect. Confidentiality, integrity, and availability (CIA) are the unifying attributes of an information security program. Collectively referred to as the CIA triad or CIA security model, each attribute represents a fundamental objective of information security.

You may be wondering which is most important: confidentiality, integrity, or availability? The answer requires an organization to assess its mission, evaluate its services, and consider regulations and contractual agreements. As Figure 1-7 illustrates, organizations might consider all three components of the CIA triad equally important, in which case resources must be allocated proportionally.

What Is Confidentiality?

When you tell a friend something in confidence, you expect them to keep the information private and not share what you told them with anyone else without your permission. You also hope that they will never use this against you. Likewise, confidentiality is the requirement that private or confidential information not be disclosed to unauthorized individuals.

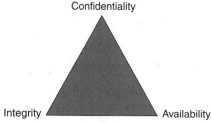

Figure 1-7 *The CIA Triad*

There are many attempts to define what confidentiality is. As an example, the ISO 2700 standard provides a good definition of confidentiality as "the property that information is not made available or disclosed to unauthorized individuals, entities, or processes."

Confidentiality relies on three general concepts, as illustrated in Figure 1-8.

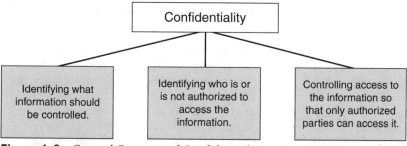

Figure 1-8 *General Concepts of Confidentiality*

There are several ways to protect the confidentiality of a system or its data; one of the most common is to use encryption. This includes encryption of data in transit with the use of site-to-site and remote access virtual private networks (VPNs) or by deploying server- and client-side encryption using Transport Layer Security (TLS).

Another important element of confidentiality is that all sensitive data needs to be controlled, audited, and monitored at all times. This is often done by encrypting data at rest. Here are some examples of sensitive data:

- Social security numbers

- Bank and credit card account information

- Criminal records

- Patient and health records

- Trade secrets

- Source code

- Military secrets

Data often is protected by law, regulation, memorandum of agreement, contractual obligation, or management discretion. Examples include nonpublic personal information (NPPI) and personally identifiable information (PII), such as Social Security number, driver's license or state-issued identification number, bank account or financial account numbers, payment card information (PCI), which is credit or debit cardholder information, and personal health information (PHI).

The following are examples of security mechanisms designed to preserve confidentiality:

- Logical and physical access controls
- Encryption (in motion and at rest)
- Database views
- Controlled traffic routing

Data classification is important when you're deciding how to protect data. By having a good data classification methodology, you can enhance the way you secure your data across your network and systems.

Not only has the amount of information stored, processed, and transmitted on privately owned networks and the public Internet increased dramatically, so has the number of ways to potentially access the data. The Internet, its inherent weaknesses, and those willing (and able) to exploit vulnerabilities are the main reasons why protecting confidentiality has taken on a new urgency. The technology and accessibility we take for granted would have been considered magic just 10 years ago. The amazing speed at which we arrived here is also the reason we have such a gap in security. The race to market often means that security is sacrificed. So although it might seem that information security requirements are a bit extreme at times, it is really a reaction to the threat environment.

Because there is value in confidential information, it is often a target of cybercriminals. For instance, many breaches involve the theft of credit card information or other personal information useful for identity theft. Criminals look for and are prepared to exploit weaknesses in network designs, software, communication channels, and people to access confidential information. The opportunities are plentiful.

Criminals are not always outsiders. Insiders can be tempted to "make copies" of information they have access to for financial gain, notoriety, or to "make a statement." The most recent threat to confidentiality is hacktivism, which is a combination of the terms "hack" and "activism." Hacktivism has been described as the fusion of hacking and activism, politics, and technology. Hacktivist groups or collectives expose or hold hostage illegally obtained information to make a political statement or for revenge.

What Is Integrity?

Whenever the word integrity comes to mind, so does Brian De Palma's classic 1987 film *The Untouchables*, starring Kevin Costner and Sean Connery. The film is about a group of police officers who could not be "bought off" by organized crime. They were incorruptible. Integrity is certainly one of the highest ideals of personal character. When we say someone has integrity, we mean she lives her life according to a code of ethics; she can be trusted to behave in certain ways in certain situations. It is interesting to note that those to whom we ascribe the quality of integrity can be trusted with our confidential information. As for information security, integrity has a very similar meaning. Integrity is basically the ability to make sure that a system and its data has not been altered or compromised. It ensures that the data is an accurate and unchanged representation of the original secure data. Integrity applies not only to data, but also to systems. For instance, if a threat actor changes the configuration of a server, firewall, router, switch, or any other infrastructure device, it is considered that he or she impacted the integrity of the system.

Data integrity is a requirement that information and programs are changed only in a specified and authorized manner. In other words, is the information the same as it was intended to be?

System integrity is a requirement that a system performs its intended function in an unimpaired manner, free from deliberate or inadvertent unauthorized manipulation of the system. Malware that corrupts some of the system files required to boot the computer is an example of deliberate unauthorized manipulation.

Errors and omissions are an important threat to data and system integrity. These errors are caused not only by data entry clerks processing hundreds of transactions per day, but also by all types of users who create and edit data and code. Even the most sophisticated programs cannot detect all types of input errors or omissions. In some cases, the error is the threat, such as a data entry error or a programming error that crashes a system. In other cases, the errors create vulnerabilities. Programming and development errors, often called "bugs," can range in severity from benign to catastrophic.

To make this a bit more personal, let's talk about medical and financial information. What if you are injured, unconscious, and taken to the emergency room of a hospital, and the doctors need to look up your health information? You would want it to be correct, wouldn't you? Consider what might happen if you had an allergy to some very common treatment, and this critical information had been deleted from your medical records. Or think of your dismay if you check your bank balance after making a deposit and find that the funds have not been credited to your account!

Integrity and confidentiality are interrelated. If a user password is disclosed to the wrong person, that person could in turn manipulate, delete, or destroy data after gaining access to the system with the password he obtained. Many of the same vulnerabilities that threaten integrity also threaten confidentiality. Most notable, though, is human error. Safeguards that protect against the loss of integrity include access controls, such as encryption and digital signatures; process controls, such as code testing; monitoring controls, such as file integrity monitoring and log analysis; and behavioral controls, such as separation of duties, rotation of duties, and training.

What Is Availability?

The last component of the CIA triad is availability, which states that systems, applications, and data must be available to authorized users when needed and requested. The most common attack against availability is a denial-of-service (DoS) attack. User productivity can be greatly affected, and companies can lose a lot of money if data is not available. For example, if you are an online retailer or a cloud service provider and your ecommerce site or service is not available to your users, you could potentially lose current or future business, thus impacting revenue.

In fact, availability is generally one of the first security issues addressed by Internet service providers (ISPs). You might have heard the expressions "uptime" and "5-9s" (99.999% uptime). This means the systems that serve Internet connections, web pages, and other such services will be available to users who need them when they need them. Service providers frequently use service level agreements (SLAs) to assure their customers of a certain level of availability.

Just like confidentiality and integrity, we prize availability. Not all threats to availability could be malicious. For example, human error or a misconfigured server or infrastructure device can cause a network outage that will have a direct impact to availability. We are more vulnerable to availability threats than to the other components of the CIA triad. We are certain to face some of them. Safeguards that address availability include access controls, monitoring, data redundancy, resilient systems, virtualization, server clustering, environmental controls, continuity of operations planning, and incident response preparedness.

Talking About Availability, What Is a Denial-of-Service (DoS) Attack?

Denial-of-service (DoS) and distributed DoS (DDoS) attacks have been around for quite some time now, but there has been heightened awareness of them over the past few years. A DoS attack typically uses one system and one network connection to perform a denial-of-service condition to a targeted system, network, or resource. DDoS attacks use multiple computers and network connections that can be geographically dispersed (that is, distributed) to perform a denial-of-service condition against the victim.

DDoS attacks can generally be divided into the following three categories:

- Direct DDoS attacks

- Reflected DDoS attacks

- Amplification DDoS attacks

Direct denial-of-service attacks occur when the source of the attack generates the packets, regardless of protocol, application, and so on, that are sent directly to the victim of the attack.

Figure 1-9 illustrates a direct denial-of-service attack.

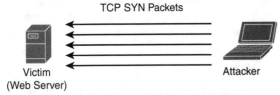

Figure 1-9 *Direct Denial-of-Service Attack*

In Figure 1-9 the attacker launches a direct DoS to the victim (a web server) by sending numerous TCP SYN packets. This type of attack is aimed at flooding the victim with an overwhelming number of packets, oversaturating its connection bandwidth, or depleting the target's system resources. This type of attack is also known as a "SYN flood attack."

Reflected DDoS attacks occur when the sources of the attack are sent spoofed packets that appear to be from the victim, and then the sources become unwitting participants in the DDoS attacks by sending the response traffic back to the intended victim. UDP is often used as the transport mechanism because it is more easily spoofed due to the lack of a three-way handshake. For example, if the attacker (A) decides he wants to attack a victim (V), he will send packets (for example, Network Time Protocol [NTP] requests) to a source (S) that thinks these packets are legitimate. The source then responds to the NTP requests by sending the responses to the victim, who was never expecting these NTP packets from the source, as shown in Figure 1-10.

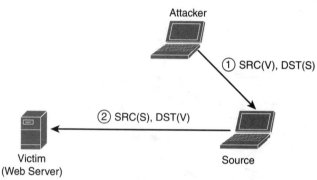

Figure 1-10 *Reflected Denial-of-Service Attack*

An amplification attack is a form of reflected attack in which the response traffic (sent by the unwitting participant) is made up of packets that are much larger than those that were initially sent by the attacker (spoofing the victim). An example is when DNS queries are sent, and the DNS responses are much larger in packet size than the initial query packets. The end result is that the victim's machine gets flooded by large packets for which it never actually issued queries.

Another type of DoS is caused by exploiting vulnerabilities such as buffer overflows to cause a server or even network infrastructure device to crash, subsequently causing a denial-of-service condition.

Many attackers use botnets to launch DDoS attacks. A botnet is a collection of compromised machines that the attacker can manipulate from a command-and-control (often referred to as a C2 or CnC) system to participate in a DDoS, send spam emails, and perform other illicit activities. Figure 1-11 shows how a botnet is used by an attacker to launch a DDoS attack.

In Figure 1-11, the attacker sends instructions to compromised systems. These compromised systems can be end-user machines or IoT devices such as cameras, sensors, and so on.

Access Control Management

Access controls are security features that govern how users and processes communicate and interact with systems and resources. The objective of implementing access controls is to ensure that authorized users and processes are able to access information and resources while unauthorized users and processes are prevented from access to the same. Access control models refer to the active entity that requests access to an object or data as the subject and the passive entity being accessed or being acted upon as the object.

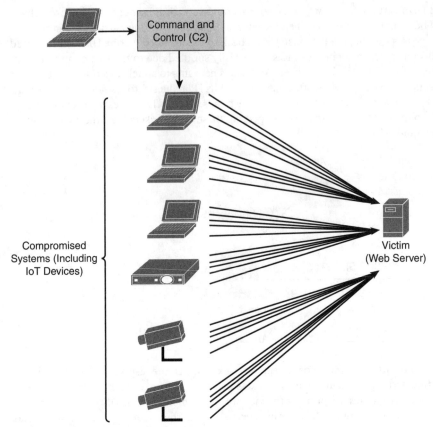

Figure 1-11 *Example of a Botnet*

An organization's approach to access controls is referred to as its security posture. There are two fundamental approaches—open and secure. Open, also referred to as *default allow*, means that access not explicitly forbidden is permitted. Secure, also referred to as *default deny*, means that access not explicitly permitted is forbidden. Access decisions should consider the security principles of *need to know* and *least privilege*. Need to know means having a demonstrated and authorized reason for being granted access to information. Least privilege means granting subjects the minimum level of access required to perform their job or function.

Gaining access is a three-step process:

1. The object recognizes the subject. Identification is the process of the subject supplying an identifier such as a username to the object.

2. Providing proof that the subjects are who they say they are. Authentication is the process of the subject supplying verifiable credentials to the object.

3. Determining the actions a subject can take. Authorization is the process of assigning authenticated subjects the rights and permissions needed to carry out a specific operation.

Authentication credentials are called factors. There are three categories of factors:

- Knowledge (something the user knows)

- Possession (something a user has)

- Inherence (something the user is)

Single-factor authentication is when only one factor is presented. Multifactor authentication is when two or more factors are presented. Multilayer authentication is when two or more of the same type of factor are presented. Out-of-band authentication requires communication over a channel that is distinct from the first factor. Data classification, regulatory requirement, the impact of unauthorized access, and the likelihood of a threat being exercised must all be considered when deciding on the level of authentication required.

Once authentication is complete, an authorization model defines how subjects access objects. Mandatory access controls (MACs) are defined by policy and cannot be modified by the information owner. Discretionary access controls (DACs) are defined by the owner of the object. Role-based access controls (RBACs, also called nondiscretionary) are access permissions based on a specific role or function. In a rule-based access controls environment, access is based on criteria independent of the user or group account, such as time of day or location.

Cloud Security Threats

Many organizations are moving to the cloud or deploying hybrid solutions to host their applications. Organizations moving to the cloud are almost always looking to transition from capital expenditure (CapEx) to operational expenditure (OpEx). Most of Fortune 500 companies operate in a multicloud environment. It is obvious that cloud computing security is more important than ever. Cloud computing security includes many of the same functionalities as traditional IT security. This includes protecting critical information from theft, data exfiltration, and deletion, as well as privacy.

The National Institute of Standards and Technology (NIST) authored Special Publication (SP) 800-145, "The NIST Definition of Cloud Computing," to provide a standard set of definitions for the different aspects of cloud computing. The SP 800-145 document also compares the different cloud services and deployment strategies.

The advantages of using a cloud-based service include the following:

- Distributed storage

- Scalability

- Resource pooling

- Access from any location

- Measured service

- Automated management

According to NIST, the essential characteristics of cloud computing include the following:

- On-demand self-service
- Broad network access
- Resource pooling
- Rapid elasticity
- Measured service

Cloud deployment models include the following:

- **Public cloud:** Open for public use
- **Private cloud:** Used just by the client organization on the premises (on-prem) or at a dedicated area in a cloud provider
- **Community cloud:** Shared between several organizations
- **Hybrid cloud:** Composed of two or more clouds (including on-prem services).

Cloud computing can be broken into the following three basic models:

- **Infrastructure as a Service (IaaS):** IaaS describes a cloud solution where you are renting infrastructure. You purchase virtual power to execute your software as needed. This is much like running a virtual server on your own equipment, except you are now running a virtual server on a virtual disk. This model is similar to a utility company model because you pay for what you use.

- **Platform as a Service (PaaS):** PaaS provides everything except applications. Services provided by this model include all phases of the system development life cycle (SDLC) and can use application programming interfaces (APIs), website portals, or gateway software. These solutions tend to be proprietary, which can cause problems if the customer moves away from the provider's platform.

- **Software as a Service (SaaS):** SaaS is designed to provide a complete packaged solution. The software is rented out to the user. The service is usually provided through some type of front end or web portal. While the end user is free to use the service from anywhere, the company pays a per-use fee.

> **NOTE**   NIST Special Publication 500-292, "NIST Cloud Computing Reference Architecture," is another resource for learning more about cloud architecture.

Cloud Computing Issues and Concerns

There are many potential threats when organizations move to a cloud model. For example, although your data is in the cloud, it must reside in a physical location somewhere. Your cloud provider should agree in writing to provide the level of security required for your customers. The following are questions to ask a cloud provider before signing a contract for its services:

- **Who has access?** Access control is a key concern because insider attacks are a huge risk. Anyone who has been approved to access the cloud is a potential hacker, so you want to know who has access and how they were screened. Even if it was not done with malice, an employee can leave, and then you find out that you don't have the password, or the cloud service gets canceled because maybe the bill didn't get paid.

- **What are your regulatory requirements?** Organizations operating in the United States, Canada, or the European Union have many regulatory requirements that they must abide by (for example, ISO/IEC 27002, EU-U.S. Privacy Shield Framework, ITIL, and COBIT). You must ensure that your cloud provider can meet these requirements and is willing to undergo certification, accreditation, and review.

- **Do you have the right to audit?** This particular item is no small matter in that the cloud provider should agree in writing to the terms of the audit. With cloud computing, maintaining compliance could become more difficult to achieve and even harder to demonstrate to auditors and assessors. Of the many regulations touching upon information technology, few were written with cloud computing in mind. Auditors and assessors might not be familiar with cloud computing generally or with a given cloud service in particular.

NOTE   Division of compliance responsibilities between cloud provider and cloud customer must be determined before any contracts are signed or service is started.

- **What type of training does the provider offer its employees?** This is a rather important item to consider because people will always be the weakest link in security. Knowing how your provider trains its employees is an important item to review.

- **What type of data classification system does the provider use?** Questions you should be concerned with here include what data classification standard is being used and whether the provider even uses data classification.

- **How is your data separated from other users' data?** Is the data on a shared server or a dedicated system? A dedicated server means that your information is the only thing on the server. With a shared server, the amount of disk space, processing power, bandwidth, and so on is limited because others are sharing this device. If it is shared, the data could potentially become comingled in some way.

- **Is encryption being used?** Encryption should be discussed. Is it being used while the data is at rest and in transit? You will also want to know what type of encryption is being used. For example, there are big technical difference between DES and AES. For both of these algorithms, however, the basic questions are the same: Who maintains control of the encryption keys? Is the data encrypted at rest in the cloud? Is the data encrypted in transit, or is it encrypted at rest and in transit?

- **What are the service level agreement (SLA) terms?** The SLA serves as a contracted level of guaranteed service between the cloud provider and the customer that specifies what level of services will be provided.

- **What is the long-term viability of the provider?** How long has the cloud provider been in business, and what is its track record? If it goes out of business, what happens to your data? Will your data be returned and, if so, in what format?

- **Will the provider assume liability in the case of a breach?** If a security incident occurs, what support will you receive from the cloud provider? While many providers promote their services as being unhackable, cloud-based services are an attractive target to hackers.

- **What is the disaster recovery/business continuity plan (DR/BCP)?** Although you might not know the physical location of your services, it is physically located somewhere. All physical locations face threats such as fire, storms, natural disasters, and loss of power. In case of any of these events, how will the cloud provider respond, and what guarantee of continued services is it promising?

Even when you end a contract, you must ask what happens to the information after your contract with the cloud service provider ends.

NOTE   Insufficient due diligence is one of the biggest issues when moving to the cloud. Security professionals must verify that issues such as encryption, compliance, incident response, and so forth are all worked out before a contract is signed.

Cloud Computing Attacks

Because cloud-based services are accessible via the Internet, they are open to any number of attacks. As more companies move to cloud computing, look for hackers to follow. Some of the potential attack vectors criminals might attempt include the following:

- **Session hijacking:** This attack occurs when the attacker can sniff traffic and intercept traffic to take over a legitimate connection to a cloud service.

- **DNS attack:** This form of attack tricks users into visiting a phishing site and giving up valid credentials.

- **Cross-site scripting (XSS):** Used to steal cookies that can be exploited to gain access as an authenticated user to a cloud-based service.

- **SQL injection:** This attack exploits vulnerable cloud-based applications that allow attackers to pass SQL commands to a database for execution.

- **Session riding:** This term is often used to describe a cross-site request forgery attack. Attackers use this technique to transmit unauthorized commands by riding an active session by using an email or malicious link to trick users while they are currently logged in to a cloud service.

- **Distributed denial-of-service (DDoS) attack:** Some security professionals have argued that the cloud is more vulnerable to DDoS attacks because it is shared by many users and organizations, which also makes any DDoS attack much more damaging.

- **Man-in-the-middle cryptographic attack:** This attack is carried out when the attacker places himself in the communication path between two users. Anytime the attacker can do this, there is the possibility that he can intercept and modify communications.

■ **Side-channel attack:** An attacker could attempt to compromise the cloud by placing a malicious virtual machine in close proximity to a target cloud server and then launching a side-channel attack.

■ **Authentication attack:** Authentication is a weak point in hosted and virtual services and is frequently targeted. There are many ways to authenticate users, such as based on what a person knows, has, or is. The mechanisms used to secure the authentication process and the method of authentication used are frequent targets of attackers.

■ **API attacks:** Often APIs are configured insecurely. An attacker can take advantage of API misconfigurations to modify, delete, or append data in applications or systems in cloud environments.

Cloud Computing Security

Regardless of the model used, cloud security is the responsibility of both the client and the cloud provider. These details will need to be worked out before a cloud computing contract is signed. The contracts will vary depending on the given security requirements of the client. Considerations include disaster recovery, SLAs, data integrity, and encryption. For example, is encryption provided end to end or just at the cloud provider? Also, who manages the encryption keys: the cloud provider or the client? Overall, you want to ensure that the cloud provider has the same layers of security (logical, physical, and administrative) in place that you would have for services you control. You will learn details on how to secure cloud environments in Chapter 9, "Securing the Cloud."

IoT Security Threats

The Internet of Things (IoT) includes any computing devices (mechanical and digital machines) that can transfer data over a network without requiring human-to-human or human-to-computer interaction—for example, sensors, home appliances, connected security cameras, wearables, and numerous other devices.

The capability of distributed intelligence in the network is a core architectural component of the IoT:

■ **Data collection:** Centralized data collection presents a few challenges for an IoT environment to be able to scale. For instance, managing millions of sensors in a smart grid network cannot efficiently be done using a centralized approach.

■ **Network resource preservation:** This is particularly important because network bandwidth may be limited, and centralized IoT device data collection leads to using a large amount of the network capabilities.

■ **Closed-loop functioning:** IoT environments often require reduced reaction times.

Fog computing is a concept of a distributed intelligence architecture designed to process data and events from IoT devices as close to the source as possible. The fog-edge device then sends the required data to the cloud. For example, a router might collect information from numerous sensors and then communicate to a cloud service or application for the processing of such data.

The following are some of the IoT security challenges and considerations:

- Numerous IoT devices are inexpensive devices with little to no security capabilities.

- IoT devices are typically constrained in memory and compute resources and do not support complex and evolving security and encryption algorithms.

- Several IoT devices are deployed with no backup connectivity if the primary connection is lost.

- Numerous IoT devices require secure remote management during and after deployment (onboarding).

- IoT devices often require the management of multiparty networks. Governance of these networks is often a challenging task. For example, who will accept liability for a breach? Who is in charge of incident response? Who has provisioning access? Who has access to the data?

- Crypto resilience is a challenge in many IoT environments. These embedded devices (such as smart meters) are designed to last decades without being replaced.

- Physical protection is another challenge, because any IoT device could be stolen, moved, or tampered with.

- Administrations should pay attention to how the IoT device authenticates to multiple networks securely.

- IoT technologies like INSTEON, Zigbee, Z-Wave, LoRaWAN, and others were not designed with security in mind (however, they have improved significantly over the past few years).

IoT devices typically communicate to the cloud via a fog-edge device or directly to the cloud. Figure 1-12 shows several sensors are communicating to a fog-edge router, and subsequently the router communicates to the cloud.

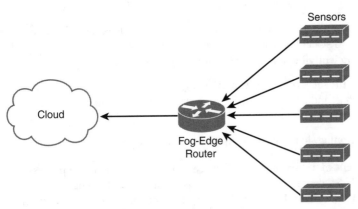

Figure 1-12 *Example of a Fog-Edge Device*

Figure 1-13 shows how a smart thermostat communicates directly to the cloud using a RESTful API via a Transport Layer Security (TLS) connection. The IoT device (a smart thermostat in this example) sends data to the cloud, and an end user checks the temperature and manages the thermostat using a mobile application.

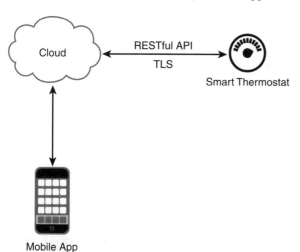

Figure 1-13   *IoT, Cloud Applications, and APIs*

In the example illustrated in Figure 1-13, securing the thermostat, the RESTful API, the cloud application, and the mobile application is easier said than done.

IoT Protocols

The following are some of the most popular IoT protocols:

- **Zigbee:** One of the most popular protocols supported by many consumer IoT devices. Zigbee takes advantage of the underlying security services provided by the IEEE 802.15.4 MAC layer. The 802.15.4 MAC layer supports the AES algorithm with a 128-bit key for both encryption and decryption. Additional information about Zigbee can be obtained from the Zigbee Alliance at https://www.zigbee.org.

- **Bluetooth Low Energy (BLE) and Bluetooth Smart:** BLE is an evolution of the Bluetooth protocol that is designed for enhanced battery life for IoT devices. Bluetooth Smart–enabled devices default to "sleep mode" and "wake up" only when needed. Both operate in the 2.4 GHz frequency range. Bluetooth Smart implements high-rate frequency-hopping spread spectrum and supports AES encryption. Additional information about BLE and Bluetooth Smart can be found at https://www.bluetooth.com.

- **Z-Wave:** Another popular IoT communication protocol. It supports unicast, multicast, and broadcast communication. Z-Wave networks consist of controllers and slaves. Some Z-Wave devices can be both primary and secondary controllers. Primary controllers are allowed to add and remove nodes form the network. Z-Wave devices operate at a frequency of 908.42 MHz (North America) and 868.42 MHz (Europe) with data rates of 100Kbps over a range of about 30 meters. Additional information about Z-Wave can be obtained from the Z-Wave Alliance at https://z-wavealliance.org.

■ **INSTEON:** A protocol that allows IoT devices to communicate wirelessly and over the power lines. It provides support for dual-band, mesh, and peer-to-peer communication. Additional information about INSTEON can be found at https://www.insteon.com/technology/.

■ **Long Range Wide Area Network (LoRaWAN):** A networking protocol designed specifically for IoT implementations. LoRaWAN has three classes of endpoint devices: Class A (lowest power, bidirectional end devices), Class B (bidirectional end devices with deterministic downlink latency), and Class C (lowest latency, bidirectional end devices). Additional information about LoRaWAN can be found at the Lora Alliance at https://lora-alliance.org.

■ **Wi-Fi:** Still one of the most popular communication methods for IoT devices.

■ **Low Rate Wireless Personal Area Networks (LRWPAN) and IPv6 over Low Power Wireless Personal Area Networks (6LoWPAN):** IPv4 and IPv6 both play a role at various points within many IoT systems. IPv6 over Low Power Wireless Personal Area Networks (6LoWPAN) supports the use of IPv6 in the network-constrained IoT implementations. 6LoWPan was designed to support wireless Internet connectivity at lower data rates. 6LoWPAN builds upon the 802.15.4 Low Rate Wireless Personal Area Networks (LRWPAN) specification to create an adaptation layer that supports the use of IPv6.

■ **Cellular Communication:** Also a popular communication method for IoT devices, including connected cars, retail machines, sensors, and others. 4G and 5G are used to connect many IoT devices nowadays.

IoT devices often communicate to applications using REST and MQTT on top of lower-layer communication protocols. These messaging protocols provide the ability for both IoT clients and application servers to efficiently agree on data to exchange. The following are some of the most popular IoT messaging protocols:

■ MQTT

■ Constrained Application Protocol (CoAP)

■ Data Distribution Protocol (DDP)

■ Advanced Message Queuing Protocol (AMQP)

■ Extensible Messaging and Presence Protocol (XMPP)

Hacking IoT Implementations

Many of the tools and methodologies for hacking applications and network apply to IoT hacking; however, several specialized tools perform IoT hardware and software hacking.

The following are a few examples of tools and methods to hack IoT devices.

■ **Hardware tools:**

  ■ Multimeters

  ■ Oscilloscopes

- Soldering tools

- UART debuggers and tools

- Universal interface tools like JTAG, SWD, I2C, and SPI tools

- Logic analyzers

- **Reverse engineering tools, such as disassemblers and debuggers:**

  - IDA

  - Binary Ninja

  - Radare2

  - Ghidra

  - Hopper

- **Wireless communication interfaces and tools:**

  - Ubertooth One (for Bluetooth hacking)

  - Software-defined radio (SDR), such as HackRF and BladeRF, to perform assessments of Z-Wave and Zigbee implementations

An Introduction to Digital Forensics and Incident Response

Cybersecurity-related incidents have become not only more numerous and diverse, but also more damaging and disruptive. A single incident can cause the demise of an entire organization. In general terms, incident management is defined as a predictable response to damaging situations. It is vital that organizations have the practiced capability to respond quickly, minimize harm, comply with breach-related state laws and federal regulations, and maintain their composure in the face of an unsettling and unpleasant experience.

ISO/IEC 27002:2013 and NIST Incident Response Guidance

Section 16 of ISO 27002:2013, "Information Security Incident Management," focuses on ensuring a consistent and effective approach to the management of information security incidents, including communication on security events and weaknesses.

Corresponding NIST guidance is provided in the following documents:

- **SP 800-61 Revision 2:** "Computer Security Incident Handling Guide"

- **SP 800-83:** "Guide to Malware Incident Prevention and Handling"

- **SP 800-86:** "Guide to Integrating Forensic Techniques into Incident Response"

Incidents drain resources, can be very expensive, and can divert attention from the business of doing business. Keeping the number of incidents as low as possible should be an organizational priority. That means as much as possible identifying and remediating weaknesses and vulnerabilities before they are exploited. A sound approach to improving an organizational security posture and preventing incidents is to conduct periodic risk assessments of systems and applications. These assessments should determine what risks are posed by combinations of threats, threat sources, and vulnerabilities. Risks can be mitigated, transferred, or avoided until a reasonable overall level of acceptable risk is reached. However, it is important to realize that users will make mistakes, external events may be out of an organization's control, and malicious intruders are motivated. Unfortunately, even the best prevention strategy isn't always enough, which is why preparation is key.

Incident preparedness includes having policies, strategies, plans, and procedures. Organizations should create written guidelines, have supporting documentation prepared, train personnel, and engage in mock exercises. An actual incident is not the time to learn. Incident handlers must act quickly and make far-reaching decisions—often while dealing with uncertainty and incomplete information. They are under a great deal of stress. The more prepared they are, the better the chance that sound decisions will be made.

Computer security incident response is a critical component of information technology (IT) programs. The incident response process and incident handling activities can be very complex. To establish a successful incident response program, you must dedicate substantial planning and resources. Several industry resources were created to help organizations establish a computer security incident response program and learn how to handle cybersecurity incidents efficiently and effectively. One of the best resources available is NIST Special Publication 800-61, which can be obtained from the following URL:

http://nvlpubs.nist.gov/nistpubs/SpecialPublications/NIST.SP.800-61r2.pdf

NIST developed Special Publication 800-61 due to statutory responsibilities under the Federal Information Security Management Act (FISMA) of 2002, Public Law 107-347.

The benefits of having a practiced incident response capability include the following:

- Calm and systematic response

- Minimization of loss or damage

- Protection of affected parties

- Compliance with laws and regulations

- Preservation of evidence

- Integration of lessons learned

- Lower future risk and exposure

What Is an Incident?

A cybersecurity incident is an adverse event that threatens business security and/or disrupts service. Sometimes confused with a disaster, an information security incident is related to loss of confidentiality, integrity, or availability (CIA), whereas a disaster is an event that results in widespread damage or destruction, loss of life, or drastic change to the environment. Examples of incidents include exposure of or modification of legally protected data, unauthorized access to intellectual property, or disruption of internal or external services. The starting point of incident management is to create an organization-specific definition of the term *incident* so that the scope of the term is clear. Declaration of an incident should trigger a mandatory response process.

Not all security incidents are the same. For example, a breach of personally identifiable information (PII) typically requires strict disclosure under many circumstances.

Before you learn the details about how to create a good incident response program within your organization, you must understand the difference between security "events" and security "incidents." The following is from NIST Special Publication 800-61:

> "An event is any observable occurrence in a system or network. Events include a user connecting to a file share, a server receiving a request for a web page, a user sending email, and a firewall blocking a connection attempt. Adverse events are events with a negative consequence, such as system crashes, packet floods, unauthorized use of system privileges, unauthorized access to sensitive data, and execution of malware that destroys data."

According to the same document, "a computer security incident is a violation or imminent threat of violation of computer security policies, acceptable use policies, or standard security practices."

The definition and criteria should be codified in policy. Incident management extends to third-party environments. Business partners and vendors should be contractually obligated to notify the organization if an actual or suspected incident occurs.

The following are a few examples of cybersecurity incidents:

- Attacker sends a crafted packet to a router and causes a denial-of-service condition.

- Attacker compromises a point-of-sale (POS) system and steals credit card information.

- Attacker compromises a hospital database and steals thousands of health records.

- Ransomware is installed in a critical server and all files are encrypted by the attacker.

False Positives, False Negatives, True Positives, and True Negatives

The term *false positive* is a broad term that describes a situation in which a security device triggers an alarm but there is no malicious activity or an actual attack taking place. In other words, false positives are "false alarms," and they are also called "benign triggers." False positives are problematic because by triggering unjustified alerts, they diminish the value and urgency of real alerts. If you have too many false positives to investigate, it becomes an operational nightmare, and you most definitely will overlook real security events.

There are also *false negatives*, which is the term used to describe a network intrusion device's inability to detect true security events under certain circumstances. In other words, a malicious activity that is not detected by the security device.

A *true positive* is a successful identification of a security attack or a malicious event. A *true negative* is when the intrusion detection device identifies an activity as acceptable behavior and the activity is actually acceptable.

Traditional IDS and IPS devices need to be tuned to avoid false positives and false negatives. Next-generation IPSs do not need the same level of tuning compared to a traditional IPS.

Also, you can obtain much deeper reports and functionality, including advanced malware protection and retrospective analysis to see what happened after an attack took place.

Traditional IDS and IPS devices also suffer from many evasion attacks. The following are some of the most common evasion techniques against traditional IDS and IPS devices:

- **Fragmentation:** When the attacker evades the IPS box by sending fragmented packets.

- **Using low-bandwidth attacks:** When the attacker uses techniques that use low bandwidth or a very small number of packets in order to evade the system.

- **Address spoofing/proxying:** Using spoofed IP addresses or sources, as well as using intermediary systems such as proxies to evade inspection.

- **Pattern change evasion:** Attackers may use polymorphic techniques to create unique attack patterns.

- **Encryption:** Attackers can use encryption to hide their communication and information.

Incident Severity Levels

Not all incidents are equal in severity. Included in the incident definition should be severity levels based on the operational, reputational, and legal impact to the organization. Corresponding to the level should be required response times as well as minimum standards for internal notification.

A cybersecurity incident is any adverse event whereby some aspect of an information system or information itself is threatened. Incidents are classified by severity relative to the impact they have on an organization. This severity level is typically assigned by an incident manager or a cybersecurity investigator. How it is validated depends on the organizational structure and the incident response policy. Each level has a maximum response time and minimum internal notification requirements.

How Are Incidents Reported?

Incident reporting is best accomplished by implementing simple, easy-to-use mechanisms that can be used by all employees to report the discovery of an incident. Employees should be required to report all actual and suspected incidents. They should not be expected to assign a severity level, because the person who discovers an incident may not have the skill, knowledge, or training to properly assess the impact of the situation.

People frequently fail to report potential incidents because they are afraid of being wrong and looking foolish, they do not want to be seen as a complainer or whistleblower, or they simply don't care enough and would prefer not to get involved. These objections must be countered by encouragement from management. Employees must be assured that even if they were to report a perceived incident that ended up being a false positive, they would not be ridiculed or met with annoyance. On the contrary, their willingness to get involved for the greater good of the company is exactly the type of behavior the company needs! They should be supported for their efforts and made to feel valued and appreciated for doing the right thing.

Digital forensic evidence is information in digital form found on a wide range of endpoint, server, and network devices—basically, any information that can be processed by a computing device or stored on other media. Evidence tendered in legal cases, such as criminal trials, is classified as witness testimony or direct evidence, or as indirect evidence in the form of an object, such as a physical document, the property owned by a person, and so forth.

Cybersecurity forensic evidence can take many forms, depending on the conditions of each case and the devices from which the evidence was collected. To prevent or minimize contamination of the suspect's source device, you can use different tools, such as a piece of hardware called a write blocker, on the specific device so you can copy all the data (or an image of the system).

The imaging process is intended to copy all blocks of data from the computing device to the forensics professional evidentiary system. This is sometimes referred to as a "physical copy" of all data, as distinct from a logical copy, which copies only what a user would normally see. Logical copies do not capture all the data, and the process will alter some file metadata to the extent that its forensic value is greatly diminished, resulting in a possible legal challenge by the opposing legal team. Therefore, a full bit-for-bit copy is the preferred forensic process. The file created on the target device is called a forensic image file.

Chain of custody is the way you document and preserve evidence from the time that you started the cyber-forensics investigation to the time the evidence is presented in court. It is extremely important to be able to show clear documentation of the following:

- How the evidence was collected

- When it was collected

- How it was transported

- How is was tracked

- How it was stored

- Who had access to the evidence and how it was accessed

A method often used for evidence preservation is to work only with a copy of the evidence—in other words, you do not want to work directly with the evidence itself. This involves creating an image of any hard drive or any storage device. Additionally, you must prevent electronic static or other discharge from damaging or erasing evidentiary data. Special evidence bags that are antistatic should be used to store digital devices. It is very important that you prevent electrostatic discharge (ESD) and other electrical discharges from damaging your evidence. Some organizations even have cyber-forensic labs that control access to only authorized users and investigators. One method often used involves constructing what is called a Faraday cage. This "cage" is often built out of a mesh of conducting material that prevents electromagnetic energy from entering into or escaping from the cage. Also, this prevents devices from communicating via Wi-Fi or cellular signals.

What's more, transporting the evidence to the forensics lab or any other place, including the courthouse, has to be done very carefully. It is critical that the chain of custody be maintained during this transport. When you transport the evidence, you should strive to secure it in a lockable container. It is also recommended that the responsible person stay with the evidence at all times during transportation.

What Is an Incident Response Program?

An incident response program is composed of policies, plans, procedures, and people. Incident response policies codify management directives. Incident response plans (IRPs) provide a well-defined, consistent, and organized approach for handling internal incidents as well as taking appropriate action when an external incident is traced back to the organization. Incident response procedures are detailed steps needed to implement the plan.

The Incident Response Plan

Having a good incident response plan and incident response process will help you minimize loss or theft of information and disruption of services caused by incidents. It will also help you enhance your incident response program by using lessons learned and information obtained during the security incident.

Section 2.3 of NIST Special Publication 800-61 Revision 2 goes over the incident response policies, plans, and procedures, including information on how to coordinate incidents and interact with outside parties. The policy elements described in NIST Special Publication 800-61 Revision 2 include the following:

- Statement of management commitment

- Purpose and objectives of the incident response policy

- The scope of the incident response policy

- Definition of computer security incidents and related terms

- Organizational structure and definition of roles, responsibilities, and levels of authority

- Prioritization or severity ratings of incidents

- Performance measures

- Reporting and contact forms

NIST's incident response plan elements include the following:

- Incident response plan's mission

- Strategies and goals of the incident response plan

- Senior management approval of the incident response plan

- Organizational approach to incident response

- How the incident response team will communicate with the rest of the organization and with other organizations

- Metrics for measuring the incident response capability and its effectiveness

- Roadmap for maturing the incident response capability

- How the program fits into the overall organization

NIST also defines standard operating procedures (SOPs) as "a delineation of the specific technical processes, techniques, checklists, and forms used by the incident response team.

SOPs should be reasonably comprehensive and detailed to ensure that the priorities of the organization are reflected in response operations."

The Incident Response Process

NIST Special Publication 800-61 goes over the major phases of the incident response process in detail. You should become familiar with that publication because it provides additional information that will help you succeed in your security operations center (SOC). The important key points are summarized here.

NIST defines the major phases of the incident response process as illustrated in Figure 1-14.

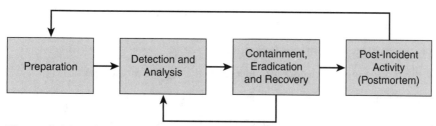

Figure 1-14 *The Phases of the Incident Response Process*

The preparation phase includes creating and training the incident response team, as well as deploying the necessary tools and resources to successfully investigate and resolve cybersecurity incidents. In this phase, the incident response team creates a set of controls based on the results of risk assessments. The preparation phase also includes the following tasks:

- Creating processes for incident handler communications and the facilities that will host the security operation center (SOC) and incident response team

- Making sure that the organization has appropriate incident analysis hardware and software as well as incident mitigation software

- Creating risk assessment capabilities within the organization

- Making sure the organization has appropriately deployed host security, network security, and malware prevention solutions

- Developing user awareness training

The detection and analysis phase is one of the most challenging phases. Although some incidents are easy to detect (for example, a denial-of-service attack), many breaches and attacks are left undetected for weeks or even months. This is why detection might be the most difficult task in incident response. The typical network is full of "blind spots" where anomalous traffic goes undetected. Implementing analytics and correlation tools is critical to eliminating these network blind spots. As a result, the incident response team must react quickly to analyze and validate each incident. This is done by following a predefined process while documenting each step the analyst takes. NIST provides various recommendations for making incident analysis easier and more effective:

- Profile networks and systems.

- Understand normal behaviors.

- Create a log retention policy.

- Perform event correlation.

- Maintain and use a knowledge base of information.

- Use Internet search engines for research.

- Run packet sniffers to collect additional data.

- Filter the data.

- Seek assistance from others.

- Keep all host clocks synchronized.

- Know the different types of attacks and attack vectors.

- Develop processes and procedures to recognize the signs of an incident.

- Understand the sources of precursors and indicators.

- Create appropriate incident documentation capabilities and processes.

- Create processes to effectively prioritize security incidents.

- Create processes to effectively communicate incident information (internal and external communications).

The containment, eradication, and recovery phase includes the following activities:

- Evidence gathering and handling

- Identifying the attacking hosts

- Choosing a containment strategy to effectively contain and eradicate the attack, as well as to successfully recover from it

NIST Special Publication 800-61 Revision 2 also defines the following criteria for determining the appropriate containment, eradication, and recovery strategy:

- The potential damage to and theft of resources

- The need for evidence preservation

- Service availability (for example, network connectivity as well as services provided to external parties)

- Time and resources needed to implement the strategy

- Effectiveness of the strategy (for example, partial containment or full containment)

- Duration of the solution (for example, emergency workaround to be removed in four hours, temporary workaround to be removed in two weeks, or permanent solution)

The post-incident activity phase includes lessons learned, how to use collected incident data, and evidence retention. NIST Special Publication 800-61 Revision 2 includes several questions that can be used as guidelines during the lessons learned meeting(s):

- Exactly what happened, and at what times?

- How well did the staff and management perform while dealing with the incident?

- Were the documented procedures followed? Were they adequate?

- What information was needed sooner?

- Were any steps or actions taken that might have inhibited the recovery?

- What would the staff and management do differently the next time a similar incident occurs?

- How could information sharing with other organizations be improved?

- What corrective actions can prevent similar incidents in the future?

- What precursors or indicators should be watched for in the future to detect similar incidents?

- What additional tools or resources are needed to detect, analyze, and mitigate future incidents?

Tabletop Exercises and Playbooks

Many organizations take advantage of tabletop (simulated) exercises to further test their capabilities. These tabletop exercises are an opportunity to practice and also perform gap analysis on their incident response processes and procedures. In addition, these exercises may allow them to create playbooks for incident response. Developing a playbook framework makes future analysis modular and extensible. A good playbook typically contains the following information:

- Report identification

- Objective statement

- Result analysis

- Data query/code

- Analyst comments/notes

There are significant long-term advantages for having relevant and effective playbooks. When developing playbooks, focus on organization and clarity within your own framework. Having a playbook and detection logic is not enough. The playbook is only a proactive plan. Your plays must actually run to generate results, those results must be analyzed, and remedial actions must be taken for malicious events. This is why tabletop exercises are very important.

Tabletop exercises could be technical and also at the executive level. You can create technical simulations for your incident response team and also risk-based exercises for your executive and management staff. A simple methodology for an incident response tabletop exercise includes the following steps:

1. **Preparation:** Identify the audience, what you want to simulate, and how the exercise will take place.

2. **Execution:** Execute the simulation and record all findings to identify all areas for improvement in your program.

3. **Report:** Create a report and distribute it to all the respective stakeholders. Narrow your assessment to specific facets of incident response. You can compare the results with the existing incident response plans. You should also measure the coordination among different teams within the organization and/or external to the organization. Provide a good technical analysis and identify gaps.

Information Sharing and Coordination

During the investigation and resolution of a security incident, you might also need to communicate with outside parties regarding the incident. Examples include, but are not limited to, contacting law enforcement, fielding media inquiries, seeking external expertise, and working with Internet service providers (ISPs), the vendor of your hardware and software products, threat intelligence vendor feeds, coordination centers, and members of other incident response teams. You can also share relevant incident indicator of compromise (IoC) information and other observables with industry peers. A good example of information-sharing communities is the Financial Services Information Sharing and Analysis Center (FS-ISAC).

Your incident response plan should account for these types of interactions with outside entities. It should also include information about how to interact with your organization's public relations (PR) department, legal department, and upper management. You should also get their buy-in when sharing information with outside parties to minimize the risk of information leakage. In other words, avoid leaking sensitive information regarding security incidents with unauthorized parties. These actions could potentially lead to additional disruption and financial loss. You should also maintain a list of all the contacts at those external entities, including a detailed list of all external communications for liability and evidentiary purposes.

Computer Security Incident Response Teams

There are different types of incident response teams. The most popular is the computer security incident response team (CSIRT). Others include the following:

- Product security incident response team (PSIRT)

- National CSIRT and computer emergency response team (CERT)

- Coordination center

- The incident response team of a security vendor and managed security service provider (MSSP)

The CSIRT is typically the team that works hand in hand with the information security teams (often called InfoSec). In smaller organizations, InfoSec and CSIRT functions may be combined and provided by the same team. In large organizations, the CSIRT focuses on the investigation of computer security incidents, whereas the InfoSec team is tasked with the implementation of security configurations, monitoring, and policies within the organization.

Establishing a CSIRT involves the following steps:

Step 1.   Defining the CSIRT constituency

Step 2.   Ensuring management and executive support

Step 3.   Making sure that the proper budget is allocated

Step 4.   Deciding where the CSIRT will reside within the organization's hierarchy

Step 5.   Determining whether the team will be central, distributed, or virtual

Step 6.   Developing the process and policies for the CSIRT

It is important to recognize that every organization is different, and these steps can be accomplished in parallel or in sequence. However, defining the constituency of a CSIRT is certainly one of the first steps in the process. When defining the constituency of a CSIRT, one should answer the following questions:

- Who will be the "customer" of the CSIRT?

- What is the scope? Will the CSIRT cover only the organization or also entities external to the organization? For example, at Cisco, all internal infrastructure and Cisco's websites and tools (that is, cisco.com) are the responsibility of the Cisco CSIRT, and any incident or vulnerability concerning a Cisco product or service is the responsibility of the Cisco PSIRT.

- Will the CSIRT provide support for the complete organization or only for a specific area or segment? For example, an organization may have a CSIRT for traditional infrastructure and IT capabilities and a separate one dedicated to cloud security.

- Will the CSIRT be responsible for part of the organization or all of it? If external entities will be included, how will they be selected?

Determining the value of a CSIRT can be challenging. One of the main questions that executives will ask is, what is the return on investment for having a CSIRT? The main goals of the CSIRT are to minimize risk, contain cyber damage, and save money by preventing incidents from happening—and when they do occur, to mitigate them efficiently. For example, the smaller the scope of the damage, the less money you need to spend to recover from a compromise (including brand reputation). Many studies in the past have covered the cost of security incidents and the cost of breaches. Also, the Ponemon Institute periodically publishes reports covering these costs. It is a good practice to review and calculate the "value add" of the CSIRT. This calculation can be used to determine when to invest more, not only in a CSIRT, but also in operational best practices. In some cases, an organization might even outsource some of the cybersecurity functions to a managed service provider, if the organization cannot afford or retain security talent.

Incident response teams must have several basic policies and procedures in place to operate satisfactorily, including the following:

- Incident classification and handling

- Information classification and protection

- Information dissemination

- Record retention and destruction

- Acceptable usage of encryption

- Engaging and cooperating with external groups (other IRTs, law enforcement, and so on)

Also, some additional policies or procedures can be defined, such as the following:

- Hiring policy

- Using an outsourcing organization to handle incidents

- Working across multiple legal jurisdictions

Even more policies can be defined depending on the team's circumstances. The important thing to remember is that not all policies need to be defined on the first day.

The following are great sources of information from the International Organization for Standardization/International Electrotechnical Commission (ISO/IEC) that you can leverage when you are conscripting your policy and procedure documents:

- **ISO/IEC 27001:2005:** "Information Technology—Security Techniques—Information Security Management Systems—Requirements"

- **ISO/IEC 27002:2005:** Information Technology—Security Techniques—Code of Practice for Information Security Management"

- **ISO/IEC 27005:2008:** "Information Technology—Security techniques—Information Security Risk Management"

- **ISO/PAS 22399:2007:** "Societal Security—Guidelines for Incident Preparedness and Operational Continuity Management"

- **ISO/IEC 27033:** Information Technology—Security Techniques—Information Security Incident Management

CERT provides a good overview of the goals and responsibilities of a CSIRT at the following site: https://www.cert.org/incident-management/csirt-development/csirt-faq.cfm.

Product Security Incident Response Teams (PSIRTs)

Software and hardware vendors may have separate teams that handle the investigation, resolution, and disclosure of security vulnerabilities in their products and services. Typically, these teams are called product security incident response teams (PSIRTs). Before you can understand how a PSIRT operates, you must understand what constitutes security vulnerability.

TIP   The following article outlines the PSIRT services framework and additional information about PSIRTs: https://blogs.cisco.com/security/psirt-services.

The Common Vulnerability Scoring System (CVSS)

Each vulnerability represents a potential risk that threat actors can use to compromise your systems and your network. Each vulnerability carries an associated amount of risk with it. One of the most widely adopted standards to calculate the severity of a given vulnerability is the Common Vulnerability Scoring System (CVSS), which has three components: base, temporal, and environmental scores. Each component is presented as a score on a scale from 0 to 10.

CVSS is an industry standard maintained by the Forum of Incident Response and Security Teams (FIRST) that is used by many PSIRTs to convey information about the severity of vulnerabilities they disclose to their customers.

In CVSS, a vulnerability is evaluated under three aspects and a score is assigned to each of them:

- The base group represents the intrinsic characteristics of a vulnerability that are constant over time and do not depend on a user-specific environment. This is the most important information and the only one that's mandatory to obtain a vulnerability score.

- The temporal group assesses the vulnerability as it changes over time.

- The environmental group represents the characteristics of a vulnerability, taking into account the organizational environment.

The score for the base group is between 0 and 10, where 0 is the least severe and 10 is assigned to highly critical vulnerabilities. For example, a highly critical vulnerability could allow an attacker to remotely compromise a system and get full control. Additionally, the score comes in the form of a vector string that identifies each of the components used to make up the score.

The formula used to obtain the score takes into account various characteristics of the vulnerability and how the attacker is able to leverage these characteristics.

CVSSv3 defines several characteristics for the base, temporal, and environmental groups.

The base group defines Exploitability metrics that measure how the vulnerability can be exploited, as well as Impact metrics that measure the impact on confidentiality, integrity, and availability. In addition to these two metrics, a metric called Scope Change (S) is used to convey the impact on other systems that may be impacted by the vulnerability but do not contain the vulnerable code. For instance, if a router is susceptible to a denial-of-service vulnerability and experiences a crash after receiving a crafted packet from the attacker, the scope is changed, since the devices behind the router will also experience the denial-of-service condition. FIRST has additional examples at https://www.first.org/cvss/v3.1/examples.

The Exploitability metrics include the following:

- Attack Vector (AV) represents the level of access an attacker needs to have to exploit a vulnerability. It can assume four values:

  - Network (N)

  - Adjacent (A)

  - Local (L)

  - Physical (P)

- Attack Complexity (AC) represents the conditions beyond the attacker's control that must exist in order to exploit the vulnerability. The values can be the following:

  - Low (L)

  - High (H)

- Privileges Required (PR) represents the level of privileges an attacker must have to exploit the vulnerability. The values are as follows:

  - None (N)

  - Low (L)

  - High (H)

- User Interaction (UI) captures whether a user interaction is needed to perform an attack. The values are as follows:

  - None (N)

  - Required (R)

- Scope (S) captures the impact on systems other than the system being scored. The values are as follows:

  - Unchanged (U)

  - Changed (C)

The Impact metrics include the following:

- Confidentiality (C) measures the degree of impact to the confidentiality of the system. It can assume the following values:

  - Low (L)

  - Medium (M)

  - High (H)

- Integrity (I) measures the degree of impact to the integrity of the system. It can assume the following values:

  - Low (L)

  - Medium (M)

  - High (H)

- Availability (A) measures the degree of impact to the availability of the system. It can assume the following values:

  - Low (L)

  - Medium (M)

  - High (H)

The temporal group includes three metrics:

- Exploit Code Maturity (E), which measures whether or not public exploit is available

- Remediation Level (RL), which indicates whether a fix or workaround is available

- Report Confidence (RC), which indicates the degree of confidence in the existence of the vulnerability

The environmental group includes two main metrics:

- Security Requirements (CR, IR, AR), which indicate the importance of confidentiality, integrity, and availability requirements for the system

- Modified Base Metrics (MAV, MAC, MAPR, MUI, MS, MC, MI, MA), which allow the organization to tweak the base metrics based on specific characteristics of the environment

For example, a vulnerability that might allow a remote attacker to crash the system by sending crafted IP packets would have the following values for the base metrics:

- Access Vector (AV) would be Network because the attacker can be anywhere and can send packets remotely.

- Attack Complexity (AC) would be Low because it is trivial to generate malformed IP packets (for example, via the Scapy Python tool).

- Privilege Required (PR) would be None because there are no privileges required by the attacker on the target system.

- User Interaction (UI) would also be None because the attacker does not need to interact with any user of the system to carry out the attack.

- Scope (S) would be Unchanged if the attack does not cause other systems to fail.

- Confidentiality Impact (C) would be None because the primary impact is on the availability of the system.

- Integrity Impact (I) would be None because the primary impact is on the availability of the system.

- Availability Impact (A) would be High because the device could become completely unavailable while crashing and reloading.

Additional examples of CVSSv3 scoring are available at the FIRST website (https://www.first.org/cvss).

In numerous instances, security vulnerabilities are not exploited in isolation. Threat actors exploit more than one vulnerability "in a chain" to carry out their attack and compromise their victims. By leveraging different vulnerabilities in a chain, attackers can infiltrate progressively further into the system or network and gain more control over it. This is something that PSIRT teams must be aware of. Developers, security professionals, and users must be aware of this because chaining can change the order in which a vulnerability needs to be fixed or patched in the affected system. For instance, multiple low-severity vulnerabilities can become a severe one if they are combined.

Performing vulnerability chaining analysis is not a trivial task. Although several commercial companies claim that they can easily perform chaining analysis, in reality the methods and procedures that can be included as part of a chain vulnerability analysis are pretty much endless. PSIRT teams should utilize an approach that works for them to achieve the best end result.

Exploits cannot exist without a vulnerability. However, there isn't always an exploit for a given vulnerability. Earlier in this chapter you were reminded of the definition of a vulnerability. As another reminder, an exploit is not a vulnerability. An exploit is a concrete manifestation, either a piece of software or a collection of reproducible steps, that leverages a given vulnerability to compromise an affected system.

In some cases, users call vulnerabilities without exploits "theoretical vulnerabilities." One of the biggest challenges with "theoretical vulnerabilities" is that there are many smart people out there capable of exploiting them. If you do not know how to exploit a vulnerability today, it does not mean that someone else will not find a way in the future. In fact, someone else may already have found a way to exploit the vulnerability and perhaps is even selling the exploit of the vulnerability in underground markets without public knowledge.

PSIRT personnel should understand there is no such thing as an "entirely theoretical" vulnerability. Sure, having a working exploit can ease the reproducible steps and help to verify whether that same vulnerability is present in different systems. However, because an exploit may not come as part of a vulnerability, you should not completely deprioritize it.

A PSIRT can learn about a vulnerability in a product or service during internal testing or during the development phase. However, vulnerabilities can also be reported by external entities, such as security researchers, customers, and other vendors.

The dream of any vendor is to be able to find and patch all security vulnerabilities during the design and development phases. However, that is close to impossible. On the other hand, that is why a secure development life cycle (SDL) is extremely important for any organization that produces software and hardware. Cisco has an SDL program that is documented at the following URL:

www.cisco.com/c/en/us/about/security-center/security-programs/secure-development-lifecycle.html

Cisco defines its SDL as "a repeatable and measurable process we've designed to increase the resiliency and trustworthiness of our products." Cisco's SDL is part of Cisco Product Development Methodology (PDM) and ISO9000 compliance requirements. It includes, but is not limited to, the following:

- Base product security requirements

- Third-party software (TPS) security

- Secure design

- Secure coding

- Secure analysis

- Vulnerability testing

The goal of the SDL is to provide tools and processes that are designed to accelerate the product development methodology, by developing secure, resilient, and trustworthy systems. TPS security is one of the most important tasks for any organization. Most of today's organizations use open source and third-party libraries. This approach creates two requirements for the product security team. The first is to know what TPS libraries are used, reused, and where. The second is to patch any vulnerabilities that affect such libraries or TPS components. For example, if a new vulnerability in OpenSSL is disclosed, what do you have to do? Can you quickly assess the impact of such a vulnerability in all your products?

If you include commercial TPS, is the vendor of such software transparently disclosing all the security vulnerabilities, including in its software? Nowadays, many organizations are including security vulnerability disclosure SLAs in their contracts with third-party vendors. This is very important because many TPS vulnerabilities (both commercial and open source) go unpatched for many months—or even years.

Many tools are available on the market today to enumerate all open source components used in a product. These tools either interrogate the product source code or scan binaries for the presence of TPS.

National CSIRTs and Computer Emergency Response Teams (CERTs)

Numerous countries have their own computer emergency response (or readiness) teams. Examples include the US-CERT (https://www.us-cert.gov), Indian Computer Emergency Response Team (http://www.cert-in.org.in), CERT Australia (https://cert.gov.au), and the Australian Computer Emergency Response Team (https://www.auscert.org.au/). The Forum of Incident Response and Security Teams (FIRST) website includes a list of all the national CERTs and other incident response teams at https://www.first.org/members/teams.

These national CERTs and CSIRTs aim to protect their citizens by providing security vulnerability information, security awareness training, best practices, and other information. For example, the following is the US-CERT mission posted at https://www.us-cert.gov/about-us:

"US-CERT's critical mission activities include:

- Providing cybersecurity protection to Federal civilian executive branch agencies through intrusion detection and prevention capabilities.

- Developing timely and actionable information for distribution to federal departments and agencies; state, local, tribal and territorial (SLTT) governments; critical infrastructure owners and operators; private industry; and international organizations.

- Responding to incidents and analyzing data about emerging cyber threats.

- Collaborating with foreign governments and international entities to enhance the nation's cybersecurity posture."

Coordination Centers

Several organizations around the world also help with the coordination of security vulnerability disclosures to vendors, hardware and software providers, and security researchers.

One of the best examples is the CERT Division of the Software Engineering Institute (SEI). Their website can be accessed at cert.org, and their "About Us" page summarizes well their role and the role of many coordination centers alike:

"CERT Division of the Software Engineering Institute (SEI), we study and solve problems with widespread cybersecurity implications, research security vulnerabilities in software products, contribute to long-term changes in networked systems, and develop cutting-edge information and training to help improve cybersecurity.

"We are more than a research organization. Working with software vendors, we help resolve software vulnerabilities. We develop tools, products, and methods to help organizations conduct forensic examinations, analyze vulnerabilities, and monitor large-scale networks. We help organizations determine how effective their security-related practices are. And we share our work at conferences; in blogs, webinars, and podcasts; and through our many articles, technical reports, and white papers. We collaborate with high-level government organizations, such as the U.S. Department of Defense and the Department of Homeland Security (DHS); law enforcement, including the FBI; the intelligence community; and many industry organizations.

"Working together, DHS and the CERT Division meet mutually set goals in areas such as data collection and mining, statistics and trend analysis, computer and network security, incident management, insider threat, software assurance, and more. The results of this work include exercises, courses, and systems that were designed, implemented, and delivered to DHS and its customers as part of the SEI's mission to transition SEI capabilities to the public and private sectors and improve the practice of cybersecurity."

Incident Response Providers and Managed Security Service Providers (MSSPs)

Cisco, along with several other vendors, provides incident response and managed security services to its customers. These incident response teams and outsourced CSIRTs operate a bit differently because their task is to provide support to their customers. However, they practice the tasks outlined earlier in this chapter for incident response and CSIRTs.

Outsourcing has been a long practice for many companies, but the onset of the complexity of cybersecurity has allowed it to bloom and become bigger as the years go by in the world of incident response.

Key Incident Management Personnel

Key incident management personnel include incident response coordinators, designated incident handlers, incident response team members, and external advisors. In various organizations, they may have different titles, but the roles are essentially the same.

The incident response coordinator (IRC) is the central point of contact for all incidents. Incident reports are directed to the IRC. The IRC verifies and logs the incident. Based on predefined criteria, the IRC notifies appropriate personnel, including the designated incident handler (DIH). The IRC is a member of the incident response team (IRT) and is responsible for maintaining all non-evidence-based incident-related documentation.

DIHs are senior-level personnel who have the crisis management and communication skills, experience, knowledge, and stamina to manage an incident. DIHs are responsible for three critical tasks: incident declaration, liaison with executive management, and managing the IRT.

The IRT is a carefully selected and well-trained team of professionals that provides services throughout the incident life cycle. Depending on the size of the organization, there may be a single team or multiple teams, each with its own specialty. The IRT members generally represent a cross-section of functional areas, including senior management, information security, information technology (IT), operations, legal, compliance, HR, public affairs and media relations, customer service, and physical security. Some members may be expected to participate in every response effort, whereas others (such as compliance) may restrict involvement to relevant events. The team as directed by the DIH is responsible for further analysis, evidence handling and documentation, containment, eradication and recovery, notification (as required), and post-incident activities.

Tasks assigned to the IRT include but are not limited to the following:

- Overall management of the incident
- Triage and impact analysis to determine the extent of the situation
- Development and implementation of containment and eradication strategies
- Compliance with government and/or other regulations
- Communication and follow-up with affected parties and/or individuals

- Communication and follow-up with other external parties, including the board of directors, business partners, government regulators (including federal, state, and other administrators), law enforcement, representatives of the media, and so on, as needed

- Root cause analysis and lessons learned

- Revision of policies/procedures necessary to prevent any recurrence of the incident

Establishing a robust response capability ensures that the organization is prepared to respond to an incident swiftly and effectively. Responders should receive training specific to their individual and collective responsibilities. Recurring tests, drills, and challenging incident response exercises can make a huge difference in responder ability. Knowing what is expected decreases the pressure on the responders and reduces errors. It should be stressed that the objective of incident response exercises isn't to get an "A" but rather to honestly evaluate the plan and procedures, to identify missing resources, and to learn to work together as a team.

Summary

This chapter started with an introduction to cybersecurity and then moved into defining what are threats, vulnerabilities, and exploits. You learned about different common threats that can affect any organization, individual, system, or network. This chapter also covered the most common software and hardware vulnerabilities such as cross-site scripting, cross-site request forgery, SQL injection, buffer overflows, and many others.

This chapter also defined what is confidentiality, integrity, and availability (the CIA triad). You also learned about different cloud and IoT security threats. At the end, this chapter provided an introduction to digital forensics and incident response.

Exam Preparation Tasks

As mentioned in the section "How to Use This Book" in the Introduction, you have a couple of choices for exam preparation: the exercises here, Chapter 12, "Final Preparation," and the exam simulation questions in the Pearson Test Prep Software Online.

Review All Key Topics

Review the most important topics in this chapter, noted with the Key Topic icon in the outer margin of the page. Table 1-2 lists these key topics and the page numbers on which each is found.

Table 1-2 Key Topics for Chapter 1

Key Topic Element	Description	Page Number
Paragraph	Understand the difference between InfoSec and Cybersecurity	7
Section	What Is a Threat?	9
Section	What Is a Vulnerability?	9
Section	What Is an Exploit?	10

Key Topic Element	Description	Page Number
List	Understand the difference between a white hat, gray hat, and black hat hacker	14
Section	Understanding What Threat Intelligence Is	14
Section	Viruses and Worms	16
Section	Trojans	18
Section	Distributing Malware	22
Section	Ransomware	23
Section	Keyloggers	25
Section	Spyware	26
Section	SQL Injection	30
Section	Command Injection	32
Section	Authentication-based Vulnerabilities	32
Section	Cross-site Scripting (XSS)	35
Section	Cross-site Request Forgery	37
Section	OWASP Top 10	40
Paragraph	The NIST Definition of Cloud Computing	47
List	Understand the different cloud models	48
List	Identifying common cloud computing security concerns	49
List	Identify cloud computing attacks	50
Section	IoT Protocols	53
Section	ISO/IEC 27002:2013 and NIST Incident Response Guidance	55
Section	What Is an Incident?	56
Section	False Positives, False Negatives, True Positives, and True Negatives	57
Section	The Incident Response Plan	60
Section	The Incident Response Process	61
Section	Information Sharing and Coordination	64
Section	Computer Security Incident Response Teams	64
Section	The Common Vulnerability Scoring System (CVSS)	67

Define Key Terms

Define the following key terms from this chapter and check your answers in the glossary:

Threat, vulnerability, exploit, white hat hackers, black hat hackers, gray hat hackers, threat intelligence, wrappers, packers, droppers, crypters, ransomware, IaaS, PaaS, SaaS

Review Questions

1. Which of the following are standards being developed for disseminating threat intelligence information?

    a. STIX

    b. TAXII

    c. CybOX

    d. All of these answers are correct.

2. Which type of hacker is considered a good guy?

    a. White hat

    b. Black hat

    c. Gray hat

    d. All of these answers are correct.

3. Which of the following is not an example of ransomware?

    a. WannaCry

    b. Pyeta

    c. Nyeta

    d. Bad Rabbit

    e. Ret2Libc

4. Which of the following is the way you document and preserve evidence from the time that you started the cyber-forensics investigation to the time the evidence is presented in court?

    a. Chain of custody

    b. Best evidence

    c. Faraday

    d. None of these answers is correct.

5. Software and hardware vendors may have separate teams that handle the investigation, resolution, and disclosure of security vulnerabilities in their products and services. Typically, these teams are called _____.

    a. CSIRT

    b. Coordination Center

    c. PSIRT

    d. MSSP

6. Which of the following are the three components in CVSS?

    a. Base, temporal, and environmental groups

    b. Base, temporary, and environmental groups

    c. Basic, temporal, and environmental groups

    d. Basic, temporary, and environmental groups

7. Which of the following are IoT technologies?

   a. Z-Wave

   b. INSTEON

   c. LoRaWAN

   d. A and B

   e. A, B, and C

   f. None of these answers is correct.

8. Which of the following is a type of cloud deployment model where the cloud environment is shared among different organizations?

   a. Community cloud

   b. IaaS

   c. PaaS

   d. None of these answers is correct.

9. _____ attacks occur when the sources of the attack are sent spoofed packets that appear to be from the victim, and then the sources become unwitting participants in the DDoS attacks by sending the response traffic back to the intended victim.

   a. Reflected DDoS

   b. Direct DoS

   c. Backtrack DoS

   d. SYN flood

10. Which of the following is a nonprofit organization that leads several industry-wide initiatives to promote the security of applications and software?

   a. CERT/cc

   b. OWASP

   c. AppSec

   d. FIRST

## CHAPTER 2

Cryptography

This chapter covers the following topics:

Introduction to Cryptography: Cryptography is used pervasively nowadays. You need to understand the concepts of secure communications. You also need to know the different encryption and hashing protocols, as well as their use.

Fundamentals of PKI: Public key infrastructure (PKI) is a set of identities, roles, policies, and actions for the creation, use, management, distribution, and revocation of public and private keys.

This chapter introduces you to the world of cryptography. You will learn what ciphers are and different encryption and hashing algorithms. In this chapter, you will also learn details about PKI implementations.

The following SCOR 350-701 exam objectives are covered in this chapter:

- Domain 1: Security Concepts

  - 1.3 Describe functions of the cryptography components such as hashing, encryption, PKI, SSL, IPsec, NAT-T IPv4 for IPsec, pre-shared key, and certificate-based authorization

"Do I Know This Already?" Quiz

The "Do I Know This Already?" quiz allows you to assess whether you should read this entire chapter thoroughly or jump to the "Exam Preparation Tasks" section. If you are in doubt about your answers to these questions or your own assessment of your knowledge of the topics, read the entire chapter. Table 2-1 lists the major headings in this chapter and their corresponding "Do I Know This Already?" quiz questions. You can find the answers in Appendix A, "Answers to the 'Do I Know This Already?' Quizzes and Q&A Sections."

Table 2-1 "Do I Know This Already?" Section-to-Question Mapping

Foundation Topics Section	Questions
Introduction to Cryptography	1–5
Fundamentals of PKI	6–10

CAUTION The goal of self-assessment is to gauge your mastery of the topics in this chapter. If you do not know the answer to a question or are only partially sure of the answer, you should mark that question as wrong for purposes of the self-assessment. Giving yourself credit for an answer you incorrectly guess skews your self-assessment results and might provide you with a false sense of security.

1. Which of the following is a good example of a key that is only used once?
   a. OTP
   b. ISAKMP
   c. Multifactor key
   d. None of these answers are correct
2. Which of the following is a type of cipher that uses the same key to encrypt and decrypt?
   a. Symmetric
   b. Asymmetric
   c. Ciphertext
   d. RSA
3. Which of the following is a symmetric key cipher where the plaintext data to be encrypted is done a bit at a time against the bits of the key stream, also called a cipher digit stream?
   a. Asymmetric cipher
   b. Block cipher
   c. Stream cipher
   d. None of these answers is correct.
4. Which of the following is not an example of a symmetric encryption algorithm?
   a. AES
   b. 3DES
   c. RC4
   d. RSA
5. Which of the following is an algorithm that allows two devices to negotiate and establish shared secret keying material (keys) over an untrusted network?
   a. Diffie-Hellman
   b. RSA
   c. RC4
   d. IKE
6. Assume that Mike is trying to send an encrypted email to Chris using PGP or S/MIME. What key will Mike use to encrypt the email to Chris?
   a. Chris's private key
   b. Chris's public key
   c. Mike's private key
   d. Mike's public key
7. Which of the following implementations uses a key pair?
   a. PGP
   b. Digital certificates on a web server running TLS
   c. S/MIME
   d. All of these answers are correct.

8. Which of the following is an entity that creates and issues digital certificates?

   a. Certificate Registry (CR)

   b. Certificate Authentication Server (CAS)

   c. Certificate Authority (CA)

   d. None of these answers is correct.

9. Which of the following statements is true?

   a. Subordinate CA servers can be invalidated.

   b. Subordinate certificates cannot be invalidated.

   c. Root certificates cannot be invalidated.

   d. Root CAs cannot be invalidated.

10. Which of the following is a series of standards focused on directory services and how those directories are organized?

    a. 802.1X

    b. X.500

    c. X.11

    d. X.409

Foundation Topics

Introduction to Cryptography

Cryptography or *cryptology* is the study of the techniques used for encryption and secure communications. Cryptographers are the people who study and analyze cryptography. Cryptographers are always constructing and analyzing protocols for preventing unauthorized users from reading private messages as well as the following areas of information security:

- Data confidentiality

- Data integrity

- Authentication

- Nonrepudiation

Cryptography is a combination of disciplines, including mathematics and computer science. Examples of the use of cryptography include virtual private networks (VPNs), ecommerce, secure email transfer, and credit card chips. Every time you check Facebook, Instagram, or watch a YouTube video in your web browser, you are using cryptographic algorithms and implementations, since those sites use Transport Layer Security (TLS). You will learn about different cryptographic algorithms later in this chapter.

You may also often hear the term *cryptanalysis*, which is the study of how to crack encryption algorithms or their implementations.

Ciphers

A *cipher* is a set of rules, which can also be called an *algorithm*, about how to perform encryption or decryption. Literally hundreds of encryption algorithms are available, and there are likely many more that are proprietary and used for special purposes, such as for governmental use and national security.

Common methods that ciphers use include the following:

- **Substitution:** This type of cipher substitutes one character for another. For instance, a simple substitution cipher can change a letter from the alphabet with another letter. To make it more challenging, we can shift more than just a single character and only chose certain letters to substitute. The exact method of substitution could be referred to as the "key." If both parties involved in the encrypted communication channel understand the key, they can both encrypt and decrypt data. An example of substitution is shown in Figure 2-1.

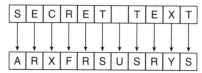

Figure 2-1   *Example of Substitution*

- **Polyalphabetic:** This is similar to substitution, but instead of using a single alphabet, it can use multiple alphabets and switch between them by some trigger character in the encoded message.

- **Transposition:** This method uses many different options, including the rearrangement of letters. For example, if we have the message "This is secret," we could write it out (top to bottom, left to right) as shown in Figure 2-2.

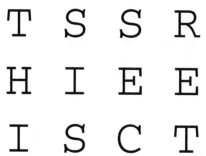

Figure 2-2   *Example of Transposition*

We then encrypt it as RETCSIHTSSEI, which involves starting at the top right and going around like a clock, spiraling inward. In order for someone to know how to encrypt/decrypt this correctly, the correct key is needed.

Keys

The key in Figure 2-2 refers to the instructions for how to reassemble the characters. In this case, it begins at the top-right corner and moves clockwise and spirals inward.

A one-time pad (OTP) is a good example of a key that is only used once. Using this method, if we want to encrypt a 32-bit message, we use a 32-bit key, also called the *pad*, which is used one time only. Each bit from the pad is mathematically computed with a corresponding bit from our message, and the results are our cipher text, or encrypted content. The key in this case is the one-time-use pad. The pad must also be known by the receiver if he or she wants to decrypt the message. (Another use of the acronym OTP is for a user's *one-time password*, which is a different topic altogether.)

Block and Stream Ciphers

Encryption algorithms can operate on blocks of data at a time, or bits and bytes of data, based on the type of cipher. Let's compare the two methods.

A block cipher is a symmetric key cipher (meaning the same key is used to encrypt and decrypt) that operates on a group of bits called a *block*. A block cipher encryption algorithm may take a 64-bit block of plaintext and generate a 64-bit block of ciphertext. With this type of encryption, the key to encrypt is also used to decrypt. Examples of symmetric block cipher algorithms include the following:

- Advanced Encryption Standard (AES)

- Triple Digital Encryption Standard (3DES)

- Blowfish

- Digital Encryption Standard (DES)

- International Data Encryption Algorithm (IDEA)

Block ciphers might add padding in cases where there is not enough data to encrypt to make a full block size. This might result in a very small amount of wasted overhead, because the small padding would be processed by the cipher along with the real data.

A stream cipher is a symmetric key cipher (meaning the same key is used to encrypt and decrypt), where the plaintext data to be encrypted is done a bit at a time against the bits of the key stream, also called a *cipher digit stream*. The resulting output is a *ciphertext stream*. Because a given algorithm ciphertext stream does not have to fit in a given block size, there may be slightly less overhead than with a block cipher that requires padding to complete a block size.

Symmetric and Asymmetric Algorithms

As you build your vocabulary, the words *symmetric* and *asymmetric* are important ones to differentiate. Let's look at the options of each and identify which of these requires the most CPU overhead and which one is used for bulk data encryption.

As mentioned previously, a symmetric encryption algorithm, also known as a symmetric cipher, uses the same key to encrypt the data and decrypt the data. Two devices connected via a VPN both need the key (or keys) to successfully encrypt and decrypt the data protected using a symmetric encryption algorithm.

> **NOTE**   You will learn the details about different VPN solutions and deployment scenarios in Chapter 8, "Virtual Private Networks (VPNs)."

Common examples of symmetric encryption algorithms include the following:

- DES

- 3DES

- AES

- IDEA

- RC2, RC4, RC5, RC6

- Blowfish

Symmetric encryption algorithms are used for most of the data we protect in VPNs today because they are much faster to use and take less CPU than asymmetric algorithms. As with all encryption, the more difficult the key, the more difficult it is for someone who does not have the key to intercept and understand the data. We usually refer to keys with VPNs by their length. A longer key means better security. A typical key length is 112 bits to 256 bits. The minimum key length should be at least 128 bits for symmetric encryption algorithms to be considered fairly safe. Again, bigger is better.

An example of an asymmetric algorithm is a public key algorithm. There is something magical about asymmetric algorithms because instead of using the same key for encrypting and decrypting, they use two different keys that mathematically work together as a pair. Let's call these keys the *public key* and the *private key*. Together they make a key pair. Let's put these keys to use with an analogy.

Imagine a huge shipping container that has a special lock with two keyholes (one large keyhole and one smaller keyhole). With this magical shipping container, if we use the small keyhole with its respective key to lock the container, the only way to unlock it is to use the big keyhole with its larger key. Another option is to initially lock the container using the big key in the big keyhole, and then the only way to unlock it is to use the small key in the small keyhole. (I told you it was magic.) This analogy explains the interrelationship between the public key and its corresponding private key. (I'll let you decide which one you want to call the big key and which one you want to call the little key.) There is a very high CPU cost when using key pairs to lock and unlock data. For that reason, we use asymmetric algorithms sparingly. Instead of using them to encrypt our bulk data, we use asymmetric algorithms for things such as authenticating a VPN peer or generating keying material that we can use for our symmetric algorithms. Both of these tasks are infrequent compared to encrypting all the user packets (which happens consistently).

With public key cryptography, one of the keys in the key pair is published and available to anyone who wants to use it (the public key). The other key in the key pair is the private key, which is known only to the device that owns the public-private key pair. An example of when a public-private key pair is used is visiting a secure website. In the background, the public-private key pair of the server is being used for the security of the session. Your PC has access to the public key, and the server is the only one that knows its private key.

Here are some examples of asymmetric algorithms:

- **RSA:** Named after Rivest, Shamir, and Adleman, who created the algorithm. The primary use of this asymmetric algorithm today is for authentication. It is also known as Public Key Cryptography Standard (PKCS) #1. The key length may be from 512 to 2048, and a minimum size for good security is 1024. Regarding security, bigger is better.

- **DH:** The Diffie-Hellman key exchange protocol is an asymmetric algorithm that allows two devices to negotiate and establish shared secret keying material (keys) over an untrusted network. The interesting thing about DH is that although the algorithm itself is asymmetric, the keys generated by the exchange are symmetric keys that can then be used with symmetric algorithms such as Triple Digital Encryption Standard (3DES) and Advanced Encryption Standard (AES).

- **ElGamal:** This asymmetric encryption system is based on the DH exchange.

- **DSA:** The Digital Signature Algorithm was developed by the U.S. National Security Agency.

- **ECC:** Elliptic Curve Cryptography.

Asymmetric algorithms require more CPU processing power than symmetric algorithms. Asymmetric algorithms, however, are more secure. A typical key length used in asymmetric algorithms can be anywhere between 2048 and 4096. A key length that is shorter than 2048 is considered unreliable and not as secure as a longer key.

A commonly asymmetric algorithm used for authentication is RSA (as in RSA digital signatures).

Hashes

Hashing is a method used to verify data integrity. For example, you can verify the integrity of a downloaded software image file from Cisco, and then verify its integrity using a tool such as the **verify md5** command in a Cisco IOS device or a checksum verification in an operating system such as Microsoft Windows, Linux, or macOS.

SHA512 checksum (512 bits) output is represented by a 128-digit hexadecimal number, whereas MD5 produces a 128-bit (16-byte) hash value, typically expressed in text format as a 32-digit hexadecimal number.

An example of using a hash to verify integrity is the sender running a hash algorithm on a packet and attaching that hash to it. The receiver runs the same hash against the packet and compares his results against the results the sender had (which are attached to the packet as well). If the hash generated matches the hash that was sent, they know that the entire packet is intact. If a single bit of the hashed portion of the packet is modified, the hash calculated by the receiver will not match, and the receiver will know that the packet had a problem— specifically with the integrity of the packet.

Let's take a look at Example 2-1. In Example 2-1, the contents of three files (file1.txt, file2.txt, file3.txt) are shown using the **cat** Linux command. As you can see, the contents of file1.txt and file3.txt are the same. The contents of file2.txt is different from file1.txt and file3.txt.

Example 2-1   *The Contents of file1.txt, file2.txt, and file3.txt*

```
$cat file1.txt
The rabbit is fast.
The turtle is slow.
$cat file2.txt
This is file 2. File two. File dos...
$cat file3.txt
The rabbit is fast.
The turtle is slow.
```

> **NOTE** You can find a blog post explaining hash verification of Cisco software at http://blogs.cisco.com/security/sha512-checksums-for-all-cisco-software.

In Example 2-2, the **shasum** Linux command is used to obtain the SHA checksum of each file and the **md5sum** Linux command is used to obtain the MD5 checksum of each file. As you can see, the checksums of files file1.txt and file3.txt are identical, since the contents of the files are the same (despite their different filenames).

Example 2-2 *The SHA and MD5 Checksums*

```
$shasum file1.txt file2.txt file3.txt
9a88d5f65e6136d351b20bc750038254c2ff6533 file1.txt
04aebe7d04587fad725588c1a8a49f7c7558252c file2.txt
9a88d5f65e6136d351b20bc750038254c2ff6533 file3.txt
$ md5sum file1.txt file2.txt file3.txt
738316215d0085ddb4c3ca4af2c928a1 file1.txt
3bb237acac848be493cdfb005d27c0f0 file2.txt
738316215d0085ddb4c3ca4af2c928a1 file3.txt
```

A cryptographic hash function is a process that takes a block of data and creates a small fixed-sized hash value. It is a one-way function, meaning that if two different computers take the same data and run the same hash function, they should get the same fixed-sized hash value (for example, a 12-bit long hash). The Message Digest 5 (MD5) algorithm is an example of a cryptographic hash function. It is not possible (at least not realistically) to generate the same hash from a different block of data. This is referred to as *collision resistance*. The result of the hash is a fixed-length small string of data, and is sometimes referred to as the digest, message digest, or simply the hash.

Hashes are also used when security experts are analyzing, searching, and comparing malware. A hash of the piece of malware is typically exchanged instead of the actual file, in order to avoid infection and collateral damage. For example, Cisco Advanced Malware Protection (AMP) use malware hashes in many of its different functions and capabilities.

The three most popular types of hashes are as follows:

- **Message Digest 5 (MD5):** This hash creates a 128-bit digest.

- **Secure Hash Algorithm 1 (SHA-1):** This hash creates a 160-bit digest.

- **Secure Hash Algorithm 2 (SHA-2):** Options include a digest between 224 bits and 512 bits.

With encryption and cryptography, and now hashing, bigger is better, and more bits equals better security. There are several vulnerabilities in the MD5 hashing protocol, including collision and pre-image vulnerabilities. Attackers use collision attacks in order to find two input strings of a hash function that produce the same hash result. This is because hash functions have infinite input length and a predefined output length. Consequently, there is the possibility of two different inputs producing the same output hash.

There are also several vulnerabilities and attacks against SHA-1. Consequently it is recommended that SHA-2 with 512 bits be used when possible.

During the last few years there has been a lot of discussion on quantum computers and their potential impact on current cryptography standards. This is an area of active research and growing interest. The industry is trying to label what are the post-quantum-ready and next-generation cryptographic algorithms. AES-256, SHA-384, and SHA-512 are believed to have post-quantum security. Other public key algorithms are believed to also be resistant to post-quantum security attacks; however, not many standards support them.

> **TIP**   Cisco provides a great resource that explains the next-generation encryption protocols and hashing protocols at http://www.cisco.com/c/en/us/about/security-center/next-generation-cryptography.html. You can use the references in that website to become familiar with encryption and hashing algorithms and their use. You can also learn about the algorithms that you should avoid and those that are recommended.

Hashed Message Authentication Code

Hashed Message Authentication Code (HMAC) uses the mechanism of hashing, but kicks it up a notch. Instead of using a hash that anyone can calculate, it includes in its calculation a secret key of some type. Thus, only the other party who also knows the secret key and can calculate the resulting hash can correctly verify the hash. When this mechanism is used, an attacker who is eavesdropping and intercepting packets cannot inject or remove data from those packets without being noticed because he cannot recalculate the correct hash for the modified packet because he does not have the key or keys used for the calculation.

Once again, MD5 is a hash function that is insecure and should be avoided. SHA-1 is a legacy algorithm and therefore is adequately secure. SHA-256 provides adequate protection for sensitive information. On the other hand, SHA-384 is required to protect classified information of higher importance.

HMAC is a construction that uses a secret key and a hash function to provide a message authentication code (MAC) for a message. HMAC is used for integrity verification. HMAC-MD5, which uses MD5 as its hash function, is a legacy algorithm. MD5 as a hash function itself is not secure. It provides adequate security today, but its keys should be renewed relatively often. Alternatively, the NIST-recommended HMAC function is HMAC-SHA-1.

Digital Signatures

When you sign something, this often represents a commitment to follow through, or at least proves that you are who you say you are. In the world of cryptography, a digital signature provides three core benefits:

- Authentication
- Data integrity
- Nonrepudiation

One of the best ways to understand how a digital signature operates is to remember what you learned in the previous sections about public and private key pairs, hashing, and encryption. Digital signatures involve each of these elements.

In most security books, three fictional characters are used to explain encryption and PKI: Bob, Alice, and Eve. Bob and Alice typically are the two entities that exchange a secured message over a public or untrusted network, and Eve is the person who tries to "eavesdrop" and steal the information being exchanged. In this book, let's make it more entertaining and use Thor, Ironman, and Thanos. In Figure 2-3, all three entities are illustrated. Thor wants to send an encrypted message to Ironman without Thanos being able to read it.

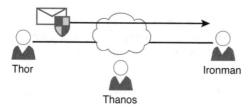

Figure 2-3 *Fundamentals of Public Key Cryptography (PKI)*

Thor and Ironman also want to verify each other to make sure they are talking to the right entity by using digital signatures. This process is illustrated in Figure 2-4.

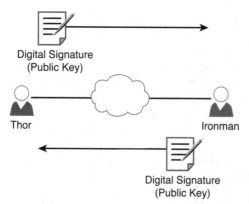

Figure 2-4 *Digital Signature Verification*

Both Thor and Ironman want to verify each other, but for simplicity let's focus on one entity: Thor wanting to prove its identity to the other device, Ironman. (This could also be phrased as Ironman asking Thor to prove Thor's identity.)

As a little setup beforehand, you should know that both Thor and Ironman have generated public-private key pairs, and they both have been given digital certificates from a common certificate authority (CA).

A CA is a trusted entity that hands out digital certificates. This concept is illustrated in Figure 2-5.

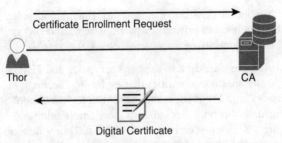

Certificate Enrollment Request

Thor

CA

Digital Certificate

Figure 2-5 *Digital Certificate Enrollment with a CA*

In Figure 2-5, Thor requests a digital certification from (enrolls with) a CA, and the CA assigns one to Thor. If you and I were to open the digital certificate, we would find the name of the entity (in this case, Thor). We would also find Thor's public key (which Thor gave to the CA when applying for the digital certificate).

Both Thor and Ironman trust the CA and have received their certificates.

Let's assume that now Thor and Ironman want to create a VPN tunnel. Thor takes a packet and generates a hash. Thor then takes this small hash and encrypts it using Thor's private key. (Think of this as a shipping container, and Thor is using the small key in the small keyhole to lock the data.) Thor attaches this encrypted hash to the packet and sends it to Ironman. The fancy name for this encrypted hash is *digital signature*.

When Ironman receives this packet, it looks at the encrypted hash that was sent and decrypts it using Thor's public key. (Think of this as a big keyhole and the big key being used to unlock the data.) Ironman then sets the decrypted hash off to the side for one moment and runs the same hash algorithm on the packet it just received. If the hash Ironman just calculated matches the hash just received (after Ironman decrypted it using the sender's public key), then Ironman knows two things: that the only person who could have encrypted it was Thor with Thor's private key, and that the data integrity on the packet is solid, because if one bit had been changed, the hashes would not have matched. This process is called authentication, using digital signatures, and it normally happens in both directions with an IPsec VPN tunnel if the peers are using digital signatures for authentication (sometimes this is referred to as *rsa-signatures* in the configuration).

At this point you might be wondering how Ironman got Thor's key (Thor's public key) to begin with. The answer is that Thor and Ironman also exchanged digital certificates that contained each other's public keys. Thor and Ironman do not trust just any certificates, but they do trust certificates that are digitally signed by a CA they trust. This also implies that to verify digital signatures from the CA, both Thor and Ironman also need the CA's public key.

Most browsers and operating systems today have the built-in certificates and public keys for the mainstream CAs on the Internet. Figure 2-6 shows the "System Roots" keychain on macOS.

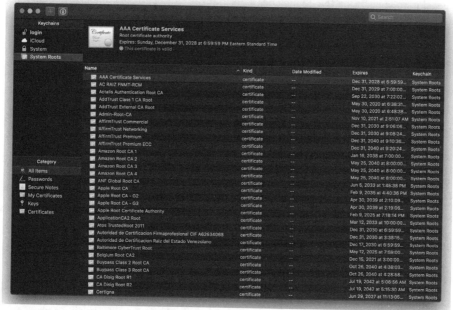

Figure 2-6 *System Roots CA certificates*

Key Management

Key management is huge in the world of cryptography. We have symmetric keys that can be used with symmetric algorithms for hashing and encryption. We have asymmetric keys such as public-private key pairs that can be used with asymmetric algorithms such as digital signatures, among other things. You could say that the key to security with all these algorithms we have taken a look at is the keys themselves.

Key management deals with generating keys, verifying keys, exchanging keys, storing keys, and at the end of their lifetime, destroying keys. An example of why this is critical is when two devices that want to establish a VPN session send their encryption keys over at the beginning of their session in plaintext. If that happens, an eavesdropper who sees the keys could use them to change ciphertext into understandable data, which would result in a lack of confidentiality within the VPN.

Keyspace refers to all the possible values for a key. The bigger the key, the more secure the algorithm will be. The only negative of having an extremely long key is that the longer the key, the more the CPU is used for the decryption and encryption of data.

Next-Generation Encryption Protocols

The industry is always looking for new algorithms for encryption, authentication, digital signatures, and key exchange to meet escalating security and performance requirements. The U.S. government selected and recommended a set of cryptographic standards called Suite B because it provides a complete suite of algorithms designed to meet future security needs. Suite B has been approved for protecting classified information at both the secret and top-secret levels. Cisco participated in the development of some of these standards. The Suite B next-generation encryption (NGE) includes algorithms for authenticated encryption, digital signatures, key establishment, and cryptographic hashing, as listed here:

- Elliptic Curve Cryptography (ECC) replaces RSA signatures with the ECDSA algorithm and replaces the DH key exchange with ECDH. ECDSA is an elliptic curve variant of the DSA algorithm, which has been a standard since 1994. The new key exchange uses DH with P-256 and P-384 curves.

- AES in the Galois/Counter Mode (GCM) of operation.

- ECC digital signature algorithm.

- SHA-256, SHA-384, and SHA-512.

IPsec

IPsec is a suite of protocols used to protect IP packets and has been around for decades. It is in use today for both remote-access VPNs and site-to-site VPNs. SSL is the new kid on the block in its application with remote-access VPNs. Let's take a closer look at both these options. IPsec use a collection of protocols and algorithms used to protect IP packets at Layer 3—hence the name IP Security (IPsec).

IPsec provides the core benefits of confidentiality through encryption, data integrity through hashing and HMAC, and authentication using digital signatures or using a pre-shared key (PSK) that is just for the authentication, similar to a password. IPsec also provides anti-replay support. The following is a high-level explanation of IPsec components (protocols, algorithms, and so on):

- **ESP and AH:** These are the two primary methods for implementing IPsec. ESP stands for Encapsulating Security Payload, which can perform all the features of IPsec, and AH stands for Authentication Header, which can do many parts of the IPsec objectives, except for the important one (the encryption of the data). For that reason, we do not frequently see AH being used.

- **Encryption algorithms for confidentiality:** DES, 3DES, and AES.

- **Hashing algorithms for integrity:** MD5 and SHA.

- **Authentication algorithms:** Pre-shared keys (PSKs) and RSA digital signatures.

- **Key management:** Examples of key management include Diffie-Hellman (DH), which can be used to dynamically generate symmetric keys to be used by symmetric algorithms; PKI, which supports the function of digital certificates issued by trusted CAs; and Internet Key Exchange (IKE), which does a lot of the negotiating and management needed for IPsec to operate.

NOTE Chapter 8 provides detailed explanations of IPsec site-to-site and remote-access VPNs, as well as other technical details about the underlying protocols.

SSL and TLS

Information transmitted over a public network needs to be secured through encryption to prevent unauthorized access to that data. An example is online banking. Not only do you want to avoid an attacker seeing your username, password, and codes, you also do not want an attacker to be able to modify the packets in transit during a transaction with the bank. This would seem to be a perfect opportunity for IPsec to be used to encrypt the data and perform integrity checking and authentication of the server you are connected to. Although it is true that IPsec can do all this, not everyone has an IPsec client or software running on their computer. What's more, not everyone has a digital certificate or a PSK they could successfully use for authentication.

You can still benefit from the concepts of encryption and authentication by using a different type of technology called Secure Sockets Layer (SSL). The convenient thing about SSL is that almost every web browser on every computer supports it, so almost anyone who has a computer can use it.

To use SSL, the user connects to an SSL server (that is, a web server that supports SSL) by using HTTPS rather than HTTP (the *S* in HTTPS stands for Secure). Depending on whom you talk to, SSL may also be called Transport Layer Security, or TLS. To the end user, it represents a secure connection to the server, and to the correct server.

TLS is the preferred method of encrypted communication. At the time of writing, TLS version 1.3 is the latest version of TLS. TLS version 1.3 is defined in RFC 8446 (https://tools. ietf.org/html/rfc8446).

TLS 1.3 provides several benefits and improvements in comparison to previous versions, such as separating key agreement and authentication algorithms from the cipher suites. TLS 1.3 also removes support for weak and less-used named elliptic curves. It also removes the use of MD5 and SHA-224 cryptographic hash functions.

The NIST Special Publication (SP) 800-52 Revision 2 (https://csrc.nist.gov/publications/detail/ sp/800-52/rev-2/final) provides detailed guidance for the selection, configuration, and use of TLS implementations. It also provides the minimum requirements for TLS implementations. For instance, it dictates that the TLS server should be configured with one or more public-key certificates and the associated private keys. TLS server implementations should also use multiple server certificates with their associated private keys to support algorithm and key size agility. TLS servers must be deployed with certificates issued by a CA that publishes revocation information in Online Certificate Status Protocol (OCSP) responses. The certificate authority may also publish revocation information in a certificate revocation list (CRL). The CRL includes a list of the serial numbers of the certificates that have been revoked.

TIP   Several sites, such as https://www.checktls.com and https://www.ssllabs.com/ssltest, allow you to test the TLS implementation of a given website. There are also tools such as https://testssl.sh that allow you to test any web server against known cryptographic weaknesses.

Even if the user does not type in HTTPS, the website can redirect the user behind the scenes to the correct URL. Once there, the browser requests that the web server identify itself. (Be aware that everything that is about to happen is occurring in the background and does not require user intervention.) The server sends the browser a copy of its digital certificate, which may also be called an SSL certificate. When the browser receives the certificate, it checks whether it trusts the certificate. Using the method for verifying a digital signature discussed earlier, the browser determines whether the certificate is valid based on the signature of the CA. Assuming the certificate is trusted, the browser now has access to the server's public key contained in the certificate.

If the signature is not valid, or at least if the browser does not think the certificate is valid, a pop-up is usually presented to the user asking whether he or she wants to proceed. This is where user training is important. Users should be trained never to accept a certificate that the browser does not trust.

Most of the time, the server does not require the browser to prove who it is. Instead, the web server uses some type of user authentication, such as a username and password, as required, to verify who the user is.

After the authentication has been done, several additional exchanges occur between the browser and the server as they establish the encryption algorithm they will use as well as the keys they will use to encrypt and decrypt the data.

As mentioned previously, understanding the terminology is important for you in mastering encryption and VPN technologies. Figure 2-7 explains the key components and their functions as well as provides examples of their implementation.

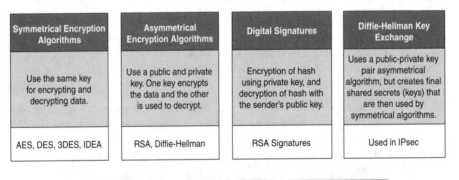

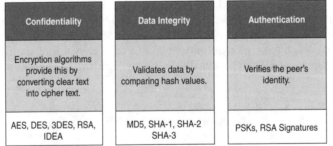

Figure 2-7 *Fundamental Encryption Components*

Fundamentals of PKI

Public key infrastructure (PKI) is a set of identities, roles, policies, and actions for the creation, use, management, distribution, and revocation of public and private keys. The reason that PKI exists is to enable the secure electronic transfer of information for many different purposes. You probably know that using simple passwords is an inadequate authentication method. PKI provides a more rigorous method to confirm the identity of the parties involved in the communication and to validate the information being transferred.

PKI binds public keys with the identities of people, applications, and organizations. This "binding" is maintained by the issuance and management of digital certificates by a certificate authority (CA).

Public and Private Key Pairs

A key pair is a set of two keys that work in combination with each other as a team. In a typical key pair, you have one public key and one private key. The public key may be shared with everyone, and the private key is not shared with anyone. For example, the private key for a web server is known only to that specific web server. If you use the public key to encrypt data using an asymmetric encryption algorithm, the corresponding private key is used to decrypt the data. The inverse is also true. If you encrypt with the private key, you then decrypt with the corresponding public key. Another name for this asymmetric encryption is *public key cryptography* or *asymmetric key cryptography*. The uses for asymmetric algorithms are not just limited to authentication, as in the case of digital signatures discussed in the previous sections, but that is one example of an asymmetric algorithm.

Let's take a look at another example. Thor and Ironman now want to send an encrypted email to each other using Pretty Good Privacy (PGP). In order for Thor to encrypt the email to Ironman, he needs Ironman's public key. Ironman can just send the key to Thor or alternatively publish his public key to a public repository or server.

> **TIP** The following are examples of public PGP key servers: pgp.mit.edu, pgp.key-server. io, and keybase.io. For example, my personal PGP key can be found at https://keybase.io/ santosomar.

Once Thor encrypts and sends the message to Ironman, Ironman decrypts the message using his private key.

More About Keys and Digital Certificates

Keys are the secrets that allow cryptography to provide confidentiality. Let's take a closer look at the keys involved with RSA and how they are used.

With RSA digital signatures, each party has a public-private key pair because both parties intend on authenticating the other side. Going back to the analogy in the previous sections, let's use the two users Thor and Ironman. They both generated their own public-private key pair, and they both enrolled with a certificate authority (CA). That CA took each of their public keys as well as their names and IP addresses and created individual digital certificates, and the CA issued these certificates back to Thor and Ironman, respectively. The CA also digitally signed each certificate.

When Thor and Ironman want to authenticate each other, they send each other their digital certificates (or at least a copy of them). Upon receiving the other party's digital certificate, they both verify the authenticity of the certificate by checking the signature of a CA they currently trust. (When you *trust* a certificate authority, it means that you know who the CA is and can verify that CA's digital signature by knowing the public key of that CA.)

Now that Thor and Ironman have each other's public keys, they can authenticate each other. This normally happens inside of a VPN tunnel in both directions (when RSA signatures are used for authentication). For the purpose of clarity, we focus on just one of these parties (for example, the computer Thor) and proving its identity to the other computer (in this case, Ironman).

Thor takes some data, generates a hash, and then encrypts the hash with Thor's private key. (Note that the private key is not shared with anyone else—not even Thor's closest friends have it.) This encrypted hash is inserted into the packet and sent to Ironman. This encrypted hash is Thor's digital signature.

Ironman, having received the packet with the digital signature attached, first decodes or decrypts the encrypted hash using Thor's public key. It then sets the decrypted hash to the side for a moment and runs a hash against the same data that Thor did previously. If the hash that Ironman generates matches the decrypted hash, which was sent as a digital signature from Thor, then Ironman has just authenticated Thor—because only Thor has the private key used for the creation of Thor's digital signature.

Certificate Authorities

A certificate authority is a computer or entity that creates and issues digital certificates. Inside of a digital certificate is information about the identity of a device, such as its IP address, fully qualified domain name (FQDN), and the public key of that device. The CA takes requests from devices that supply all of that information (including the public key generated by the computer making the request) and generates a digital certificate, which the CA assigns a serial number to. The CA then signs the certificate with its own digital signature. Also included in the final certificate is a URL that other devices can check to see whether this certificate has been revoked and the certificate's validity dates (the time window during which the certificate is considered valid). Also in the certificate is the information about the CA that issued the certificate and several other parameters used by PKI.

Figure 2-8 shows an example of a digital certificate, which, in this case, is for the website ccnpsecurity.org. You also see the digital signature of the CA.

Now let's go back to our scenario. Thor and Ironman's computers can receive and verify identity certificates from each other (and thousands of others) by using a third-party trusted certificate authority, as long as the certificates are signed by a CA that is trusted by Thor and Ironman. Commercial CAs charge a fee to issue and maintain digital certificates. One benefit of using a commercial CA server to obtain digital certificates for your devices is that most web browsers maintain a list of the more common trusted public CA servers, and as a result, anyone using a browser can verify the identity of your web server by default without having to modify their web browser at all. If a company wants to set up its own internal CA and then configure each of the end devices to trust the certificates issued by this internal CA, no commercial certificate authority is required, but the scope of that CA is limited to the company and its managed devices, because any devices outside of the company would not trust the company's internal CA by default.

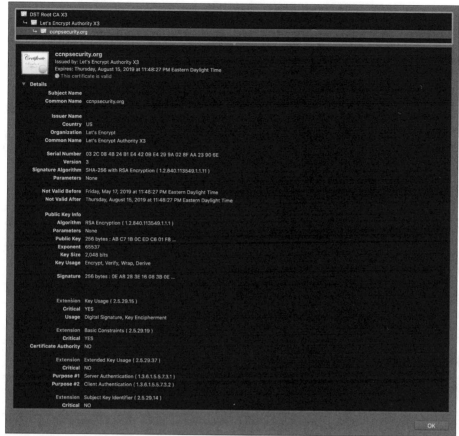

Figure 2-8 *An Example of a Digital Certificate*

Root Certificates

A digital certificate can be thought of as an electronic document that identifies a device or person. It includes information such as the name of the person or organization, their address, and the public key of that person or device. There are different types of certificates, including root certificates (which identify the CA), and identity certificates, which identify devices such as servers and other devices that want to participate in PKI.

A root certificate contains the public key of the CA server and the other details about the CA server. Figure 2-9 shows an example of one.

The output in Figure 2-9 can be seen on most browsers, although the location of the information might differ a bit depending on the browser vendor and version.

Here are the relevant parts of the certificate:

- **Serial number:** Issued and tracked by the CA that issued the certificate.

- **Issuer:** The CA that issued this certificate. (Even root certificates need to have their certificates issued from someone, perhaps even themselves.)

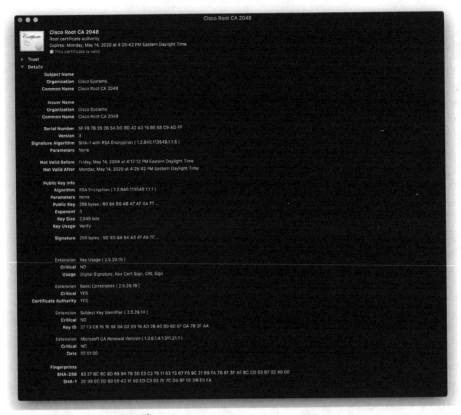

Figure 2-9   *A Root Certificate*

- **Validity dates:** The time window during which the certificate is considered valid. If a local computer believes the date to be off by a few years, that same PC may consider the certificate invalid due to its own error about the time. Using the Network Time Protocol (NTP) is a good idea to avoid this problem.

- **Subject of the certificate:** This includes the organizational unit (OU), organization (O), country (C), and other details commonly found in an X.500 structured directory. The subject of the root certificate is the CA itself. The subject for a client's identity certificate is the client.

- **Public key:** The contents of the public key and the length of the key are often both shown. After all, the public key is public.

- **Thumbprint algorithm and thumbprint:** This is the hash for the certificate. On a new root certificate, you could use a phone to call and ask for the hash value and compare it to the hash value you see on the certificate. If it matches, you have just performed out-of-band verification (using the telephone) of the digital certificate.

Identity Certificates

An identity certificate is similar to a root certificate, but it describes the client and contains the public key of an individual host (the client). An example of a client is a web server that wants to support SSL or a router that wants to use digital signatures for authentication of a VPN tunnel.

Basically, any device that wants to verify a digital signature must have the public key of the sender. So, as an example, let's say that you and I want to authenticate each other, and we both trust a common CA and have previously requested and received digital certificates (identity certificates) from the CA server. We exchange our identity certificates, which contain our public keys. We both verify the CA's signature on the digital certificate we just received from each other using the public key of the CA. In practice, this public key for the CA is built in to most of the browsers today for public CA servers. Once we verify each other's certificates, we can then trust the contents of those certificates (and most important, the public key). Now that you and I both have each other's public key, we can use those public keys to verify each other's digital signatures.

X.500 and X.509v3

X.500 is a series of standards focused on directory services and how those directories are organized. Many popular network operating systems have been based on X.500, including Microsoft Active Directory. This X.500 structure is the foundation from which you see common directory elements such as CN=Thor (CN stands for common name), OU=engineering (OU stands for organizational unit), O=cisco.com (O stands for organization), and so on, all structured in an "org chart" way (that is, shaped like a pyramid). X.509 Version 3 is a standard for digital certificates that is widely accepted and incorporates many of the same directory and naming standards. A common protocol used to perform lookups from a directory is the Lightweight Directory Access Protocol (LDAP). A common use for this protocol is having a digital certificate that's used for authentication, and then based on the details of that certificate (for example, OU=sales in the certificate itself), the user can be dynamically assigned the access rights associated with that group in Active Directory or some other LDAP-accessible database. The concept is to define the rights in one place and then leverage them over and over again. An example is setting up Active Directory for the network and then using that to control what access is provided to each user after he or she authenticates.

As a review, most digital certificates contain the following information:

- **Serial number:** Assigned by the CA and used to uniquely identify the certificate
- **Subject:** The person or entity that is being identified
- **Signature algorithm:** The specific algorithm used for signing the digital certificate
- **Signature:** The digital signature from the certificate authority, which is used by devices that want to verify the authenticity of the certificate issued by that CA
- **Issuer:** The entity or CA that created and issued the digital certificate
- **Valid from:** The date the certificate became valid
- **Valid to:** The expiration date of the certificate
- **Key usage:** The functions for which the public key in the certificate may be used
- **Public key:** The public portion of the public and private key pair generated by the host whose certificate is being looked at

■ **Thumbprint algorithm:** The hash algorithm used for data integrity

■ **Thumbprint:** The actual hash

■ **Certificate revocation list location:** The URL that can be checked to see whether the serial number of any certificates issued by the CA have been revoked

Authenticating and Enrolling with the CA

Using a new CA as a trusted entity, as well as requesting and receiving your own identity certificate from this CA, is really a two-step process, as demonstrated in Figure 2-10.

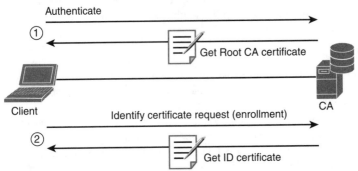

Figure 2-10 *Authenticating and Enrolling with the CA*

The following are the steps illustrated in Figure 2-10:

Step 1. The first step is to authenticate the CA server (in other words, to *trust* the CA server). Unfortunately, if you do not have the public key for a CA server, you cannot verify the digital signature of that CA server. This is sort of like the chicken and the egg story, because you need the public key, which can be found in the root's CA certificate, but you cannot verify the signature on a certificate until you have the public key.

To get the ball rolling, you could download the root certificate and then use an out-of-band method, such as making a telephone call, to validate the root certificate. This can be done after downloading the root certificate and looking at the hash value by calling the administrators for the root CA and asking them to verbally tell you what the hash is. If the hash that they tell you over the phone matches the hash you see on the digital certificate (and assuming that you called the right phone number and talked with the right people), you know that the certificate is valid, and you can then use the public key contained in a certificate to verify future certificates signed by that CA. This process of getting the root CA certificate installed is often referred to as *authenticating the CA*. Current web browsers automate this process for well-known CAs.

Step 2. After you have authenticated the root CA and have a known-good root certificate for that CA, you can then request your own identity certificate. This involves generating a public-private key pair and including the public key portion in any requests for your own identity certificate. An identity certificate could be for a device or person. Once you make this request, the CA can take all of your information and generate an identity certificate for you, which includes your public key, and then send this certificate back to you. If this is done electronically, how do you verify the identity certificate you got is really from the CA server that you trust? The answer is simple, because the CA has not only issued the certificate but has also signed the certificate. Because you authenticated the CA server earlier and you have a copy of its digital certificate with its public key, you can now verify the digital signature it has put on your own identity certificate. If the signature from the CA is valid, you also know that your certificate is valid and therefore you can install it and use it.

Public Key Cryptography Standards

Many standards are in use for the PKI. Many of them have Public Key Cryptography Standards (PKCS) numbers. Some of these standards control the format and use of certificates, including requests to a CA for new certificates, the format for a file that is going to be the new identity certificate, and the file format and usage access for certificates. Having the standards in place helps with interoperability between different CA servers and many different CA clients.

Here are a few standards you should become familiar with; these include protocols by themselves and protocols used for working with digital certificates:

- **PKCS #10:** This is a format of a certificate request sent to a CA that wants to receive its identity certificate. This type of request would include the public key for the entity desiring a certificate.

- **PKCS #7:** This is a format that can be used by a CA as a response to a PKCS #10 request. The response itself will very likely be the identity certificate (or certificates) that had been previously requested.

- **PKCS #1:** The RSA cryptography standard.

- **PKCS #12:** A format for storing both public and private keys using a symmetric password-based key to "unlock" the data whenever the key needs to be used or accessed.

- **PKCS #3:** Diffie-Hellman key exchange.

Simple Certificate Enrollment Protocol

The process of authenticating a CA server, generating a public-private key pair, requesting an identity certificate, and then verifying and implementing the identity certificate can take several steps. Cisco, in association with a few other vendors, developed the Simple Certificate Enrollment Protocol (SCEP), which can automate most of the process for requesting and installing an identity certificate. Although it is not an open standard, it is supported by most Cisco devices and makes getting and installing both root and identity certificates convenient.

Revoking Digital Certificates

If you decommission a device that has been assigned an identity certificate, or if the device assigned a digital certificate has been compromised and you believe that the private key information is no longer "private," you could request from the CA that the previously issued certificate be revoked. This poses a unique problem. Normally when two devices authenticate with each other, they do not need to contact a CA to verify the identity of the other party. This is because the two devices already have the public key of the CA and can validate the signature on a peer's certificate without direct contact with the CA. So here's the challenge: If a certificate has been revoked by the CA, and the peers are not checking with the CA each time they try to authenticate the peers, how does a peer know whether the certificate it just received has been revoked? The answer is simple: It has to check and see. A digital certificate contains information on where an updated list of revoked certificates can be obtained. This URL could point to the CA server itself or to some other publicly available resource on the Internet. The revoked certificates are listed based on the serial number of the certificates, and if a peer has been configured to check for revoked certificates, it adds this check before completing the authentication with a peer.

If a certificate revocation list (CRL) is checked, and the certificate from the peer is on that list, the authentication stops at that moment. The three basic ways to check whether certificates have been revoked are as follows, in order of popularity:

- **Certificate revocation list (CRL):** This is a list of certificates, based on their serial numbers, that had initially been issued by a CA but have since been revoked and as a result should not be trusted. A CRL could be very large, and the client would have to process the entire list to verify a particular certificate is not on the list. A CRL can be thought of as the naughty list. CRL is the primary protocol used for this purpose, compared to OSCP and AAA. A CRL can be accessed by several protocols, including LDAP and HTTP. A CRL can also be obtained via SCEP.

- **Online Certificate Status Protocol (OCSP):** This is an alternative to CRLs. Using this method, a client simply sends a request to find the status of a certificate and gets a response without having to know the complete list of revoked certificates.

- **Authentication, authorization, and accounting (AAA):** Cisco AAA services also provide support for validating digital certificates, including a check to see whether a certificate has been revoked. Because this is a proprietary solution, it is not often used in PKI.

Digital Certificates in Practice

Digital certificates can be used for clients who want to authenticate a web server to verify they are connected to the correct server using HTTP Secure (HTTPS), Transport Layer Security (TLS), or Secure Sockets Layer (SSL). For the average user who does not have to write these protocols, but simply benefits from using them, they are all effectively the same, which is HTTP combined with TLS/SSL for the security benefits. This means that digital certificates can be used when you do online banking from your PC to the bank's website. It also means that if you use SSL technology for your remote-access VPNs, you can use digital certificates for authenticating the peers (at each end) of the VPN.

You can also use digital certificates with the protocol family of IPsec, which can also use digital certificates for the authentication portion.

In addition, digital certificates can be used with protocols such as 802.1X, which involves authentication at the edge of the network before allowing the user's packets and frames to progress through it. An example is a wireless network, controlling access and requiring authentication, using digital certificates for the PCs/users, before allowing them in on the network.

PKI Topologies

There is no one-size-fits-all solution for PKI. In small networks, a single CA server may be enough, but in a network with 30,000 devices, a single server might not provide the availability and fault tolerance required. To address these issues, let's investigate the options available to us for implementation of the PKI, using various topologies, including single and hierarchical. Let's start off with the single CA and expand from there.

Single Root CA

If you have one trusted CA, and you have tens of thousands of customers who want to authenticate that CA and request their own identity certificates, there might be too large of a demand on a single server, even though a single CA does not have to be directly involved in the day-to-day authentication that happens between peers. To offload some of the workload from a single server, you could publish CRLs on other servers. At the end of the day, it still makes sense to have at least some fault tolerance for your PKI, which means more than just a single root CA server.

Hierarchical CA with Subordinate CAs

One option for supporting fault tolerance and increased capacity is to use intermediate or subordinate CAs to assist the root CA. The root CA is the king of the hill. The root CA delegates the authority (to the subordinate CAs) to create and assign identity certificates to clients. This is called a *hierarchical PKI topology*. The root CA signs the digital certificates of its subordinate or intermediate CAs, and the subordinate CAs are the ones to issue certificates to clients. Figure 2-11 shows a hierarchical CA deployment with a root and three subordinate CAs.

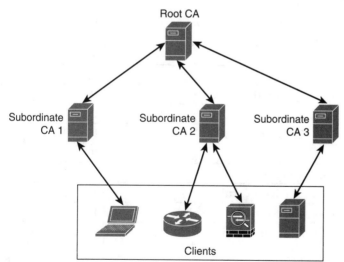

Figure 2-11   *Hierarchical CA Deployment with Subordinate CAs*

For a client to verify the "chain" of authority, the client needs both the subordinate CA's certificate and the root certificate. The root certificate (and its public key) is required to verify the digital signature of the root CA, and the subordinate CA's certificate (and its public key) is required to verify the signature of the subordinate CA. If there are multiple levels of subordinate CAs, a client needs the certificates of all the devices in the chain, from the root all the way to the CA that issued the client's certificate.

Cross-Certifying CAs

Another approach to hierarchical PKIs is called *cross-certification*. With cross-certification, you would have a CA with a horizontal trust relationship over to a second CA so that clients of either CA can trust the signatures of the other CA.

Exam Preparation Tasks

As mentioned in the section "How to Use This Book" in the Introduction, you have a couple of choices for exam preparation: the exercises here, Chapter 12, "Final Preparation," and the exam simulation questions in the Pearson Test Prep Software Online.

Review All Key Topics

Review the most important topics in this chapter, noted with the Key Topic icon in the outer margin of the page. Table 2-2 lists these key topics and the page numbers on which each is found.

Table 2-2   Key Topics for Chapter 2

Key Topic Element	Description	Page Number
Paragraph	Define and identify block ciphers	82
Paragraph	Define and identify stream ciphers	82
Section	Symmetric and Asymmetric Algorithms	82
Paragraph	Understand that asymmetric algorithms use a public and private key (key pair)	83
Section	Hashes	84
List	Identify hashing algorithms	85
Section	Hashed Message Authentication Code	86
List	Define digital signatures	86
Section	Next-Generation Encryption Protocols	89
Section	IPSec	90
Section	Public and Private Key Pairs	93
Section	More About Keys and Digital Certificates	93
Section	Certificate Authorities	94
List	Define what is a root certificate and the relevant parts of a certificate	95
Section	Identity Certificates	96
List	Understand the Public Key Cryptography Standards (PKCS)	99
Section	Simple Certificate Enrollment Protocol	99
List	Define what is a CRL	100

Define Key Terms

Define the following key terms from this chapter and check your answers in the glossary:

Block ciphers, symmetric algorithms, asymmetric algorithms, stream cipher, hashing, Hashed Message Authentication Code (HMAC), hashing algorithms, digital certificates, certificate authority, root certificate, identity certificate

Review Questions

1. Which of the following are examples of common methods used by ciphers?
   a. Transposition
   b. Substitution
   c. Polyalphabetic
   d. Polynomial

2. Which of the following are examples of symmetric block cipher algorithms?
   a. Advanced Encryption Standard (AES)
   b. Triple Digital Encryption Standard (3DES)
   c. DSA
   d. Blowfish
   e. ElGamal

3. Which of the following are examples of hashes?
   a. ASH-160
   b. SHA-1
   c. SHA-2
   d. MD5

4. Which of the following are benefits of digital signatures?
   a. Authentication
   b. Nonrepudiation
   c. Encryption
   d. Hashing

5. Which of the following statements are true about public and private key pairs?
   a. A key pair is a set of two keys that work in combination with each other as a team.
   b. A key pair is a set of two keys that work in isolation.
   c. If you use the public key to encrypt data using an asymmetric encryption algorithm, the corresponding private key is used to decrypt the data.
   d. If you use the public key to encrypt data using an asymmetric encryption algorithm, the peer decrypts the data with that public key.

6. Which of the following entities can be found inside of a digital certificate?

   a. FQDN

   b. DNS server IP address

   c. Default gateway

   d. Public key

7. Which of the following is true about root certificates?

   a. A root certificate contains information about the user.

   b. A root certificate contains information about the network security device.

   c. A root certificate contains the public key of the CA.

   d. Root certificates never expire.

8. Which of the following are public key standards?

   a. IPsec

   b. PKCS #10

   c. PKCS #12

   d. ISO33012

   e. AES

9. Most digital certificates contain which of the following information?

   a. Serial number

   b. Signature

   c. Thumbprint (fingerprint)

   d. All of these answers are correct.

10. Which of the following is a format for storing both public and private keys using a symmetric password-based key to "unlock" the data whenever the key needs to be used or accessed?

   a. PKCS #12

   b. PKCS #10

   c. PKCS #7

   d. None of these answers is correct.

## CHAPTER 3

Software-Defined Networking Security and Network Programmability

This chapter covers the following topics:

Software-Defined Networking (SDN) and SDN Security

Network Programmability

This chapter starts with an introduction to SDN and different SDN security concepts, such as centralized policy management and micro-segmentation. This chapter also introduces SDN solutions such as Cisco ACI and modern networking environments such as Cisco DNA. You will also learn what are network overlays and what they are trying to solve.

The second part of this chapter provides an overview of network programmability and how networks are being managed using modern application programming interfaces (APIs) and other functions. This chapter also includes dozens of references that are available to enhance your learning.

The following SCOR 350-701 exam objectives are covered in this chapter:

- Domain 1: Security Concepts

  - 1.7 Explain northbound and southbound APIs in the SDN architecture

  - 1.8 Explain DNAC APIs for network provisioning, optimization, monitoring, and troubleshooting

"Do I Know This Already?" Quiz

The "Do I Know This Already?" quiz allows you to assess whether you should read this entire chapter thoroughly or jump to the "Exam Preparation Tasks" section. If you are in doubt about your answers to these questions or your own assessment of your knowledge of the topics, read the entire chapter. Table 3-1 lists the major headings in this chapter and their corresponding "Do I Know This Already?" quiz questions. You can find the answers in Appendix A, "Answers to the 'Do I Know This Already?' Quizzes and Q&A Sections."

Table 3-1 "Do I Know This Already?" Section-to-Question Mapping

Foundation Topics Section	Questions
Software-Defined Networking (SDN) and SDN Security	1–5
Network Programmability	6–10

1. Which of the following are the three different "planes" in traditional networking?

   a. The management, control, and data planes

   b. The authorization, authentication, and accountability planes

   c. The authentication, control, and data planes

   d. None of these answers is correct.

2. Which of the following is true about Cisco ACI?

   a. Spine nodes interconnect leaf devices, and they can also be used to establish connections from a Cisco ACI pod to an IP network or interconnect multiple Cisco ACI pods.

   b. Leaf switches provide the Virtual Extensible LAN (VXLAN) tunnel endpoint (VTEP) function.

   c. The APIC manages the distributed policy repository responsible for the definition and deployment of the policy-based configuration of the Cisco ACI infrastructure.

   d. All of these answers are correct.

3. Which of the following is used to create network overlays?

   a. SDN-Lane

   b. VXLAN

   c. VXWAN

   d. None of these answers is correct.

4. Which of the following is an identifier or a tag that represents a logical segment?

   a. VXLAN Network Identifier (VNID)

   b. VXLAN Segment Identifier (VSID)

   c. ACI Network Identifier (ANID)

   d. Application Policy Infrastructure Controller (APIC)

5. Which of the following is network traffic between servers (virtual servers or physical servers), containers, and so on?

   a. East-west traffic

   b. North-south traffic

   c. Micro-segmentation

   d. Network overlays

6. Which of the following is an HTTP status code message range related to successful HTTP transactions?

    a. Messages in the 100 range

    b. Messages in the 200 range

    c. Messages in the 400 range

    d. Messages in the 500 range

7. Which of the following is a Python package that can be used to interact with REST APIs?

    a. argparse

    b. requests

    c. rest_api_pkg

    d. None of these answers is correct.

8. Which of the following is a type of API that exclusively uses XML?

    a. APIC

    b. REST

    c. SOAP

    d. GraphQL

9. Which of the following is a modern framework of API documentation and is now the basis of the OpenAPI Specification (OAS)?

    a. SOAP

    b. REST

    c. Swagger

    d. WSDL

10. Which of the following can be used to retrieve a network device configuration?

    a. RESTCONF

    b. NETCONF

    c. SNMP

    d. All of these answers are correct.

Foundation Topics

Introduction to Software-Defined Networking

In the last decade there have been several shifts in networking technologies. Some of these changes are due to the demand of modern applications in very diverse environments and the cloud. This complexity introduces risks, including network configuration errors that can cause significant downtime and network security challenges.

Subsequently, networking functions such as routing, optimization, and security have also changed. The next generation of hardware and software components in enterprise networks must support both the rapid introduction and the rapid evolution of new technologies and solutions. Network infrastructure solutions must keep pace with the business environment and support modern capabilities that help drive simplification within the network.

These elements have fueled the creation of software-defined networking (SDN). SDN was originally created to decouple control from the forwarding functions in networking equipment. This is done to use software to centrally manage and "program" the hardware and virtual networking appliances to perform forwarding.

Traditional Networking Planes

In traditional networking, there are three different "planes" or elements that allow network devices to operate: the management, control, and data planes. Figure 3-1 shows a high-level explanation of each of the planes in traditional networking.

TRADITIONAL ROUTING AND SWITCHING PLANES		
Management Plane	**Control Plane**	**Data Plane**
• Configuration and monitoring • Typically done via the traditional CLI or GUI • Each vendor has its proprietary way to configure its devices	• Layer 2 protocols and control • Layer 3 protocols (e.g., OSPF, RIP, BGP, etc.)	• Institutes how data is forwarded inside the hardware from interface to interface

Figure 3-1   *The Management, Control, and Data Planes*

The control plane has always been separated from the data plane. There was no central brain (or controller) that controlled the configuration and forwarding. Let's take a look at the example shown in Figure 3-2. Routers, switches, and firewalls were managed by the command-line interface (CLI), graphical user interfaces (GUIs), and custom Tcl scripts. For instance, the firewalls were managed by the Adaptive Security Device Manager (ASDM), while the routers were managed by the CLI.

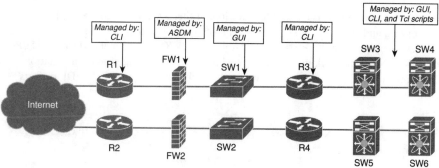

Figure 3-2   *Traditional Network Management Solutions*

Each device in Figure 3-2 has its "own brain" and does not really exchange any intelligent information with the rest of the devices.

So What's Different with SDN?

SDN introduced the notion of a centralized controller. The SDN controller has a global view of the network, and it uses a common management protocol to configure the network infrastructure devices. The SDN controller can also calculate reachability information from many systems in the network and pushes a set of flows inside the switches. The flows are used by the hardware to do the forwarding. Here you can see a clear transition from a distributed "semi-intelligent brain" approach to a "central and intelligent brain" approach.

TIP An example of an open source implementation of SDN controllers is the Open vSwitch (OVS) project using the OVS Database (OVSDB) management protocol and the OpenFlow protocol. Another example is the Cisco Application Policy Infrastructure Controller (Cisco APIC). Cisco APIC is the main architectural component and the brain of the Cisco Application Centric Infrastructure (ACI) solution. A great example of this is Cisco ACI, which is discussed in the next section of the chapter.

SDN changed a few things in the management, control, and data planes. However, the big change was in the control and data planes in software-based switches and routers (including virtual switches inside of hypervisors). For instance, the Open vSwitch project started some of these changes across the industry.

SDN provides numerous benefits in the area of management plane. These benefits are in both physical switches and virtual switches. SDN is now widely adopted in data centers. A great example of this is Cisco ACI.

Introduction to the Cisco ACI Solution

Cisco ACI provides the ability to automate setting networking policies and configurations in a very flexible and scalable way. Figure 3-3 illustrates the concept of a centralized policy and configuration management in the Cisco ACI solution.

The Cisco ACI scenario shown in Figure 3-3 uses a leaf-and-spine topology. Each leaf switch is connected to every spine switch in the network with no interconnection between leaf switches or spine switches.

The leaf switches have ports connected to traditional Ethernet devices (for example, servers, firewalls, routers, and so on). Leaf switches are typically deployed at the edge of the fabric. These leaf switches provide the Virtual Extensible LAN (VXLAN) tunnel endpoint (VTEP) function. VXLAN is a network virtualization technology that leverages an encapsulation technique (similar to VLANs) to encapsulate Layer 2 Ethernet frames within UDP packets (over UDP port 4789, by default).

NOTE The section "VXLAN and Network Overlays," later in the chapter, will discuss VXLAN and overlays in more detail.

In Cisco ACI, the IP address that represents the leaf VTEP is called the physical tunnel endpoint (PTEP). The leaf switches are responsible for routing or bridging tenant packets and for applying network policies.

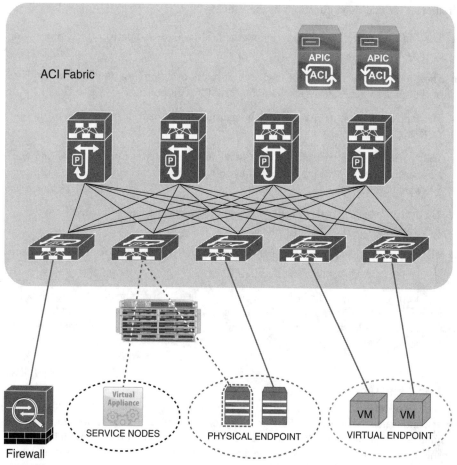

Figure 3-3 *Cisco APIC Configuration and Policy Management*

Spine nodes interconnect leaf devices, and they can also be used to establish connections from a Cisco ACI pod to an IP network or to interconnect multiple Cisco ACI pods. Spine switches store all the endpoint-to-VTEP mapping entries. All leaf nodes connect to all spine nodes within a Cisco ACI pod. However, no direct connectivity is allowed between spine nodes or between leaf nodes.

NOTE   All workloads in Cisco ACI connect to leaf switches. The leaf switches used in a Cisco ACI fabric are Top-of-the-Rack (ToR) switches. The acronym "ToR" here is not the same as "The Onion Router" (a solution used for anonymity and to access the "deep web").

The APIC can be considered a policy and a topology manager. APIC manages the distributed policy repository responsible for the definition and deployment of the policy-based configuration of the Cisco ACI infrastructure. APIC also manages the topology and inventory information of all devices within the Cisco ACI pod.

The following are additional functions of the APIC:

- The APIC "observer" function monitors the health, state, and performance information of the Cisco ACI pod.

- The "boot director" function is in charge of the booting process and firmware updates of the spine switches, leaf switches, and the APIC components.

- The "appliance director" APIC function manages the formation and control of the APIC appliance cluster.

- The "virtual machine manager (VMM)" is an agent between the policy repository and a hypervisor. The VMM interacts with hypervisor management systems (for example, VMware vCenter).

- The "event manager" manages and stores all the events and faults initiated from the APIC and the Cisco ACI fabric nodes.

- The "appliance element" maintains the inventory and state of the local APIC appliance.

> **TIP**   The Cisco ACI Design Guide provides comprehensive information about the design, deployment, and configuration of the ACI solution. The design guide can be found here: https://www.cisco.com/c/en/us/solutions/collateral/data-center-virtualization/application-centric-infrastructure/white-paper-c11-737909.pdf.

VXLAN and Network Overlays

Modern networks and data centers need to provide load balancing, better scalability, elasticity, and faster convergence. Many organizations use the overlay network model. Deploying an overlay network allows you to tunnel Layer 2 Ethernet packets with different encapsulations over a Layer 3 network. The overlay network uses "tunnels" to carry the traffic across the Layer 3 fabric. This solution also needs to allow the "underlay" to separate network flows between different "tenants" (administrative domains). The solution also needs to switch packets within the same Layer 2 broadcast domain, route traffic between Layer 3 broadcast domains, and provide IP separation, traditionally done via virtual routing and forwarding (VRF).

There have been multiple IP tunneling mechanisms introduced throughout the years. The following are a few examples of tunneling mechanisms:

- Virtual Extensible LAN (VXLAN)

- Network Virtualization using Generic Routing Encapsulation (NVGRE)

- Stateless Transport Tunneling (STT)

- Generic Network Virtualization Encapsulation (GENEVE)

All of the aforementioned tunneling protocols carry an Ethernet frame inside an IP frame. The main difference between them is in the type of the IP frame used. For instance, VXLAN uses UDP, and STT uses TCP.

The use of UDP in VXLAN enables routers to apply hashing algorithms on the outer UDP header to load balance network traffic. Network traffic that is riding the overlay network tunnels is load balanced over multiple links using equal-cost multi-path routing (ECMP). This introduces a better solution compared to traditional network designs. In traditional network designs, access switches connect to distribution switches. This causes redundant links to block due to spanning tree.

VXLAN uses an identifier or a tag that represents a logical segment that is called the VXLAN Network Identifier (VNID). The logical segment identified with the VNID is a Layer 2 broadcast domain that is tunneled over the VTEP tunnels.

Figure 3-4 shows an example of an overlay network that provides Layer 2 capabilities.

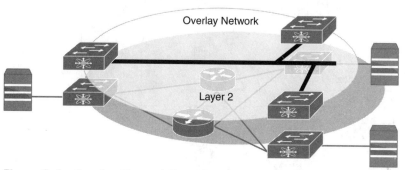

Figure 3-4   *Overlay Network Providing Layer 2 Capabilities*

Figure 3-5 shows an example of an overlay network that provides Layer 3 routing capabilities.

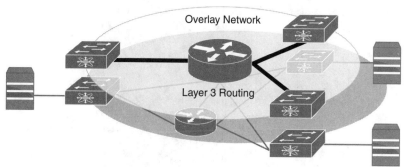

Figure 3-5   *Overlay Network Providing Layer 3 Routing Capabilities*

Figure 3-6 illustrates the VXLAN frame format for your reference.

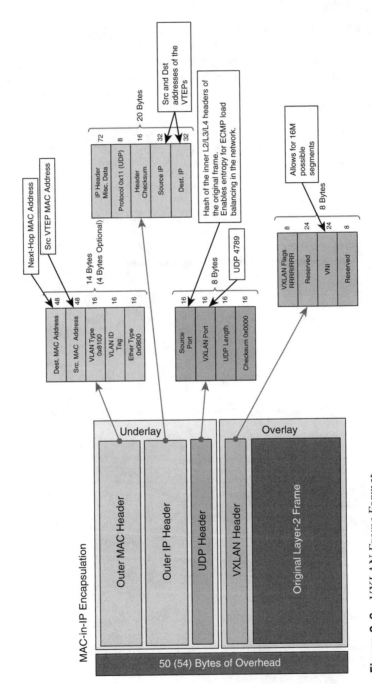

Figure 3-6 *VXLAN Frame Format*

Micro-Segmentation

For decades, servers were assigned subnets and VLANs. Sounds pretty simple, right? Well, this introduced a lot of complexities because application segmentation and policies were physically restricted to the boundaries of the VLAN within the same data center (or even in "the campus"). In virtual environments, the problem became harder. Nowadays applications can move around between servers to balance loads for performance or high availability upon failures. They also can move between different data centers and even different cloud environments.

Traditional segmentation based on VLANs constrains you to maintain the policies of which application needs to talk to which application (and who can access such applications) in centralized firewalls. This is ineffective because most traffic in data centers is now "East-West" traffic. A lot of that traffic does not even hit the traditional firewall. In virtual environments, a lot of the traffic does not even leave the physical server.

Let's define what people refer to as "East-West" traffic and "North-South" traffic. "East-West" traffic is network traffic between servers (virtual servers or physical servers, containers, and so on).

"North-South" traffic is network traffic flowing in and outside the data center. Figure 3-7 illustrates the concepts of "East-West" and "North-South" traffic.

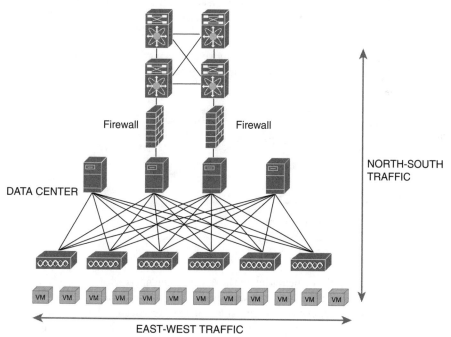

Figure 3-7   *"East-West" and "North-South" Traffic*

Many vendors have created solutions where policies applied to applications are independent from the location or the network tied to the application.

For example, let's suppose that you have different applications running in separate VMs and those applications also need to talk to a database (as shown in Figure 3-8).

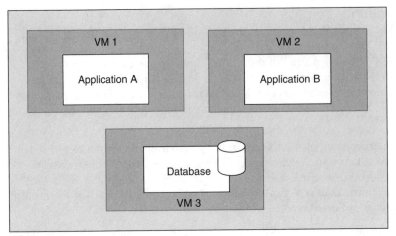

Figure 3-8 *Applications in VMs*

You need to apply policies to restrict if application A needs or does not need to talk to application B, or which application should be able to talk to the database. These policies should not be bound by which VLAN or IP subnet the application belongs to and whether it is in the same rack or even in the same data center. Network traffic should not make multiple trips back and forth between the applications and centralized firewalls to enforce policies between VMs.

Containers make this a little harder because they move and change more often. Figure 3-9 illustrates a high-level representation of applications running inside of containers (for example, Docker containers).

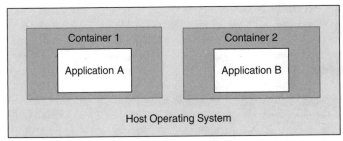

Figure 3-9 *Applications in Containers*

The ability to enforce network segmentation in those environments is what's called "micro-segmentation." Micro-segmentation is at the VM level or between containers regardless of a VLAN or a subnet. Micro-segmentation segmentation solutions need to be "application aware." This means that the segmentation process starts and ends with the application itself.

Most micro-segmentation environments apply a "zero-trust model." This model dictates that users cannot talk to applications, and applications cannot talk to other applications unless a defined set of policies permits them to do so.

Open Source Initiatives

There are several open source projects that are trying to provide micro-segmentation and other modern networking benefits. Examples include the following:

- Neutron from OpenStack

- Open vSwitch (OVS)

- Open Virtual Network (OVN)

- OpenDaylight (ODL)

- Open Platform for Network Function Virtualization (OPNFV)

- Contiv

The concept of SDN is very broad, and every open source provider and commercial vendor takes it in a different direction. The networking component of OpenStack is called Neutron. Neutron is designed to provide "networking as a service" in private, public, and hybrid cloud environments. Other OpenStack components, such as Horizon (Web UI) and Nova (compute service), interact with Neutron using a set of APIs to configure the networking services. Neutron uses plug-ins to deliver advanced networking capabilities and allow third-party vendor integration. Neutron has two main components: the neutron server and a database that handles persistent storage and plug-ins to provide additional services. Additional information about Neutron and OpenStack can be found at https://docs.openstack.org/neutron/latest.

OVN was originally created by the folks behind Open vSwitch (OVS) for the purpose of bringing an open source solution for virtual network environments and SDN. Open vSwitch is an open source implementation of a multilayer virtual switch inside the hypervisor.

> **NOTE** You can download Open vSwitch and access its documentation at https://www.openvswitch.org.

OVN is often used in OpenStack implementations with the use of OVS. You can also use OVN with the OpenFlow protocol. OpenStack Neutron uses OVS as the default "control plane."

> **NOTE** You can access different tutorials about OVN and OVS at http://docs.openvswitch.org/en/latest/tutorials/.

OpenDaylight (ODL) is another popular open source project that is focused on the enhancement of SDN controllers to provide network services across multiple vendors. OpenDaylight participants also interact with the OpenStack Neutron project and attempt to solve the existing inefficiencies.

OpenDaylight interacts with Neutron via a northbound interface and manages multiple interfaces southbound, including the Open vSwitch Database Management Protocol (OVSDB) and OpenFlow.

TIP   You can find more information about OpenDaylight at https://www.opendaylight. org. Cisco has several tutorials and additional information about OpenDaylight in DevNet at https://developer.cisco.com/site/opendaylight/.

 So, what is a northbound and southbound API? In an SDN architecture, southbound APIs are used to communicate between the SDN controller and the switches and routers within the infrastructure. These APIs can be open or proprietary.

NOTE   Cisco provides detailed information about the APIs supported in all platforms in DevNet (developer.cisco.com). DevNet will be discussed in detail later in this chapter.

Southbound APIs enable SDN controllers to dynamically make changes based on real-time demands and scalability needs. OpenFlow and Cisco OpFlex provide southbound API capabilities.

Northbound APIs (SDN northbound APIs) are typically RESTful APIs that are used to communicate between the SDN controller and the services and applications running over the network. Such northbound APIs can be used for the orchestration and automation of the network components to align with the needs of different applications via SDN network programmability. In short, northbound APIs are basically the link between the applications and the SDN controller. In modern environments, applications can tell the network devices (physical or virtual) what type of resources they need and, in turn, the SDN solution can provide the necessary resources to the application.

Cisco has the concept of intent-based networking. On different occasions, you may see northbound APIs referred to as "intent-based APIs."

 ## More About Network Function Virtualization

Network virtualization is used for logical groupings of nodes on a network. The nodes are abstracted from their physical locations so that VMs and any other assets can be managed as if they are all on the same physical segment of the network. This is not a new technology. However, it is still one that is key in virtual environments where systems are created and moved despite their physical location.

Network Functions Virtualization (NFV) is a technology that addresses the virtualization of Layer 4 through Layer 7 services. These include load balancing and security capabilities such as firewall-related features. In short, with NFV, you convert certain types of network appliances into VMs. NFV was created to address the inefficiencies that were introduced by virtualization.

NFV allows you to create a virtual instance of a virtual node such as a firewall that can be deployed where it is needed, in a flexible way that's similar to how you do with a traditional VM.

Open Platform for Network Function Virtualization (OPNFV) is an open source solution for NFV services. It aims to be the base infrastructure layer for running virtual network functions. You can find detailed information about OPNFV at opnfv.org.

NFV nodes such as virtual routers and firewalls need an underlying infrastructure:

■ A hypervisor to separate the virtual routers, switches, and firewalls from the underlying physical hardware. The hypervisor is the underlying virtualization platform that allows the physical server (system) to operate multiple VMs (including traditional VMs and network-based VMs).

■ A virtual forwarder to connect individual instances.

■ A network controller to control all of the virtual forwarders in the physical network.

■ A VM manager to manage the different network-based VMs.

Figure 3-10 demonstrates the high-level components of the NFV architecture.

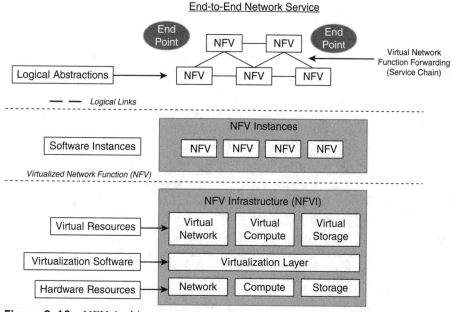

Figure 3-10 *NFV Architecture*

Several NFV infrastructure components have been created in open community efforts. On the other hand, traditionally, the actual integration has so far remained a "private" task. You've either had to do it yourself, outsource it, or buy a pre-integrated system from some vendor, keeping in mind that the systems integration undertaken is not a one-time task. OPNFV was created to change the NFV ongoing integration task from a private solution into an open community solution.

NFV MANO

NFV changes the way networks are managed. NFV management and network orchestration (MANO) is a framework and working group within the European Telecommunications Standards Institute (ETSI) Industry Specification Group for NFV (ETSI ISG NFV). NFV MANO is designed to provide flexible on-boarding of network components. NFV MANO is divided into the three functional components listed in Figure 3-11.

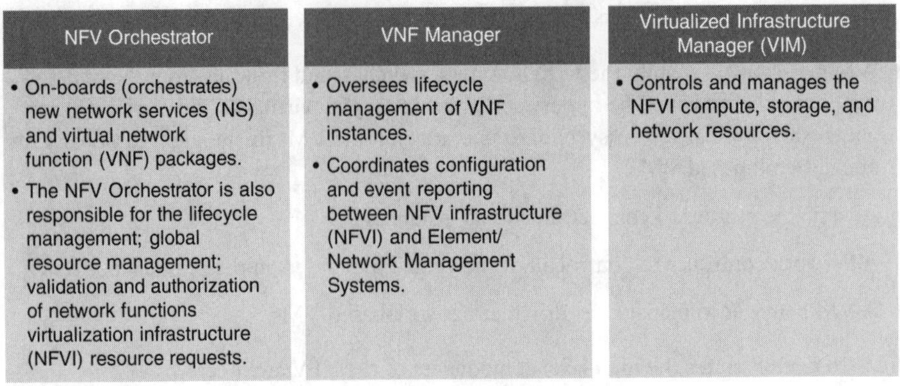

NFV Orchestrator	VNF Manager	Virtualized Infrastructure Manager (VIM)
• On-boards (orchestrates) new network services (NS) and virtual network function (VNF) packages. • The NFV Orchestrator is also responsible for the lifecycle management; global resource management; validation and authorization of network functions virtualization infrastructure (NFVI) resource requests.	• Oversees lifecycle management of VNF instances. • Coordinates configuration and event reporting between NFV infrastructure (NFVI) and Element/ Network Management Systems.	• Controls and manages the NFVI compute, storage, and network resources.

Figure 3-11 *NFV MANO Functional Components*

The NFV MANO architecture is integrated with open application program interfaces (APIs) in the existing systems. The MANO layer works with templates for standard VNFs. It allows implementers to pick and choose from existing NFV resources to deploy their platform or element.

Contiv

Contiv is an open source project that allows you to deploy micro-segmentation policy-based services in container environments. It offers a higher level of networking abstraction for microservices by providing a policy framework. Contiv has built-in service discovery and service routing functions to allow you to scale out services.

NOTE You can download Contiv and access its documentation at https://contiv.io.

With Contiv you can assign an IP address to each container. This feature eliminates the need for host-based port NAT. Contiv can operate in different network environments such as traditional Layer 2 and Layer 3 networks, as well as overlay networks.

Contiv can be deployed with all major container orchestration platforms (or schedulers) such as Kubernetes and Docker Swarm. For instance, Kubernetes can provide compute resources to containers and then Contiv provides networking capabilities.

NOTE Contiv supports Layer 2, Layer 3 (BGP), VXLAN for overlay networks, and Cisco ACI mode. It also provides built-in east-west service load balancing and traffic isolation.

The Netmaster and Netplugin (Contiv host agent) are the two major components in Contiv. Figure 3-12 illustrates how the Netmaster and the Netplugin interact with all the underlying components of the Contiv solution.

TIP The Contiv website includes several tutorials and step-by-step integration documentation at https://contiv.io/documents/tutorials/index.html.

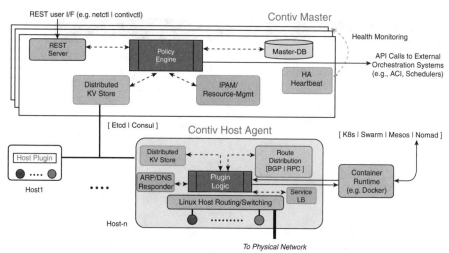

Figure 3-12 *Contiv Netmaster and Netplugin (Contiv Host Agent) Components*

Cisco Digital Network Architecture (DNA)

Cisco DNA is a solution created by Cisco that is often referred to as the "intent-based networking" solution. Cisco DNA provides automation and assurance services across campus networks, wide area networks (WANs), and branch networks. Cisco DNA is based on an open and extensible platform and provides the policy, automation, and analytics capabilities, as illustrated in Figure 3-13.

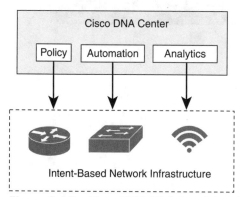

Figure 3-13 *Cisco DNA High-Level Architecture*

The heart of the Cisco DNA solution is Cisco DNA Center (DNAC). DNAC is a command-and-control element that provides centralized management via dashboards and APIs. Figure 3-14 shows one of the many dashboards of Cisco DNA Center (the Network Hierarchy dashboard).

Cisco DNA Center can be integrated with external network and security services such as the Cisco Identity Services Engine (ISE). Figure 3-15 shows how the Cisco ISE is configured as an authentication, authorization, and accounting (AAA) server in the Cisco DNA Center Network Settings screen.

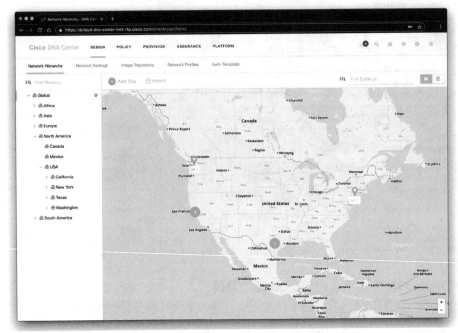

Figure 3-14 *Cisco DNA Center Network Hierarchy Dashboard*

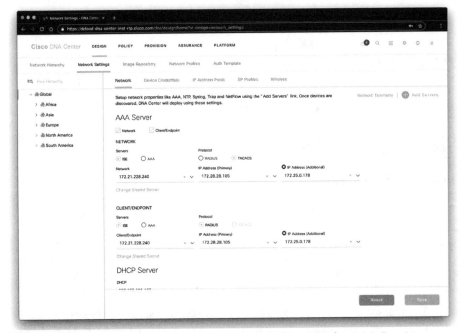

Figure 3-15 *Cisco DNA Center Integration with Cisco ISE for AAA Services*

Cisco DNA Policies

The following are the policies you can create in the Cisco DNA Center:

- Group-based access control policies

- IP-based access control policies

- Application access control policies

- Traffic copy policies

Figure 3-16 shows the Cisco DNA Center Policy Dashboard. There you can see the number of virtual networks, group-based access control policies, IP-based access control policies, traffic copy policies, scalable groups, and IP network groups that have been created. The Policy Dashboard will also show any policies that have failed to deploy.

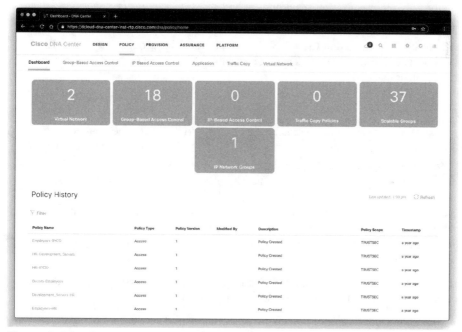

Figure 3-16 *Cisco DNA Center Policy Dashboard*

The Policy Dashboard window also provides a list of policies and the following information about each policy:

- **Policy Name:** The name of the policy.

- **Policy Type:** The type of policy.

- **Policy Version:** The version number is incremented by one version each time you change a policy.

- **Modified By:** The user who created or modified the policy.

- **Description:** The policy description.

- **Policy Scope:** The policy scope defines the users and device groups or applications that a policy affects.

- **Timestamp:** The date and time when a particular version of a policy was saved.

Cisco DNA Group-Based Access Control Policy

When you configure group-based access control policies, you need to integrate the Cisco ISE with Cisco DNA Center, as you learned previously in this chapter. In Cisco ISE, you configure the work process setting as "Single Matrix" so that there is only one policy matrix for all devices in the TrustSec network. You will learn more about Cisco TrustSec and Cisco ISE in Chapter 4, "Authentication, Authorization, Accounting (AAA) and Identity Management."

Depending on your organization's environment and access requirements, you can segregate your groups into different virtual networks to provide further segmentation.

After Cisco ISE is integrated in Cisco DNA Center, the scalable groups that exist in Cisco ISE are propagated to Cisco DNA Center. If a scalable group that you need does not exist, you can create it in Cisco ISE.

NOTE   You can access Cisco ISE through the Cisco DNA Center interface to create scalable groups. After you have added a scalable group in Cisco ISE, it is synchronized with the Cisco DNA Center database so that you can use it in an access control policy. You cannot edit or delete scalable groups from Cisco DNA Center; you need to perform these tasks from Cisco ISE.

Cisco DNA Center has the concept of access control contracts. A contract specifies a set of rules that allow or deny network traffic based on such traffic matching particular protocols or ports. Figure 3-17 shows a new contract being created in Cisco DNA Center to allow SSH access (TCP port 22).

To create a contract, navigate to **Policy > Group-Based Access Control > Access Contract** and click **Add Contract**. The dialog box shown in Figure 3-17 will be displayed.

Figure 3-18 shows an example of how to create a group-based access control policy.

In Figure 3-18, an access control policy named **omar_policy_1** is configured to **deny** traffic from all users and related devices in the group called **Guests** to any user or device in the **Finance** group.

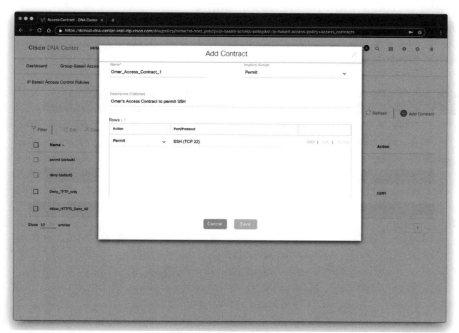

Figure 3-17 *Adding a Cisco DNA Center Contract*

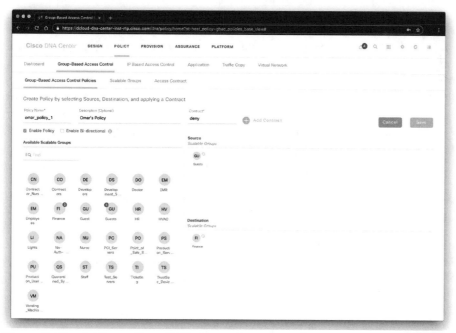

Figure 3-18 *Adding a Cisco DNA Center Group-Based Access Control Policy*

Cisco DNA IP-Based Access Control Policy

You can also create IP-based access control policies in Cisco DNA Center. To create IP-based access control policies, navigate to **Policy > IP Based Access Control > IP Based Access Control Policies,** as shown in Figure 3-19.

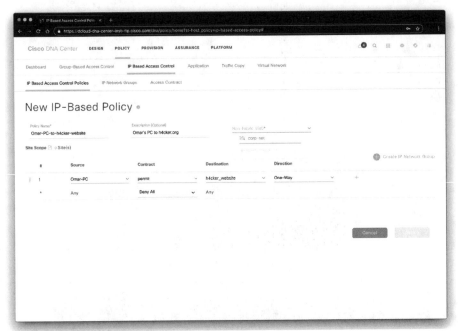

Figure 3-19 *Adding a Cisco DNA Center IP-Based Access Control Policy*

In the example shown in Figure 3-19, a policy is configured to permit Omar's PC to communicate with h4cker.org.

> **NOTE**   An IP network group named h4cker_website is already configured. To configure IP network groups, navigate to **Policy > IP Based Access Control > IP Network Groups.** These IP network groups can also be automatically populated from Cisco ISE.

You can also associate these policies to specific wireless SSIDs. The **corp-net** SSID is associated to the policy entry in Figure 3-19.

Cisco DNA Application Policies

Application policies can be configured in Cisco DNA Center to provide Quality of Service (QoS) capabilities. The following are the Application Policy components you can configure in Cisco DNA Center:

■ Applications

■ Application sets

- Application policies

- Queuing profiles

Applications in Cisco DNA Center are the software programs or network signaling protocols that are being used in your network.

> **NOTE** Cisco DNA Center supports all of the applications in the Cisco Next Generation Network-Based Application Recognition (NBAR2) library.

Applications can be grouped into logical groups called *application sets*. These application sets can be assigned a business relevance within a policy.

You can also map applications to industry standard-based traffic classes, as defined in RFC 4594.

Cisco DNA Traffic Copy Policy

You can also use an Encapsulated Remote Switched Port Analyzer (ERSPAN) configuration in Cisco DNA Center so that the IP traffic flow between two entities is copied to a given destination for monitoring or troubleshooting. In order for you to configure ERSPAN using Cisco DNA Center, you need to create a traffic copy policy that defines the source and destination of the traffic flow you want to copy. To configure a traffic copy policy, navigate to **Policy > Traffic Copy > Traffic Copy Policies**, as shown in Figure 3-20.

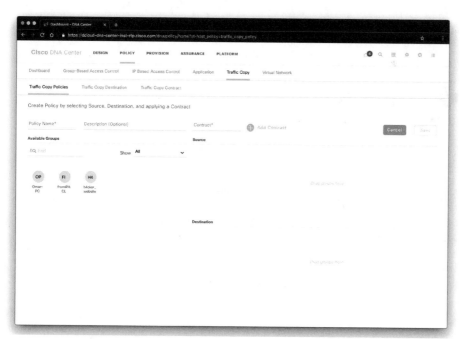

Figure 3-20 *Adding a Traffic Copy Policy*

You can also define a traffic copy contract that specifies the device and interface where the copy of the traffic is sent.

Cisco DNA Center Assurance Solution

The Cisco DNA Center Assurance solution allows you to get contextual visibility into network functions with historical, real-time, and predictive insights across users, devices, applications, and the network. The goal is to provide automation capabilities to reduce the time spent on network troubleshooting.

Figure 3-21 shows the Cisco DNA Center Assurance Overall Health dashboard.

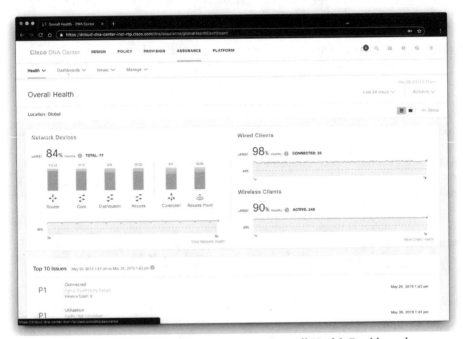

Figure 3-21 *The Cisco DNA Center Assurance Overall Health Dashboard*

The Cisco DNA Center Assurance solution allows you to investigate different networkwide (global) issues, as shown in Figure 3-22.

The Cisco DNA Center Assurance solution also allows you to configure sensors to test the health of wireless networks. A wireless network includes access point (AP) radios, WLAN configurations, and wireless network services. Sensors can be dedicated or on-demand sensors. A dedicated sensor is when an AP is converted into a sensor, and it stays in sensor mode (is not used by wireless clients) unless it is manually converted back into AP mode. An on-demand sensor is when an AP is temporarily converted into a sensor to run tests. After the tests are complete, the sensor goes back to AP mode. Figure 3-23 shows the Wireless Sensor dashboard in Cisco DNA Center.

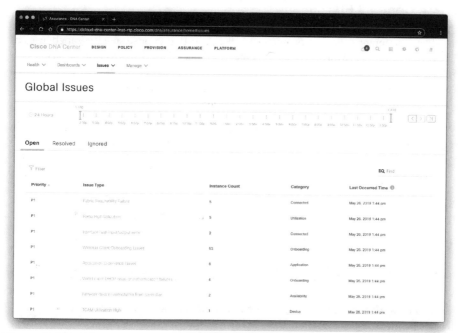

Figure 3-22 *The Cisco DNA Center Assurance Global Issues Dashboard*

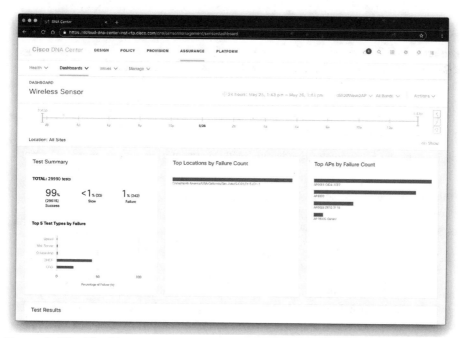

Figure 3-23 *The Cisco DNA Center Assurance Wireless Sensor Dashboard*

Cisco DNA Center APIs

One of the key benefits of the Cisco DNA Center is the comprehensive available APIs (aka Intent APIs). The Intent APIs are northbound REST APIs that expose specific capabilities of the Cisco DNA Center platform. These APIs provide policy-based abstraction of business intent, allowing you to focus on an outcome to achieve instead of struggling with the mechanisms that implement that outcome. The APIs conform to the REST API architectural style and are simple, extensible, and secure to use.

Cisco DNA Center also has several integration APIs. These integration capabilities are part of westbound interfaces. Cisco DNA Center also allows administrators to manage their non-Cisco devices. Multivendor support comes to Cisco DNA Center through the use of an SDK that can be used to create device packages for third-party devices. A device package enables Cisco DNA Center to communicate with third-party devices by mapping Cisco DNA Center features to their southbound protocols.

TIP   Cisco has very comprehensive documentation and tutorials about the Cisco DNA Center APIs at DevNet (https://developer.cisco.com/dnacenter).

Cisco DNA Center also has several events and notifications services that allow you to capture and forward Cisco DNA Assurance and Automation (SWIM) events to third-party applications via a webhook URL.

All Cisco DNA Center APIs conform to the REST API architectural styles.

NOTE   A REST endpoint accepts and returns HTTPS messages that contain JavaScript Object Notation (JSON) documents. You can use any programming language to generate the messages and the JSON documents that contain the API methods. These APIs are governed by the Cisco DNA Center Role-Based Access Control (RBAC) rules and as a security measure require the user to authenticate successfully prior to using the API.

You can view information about all the Cisco DNA Center APIs by clicking the **Platform** tab and navigating to **Developer Toolkit > APIs**, as shown in Figure 3-24.

Figure 3-25 shows an example of the detailed API documentation within Cisco DNA Center.

TIP   All REST requests in Cisco DNA Center require authentication. The Authentication API generates a security token that encapsulates the privileges of an authenticated REST caller. All requested operations are authorized by Cisco DNA Center according to the access privileges associated with the security token that is sent in the request.

Cisco is always expanding the capabilities of the Cisco DNA Center APIs. Please study and refer to the following API documentation and tutorials for the most up-to-date capabilities: https://developer.cisco.com/docs/dna-center and https://developer.cisco.com/site/dna-center-rest-api.

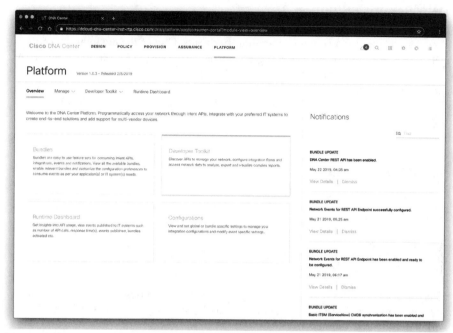

Figure 3-24 *The Cisco DNA Center APIs and Developer Toolkit*

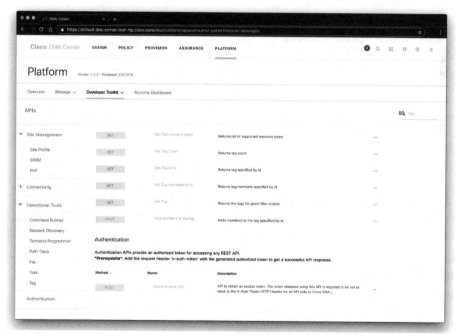

Figure 3-25 *API Developer Toolkit Documentation*

Cisco DNA Security Solution

The Cisco DNA Security solution supports several other security products and operations that allow you to detect and contain cybersecurity threats. One of the components of the Cisco DNA Security solution is the Encrypted Traffic Analytics (ETA) solution. Cisco ETA allows you to detect security threats in encrypted traffic without decrypting the packets. It is able to do this by using machine learning and other capabilities. To use Encrypted Traffic Analytics, you need one of the following network devices along with Cisco Stealthwatch Enterprise:

- Catalyst 9000 switches

- ASR 1000 Series routers

- ISR 4000 Series routers

- CSR 1000V Series virtual routers

- ISR 1000 Series routers

- Catalyst 9800 Series wireless controllers

Cisco Stealthwatch provides network visibility and security analytics to rapidly detect and contain threats. You will learn more about the Cisco Stealthwatch solution in Chapter 5, "Network Visibility and Segmentation."

As you learned in previous sections of this chapter, the Cisco TrustSec solution and Cisco ISE enable you to control networkwide access, enforce security policies, and help meet compliance requirements.

Cisco DNA Multivendor Support

Cisco DNA Center now allows customers to manage their non-Cisco devices. Multivendor support comes to Cisco DNA Center through the use of an SDK that can be used to create device packages for third-party devices. A device package enables Cisco DNA Center to communicate with third-party devices by mapping Cisco DNA Center features to their southbound protocols. Multivendor support capabilities are based on southbound interfaces. These interfaces interact directly with network devices by means of CLI, SNMP, or NETCONF.

NOTE Southbound interfaces are not exposed to the consumer. Instead, the consumer uses Intent APIs, which abstract the underlying complexity of the traditional network. The user of Intent APIs need not be concerned with the particular protocols that the southbound interfaces use to implement network intent on devices that Cisco DNA Center supports.

Introduction to Network Programmability

As you were able to see in previous sections of this chapter, learning to code and work with programmable infrastructures is very important in today's environment. You saw the value of using APIs. Whether you have configured large networks in the past or are just getting started, you know that this probably involved a lot of clicking, typing, copying-and-pasting, and many repetitive tasks. Nowadays, modern APIs enable you to complete powerful tasks, reduce all the repetitive work, and save time.

Using APIs, you can make requests like the ones shown in Figure 3-26 in a very simple way.

Get the status for interface X
Get the last-change time for interface X
Shutdown interface X

REST API

Administrator   Router 88

Figure 3-26 *Using Network Infrastructure Device APIs*

Modern Programming Languages and Tools

Modern programming languages like JavaScript, Python, Go, Swift, and others are more flexible and easier to learn than their predecessors. You might wonder what programming language you should learn first. Python is one of the programming languages recommended to learn first—not only for network programmability, but for many other scenarios.

> **TIP** Many different sites allow you to get started with Python. The following are several great resources to learn Python:
>
> - Learn Python dot org: https://www.learnpython.org
> - W3 Schools Python tutorials: https://www.w3schools.com/python/
> - The Python Tutorial: https://docs.python.org/3/tutorial/

Combining programming capabilities with developer tools like Git (GitHub or GitLab repositories), package management systems, virtual environments, and integrated development environments (IDEs) allows you to create your own set of powerful tools and workflows.

Another amazing thing is the power of code reuse and online communities. In the past, when you wanted to create some program, you often had to start "from scratch." For example, if you wanted to just make an HTTPS web request, you had to create code to open a TCP connection over port 443, perform the TLS negotiation, exchange and validate certificates, and format and interpret HTTP requests and responses.

Nowadays, you can just use open source software in GitHub or simply use packages such as the Python requests package, as shown in Figure 3-27.

In Figure 3-27, the Python package called *requests* is installed using the package manager for Python called *pip* (https://pypi.org/project/pip). The requests library allows you to make HTTP/HTTPS requests in Python very easily.

Now that you have the requests package installed, you can start making HTTP requests, as shown in Figure 3-28.

Figure 3-27 *Installing the Python Requests Package Using pip*

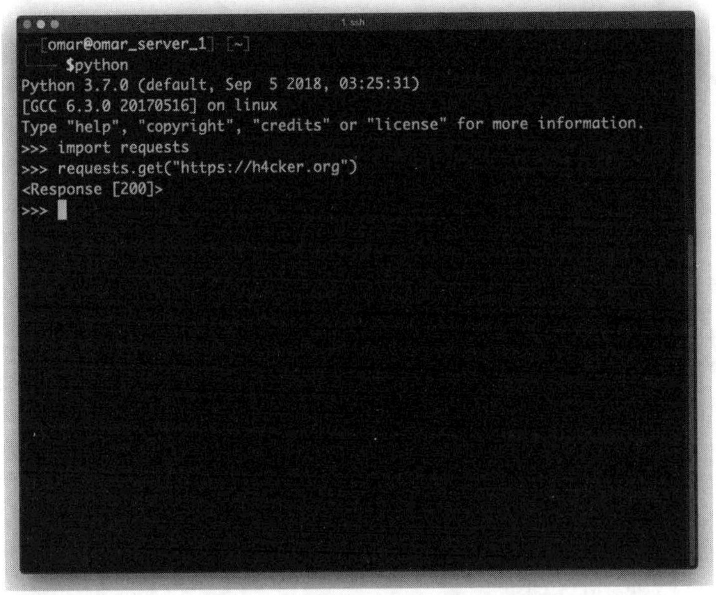

Figure 3-28 *Using the Python Requests Package*

In Figure 3-28, the interactive Python shell (interpreter) is used to use (import) the requests package and send an HTTP GET request to the website at https://h4cker.org. The HTTP GET request is successful and the 200 message/response is shown.

Additional information about the Python interpreter can be found at https://docs.python.org/3/tutorial/interpreter.html and https://www.python-course.eu/python3_interactive.php.

TIP The W3 schools website has a very good explanation of the HTTP status code messages at https://www.w3schools.com/tags/ref_httpmessages.asp.

The HTTP status code messages can be in the following ranges:

- Messages in the 100 range are informational.
- Messages in the 200 range are related to successful transactions.
- Messages in the 300 range are related to HTTP redirections.
- Messages in the 400 range are related to client errors.
- Messages in the 500 range are related to server errors.

When HTTP servers and browsers communicate with each other, they perform interactions based on headers as well as body content. The HTTP Request has the following structure:

1. The METHOD, which in this example is an HTTP GET. However, the HTTP methods can be the following:

   - **GET:** Retrieves information from the server.

   - **HEAD:** Basically, this is the same as a GET, but it returns only HTTP headers and no document body.

   - **POST:** Sends data to the server (typically using HTML forms, API requests, and the like).

   - **TRACE:** Does a message loopback test along the path to the target resource.

   - **PUT:** Uploads a representation of the specified URI.

   - **DELETE:** Deletes the specified resource.

   - **OPTIONS:** Returns the HTTP methods that the server supports.

   - **CONNECT:** Converts the request connection to a transparent TCP/IP tunnel.

2. The URI and the path-to-resource field represent the path portion of the requested URL.

3. The request version-number field specifies the version of HTTP used by the client.

4. The user agent is Chrome in this example, and it was used to access the website. In the packet capture, you see the following:

   ```
   User-Agent: Mozilla/5.0 (Macintosh; Intel Mac OS X 10_13_4)
   AppleWebKit/537.36 (KHTML, like Gecko) Chrome/66.0.3359.181
   Safari/537.36\r\n.
   ```

5. Next, you see several other fields like accept, accept-language, accept encoding, and others.

6. The server, after receiving this request, generates a response.

7. The server response has a three-digit status code and a brief human-readable explanation of the status code. Then below you see the text data (which is the HTML code coming back from the server and displaying the website contents).

TIP   The requests Python package is used often to interact with APIs. You can obtain more information about the requests Python package at https://realpython.com/python-requests and https://developer.cisco.com/learning/lab/intro-python-part1/step/1.

DevNet

DevNet is a platform created by Cisco that has numerous resources for network and application developers. DevNet is an amazing resource that includes many tutorials, free video courses, sandboxes, learning paths, and sample code to interact with many APIs. You can access DevNet at developer.cisco.com.

If you are new to programming and network programmability, you can take advantage of the following DevNet tutorials and learning paths:

- Introduction to Coding and APIs: https://developer.cisco.com/startnow

- Network Programmability Basics Video Course: https://developer.cisco.com/video/net-prog-basics/

- Parsing JSON using Python: https://developer.cisco.com/learning/lab/coding-202-parsing-json/step/1

- DevNet GitHub Repositories: https://github.com/CiscoDevNet

- DevNet Developer Videos: https://developer.cisco.com/video

- DevNet Git Tutorials: https://developer.cisco.com/learning/lab/git-intro/step/1

- DevNet ACI Programmability: https://developer.cisco.com/learning/tracks/aci-programmability

- Build Applications with Cisco: https://developer.cisco.com/learning/tracks/app-dev

- IOS-XE Programmability: https://developer.cisco.com/learning/tracks/iosxe-programmability

- Network Programmability for Network Engineers: https://developer.cisco.com/learning/tracks/netprog-eng

Getting Started with APIs

APIs are used everywhere these days. A large number of modern applications use some type of APIs because they make access available to other systems to interact with the application. There are few methods or technologies behind modern APIs:

- **Simple Object Access Protocol (SOAP):** SOAP is a standards-based web services access protocol that was originally developed by Microsoft and has been used by numerous legacy applications for many years. SOAP exclusively uses XML to provide API services. XML-based specifications are governed by XML Schema Definition (XSD) documents. SOAP was originally created to replace older solutions such as the Distributed Component Object Model (DCOM) and Common Object Request Broker Architecture (CORBA). You can find the latest SOAP specifications at https://www.w3.org/TR/soap.

- **Representational State Transfer (REST):** REST is an API standard that is easier to use than SOAP. It uses JSON instead of XML, and it uses standards like Swagger and the OpenAPI Specification (https://www.openapis.org) for ease of documentation and to help with adoption.

- **GraphQL and queryable APIs:** This is another query language for APIs that provides many developer tools. GraphQL is now used for many mobile applications and online dashboards. Many languages support GraphQL. You can learn more about GraphQL at https://graphql.org/code.

NOTE   SOAP and REST share similarities over the HTTP protocol. SOAP limits itself to a stricter set of API messaging patterns than REST.

APIs often provide a roadmap describing the underlying implementation of an application. API documentation can provide a great level of detail that can be very valuable to security professional. These types of documentation include the following:

- **Swagger (OpenAPI):** Swagger is a modern framework of API documentation and is now the basis of the OpenAPI Specification (OAS). Additional information about Swagger can be obtained at https://swagger.io. The OAS specification is available at https://github.com/OAI/OpenAPI-Specification.

- **Web Services Description Language (WSDL) documents:** WSDL is an XML-based language that is used to document the functionality of a web service. The WSDL specification can be accessed at https://www.w3.org/TR/wsdl20-primer.

- **Web Application Description Language (WADL) documents:** WADL is also an XML-based language for describing web applications. The WADL specification can be obtained from https://www.w3.org/Submission/wadl.

NOTE   Most Cisco products and services use RESTful (REST) APIs.

REST APIs

Let's take a look at a quick example of a REST API. There is a sample API you can use to perform several tests at https://deckofcardsapi.com. In Figure 3-29, the Linux **curl** utility is used to retrieve a "new deck of cards" from the Deck of Cards API. The API "shuffles" a deck of cards for you. The deck ID (**deck_id**) is **wkc12q20frlh** in this example.

NOTE   The **python -m json.tool** command is used to invoke the json.tool Python module to "pretty print" the JSON output. You can obtain more information about the json.tool Python module at https://docs.python.org/3/library/json.html#module-json.tool.

Suppose that you want to draw a random card from the deck. Since you have the deck ID, you can easily use the command shown in Figure 3-30 to draw a random card.

Figure 3-29 *Using* *curl* *to Obtain Information from an API*

Figure 3-30 *Using* *curl* *to Obtain Additional Information from the Deck of Cards API*

You can see the response (in JSON), including the remaining number of cards and the card that was retrieved (the 9 of spades). Other information, such as the code, suit, value, and images of the card, is also included in the JSON output.

> **NOTE**   The DevNet tutorial at the following link shows how to interact with this sample API using Postman: https://developer.cisco.com/learning/lab/hands-on-postman/step/1.

Using Network Device APIs

Earlier in this chapter you learned that there are several API resources available in many Cisco solutions such as the Cisco DNA Center. The following are a few basic available API resources on the Cisco DNA Center Platform (10.1.1.1 is the IP address of the Cisco DNA Center):

- **https://10.1.1.1/api/system/v1/auth/token**: Used to get and encapsulate user identity and role information as a single value.

- **https://10.1.1.1/api/v1/network-device**: Used to get the list of first 500 network devices sorted lexicographically based on host name.

- **https://10.1.1.1/api/v1/interface**: Used to get information about every interface on every network device.

- **https://10.1.1.1/api/v1/host**: Used to get the name of a host, the ID of the VLAN that the host uses, the IP address of the host, the MAC address of the host, the IP address of the network device to which the host is connected, and more.

- **https://10.1.1.1/api/v1/flow-analysis**: Used to trace a path between two IP addresses. The function will wait for analysis to complete, and return the results.

There are a dozen (or dozens?) more APIs that you can use and interact with Cisco DNA Center at https://developer.cisco.com/dnacenter. Many other Cisco products include APIs that can be used for integrating third-party applications, obtain information similar to the preceding examples, as well as change the configuration of the device, apply policies, and more. Many of those APIs are also documented in DevNet (developer.cisco.com).

Modern networking devices support programmable capabilities such as NETCONF, RESTCONF, and YANG models. The following sections provide details about these technologies.

YANG Models

YANG is an API contract language used in many networking devices. In other words, you can use YANG to write a specification for what the interface between a client and networking device (server) should be on a particular topic. YANG was originally defined in RFC 6020 (https://tools.ietf.org/html/rfc6020).

> **TIP**   A specification written in YANG is referred to as a "YANG module." A collection (or set) of YANG modules are often called a "YANG model."

A YANG model typically concentrates on the data that a client processes using standardized operations.

NOTE Keep in mind that in NETCONF and RESTCONF implementations, the YANG controller is the client and the network elements are the server. You will learn more about NETCONF and RESTCONF later in this chapter.

Figure 3-31 shows an example of a network management application (client) interacting with a router (server) using YANG as the API contract.

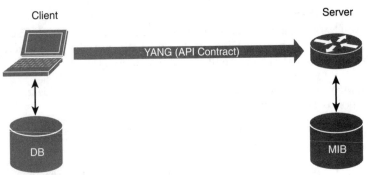

Figure 3-31 *A Basic YANG Example*

A YANG-based server (as shown in Figure 3-31) publishes a set of YANG modules, which taken together form the system's YANG model. The YANG modules specify what a client can do. The following are a few examples of what a client can do using different YANG models:

■ **Configure:** For example, enabling a routing protocol or a particular interface.

■ **Receive notifications:** An example of notifications can be repeated login failures, interface failures, and so on.

■ **Monitor status:** For example, retrieving information about CPU and memory utilization, packet counters, and so on.

■ **Invoke actions:** For instance, resetting packet counters, rebooting the system, and so on.

NOTE The YANG model of a device is often called a "schema" defining the structure and content of messages exchanged between the application and the device.

The YANG language provides flexibility and extensibility capabilities that are not present in other model languages. When you create new YANG modules, you can leverage the data hierarchies defined in other modules. YANG also permits new statements to be defined, allowing the language itself to be expanded in a consistent way.

TIP DevNet has a series of videos that demonstrate how YANG works at https://developer.cisco.com/video/net-prog-basics/02-network_device_apis/yang.

NETCONF

NETCONF is defined in RFCs 6241 and 6242. NETCONF was created to overcome the challenges in legacy Simple Network Management Protocol (SNMP) implementations.

A NETCONF client typically has the role of a network management application. The NETCONF server is a managed network device (router, switch, and so on). You can also have intermediate systems (often called "controllers") that control a particular aspect or domain. Controllers can act as a server to its managers and as a client to its networking devices, as shown in Figure 3-32.

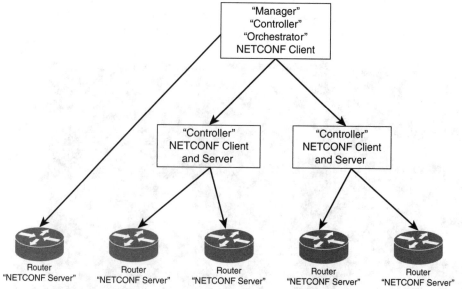

Figure 3-32 *NETCONF Clients, Servers, and Controllers*

In Figure 3-32, a node called a "Manager" manages a NETCONF server (router) and two "Controllers," which are both a server for the Manager and a client for the other network devices (routers).

> **NOTE** NETCONF was created before YANG. Other languages were used for NETCONF operations. On the other hand, YANG is the only language widely used for NETCONF nowadays.

NETCONF sessions established from a NETCONF client to a NETCONF server consist of a sequence of messages. Both parties send a "hello" message when they initially connect. All message exchanges are initiated by the NETCONF client. The hello message includes which NETCONF protocol version(s) the devices support. The server states which optional capabilities it supports.

NETCONF messages are either a remote procedure call (RPC) or an "rpc-reply." Each RPC is a request from the client to the server to execute a given operation. The NETCONF rpc-reply is sent by the server when it has completed or failed to complete the request. Some NETCONF rpc-replies are short answers to a simple query, or just an OK that the order was executed. Some are long and may contain the entire device configuration or status. NETCONF rpc-replies to subscriptions consist of a message that technically never ends. Other information of the rpc-reply is generated by the server. A NETCONF rpc-reply may also be a NETCONF rpc-error, indicating that the requested operation failed.

NETCONF messages are encoded in an XML-based structure defined by the NETCONF standard. The NETCONF communication is done over Secure Shell (SSH), but using a default TCP port 830. This can be configured to a different port.

SSH supports a subsystem concept. NETCONF has its own subsystem: netconf. Figure 3-33 shows how you can connect to a networking device (in this case, a CSR-1000v router configured with the hostname **ios-xe-mgmt.cisco.com**). The username of the router is **root**. You are also asked to provide a password. The router is configured for NETCONF over TCP port 10000.

```
[omar@omar_server_1] [~]
    $ssh root@ios-xe-mgmt.cisco.com -p 10000 -s netconf
root@ios-xe-mgmt.cisco.com's password:
<?xml version="1.0" encoding="UTF-8"?>
<hello xmlns="urn:ietf:params:xml:ns:netconf:base:1.0">
<capabilities>
<capability>urn:ietf:params:netconf:base:1.0</capability>
<capability>urn:ietf:params:netconf:base:1.1</capability>
<capability>urn:ietf:params:netconf:capability:writable-running:1.0</capability>
<capability>urn:ietf:params:netconf:capability:xpath:1.0</capability>
<capability>urn:ietf:params:netconf:capability:validate:1.0</capability>
<capability>urn:ietf:params:netconf:capability:validate:1.1</capability>
<capability>urn:ietf:params:netconf:capability:rollback-on-error:1.0</capability>
<capability>urn:ietf:params:netconf:capability:notification:1.0</capability>
<capability>urn:ietf:params:netconf:capability:interleave:1.0</capability>
<capability>urn:ietf:params:netconf:capability:with-defaults:1.0?basic-mode=explicit&amp;also-supported=report-all-tagge
d</capability>
<capability>urn:ietf:params:netconf:capability:yang-library:1.0?revision=2016-06-21&amp;module-set-id=88c694c75e847aba17
e8ab19254ad090</capability>
<capability>http://tail-f.com/ns/netconf/actions/1.0</capability>
<capability>http://tail-f.com/ns/netconf/extensions</capability>
<capability>http://cisco.com/ns/cisco-xe-ietf-ip-deviation?module=cisco-xe-ietf-ip-deviation&amp;revision=2016-08-10</ca
pability>
<capability>http://cisco.com/ns/cisco-xe-ietf-ipv4-unicast-routing-deviation?module=cisco-xe-ietf-ipv4-unicast-routing-d
eviation&amp;revision=2015-09-11</capability>
<capability>http://cisco.com/ns/cisco-xe-ietf-ipv6-unicast-routing-deviation?module=cisco-xe-ietf-ipv6-unicast-routing-d
eviation&amp;revision=2015-09-11</capability>
<capability>http://cisco.com/ns/cisco-xe-ietf-ospf-deviation?module=cisco-xe-ietf-ospf-deviation&amp;revision=2018-02-09
</capability>
<capability>http://cisco.com/ns/cisco-xe-ietf-routing-deviation?module=cisco-xe-ietf-routing-deviation&amp;revision=2016
-07-09</capability>
<capability>http://cisco.com/ns/cisco-xe-openconfig-acl-deviation?module=cisco-xe-openconfig-acl-deviation&amp;revision=
2017-08-25</capability>
<capability>http://cisco.com/ns/mpls-static/devs?module=common-mpls-static-devs&amp;revision=2015-09-11</capability>
<capability>http://cisco.com/ns/nvo/devs?module=nvo-devs&amp;revision=2015-09-11</capability>
```

Figure 3-33 *Using the NETCONF SSH Subsystem*

TIP   DevNet has several sandboxes where you can practice these concepts and more at https://devnetsandbox.cisco.com.

An open source Python library for NETCONF clients called ncclient is available on GitHub at https://github.com/ncclient/ncclient. You can install it using Python pip, as shown here:

```
pip install ncclient
```

There are several sample scripts at the DevNet GitHub repositories that can help you get started at https://github.com/CiscoDevNet/python_code_samples_network.

Figure 3-34 shows how to use a Python script that leverages ncclient to interact with the router (**ios-xe-mgmt.cisco.com**).

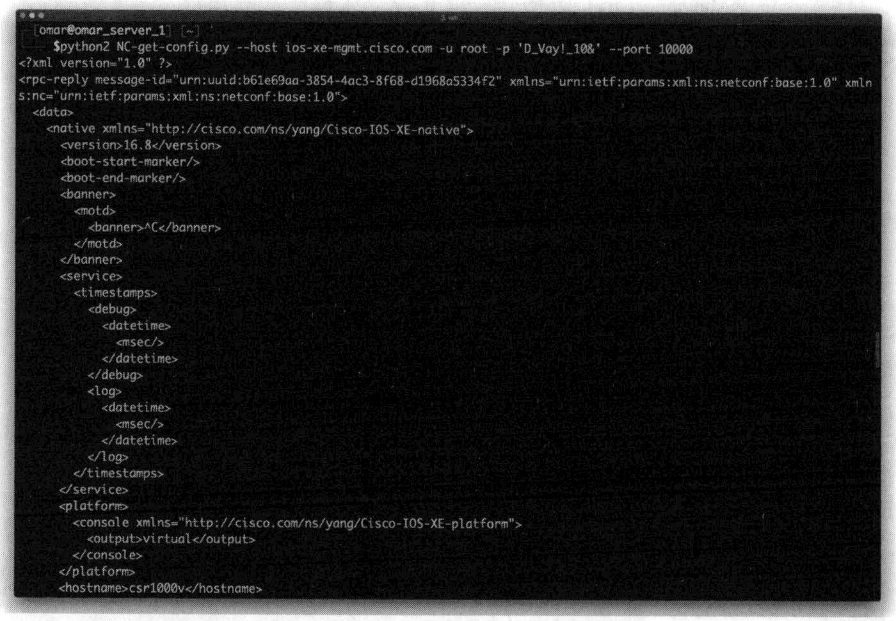

```
[omar@omar_server_1] [~]
     $python2 NC-get-config.py --host ios-xe-mgmt.cisco.com -u root -p 'D_Vay!_10&' --port 10000
<?xml version="1.0" ?>
<rpc-reply message-id="urn:uuid:b61e69aa-3854-4ac3-8f68-d1968a5334f2" xmlns="urn:ietf:params:xml:ns:netconf:base:1.0" xmln
s:nc="urn:ietf:params:xml:ns:netconf:base:1.0">
  <data>
    <native xmlns="http://cisco.com/ns/yang/Cisco-IOS-XE-native">
      <version>16.8</version>
      <boot-start-marker/>
      <boot-end-marker/>
      <banner>
        <motd>
          <banner>^C</banner>
        </motd>
      </banner>
      <service>
        <timestamps>
          <debug>
            <datetime>
              <msec/>
            </datetime>
          </debug>
          <log>
            <datetime>
              <msec/>
            </datetime>
          </log>
        </timestamps>
      </service>
      <platform>
        <console xmlns="http://cisco.com/ns/yang/Cisco-IOS-XE-platform">
          <output>virtual</output>
        </console>
      </platform>
      <hostname>csr1000v</hostname>
```

Figure 3-34   *Using Python to Obtain the Entire Configuration of a Network Device*

TIP   You can obtain NC-get-config.py from https://github.com/CiscoDevNet/
python_code_samples_network/tree/master/NC-get-config.

RESTCONF

You already learned that REST is a type of modern API. Many network administrators
wanted to have the capabilities of NETCONF over "REST." This is why a REST-based variant
of NETCONF was created. RESTCONF is now supported in many networking devices in the
industry.

RESTCONF is defined in RFC 8040 and it follows the REST principles. However, not all
REST-based APIs are compatible or even comparable to RESTCONF.

The RESTCONF interface is built around a small number of standardized requests (GET,
PUT, POST, PATCH, and DELETE). Several of the REST principles are similar to NETCONF:

- The client-server model

- The layered system principle

- The first two uniform interface principles

One of the differences between RESTCONF and NETCONF is the stateless server principle.
NETCONF is based on clients establishing a session to the server (which is not stateless).
NETCONF clients frequently connect and then manipulate the candidate datastore with a
number of *edit-config* operations. The NETCONF clients may also send a *validation* call to
NETCONF servers. This is different in RESTCONF.

RESTCONF requires the server to keep some client state. Any request the RESTCONF client sends is acted upon by the server immediately. You cannot send any transactions that span multiple RESTCONF messages. Subsequently, some of the key features of NETCONF (including networkwide transactions) are not possible in RESTCONF.

Let's take a look at a quick example of using RESTCONF. Example 3-1 shows a Python script that is used to obtain the details of all interfaces in a networking device using RESTCONF.

Example 3-1 *Python Script to Retrieve Interface Details from a Networking Device Using RESTCONF*

```
#!/usr/bin/python
import requests
import sys

# disable warnings from SSL/TLS certificates
requests.packages.urllib3.disable_warnings()

# the IP address or hostname of the networking device
HOST = 'ios-xe-mgmt.cisco.com'

# use your user credentials to access the networking device
USER = 'root'
PASS = 'supersecretpassword'

# create a main() method
def main():
    """Main method that retrieves the interface details from a
    networking device via RESTCONF."""

    # RESTCONF url of the networking device
    url="https://{h}:9443/restconf/data/ietf-
    interfaces:interfaces".format(h=HOST)

    # RESTCONF media types for REST API headers
    headers = {'Content-Type': 'application/yang-data+json',
               'Accept': 'application/yang-data+json'}

    # this statement performs a GET on the specified url
    response = requests.get(url, auth=(USER, PASS),
                            headers=headers, verify=False)

    # print the json that is returned
    print(response.text)

if __name__ == '__main__':
    sys.exit(main())
```

Figure 3-35 shows the output of the Python script, including the information of all the interfaces in that networking device (**ios-xe-mgmt.cisco.com**).

```
[ omar@omar_server_1 ] [ ~ ]
    $python3 get-interface-config.py
{
"ietf-interfaces:interfaces": {
  "interface": [
    {
      "name": "GigabitEthernet1",
      "description": "NO TOUCHY!",
      "type": "iana-if-type:ethernetCsmacd",
      "enabled": true,
      "ietf-ip:ipv4": {
        "address": [
          {
            "ip": "10.10.20.48",
            "netmask": "255.255.255.0"
          },
          {
            "ip": "10.10.20.49",
            "netmask": "255.255.255.0"
          }
        ]
      },
      "ietf-ip:ipv6": {
      }
    },
    {
      "name": "GigabitEthernet2",
      "type": "iana-if-type:ethernetCsmacd",
      "enabled": true,
      "ietf-ip:ipv4": {
      },
      "ietf-ip:ipv6": {
      }
    },
    {
```

Figure 3-35 *Using Python to Obtain Information from a Network Device Using RESTCONF*

TIP Watch the DevNet "Getting Started with Network Device APIs" video for additional step-by-step information about Network APIs, NETCONF, RESTCONF, and YANG at https://developer.cisco.com/video/net-prog-basics/02-network_device_apis.

OpenConfig and gNMI

The OpenConfig consortium (https://github.com/openconfig) is a collaborative effort to provide vendor-neutral data models (in YANG) for network devices. OpenConfig uses the gRPC Network Management Interface (gNMI). The following GitHub repository includes detailed information about gNMI, as well as sample code (https://github.com/openconfig/gnmi).

NOTE The gRPC specification (https://grpc.io) is a modern Remote Procedure Call (RPC) framework. RPC allows a client to invoke operations (also called "procedures") on a server. RPC includes an interface description language (IDL) used to state what procedures the server supports (including the input and output data from them). RPC also uses client libraries to call upon those procedures (supported in different programming languages). RPC uses a serialization, marshalling, and transport mechanism for the messages (generally called an RPC protocol).

The gNMI protocol is similar to NETCONF and RESTCONF. gNMI uses YANG models, but it can be used with other interface description languages (IDLs). The OpenConfig consortium defined several standard YANG models to go with the protocols. These YANG models describe many essential networking features such as interface configuration, routing protocols, QoS, Wi-Fi configurations, and more.

Exam Preparation Tasks

As mentioned in the section "How to Use This Book" in the Introduction, you have a couple of choices for exam preparation: the exercises here, Chapter 12, "Final Preparation," and the exam simulation questions in the Pearson Test Prep Software Online.

Review All Key Topics

Review the most important topics in this chapter, noted with the Key Topic icon in the outer margin of the page. Table 3-2 lists these key topics and the page numbers on which each is found.

Table 3-2 Key Topics for Chapter 3

Key Topic Element	Description	Page Number
Section	Traditional Networking Planes	109
Section	So What's Different with SDN?	110
Section	Introduction to the Cisco ACI Solution	110
List	Understand the functions of the APIC	112
Section	VXLAN and Network Overlays	112
Paragraph	Understand what is micro-segmentation	115
Paragraph	Understand "east-west" traffic and "north-south" traffic	115
Section	Open Source Initiatives	117
Paragraph	Understand northbound and southbound APIs	118
Section	More About Network Function Virtualization	118
Section	Cisco DNA Center APIs	130
Tip	Cisco DNA Center APIs in DevNet	130
Section	Cisco DNA Security Solution	132
Section	Modern Programming Languages and Tools	133
Section	DevNet	136
Section	Getting Started with APIs	136
Section	REST APIs	137
Section	YANG Models	139
Section	NETCONF	141
Section	RESTCONF	143

Define Key Terms

Define the following key terms from this chapter and check your answers in the glossary:

Representational State Transfer (REST), Simple Object Access Protocol (SOAP), Contiv, Network Functions Virtualization (NFV), Neutron, Open vSwitch, OpenDaylight (ODL), YANG, NETCONF, RESTCONF

Review Questions

1. The RESTCONF interface is built around a small number of standardized requests. Which of the following are requests supported by RESTCONF?

   a. GET

   b. PUT

   c. PATCH

   d. All of these answers are correct.

2. NETCONF messages are encoded in a(n) _____ structure defined by the NETCONF standard.

   a. JSON

   b. XML

   c. OWASP

   d. RESTCONF

3. Which of the following is a Cisco resource where you can learn about network programmability and obtain sample code?

   a. APIC

   b. ACI

   c. DevNet

   d. NETCONF

4. A YANG-based server publishes a set of YANG modules, which taken together form the system's _____.

   a. YANG model

   b. NETCONF model

   c. RESTCONF model

   d. gRPC model

5. Which of the following HTTP methods sends data to the server typically used in HTML forms and API requests?

   a. POST

   b. GET

   c. TRACE

   d. PUT

6. Which of the following is a solution that allows you to detect security threats in encrypted traffic without decrypting the packets?

   a. ETA

   b. ESA

   c. WSA

   d. None of these answers is correct.

7. Which of the following is an open source project that allows you to deploy micro-segmentation policy-based services in container environments?

   a. OVS

   b. Contiv

   c. ODL

   d. All of the above

8. NFV nodes such as virtual routers and firewalls need which of the following components as an underlying infrastructure?

   a. A hypervisor

   b. A virtual forwarder to connect individual instances

   c. A network controller

   d. All of these answers are correct.

9. There have been multiple IP tunneling mechanisms introduced throughout the years. Which of the following are examples of IP tunneling mechanisms?

   a. VXLAN

   b. SST

   c. NVGRE

   d. All of these answers are correct.

10. Which of the following is true about SDN?

   a. SDN provides numerous benefits in the area of management plane. These benefits are in both physical switches and virtual switches.

   b. SDN changed a few things in the management, control, and data planes. However, the big change was in the control and data planes in software-based switches and routers (including virtual switches inside of hypervisors).

   c. SDN is now widely adopted in data centers.

   d. All of these answers are correct.

CHAPTER 4

Authentication, Authorization, Accounting (AAA) and Identity Management

This chapter covers the following topics:

An Introduction to Authentication, Authorization, and Accounting: You will learn AAA concepts, models, and techniques used in different environments.

Authentication: This chapter covers what is authentication and the different types of authentication, such as authentication by knowledge, ownership, biometrics, and others. You will also learn about multifactor authentication solutions, single sign-on (SSO), and many other authentication protocols and schemes.

Authorization: Once authenticated, a subject must be authorized. In this chapter, you will learn about the process of assigning authenticated subjects permission to carry out a specific operation. You will learn about different authorization models that define how access rights and permission are granted. The three primary authorization models are object capability, security labels, and ACLs.

Accounting: You will learn what is the process of auditing and monitoring what a user does once a specific resource is accessed.

Infrastructure Access Control: Infrastructure access controls include physical and logical network design, border devices, communication mechanisms, and host security settings. Because no system is foolproof, access must be continually monitored; if suspicious activity is detected, a response must be initiated.

AAA Protocols: This chapter covers the details about the RADIUS and TACACS+ protocols and how they are used in many different implementations.

Cisco Identity Service Engine (ISE): Cisco ISE is the centralized AAA and policy engine solution from Cisco. In this chapter, you will learn how the Cisco ISE integrates with numerous Cisco products and third-party solutions to allow you to maintain visibility of who and what is accessing your network, and to enforce access control consistently.

Configuring TACACS+ Access: Each of the CCNP concentration exams and the CCIE lab exam focus on configuration and troubleshooting. However, in this chapter, you will learn the concepts of TACACS+ access configurations in infrastructure devices such as routers and switches running Cisco IOS and Cisco IOS-XE software.

Configuring RADIUS Authentication: You can configure RADIUS authentication in multiple scenarios, including Remote Access VPN, Secure Network Access, 802.1X, and more. This chapter includes several examples of how you can configure RADIUS authentication for secure network access.

Cisco ISE Design Tips: In this chapter, you will learn different best practices when designing and deploying Cisco ISE in different environments.

The following SCOR 350-701 exam objectives are covered in this chapter:

- **Domain 2: Network Security**

  - 2.7 Configure AAA for device and network access (authentication and authorization, TACACS+, RADIUS and RADIUS flows, accounting, and dACL)

- **Domain 5: Endpoint Protection and Detection**

  - 5.6 Describe the uses and importance of a multifactor authentication (MFA) strategy

- **Domain 6: Secure Network Access, Visibility, and Enforcement**

  - 6.1 Describe identity management and secure network access concepts such as guest services, profiling, posture assessment, and BYOD

  - 6.2 Configure and verify network access device functionality such as 802.1X, MAB, and WebAuth

  - 6.3 Describe network access with CoA

  - 6.4 Describe the benefits of device compliance and application control

"Do I Know This Already?" Quiz

The "Do I Know This Already?" quiz allows you to assess whether you should read this entire chapter thoroughly or jump to the "Exam Preparation Tasks" section. If you are in doubt about your answers to these questions or your own assessment of your knowledge of the topics, read the entire chapter. Table 4-1 lists the major headings in this chapter and their corresponding "Do I Know This Already?" quiz questions. You can find the answers in Appendix A, "Answers to the 'Do I Know This Already?' Quizzes and Q&A Sections."

Table 4-1 "Do I Know This Already?" Section-to-Question Mapping

Foundation Topics Section	Questions
Introduction to Authentication, Authorization, and Accounting	1
Authentication	2–4
Authorization	5–6
Accounting	7
Infrastructure Access Controls	8
AAA Protocols	9–10
Cisco Identity Service Engine (ISE)	11–12
Configuring TACACS+ Access	13
Configuring RADIUS Authentication	14
Additional Cisco ISE Design Tips	15

CAUTION The goal of self-assessment is to gauge your mastery of the topics in this chapter. If you do not know the answer to a question or are only partially sure of the answer, you should mark that question as wrong for the purposes of the self-assessment. Giving yourself credit for an answer you incorrectly guess skews your self-assessment results and might provide you with a false sense of security.

1. You were hired to configure AAA services in an organization and are asked to make sure that users in the engineering department do not have access to resources that are only meant for the finance department. What authorization principle addresses this scenario?

   a. The principle of least privilege and separation of duties

   b. Accounting and MAC Auth-bypass

   c. Deter, delay, and detect

   d. Policy-based segmentation

2. Which of the following describes the type of authentication where the user provides a secret that is only known by him or her?

   a. Authentication by password

   b. Authentication by knowledge

   c. Personal identification number (PIN) code

   d. Authentication by characteristics

3. Which of the following is a set of characteristics that can be used to prove a subject's identity one time and one time only?

   a. One-time passcode (OTP)

   b. Out-of-band (OOB)

   c. Biometrics

   d. None of these answers is correct.

4. Which of the following is an open standard for exchanging authentication and authorization data between identity providers, and is used in many single sign-on (SSO) implementations?

   a. SAML

   b. OAuth 2.0

   c. OpenConnectID

   d. DUO Security

5. Which of the following defines how access rights and permission are granted? Examples of that model include object capability, security labels, and ACLs.

   a. A mandatory access control model

   b. An authorization model

   c. An authentication model

   d. An accounting model

6. An authorization policy should always implement which of the following concepts? (Select all that apply.)

   a. Implicit deny

   b. Need to know

   c. Access control debugging logs

   d. Access control filter logs

7. Which of the following is the process of auditing and monitoring what a user does once a specific resource is accessed?

   a. CoA

   b. Authorization

   c. Accounting

   d. TACACS+ auditing

8. Access control lists classify packets by inspecting Layer 2 through Layer 7 headers for a number of parameters, including which of the following?

   a. Layer 2 protocol information such as EtherTypes

   b. Layer 3 header information such as source and destination IP addresses

   c. Layer 4 header information such as source and destination TCP or UDP ports

   d. All of these options are correct.

9. Which of the following statements are true?

   a. RADIUS uses UDP, and TACACS+ uses TCP.

   b. In RADIUS, authentication and authorization are performed with the same exchange. Accounting is done with a separate exchange.

   c. In TACACS+, authentication, authorization, and accounting are performed with separate exchanges.

   d. RADIUS provides limited support for command authorization. TACACS+ provides granular command authorization.

   e. All of these answers are correct.

10. Network access devices (such as network switches and wireless access points) can use an IEEE protocol that when enabled, will allow traffic on the port only after the device has been authenticated and authorized. Which of the following is an IEEE standard that is used to implement port-based access control?

    a. 802.11ac

    b. 802.1Q

    c. 802.1X

    d. pxGrid

11. Which of the following provides a cross-platform integration capability between security monitoring applications, threat detection systems, asset management platforms, network policy systems, and practically any other IT operations platform?

    a. pxGrid

    b. 802.1X

    c. TrustSec

    d. SGTs

12. Which of the following are examples of some of the more popular policy attributes supported by Cisco ISE?

   a. Active Directory group membership and Active Directory user-based attributes

   b. Time and date

   c. Location of the user

   d. Access method (MAB, 802.1X, wired, wireless, and so on)

   e. None of these options is correct.

   f. All of these options are correct.

13. Which of the following commands enables AAA services on a Cisco router?

   a. aaa new-model

   b. aaa authentication enable

   c. aaa authentication model

   d. aaa enable console

14. Which of the following is the default behavior of an 802.1X-enabled port?

   a. To authorize only a single MAC address per port

   b. To authorize only a single IP address per port

   c. To perform MAC auth bypass only if the MAC is registered to ISE

   d. To authenticate only a single host that has an identity certificate

15. Which of the following are Cisco ISE distributed node types?

   a. Primary Administration Node (PAN)

   b. Secondary Administration Node (SAN)

   c. Policy Service Node (PSN)

   d. All of these options are correct.

Foundation Topics

Introduction to Authentication, Authorization, and Accounting

In Chapter 1, "Cybersecurity Fundamentals," you learned about different types of authentication-based attacks and the high-level concepts of the authentication process. In the following sections, you will learn the tenets of authentication, authorization, and accounting (AAA).

An identification scheme, an authentication method, and an authorization model are common attributes of access controls.

NOTE Access controls are security features that govern how users and processes communicate and interact with systems and resources. The primary objective of access controls is to protect information and information systems from unauthorized access (confidentiality), modification (integrity), or disruption (availability). When we're discussing access controls, the active entity (that is, the user or system) that requests access to a resource or data is referred to as the *subject*, and the passive entity being accessed or being acted upon is referred to as the *object*.

An identification scheme is used to identify unique records in a set, such as a username. Identification is the process of the subject supplying an identifier to the object. The authentication method is how identification is proven to be genuine. Authentication is the process of the subject supplying verifiable credentials to the object. The authorization model defines how access rights and permission are granted. Authorization is the process of assigning authenticated subjects the permission to carry out a specific operation.

The Principle of Least Privilege and Separation of Duties

The principle of least privilege states that all users—whether they are individual contributors, managers, directors, or executives—should be granted only the level of privilege they need to do their jobs, and no more. For example, a sales account manager really has no business having administrator privileges over the network, or a call center staff member over critical corporate financial data.

The same concept can be applied to software. For example, programs or processes running on a system should have the capabilities they need to get their job done, but no root access to the system. If a vulnerability is exploited on a system that runs everything as root, the damage could extend to a complete compromise of the system. This is why you should always limit users, applications, and processes to access and run as the least privilege they need.

Somewhat related to the principle of least privilege is the concept of "need to know," which means that users should get access only to data and systems that they need to do their job, and no other.

Separation of duties is an administrative control that dictates that a single individual should not perform all critical- or privileged-level duties. Additionally, important duties must be separated or divided among several individuals within the organization. The goal is to safeguard against a single individual performing sufficiently critical or privileged actions that could seriously damage a system or the organization as a whole. For instance, security auditors responsible for reviewing security logs should not necessarily have administrative rights over the systems. Another example is that a network administrator should not have the ability to alter logs on the system. This is to prevent such individuals from carrying out unauthorized actions and then deleting evidence of such actions from the logs (in other words, covering their tracks).

Think about two software developers in the same organization ultimately working toward a common goal, but one is tasked with developing a portion of a critical application and the other is tasked with creating an application programming interface (API) for other critical applications. Each developer has the same seniority and working grade level; however, they do not know or have access to each other's work or systems.

Authentication

Identification is the process of providing the identity of a subject or user. This is the first step in the authentication, authorization, and accounting process. Providing a username, a passport, an IP address, or even pronouncing your name is a form of identification. A secure identity should be unique in the sense that two users should be able to identify themselves unambiguously. This is particularly important in the context of account monitoring. Duplication of identity is possible if the authentication systems are not connected. For example, a user can use the same user ID for his corporate account and for his personal email account. A secure identity should also be nondescriptive so that information about the user's identity cannot be inferred. For example, using "Administrator" as the user ID is generally not recommended. An identity should also be issued in a secure way. This includes all processes and steps in requesting and approving an identity request. This property is usually referred to as "secure issuance."

The following list highlights the key concepts of identification:

- Identities should be unique. Two users with the same identity should not be allowed.

- Identities should be nondescriptive, meaning it should not be possible to infer the role or function of the user from his or her identity. For example, a user called "admin" represents a descriptive identity, whereas a user called "om1337ar" represents a nondescriptive identity.

- Identities should be securely issued. A secure process for issuing an identity to a user needs to be established.

- Identities can be location-based. A process for authenticating someone based on his or her location needs to be established.

There are three categories of factors: knowledge (something the user knows), possession (something a user has), and inherence or characteristics (something the user is).

Authentication by Knowledge

Authentication by knowledge is where the user provides a secret that is only known by him or her. An example of authentication by knowledge would be a user providing a password, a personal identification number (PIN) code, or answering security questions.

The disadvantage of using this method is that once the information is lost or stolen (for example, if a user's password is stolen), an attacker would be able to successfully authenticate. Nowadays, a day does not pass without hearing about a new breach in retailers, service providers, cloud services, and social media companies.

> **NOTE** If you look at the VERIS community database, you will see hundreds of breach cases where users' passwords were exposed (https://github.com/vz-risk/VCDB). Websites like "Have I been pwned" (https://haveibeenpwned.com) include a database of billions of usernames and passwords from past breaches and even allow you to search for your email address to see if your account or information has potentially been exposed.

Something you know is knowledge-based authentication. It could be a string of characters, referred to as a password or PIN, or it could be an answer to a question. Passwords are the most commonly used single-factor network authentication method. The authentication strength of a password is a function of its length, complexity, and unpredictability. If it is easy to guess or deconstruct, it is vulnerable to attack. Once known, it is no longer useful as a verification tool. The challenge is to get users to create, keep secret, and remember secure passwords. Weak passwords can be discovered within minutes or even seconds using any number of publicly available password crackers or social engineering techniques. Best practices dictate that passwords be a minimum of eight characters in length (preferably longer), include a combination of at least three of the following categories: upper- and/or lowercase letters, punctuation, symbols, and numerals (referred to as complexity), be changed frequently, and be unique. Using the same password to log in to multiple applications and sites significantly increases the risk of exposure.

NOTE NIST Special Publication 800-63B, "Digital Identity Guidelines: Authentication and Lifecycle Management," provides guidelines for authentication and password strengths. NIST confirms that the length of a password has been found to be a primary factor in characterizing password strength. The longer the password, the better. Passwords that are too short are very susceptible to brute force and dictionary attacks using words and commonly chosen passwords. NIST suggests that "the minimum password length that should be required depends to a large extent on the threat model being addressed. Online attacks where the attacker attempts to log in by guessing the password can be mitigated by limiting the rate of login attempts permitted."

Generally, when users are granted initial access to an information system, they are given a temporary password. Most systems have a technical control that will force the user to change his or her password at first login. A password should be changed immediately if there is any suspicion that it has been compromised.

As any help desk person will tell you, users forget their passwords with amazing regularity. If a user forgets his password, there needs to be a process for reissuing passwords that includes verification that the requester is indeed who he says he is. Often cognitive passwords are used as secondary verification. A cognitive password is a form of knowledge-based authentication that requires a user to answer a question based on something familiar to them. Common examples are mother's maiden name and favorite color. The problem, of course, is that this information is very often publicly available. This weakness can be addressed using sophisticated questions that are derived from subscription databases such as credit reports. These questions are commonly referred to as out-of-wallet challenge questions. The term was coined to indicate that the answers are not easily available to someone other than the user, and that the user is not likely to carry such information in his or her wallet. Out-of-wallet question systems usually require that the user correctly answer more than one question and often include a "red herring" question that is designed to trick an imposter but which the legitimate user will recognize as nonsensical.

It might seem very convenient when a website or application offers to remember a user's logon credentials or provide an automatic logon to a system, but this practice should be strictly prohibited. If a user allows websites or software applications to automate the authentication process, unattended devices can be used by unauthorized people to gain access to information resources.

Authentication by Ownership or Possession

With this type of authentication, the user is asked to provide proof that he owns something specific—for example, a system might require an employee to use a badge to access a facility. Another example of authentication by ownership is the use of a token or smart card. Similar to the previous method, if an attacker is able to steal the object used for authentication, he will be able to successfully access the system.

Examples of authentication by ownership or possession include the following: a one-time passcode, memory card, smartcard, and out-of-band communication.

The most common of the four is the one-time passcode sent to a device in the user's possession. A *one-time passcode (OTP)* is a set of characteristics that can be used to prove a subject's identity one time and one time only. Because the OTP is valid for only one access, if it's captured, additional access would be automatically denied. OTPs are generally delivered through a hardware or software token device. The token displays the code, which must then be typed in at the authentication screen. Alternatively, the OTP may be delivered via email, text message, or phone call to a predetermined address or phone number.

A *memory card* is an authentication mechanism that holds user information within a magnetic strip and relies on a reader to process the information. The user inserts the card into the reader and enters a personal identification number (PIN). Generally, the PIN is hashed and stored on the magnetic strip. The reader hashes the inputted PIN and compares it to the value on the card itself. A familiar example of this is a bank ATM card.

A *smartcard* works in a similar fashion. Instead of a magnetic strip, it has a microprocessor and integrated circuits. The user inserts the card into a reader, which has electrical contacts that interface with the card and power the processor. The user enters a PIN that unlocks the information. The card can hold the user's private key, generate an OTP, or respond to a challenge-response.

Out-of-band authentication requires communication over a channel that is distinct from the first factor. A cellular network is commonly used for out-of-band authentication. For example, a user enters her name and password at an application logon prompt (factor 1). The user then receives a call on her mobile phone; the user answers and provides a predetermined code (factor 2). For the authentication to be compromised, the attacker would have to have access to both the computer and the phone.

Authentication by Characteristic

A system that uses authentication by characteristic authenticates the user based on some physical or behavioral characteristic, sometimes referred to as a biometric attribute. The most used physical or physiological characteristics are as follows:

- Fingerprints
- Face recognition
- Retina and iris
- Palm and hand geometry
- Blood and vascular information
- Voice recognition

Examples of behavioral characteristics are as follows:

- Signature dynamic
- Keystroke dynamic/pattern

The drawback of a system based on this type of authentication is that it's prone to accuracy errors. For example, a signature-dynamic-based system would authenticate a user by requesting that the user write his signature and then comparing the signature pattern to a record in the system. Given that the way a person signs his name differs slightly every time, the system should be designed so that the user can still authenticate, even if the signature and pattern are not exactly what are in the system. However, it should also not be too loose and unintentionally authenticate an unauthorized user attempting to mimic the pattern.

Two types of errors are associated with the accuracy of a biometric system:

■ A Type I error, also called false rejection, happens when the system rejects a valid user who should have been authenticated.

■ A Type II error, also called false acceptance, happens when the system accepts a user who should have been rejected (for example, an attacker trying to impersonate a valid user).

The crossover error rate (CER), also called the equal error rate (EER), is the point where the rate of false rejection errors (FRR) and the rate of false acceptance errors (FAR) are equal. This is generally accepted as an indicator of the accuracy (and hence the quality) of a biometric system.

Multifactor Authentication

The process of authentication requires the subject to supply verifiable credentials. The credentials are often referred to as *factors*.

Single-factor authentication is when only one factor is presented. The most common method of single-factor authentication is the use of passwords.

Multifactor authentication is when two or more factors are presented. *Multilayer authentication* is when two or more of the same type of factors are presented. Data classification, regulatory requirements, the impact of unauthorized access, and the likelihood of a threat being exercised should all be considered when you're deciding on the level of authentication required. The more factors, the more robust the authentication process.

Identification and authentication are often performed together; however, it is important to understand that they are two different operations. Identification is about establishing who you are, whereas authentication is about proving you are the entity you claim to be.

In response to password insecurity, many organizations have deployed multifactor authentication options to their users. With multifactor authentication, accounts are protected by something you know (password) and something you have (one-time verification code provided to you). Even gamers have been protecting their accounts using MFA for years.

Duo Security

Duo Security was a company acquired by Cisco that develops a very popular multifactor authentication solution that is used by many small, medium, and large organizations. Duo provides protection of on-premises and cloud-based applications. This is done by both preconfigured solutions and generic configurations via RADIUS, Security Assertion Markup Language (SAML), LDAP, and more.

> **TIP**   SAML is an open standard for exchanging authentication and authorization data between identity providers. SAML is used in many single sign-on (SSO) implementations. You will learn more about SAML and SSO later in this chapter.

Duo integrates with many different third-party applications, cloud services, and other solutions. Duo allows administrators to create policy rules around who can access applications and under what conditions. You can customize policies globally or per user group or application. User enrollment strategy will also inform your policy configuration. Duo Beyond subscribers can benefit from additional management within their environment by configuring a Trusted Endpoints policy to check the posture of the device that is trying to connect to the network, application, or cloud resource.

Duo Access Gateway is another component of the Duo solution. The Duo Access Gateway provides multifactor authentication access to cloud applications. You can use your users' existing directory credentials (such as accounts from Microsoft Active Directory or Google Apps). This is done by using the Security Assertion Markup Language (SAML) 2.0 authentication standard. SAML delegates authentication from a service provider to an identity provider.

Figure 4-1 shows a high-level overview of the Duo Access Gateway solution.

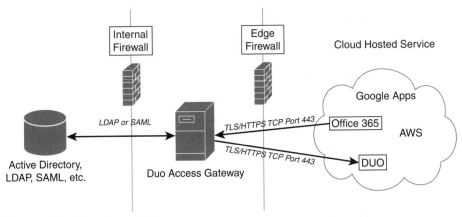

Figure 4-1 *Duo Access Gateway*

In Figure 4-1, you can see how Duo provides SAML connectors for enterprise cloud applications (Office 365, Google Apps, Amazon Web Services, and so on). The protected cloud applications redirect users to the Duo Access Gateway server that is typically deployed on-premises (that is, on your network). The Duo Access Gateway acts as a SAML identity provider (IdP). It authenticates users by leveraging existing authentication sources for credential verification and then prompting for two-factor authentication before permitting access to the SAML application.

NOTE   You can obtain detailed information on how to deploy the different solutions provided by Duo at https://duo.com/docs/getting-started.

You can also use Duo to protect virtual private network (VPN) users in your organization. For instance, you can configure a Cisco ASA or Cisco Firepower Threat Defense (FTD) device to terminate connections from remote access VPN clients and integrate Duo to provide multifactor authentication.

NOTE   You will learn more about VPNs and respective configurations in Chapter 8, "Virtual Private Networks (VPNs)."

Figure 4-2 shows an example of multifactor authentication where a user (osantos) connects to a VPN device and is prompted to verify the VPN connection on his iPhone's Duo mobile app.

Figure 4-2 *Duo Multifactor Authentication in VPN Implementations*

Figure 4-3 shows how the Duo app Security Checkup verifies the posture of the mobile device used for multifactor authentication.

> **TIP** You can use Duo for free to provide multifactor authentication for up to 10 users. That is a good way to get started and get familiar with the Duo management console. For additional information about the pricing and other service options, visit https://duo.com/pricing.

You can integrate Duo with many applications. Figure 4-4 shows the Duo administrative dashboard where you can integrate many different applications. In Figure 4-4, three applications are integrated (two Unix-based systems and a macOS system).

> **TIP** To learn how to integrate and protect a Unix/Linux-based system with Duo, visit https://duo.com/docs/duounix.

To integrate (protect) a new application, you can click the **Protect an Application** button, and the screen shown in Figure 4-5 is displayed. Then select the application you want to protect. You can access detailed information on how to integrate each application by clicking the **Read the documentation** link by the name of the respective application.

Zero Trust and BeyondCorp

Zero trust has been a buzzword in the cybersecurity industry for several years. The zero-trust concept assumes that no system or user will be "trusted" when requesting access to the corporate network, systems, and applications hosted on-premises or in the cloud. You must first

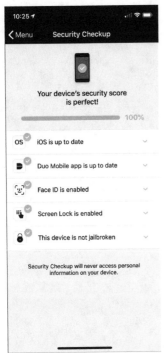

Figure 4-3 *Duo App Security Checkup Screen*

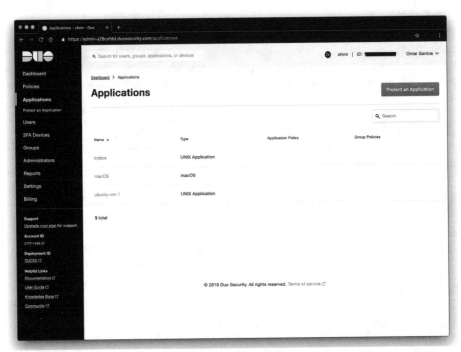

Figure 4-4 *Duo Administrative Dashboard Application Integrations*

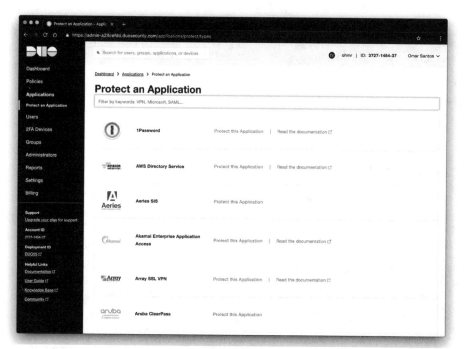

Figure 4-5 *Protecting a New Application in the Duo Administrative Dashboard*

verify their trustworthiness before granting access. Attackers can bypass different firewalls by employing different evasion techniques. In addition, a lot of attacks start from the inside (from insiders) and can spread out to compromise critical systems and steal sensitive data.

When an application or system is "protected" with access controls dependent on whether the user is "inside or outside the perimeter," an attacker can take advantage of these looser sets of controls. Furthermore, a lot of applications do not reside inside the perimeter anymore. They reside in the cloud. Consequently, external cloud-based applications and mobile users can face attacks that are outside of the perimeter protections. Users often make your organization even more vulnerable by using unmanaged and unpatched devices to connect to critical systems and applications. This is why the concept of "zero trust" was introduced in the industry.

NOTE   Google also introduced a similar concept called BeyondCorp. The BeyondCorp security model is based on Google's own implementation of a "zero-trust" model, which shifts access control from the network perimeter firewalls and other security devices to individual devices and users. The BeyondCorp model is documented at

https://research.google/pubs/pub43231/ and https://cloud.google.com/beyondcorp/.

Duo sits in the heart of the Cisco Zero Trust security framework. This framework helps you prevent unauthorized access, contain security incidents, and reduce the risk of an adversary pivoting (performing lateral movement) through your network. The Cisco Zero Trust solution allows administrators to consistently enforce policy-based controls and maintain visibility of users and systems across the entire environment. The Cisco Zero Trust solution also provides detailed logs, reports, and alerts that can help security professionals better detect and respond to threats.

Single Sign-On

Suppose you have two separate applications (Application_1 and Application_2). Both applications require user authentication, as demonstrated in Figure 4-6.

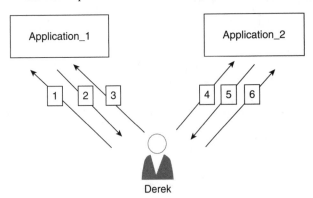

Figure 4-6   *Two Applications Without Single Sign-On*

The following steps occur in Figure 4-6:

1. Derek (the user) tries to connect to Application_1.

2. Application_1 prompts Derek for authentication.

3. Derek authenticates to Application_1.

4. Then Derek wants to connect to Application_2.

5. Application_2 prompts Derek for authentication.

6. Derek authenticates to Application_2.

As you can see, this is not very "user friendly." In most environments, you are accessing dozens of applications throughout an enterprise network or even applications hosted in the cloud. This is why single sign-on (SSO) was created.

Figure 4-7 shows an example of a typical SSO implementation.

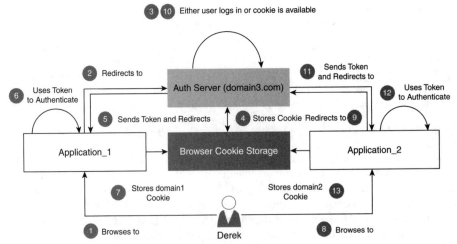

Figure 4-7   *Single Sign-On Example*

Even though the steps in Figure 4-7 are more elaborate than those in Figure 4-6, the user experience is better. This is because the user (Derek) only needs to authenticate once (to Application_1) and then he can browse (access) any other application that is a participant on the SSO implementation and that Derek is authorized to access.

TIP   The concept of a centralized identity (or linked identity) is also referred to as "federated identity." Federated identity systems handle authentication, authorization, user attributes exchange, and user management. The attributes exchange concept orchestrates data sharing across different user management systems. For example, attributes like "real name" may be present in multiple systems or applications. Federated identity systems counteract data duplication problems by linking the related attributes within all the elements that are part of the SSO environment.

SAML is used in SSO implementations. However, there are other identity technologies used in SSO implementations such as OpenID Connect, Microsoft Account (formerly known as Passport), and Facebook Connect.

NOTE   SAML is an open standard for exchanging authentication and authorization data between identity providers.

The following are several elements that are part of an SSO and federated identity implementation:

- **Delegation:** SSO implementations use delegation to call external APIs to authenticate and authorize users. Delegation is also used to make sure that applications and services do not store passwords and user information on-site.

- **Domain:** A domain in an SSO environment is the network where all resources and users are linked to a centralized database (this is, where all authentication and authorization occurs).

- **Factor:** You already learned about "multifactor" authentication. A factor in authentication is a vector through which identity can be confirmed.

- **Federated Identity Management:** A collection of shared protocols that allows user identities to be managed across organizations.

- **Federation provider:** An identity provider that offers single sign-on, consistency in authorization practices, user management, and attributes-exchange practices between identity providers (issuers) and relying parties (applications).

- **Forest:** A collection of domains managed by a centralized system.

- **Identity provider (IdP):** An application, website, or service responsible for coordinating identities between users and clients. IdPs can provide a user with identifying information and provide that information to services when the user requests access.

- **Kerberos:** A ticket-based protocol for authentication built on symmetric-key cryptography.

- **Multitenancy:** A term in computing architecture referring to the serving of many users (tenants) from a single instance of an application. Software as a Service (SaaS) offerings are examples of multitenancy. They exist as a single instance but have dedicated shares served to many companies and teams.

- **OAuth:** An open standard for authorization used by many APIs and modern applications. You can access OAuth and OAuth 2.x specifications and documentation at https://oauth.net/2.

- **OpenID (or OpenID Connect):** Another open standard for authentication. OpenID Connect allows third-party services to authenticate users without clients needing to collect, store, and subsequently become liable for a user's login information. Detailed information about OpenID can be accessed at https://openid.net/connect/.

- **Passwordless:** A type of authentication based on tokens. Passwordless authentication challenges are typically received and sent through SMS, email (magic links), or biometric sensors.

- **Social identity provider (social IdP):** A type of identity provider originating in social services like Google, Facebook, Twitter, and so on.

- **Web identity:** The identifying data is typically obtained from an HTTP request (often these are retrieved from an authenticated email address).

- **Windows identity:** This is how Active Directory in Microsoft Windows environments organizes user information.

- **WS-Federation:** A common infrastructure (federated standard) for identity, used by web services and browsers on Windows Identity Foundation. Windows Identity Foundation is a framework created by Microsoft for building identity-aware applications. You can obtain detailed information about Windows Identity Foundation and WS-Federation at https://docs.microsoft.com/en-us/dotnet/framework/security/.

Now that you have an understanding of the different elements of SSO, multifactor authentication, and user identity, let's take a look at a couple of examples of how these concepts are used together. In Figure 4-8, the Duo SSO and multifactor authentication (MFA) services provide user authentication for three different cloud services (Cisco WebEx, Cisco Meraki Cloud, and Cisco Umbrella). SSO is done using the SAML protocol.

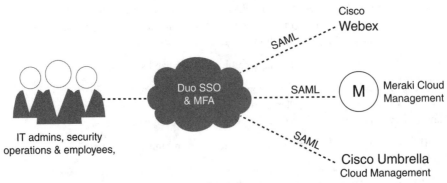

Figure 4-8 *Duo Providing SSO and Multifactor Authentication*

> **TIP**   Cisco WebEx (https://www.webex.com) is a cloud SaaS collaboration service that is extremely popular among enterprises and small business alike. Cisco Meraki (https://meraki.cisco.com) is a series of solutions for businesses of all sizes that provide networking (wired and wireless) and security products that are managed in the cloud. Cisco Umbrella (https://umbrella.cisco.com, formerly known as OpenDNS) is a cloud-based solution that protects thousands of organizations and users around the world. Cisco Umbrella blocks malicious destinations before a connection is ever established. You will learn more about Cisco Umbrella in Chapter 9, "Securing the Cloud," Chapter 10, "Content Security," and Chapter 11, "Endpoint Protection and Detection."

Figure 4-9 shows another example of how Duo provides user multifactor authentication and SSO to different services, including VPN users, Microsoft Azure applications, remote desktop protocol (RDP) implementations, Secure Shell (SSH), API authentication, and other implementations.

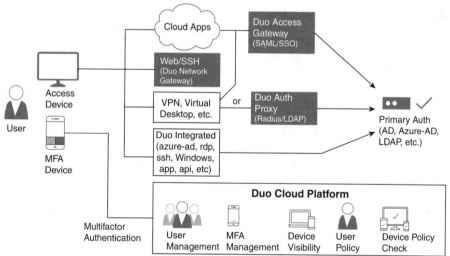

Figure 4-9   *Duo Cloud Platform Services*

The Duo cloud platform not only provides multifactor authentication and SSO services, but also device visibility, user policy, and device policy checks.

Authorization

Once authenticated, a subject must be authorized. *Authorization* is the process of assigning authenticated subjects permission to carry out a specific operation. The *authorization model* defines how access rights and permission are granted. The three primary authorization models are object capability, security labels, and ACLs. *Object capability* is used programmatically and is based on a combination of an unforgeable reference and an operational message. *Security labels* are mandatory access controls embedded in object and subject properties. Examples of security labels (based on its classification) are "confidential," "secret," and "top secret." *Access control lists (ACLs)* are used to determine access based on some combination of specific criteria, such as a user ID, group membership, classification, location, address, and date.

Additionally, when granting access, the authorization process would check the permissions associated with the subject/object pair so that the correct access right is provided. The object owner and management usually decide (or give input on) the permission and authorization policy that governs the authorization process.

The authorization policy and rule should take various attributes into consideration, such as the identity of the subject, the location from where the subject is requesting access, the subject's role within the organization, and so on. Access control models, which are described in more detail later in this chapter, provide the framework for the authorization policy implementation.

An authorization policy should implement two concepts:

- **Implicit deny:** If no rule is specified for the transaction of the subject/object, the authorization policy should deny the transaction.

- **Need to know:** A subject should be granted access to an object only if the access is needed to carry out the job of the subject. You learned about the least privilege principle and need-to-know concepts earlier in this chapter.

The sections that follow cover the details about the different access control categories.

Mandatory Access Control (MAC)

Mandatory access controls (MACs) are defined by policy and cannot be modified by the information owner. MACs are primarily used in secure military and government systems that require a high degree of confidentiality. In a MAC environment, objects are assigned a security label that indicates the classification and category of the resource. Subjects are assigned a security label that indicates a clearance level and assigned categories (based on the need to know). The operating system compares the object's security label with the subject's security label. The subject's clearance must be equal to or greater than the object's classification. The category must match. For example, for a user to access a document classified as "Secret" and categorized as "Flight Plans," the user must have either Secret or Top-Secret clearance and have been tagged to the Flight Plan category.

Discretionary Access Control (DAC)

Discretionary access controls (DACs) are defined by the owner of the object. DACs are used in commercial operating systems. The object owner builds an ACL that allows or denies access to the object based on the user's unique identity. The ACL can reference a user ID or a group (or groups) that the user is a member of. Permissions can be cumulative. For example, John belongs to the Accounting group. The Accounting group is assigned read permissions to the Income Tax folder and the files in the folder. John's user account is assigned write permissions to the Income Tax folder and the files in the folder. Because DAC permissions are cumulative, John can access, read, and write to the files in the Income Tax folder.

Role-Based Access Control (RBAC)

Role-based access controls (RBACs, also called "nondiscretionary controls") are access permissions based on a specific role or function. Administrators grant access rights and permissions to roles. Users are then associated with a single role. There is no provision for assigning rights to a user or group account.

Let's take a look at the example illustrated in Figure 4-10.

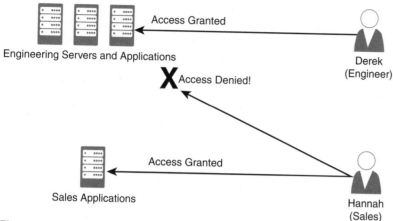

Figure 4-10 *An Example of RBAC*

The user Derek is associated with the role of "Engineer" and inherits all the permissions assigned to the Engineer role. Derek cannot be assigned any additional permissions. Hannah is associated with the role of "Sales" and inherits all the permissions assigned to the Sales role and cannot access engineering resources. Users can belong to multiple groups. RBAC enables you to control what users can do at both broad and granular levels.

Rule-Based Access Control

In a rule-based access control environment, access is based on criteria that are independent of the user or group account. The rules are determined by the resource owner. Commonly used criteria include source or destination address, geographic location, and time of day. For example, the ACL on an application requires that it be accessed from a specific workstation. Rule-based access controls can be combined with DACs and RBACs.

Attribute-Based Access Control

Attribute-based access control (ABAC) is a logical access control model that controls access to objects by evaluating rules against the attributes of entities (both subject and object), operations, and the environment relevant to a request. The characteristics of ABAC are as follows:

- ABAC supports a complex Boolean rule set that can evaluate many different attributes.

- The policies that can be implemented in an ABAC model are limited only to the degree imposed by the computational language and the richness of the available attributes.

- An example of an access control framework that is consistent with ABAC is the Extensible Access Control Markup Language (XACML).

Accounting

Accounting is the process of auditing and monitoring what a user does once a specific resource is accessed. This process is sometimes overlooked; however, as a security professional, it is important to be aware of accounting and to advocate that it be implemented because of the great help it provides during detection and investigation of cybersecurity breaches.

When accounting is implemented, an audit trail log is created and stored that details when the user has accessed the resource, what the user did with that resource, and when the user stopped using the resource. Given the potential sensitive information included in the auditing logs, special care should be taken in protecting them from unauthorized access.

Infrastructure Access Controls

A *network infrastructure* is defined as an interconnected group of hosts and devices. The infrastructure can be confined to one location or, as often is the case, widely distributed, including branch locations and home offices. Access to the infrastructure enables the use of its resources. *Infrastructure access controls* include physical and logical network designs, border devices, communication mechanisms, and host security settings. Because no system is foolproof, access must be continually monitored; if suspicious activity is detected, a response must be initiated.

Access Control Mechanisms

An access control mechanism is, in simple terms, a method for implementing various access control models. A system may implement multiple access control mechanisms. In some modern systems, this notion of an access control mechanism may be considered obsolete because the complexity of the system calls for more advanced mechanisms. Nevertheless, here are some of the most well-known methods:

- **Access control list (ACL):** This is the simplest way to implement a DAC-based system. ACLs can apply to different objects (like files) or they can also be configured statements (policies) in network infrastructure devices (routers, firewalls, etc.). For instance, an ACL, when applied to an object, will include all the subjects that can access the object and their specific permissions. Figure 4-11 shows an example of an ACL applied to a file

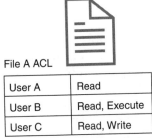

File A ACL

User A	Read
User B	Read, Execute
User C	Read, Write

Figure 4-11 *An example of an ACL applied to a file*

There is also the concept of ACLs in routers and firewalls. In those implementations ACLs provide packet filtering to protect "internal" networks from the "outside" systems and to filter traffic leaving the inside network. ACL criteria could be the source address of the traffic, the destination address of the traffic, destination port, source port, and the upper-layer protocol (otherwise known as the 5-tuple). In Cisco routers and firewalls an ACL is a collection of security rules or policies that allows or denies packets after looking at the packet headers and other attributes. Each permit or deny statement in the ACL is referred to as an access control entry (ACE). These ACEs classify packets by inspecting Layer 2 through Layer 7 headers for a number of parameters, including the following:

- Layer 2 protocol information such as EtherTypes

- Layer 3 protocol information such as ICMP, TCP, or UDP

- Layer 3 header information such as source and destination IP addresses

- Layer 4 header information such as source and destination TCP or UDP ports

- Layer 7 information such as application and system service calls

TIP   After an ACL has been properly configured, apply it to an interface to filter traffic. The security appliance filters packets in both the inbound and outbound direction on an interface. When an inbound ACL is applied to an interface, the security appliance analyzes packets against the ACEs after receiving them. If a packet is permitted by the ACL, the firewall continues to process the packet and eventually passes the packet out the egress interface.

In Cisco Next-Generation firewalls you can also configure different access control policies, that go beyond the traditional 5-tuple. You will learn detailed information about ACLs and access control policies in Chapter 7, "Cisco Next-Generation Firewalls and Cisco Next-Generation Intrusion Prevention Systems."

- **Capability table:** This is a collection of objects that a subject can access, together with the granted permissions. The key characteristic of a capability table is that it's subject centric instead of being object centric, like in the case of an access control list. Figure 4-12 shows a user (Derek) capability table.

Derek

User (Derek) Capability Table

File A	Read
File B	Read, Execute
File C	Read, Write

Figure 4-12   *User Capability Table*

- **Access control matrix (ACM):** This is an access control mechanism that is usually associated with a DAC-based system. An ACM includes three elements: the subject, the object, and the set of permissions. Each row of an ACM is assigned to a subject, while each column represents an object. The cell that identifies a subject/object pair includes the permission that subject has on the object. An ACM could be seen as a collection of access control lists or a collection of capabilities table, depending on how you want to read it. Figure 4-13 shows an example of access controls using an ACM.

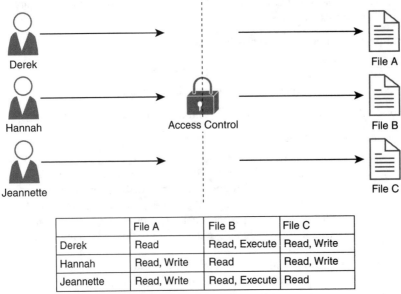

	File A	File B	File C
Derek	Read	Read, Execute	Read, Write
Hannah	Read, Write	Read	Read, Write
Jeannette	Read, Write	Read, Execute	Read

Figure 4-13 *Access Controls using an ACM*

- **Content-dependent access control:** This type of control uses the information (content) within a resource to make an authorization decision. This type of control is generally used in database access controls. A typical example is a database view.

> **TIP** A database view could also be considered a type of restricted interface because the available information is restricted depending on the identity of the user.

- **Context-dependent access control:** This type of control uses contextual information to make an access decision, together with other information such as the identity of the subject. For example, a system implementing a context-dependent control may look at events preceding an access request to make an authorization decision. A typical system that uses this type of control is a stateful firewall, such as Cisco ASA, Cisco Firepower Threat Defense (FTD), or Cisco IOS configured with the Zone-Based Firewall feature, where a packet is allowed or denied based on the information related to the session the packet belongs to.

AAA Protocols

Several protocols are used to grant access to networks or systems, provide information about access rights, and provide capabilities used to monitor, audit, and account for user actions once authenticated and authorized. These protocols are called authentication, authorization, and accounting (AAA) protocols.

The most well-known AAA protocols are RADIUS, TACACS+, and Diameter. The sections that follow provide some background information about each.

RADIUS

The Remote Authentication Dial-In User Service (RADIUS) is an AAA protocol mainly used to provide network access services. Due to its flexibility, it has been adopted in other scenarios as well. The authentication and authorization parts are specified in RFC 2865, while the accounting part is specified in RFC 2866.

RADIUS is a client-server protocol. In the context of RADIUS, the client is the access server, which is the entity to which a user sends the access request. The server is usually a machine running RADIUS services and that provides authentication and authorization responses containing all the information used by the access server to provide service to the user.

The RADIUS server can act as proxy for other RADIUS servers or other authentication systems. Also, RADIUS can support several types of authentication mechanisms, such as PPP PAP, CHAP, and EAP. It also allows protocol extension via the attribute field. For example, vendors can use the attribute "vendor-specific" (type 26) to pass vendor-specific information.

Figure 4-14 shows a typical deployment of a RADIUS server, a RADIUS client (wireless router in this example), and a laptop (wireless client or supplicant).

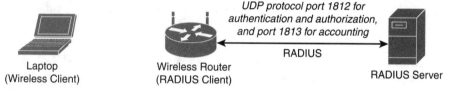

Laptop
(Wireless Client)

Wireless Router
(RADIUS Client)

UDP protocol port 1812 for
authentication and authorization,
and port 1813 for accounting

RADIUS

RADIUS Server

Figure 4-14 *RADIUS Server Implementation*

RADIUS operates in most cases over UDP protocol port 1812 for authentication and authorization, and port 1813 for accounting, which are the officially assigned ports for this service. In earlier implementations, RADIUS operated over UDP port 1645 for authentication and authorization, and port 1646 for accounting.

The authentication and authorization phase consist of two messages:

1. The access server sends an ACCESS-REQUEST to the RADIUS server that includes the user identity, the password, and other information about the requestor of the access (for example, the IP address).

2. The RADIUS server may reply with three different messages:

   - ACCESS-ACCEPT if the user is authenticated. This message will also include in the Attribute field authorization information and specific vendor information used by the access server to provide services.

   - ACCESS-REJECT if access for the user is rejected.

   - ACCESS-CHALLENGE if additional information is needed. RADIUS server needs to send an additional challenge to the access server before authenticating the user. The ACCESS-CHALLENGE will be followed by a new ACCESS-REQUEST message.

Figure 4-15 demonstrates the RADIUS exchange for authentication and authorization.

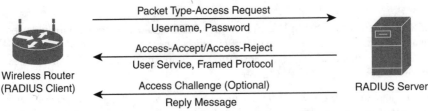

Figure 4-15 *RADIUS Exchange for Authentication and Authorization*

The accounting exchange consists of two messages: ACCOUNTING-REQUEST and ACCOUNTING-RESPONSE. Accounting can be used, for example, to specify how long a user has been connected to the network (the start and stop of a session).

The RADIUS exchange is authenticated by using a shared secret key between the access server and the RADIUS server. Only the user password information in the ACCESS-REQUEST is encrypted; the rest of the packets are sent in plaintext.

TACACS+

Terminal Access Controller Access Control System Plus (TACACS+) is a proprietary protocol developed by Cisco. It also uses a client-server model, where the TACACS+ client is the access server and the TACACS+ server is the machine providing TACACS+ services (that is, authentication, authorization, and accounting).

Similar to RADIUS, TACACS+ also supports protocol extension by allowing vendor-specific attributes and several types of authentication protocols. TACACS+ uses TCP as the transport protocol, and the TACACS+ server listens on port 49. Using TCP ensures a more reliable connection and fault tolerance.

TACACS+ has the authentication, authorization, and accounting processes as three separate steps. This allows the use of different protocols (for example, RADIUS) for authentication or accounting. Additionally, the authorization and accounting capabilities are more granular than in RADIUS (for example, allowing specific authorization of commands). This makes TACACS+ the preferred protocol for authorization services for remote device administration.

The TACACS+ exchange requires several packets:

- START, REPLY, and CONTINUE packets are used during the authentication process.

- REQUEST and RESPONSE packets are used during the authorization and accounting process.

The following is an example of an authentication exchange:

1. The access server sends a START authentication request.
2. The TACACS+ server sends a REPLY to acknowledge the message and ask the access server to provide a username.
3. The access server sends a CONTINUE with the username.
4. The TACACS+ server sends a REPLY to acknowledge the message and ask for the password.
5. The access server sends a CONTINUE with the password.
6. The TACACS+ server sends a REPLY with authentication response (pass or fail).

Figure 4-16 demonstrates the TACACS+ authentication, authorization, and accounting exchange.

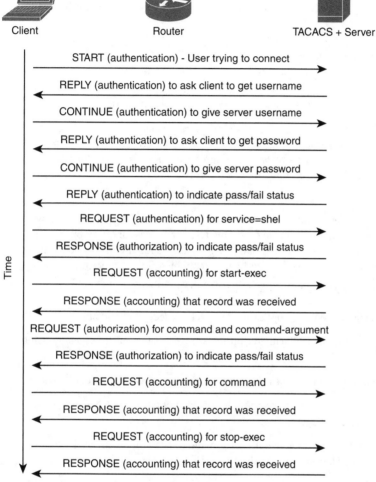

Figure 4-16 *TACACS+ Message Exchange for Authentication, Authorization, and Accounting*

TACACS+ offers better security protection compared to RADIUS. For example, the full body of the packet can be encrypted.

Table 4-2 summarizes the main differences between RADIUS and TACACS+.

Table 4-2   RADIUS vs TACACS+

	RADIUS	TACACS+
Transport Protocol	UDP.	TCP.
Security	Encrypts user password in ACCESS-REQUEST packets.	Optionally encrypts the full payload.

	RADIUS	TACACS+
AAA Phases	Authentication and authorization are performed with the same exchange. Accounting is done with a separate exchange.	Authentication, authorization, and accounting are performed with separate exchanges.
Command Authorization	Limited support for command authorization.	Provides granular command authorization.
Accounting	Supports strong accounting capabilities.	Provides basic accounting capabilities.

RADIUS authentication and authorization capabilities are defined in RFC 2865, and RADIUS accounting is defined in RFC 2866. TACACS+ is a Cisco proprietary protocol, despite the efforts through the years to have it adopted as an RFC in IETF (for example, https://www.ietf.org/id/draft-ietf-opsawg-tacacs-13.txt).

Diameter

RADIUS and TACACS+ were created with the aim of providing AAA services to network access via dial-up protocols or terminal access. Due to their success and flexibility, they have been used in several other scopes. To respond to newer access requirements and protocols, the IETF has proposed a new protocol called Diameter, which is described in RFC 6733.

Diameter has been built with the following functionality in mind:

- **Failover:** Diameter implements application-level acknowledgment and failover algorithms.

- **Transmission-level security:** Diameter protects the exchange of messages by using TLS or DTLS.

- **Reliable transport:** Diameter uses TCP or SCTP as the transport protocol.

- **Agent support:** Diameter specifies the roles of different agents such as proxy, relay, redirect, and translation agents.

- **Server-initiated messages:** Diameter makes mandatory the implementation of server-initiated messages. This enables capabilities such as on-demand re-authentication and re-authorization.

- **Transition support:** Diameter allows compatibility with systems using RADIUS.

- **Capability negotiation:** Diameter includes capability negotiations such as error handling as well as mandatory and nonmandatory attribute/value pairs (AVP).

- **Peer discovery:** Diameter enables dynamic peer discovery via DNS.

The main reason for the introduction of the Diameter protocol is the capability to work with applications that enable protocol extension. The main Diameter application is called *Diameter base* and it implements the core of the Diameter protocol. Other applications are Mobile IPv4 Application, Network Access Server Application, Diameter Credit-Control Application, and so on. Each application specifies the content of the information exchange in Diameter packets. For example, to use Diameter as AAA protocol for network access, the Diameter peers will use the Diameter Base Application and the Diameter Network Access Server Application.

The Diameter header field *Application ID* indicates the ID of the application. Each application, including the Diameter Base application, uses command code to identify specific application actions. Diameter is a peer-to-peer protocol, and entities in a Diameter context are called Diameter nodes. A Diameter node is defined as a host that implements the Diameter protocol.

The protocol is based on two main messages: a REQUEST, which is identified by setting the R bit in the header, and an ANSWER, which is identified by unsetting the R bit. Each message will include a series of attribute/value pairs (AVPs) that include application-specific information.

In its basic protocol flow, after the transport layer connection is created, the Diameter initiator peer sends a Capability-Exchange-Request (CER) to the other peer that will respond with a Capability-Exchange-Answer (CEA). The CER can include several AVPs, depending on the application that is requesting a connection. Once the capabilities are exchanged, the Diameter applications can start sending information.

Diameter also implements a keep-alive mechanism by using a Device-Watchdog-Request (DWR), which needs to be acknowledged with a Device-Watchdog-Answer (DWA). The communication is terminated by using a Disconnect-Peer-Request (DPR) and Disconnect-Peer-Answer (DPA). Both the Device-Watchdog and Disconnect-Peer can be initiated by both parties.

Figure 4-17 shows an example of a Diameter capability exchange and communication termination.

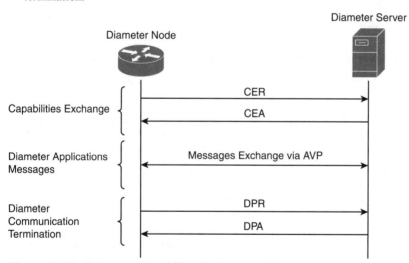

Figure 4-17 *Diameter Capability Exchange and Communication Termination*

The following is an example of protocol flows where Diameter is used to provide user authentication service for network access (as defined in the Network Access Server Application RFC 7155):

1. The initiator peer, the access server, sends a CER message with the Auth-Application-Id AVP set to 1, meaning that it supports authentication capabilities.

2. The Diameter server sends a CEA back to the access server with the Auth-Application-Id AVP set to 1.

3. The access server sends an AA-Request (AAR) to the Diameter server that includes information about the user authentication, such as username and password.

4. The access server will reply with an AA-Answer (AAA) message including the authentication results.

Figure 4-18 shows an example of a Diameter exchange for network access services.

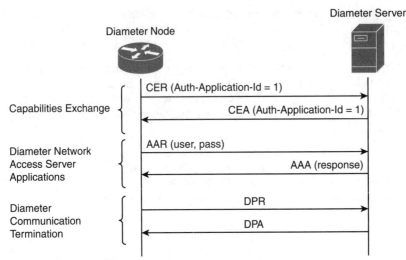

Figure 4-18   *Diameter Exchange for Network Access Services*

Diameter is a much more complex protocol and is used in some mobile service provider environments.

802.1X

802.1X is an IEEE standard that is used to implement port-based access control. In simple terms, an 802.1X access device will allow traffic on the port only after the device has been authenticated and authorized.

In an 802.1X-enabled network, three main roles are defined:

- **Authentication server:** An entity that provides an authentication service to an authenticator. The authentication server determines whether the supplicant is authorized to access the service. This is sometimes referred to as the policy decision point (PdP). The Cisco Identity Services Engine (ISE) is an example of an authentication server. You will learn more about Cisco ISE later in this chapter.

- **Supplicant:** An entity that seeks to be authenticated by an authenticator. For example, this could be a client laptop connected to a switch port. An example of a supplicant software is the Cisco AnyConnect Secure Mobility Client.

- **Authenticator:** An entity that facilitates authentication of other entities attached to the same LAN. This is sometimes referred to as the policy enforcement point (PeP). Cisco switches, wireless routers, and access points are examples of authenticators.

Other components, such as an identity database or a PKI infrastructure, may be required for a correct deployment. Figure 4-19 illustrates an example of an authentication server, supplicant, and authenticator. The supplicant is connected to the authenticator (wireless router) via a Wi-Fi connection.

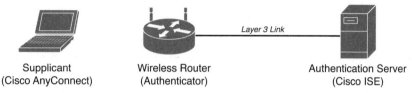

| | Supplicant
(Cisco AnyConnect) | Wireless Router
(Authenticator) | Authentication Server
(Cisco ISE) |

Figure 4-19 *802.1X Sample Topology*

802.1X uses the following protocols:

- **EAP over LAN (EAPoL):** An encapsulation defined in 802.1X that's used to encapsulate EAP packets to be transmitted from the supplicant to the authenticator.

- **Extensible Authentication Protocol (EAP):** An authentication protocol used between the supplicant and the authentication server to transmit authentication information.

- **RADIUS or Diameter:** The AAA protocol used for communication between the authenticator and authentication server.

The 802.1X port-based access control includes four phases (in this example, RADIUS is used as the protocol and a Cisco switch as the authenticator):

1. **Session initiation:** The session can be initiated either by the authenticator with an EAP-Request-Identity message or by the supplicant with an EAPoL-Start message. Before the supplicant is authenticated and the session authorized, only EAPoL, Cisco Discovery Protocol (CDP), and Spanning Tree Protocol (STP) traffic is allowed on the port from the authenticator.

2. **Session authentication:** The authenticator extracts the EAP message from the EAPoL frame and sends a RADIUS Access-Request to the authentication server, adding the EAP information in the AV pair of the RADIUS request. The authenticator and the supplicant will use EAP to agree on the authentication method (for example, EAP-TLS). Depending on the authentication method negotiated, the supplicant may provide a password, a certificate, a token, and so on.

3. **Session authorization:** If the authentication server can authenticate the supplicant, it will send a RADIUS Access-Accept to the authenticator that includes additional authorization information such as VLAN, downloadable access control list (dACL), and so on. The authenticator will send an EAP Success message to the supplicant, and the supplicant can start sending traffic.

4. **Session accounting:** This represents the exchange of accounting RADIUS packets between the authenticator and the authentication server.

Figure 4-20 illustrates an example of the 802.1X exchange.

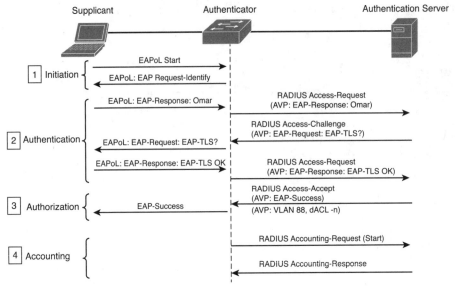

Figure 4-20   *802.1X Port Access Control Exchange*

In addition to these four phases, it is also very important that the session is correctly terminated. In the standard scenario, where the supplicant terminates the connection, it will send
an EAPoL-Logoff message.

Network Access Control List and Firewalling

You learned that one of the most basic implementations of an access control is an ACL.
When an ACL is applied to network traffic, it is called a *network ACL*. Cisco networking
devices such as routers, switches, and firewalls include network ACL capabilities to control
access to network resources. As for port-based access controls, network ACLs and firewalling are usually seen as special cases of the ABAC model and also sometimes classified as
identity-based or rule-based access control because they base the control decision on attributes such as IP or MAC addresses or Layer 4 information. Security group ACLs, on the
other hand, are access lists based on the role of the subject trying to access a resource, and
they implement role-based access control.

Network ACLs can be implemented at various levels of the OSI model:

- A Layer 2 ACL operates at the data link layer and implements filters based on Layer 2
  information. An example of this type of access list is a MAC access list, which uses
  information about the MAC address to create the filter.

- A Layer 3 ACL operates at the networking layer. Cisco devices usually allow Layer 3
  ACLs for different Layer 3 protocols, including the most used ones nowadays—IPv4
  and IPv6. In addition to selecting the Layer 3 protocol, a Layer 3 ACL allows the
  configuration of filtering for a protocol using raw IP, such as OSPF or ESP.

- A Layer 4 ACL operates at the transport layer. An example of a Layer 4 ACL is a TCP-
  or UDP-based ACL. Typically, a Layer 4 ACL includes the source and destination. This
  allows filtering of specific upper-layer packets.

VLAN ACLs

VLAN ACLs, also called VLAN maps, are not specifically Layer 2 ACLs; however, they are used to limit the traffic within a specific VLAN. A VLAN map can apply a MAC access list, a Layer 3 ACL, and a Layer 4 ACL to the inbound direction of a VLAN to provide access control.

Security Group–Based ACL

A security group–based ACL (SGACL) is an ACL that implements access control based on the security group assigned to a user (for example, based on his role within the organization) and the destination resources. SGACLs are implemented as part of Cisco TrustSec policy enforcement. Cisco TrustSec is described in a bit more detail in the sections that follow. The enforced ACL may include both Layer 3 and Layer 4 access control entries (ACEs).

Downloadable ACL

A downloadable ACL (dACL), also called a *per-user ACL*, is an ACL that can be applied dynamically to a port. The term *downloadable* stems from the fact that these ACLs are pushed from the authenticator server (for example, from a Cisco ISE) during the authorization phase.

When a client authenticates to the port (for example, by using 802.1X), the authentication server can send a dACL that will be applied to the port and that will limit the resources the client can access over the network.

> **TIP** ACLs are stateless access controls because they do not maintain the state of a session or a connection. A more advanced implementation of access control is provided by stateful firewalls, which are able to implement access control based on the state of a connection. Firewalls often implement inspection capabilities that enforce application layer protocol conformance and dynamic access control based on the state of the upper-layer protocol. Next-generation firewalls go one step further and implement context-aware controls, where not only the IP address or specific application information are taken into account, but other contextual information, such as the location, the type of device requesting access, and the sequence of events, are taken into consideration when allowing or denying a packet.

Cisco Identity Services Engine (ISE)

Cisco ISE is the centralized AAA and policy engine solution from Cisco. Cisco ISE integrates with numerous Cisco products and third-party solutions to allow you to maintain visibility of who and what is accessing your network, and to enforce access control consistently.

The following are some of the benefits of Cisco ISE:

- Centralizes network access control for wired, wireless, or VPN users.

- Helps administrators to comply with security regulations and audits by providing for easy policy creation, visibility, and reporting across the organization. Administrators can easily perform audits for regulatory requirements and mandated guidelines.

- Allows administrators to match users, endpoints, and each endpoint's security posture. It can also process attributes such as location, the time the user logged in or logged off, and the access method.

- Provides network visibility and host identification by supporting profiling capabilities. Profiling allows you to obtain real-time and historical visibility of all the devices on the network.

- Simplifies the experience of guest users or contractors when accessing the network. Cisco ISE provides self-service registration and fully customizable, branded guest portals that you can configure in minutes.

- Provides great support for bring-your-own-device (BYOD) and enterprise mobility also, with self-service device onboarding and management.

- Supports internal device certificate management and integration with enterprise mobility management (EMM) partners.

- Supports software-defined segmentation policies for users, endpoints, and other devices on your network based on security policies.

- Leverages Cisco TrustSec technology to define context-based access control policies using security group tags (SGTs). SGTs make segmentation easier when used in a security group ACL (SGACL).

- Uses the Cisco Platform Exchange Grid (pxGrid) technology to integrate with other Cisco products and third-party solutions. pxGrid allows you to maintain threat visibility and fast-tracks the capabilities to detect, investigate, contain, and recover (remediate) security incidents.

- Supports TACACS+ and RADIUS AAA services, as well as integration with Duo for multifactor authentication and secure access. Cisco ISE also supports external authentication servers such as LDAP and Active Directory servers.

Cisco ISE can be deployed in a physical appliance or in virtual machines (VMs). You can create physical or virtual ISE clusters for greater scalability, redundancy, and failover.

TIP Cisco ISE VMs are supported on VMware ESXi and on Kernel-based Virtual Machine (KVM) on Red Hat 7.x. You can obtain detailed information about the most up-to-date support and access numerous Cisco ISE resources, tutorials, and troubleshooting tips at https://community.cisco.com/t5/security-documents/ise-community-resources/ta-p/3621621#Start.

Cisco Platform Exchange Grid (pxGrid)

Cisco pxGrid provides a cross-platform integration capability among security monitoring applications, threat detection systems, asset management platforms, network policy systems, and practically any other IT operations platform. Cisco ISE supports Cisco pxGrid to provide a unified ecosystem to integrate multivendor tools to exchange information either unidirectionally or bidirectionally.

Legacy Cisco pxGrid implementations supported Extensible Messaging and Presence Protocol (XMPP) for communication. The current Cisco pxGrid implementations use the REST and Websocket protocols.

TIP   XMPP-based pxGrid implementations suffered from numerous limitations when it came to application-to-application messaging. For instance, the publish-subscribe (pubsub) model required modification of all XML messages between servers and clients. The server-side implementation of XMPP (otherwise known as the XCP server) was originally created for human chat applications. On the other hand, application messages are very different, since they require large amounts of data to be streamed quickly over a long period of time. REST and Websocket are industry-standards for application-to-application communications and they provide a better support structure for pxGrid implementations. Websocket provides bidirectional data transfer (fast and scalable) and is used for pubsub components. REST provides extensible querying web services that are used for both control and service data. All message bodies in newer pxGrid implementations are formatted in JSON.

Cisco pxGrid provides a unified method of publishing or subscribing to relevant contexts with Cisco platforms that utilize pxGrid for third-party integrations. The following links provide detailed information about pxGrid and the supported integration:

- https://www.cisco.com/c/en/us/products/security/pxgrid.html

- https://developer.cisco.com/docs/pxgrid/#!what-is-pxgrid

Figure 4-21 shows the pxGrid high-level architecture where two pxGrid servers (controllers) communicate with different participating nodes.

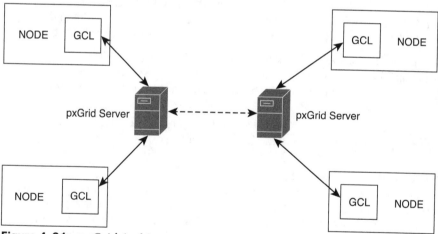

Figure 4-21   *pxGrid Architecture*

In pxGrid, participant nodes do not communicate directly with pxGrid servers. Participant nodes make programmatic calls to the Grid Client Library (GCL), and then the GCL connects and communicates with pxGrid servers. Some deployments may have only a few nodes, while large deployments may have thousands of nodes.

There are two different types of pxGrid clients: a pxGrid service consumer and a pxGrid service provider. Figure 4-22 illustrates the typical pxGrid client flow.

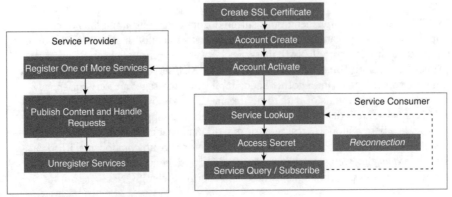

Figure 4-22 *pxGrid Client Flow*

All pxGrid clients need to authenticate using certificate-based SSL authentication or by using usernames and passwords.

> **TIP**   You can generate passwords via the pxGrid Account Create API; however, certificate-based SSL authentication is far more secure and is the recommended way to authenticate pxGrid clients. In Cisco ISE, you can generate and reuse certificates for pxGrid clients.

All pxGrid clients must request to activate their accounts on the pxGrid server (via a REST API). pxGrid clients poll on this REST API call until a "ENABLED" message is received from the server. Service providers use the Register/Unregister Service APIs to provide and update the necessary URLs from which their services are accessible for other pxGrid clients.

All pxGrid clients use the Service Lookup API to dynamically discover all available provider services and their locations. pxGrid clients can then perform REST-based queries (via the Service Query/Subscribe API) or build Websocket connections to receive information.

> **TIP**   Cisco pxGrid sample Java, Python, and Go code can be obtained from the following GitHub repository: https://github.com/cisco-pxgrid/pxgrid-rest-ws.

Cisco ISE Context and Identity Services

Cisco ISE provides the ability to maintain contextual awareness of the "who, what, where, when, and how" of network access. It does this by providing identity, visibility, and policy features. Figure 4-23 shows how the Cisco ISE maintains contextual awareness and offers identity services to firewalls, routers, wireless infrastructure devices, and switches.

The two main parts are of the figure are labelled "Context" and "Identity." Let's separate the two and start with Context. Figure 4-24 shows some of the major Context elements supported by Cisco ISE.

Cisco ISE Profiling Services

Starting with profiling services, this functionality allows you to dynamically detect and classify endpoints connected to the network. Cisco ISE uses MAC addresses as the unique identifier and captures various attributes for each network endpoint that are stored in an internal

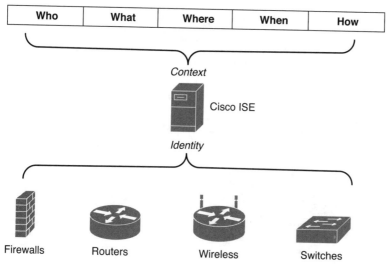

Figure 4-23   *Cisco ISE Contextual Awareness and Identity*

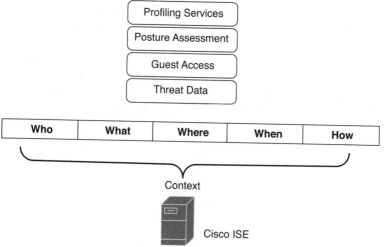

Figure 4-24   *Cisco ISE Context Elements*

endpoint database. This classification process ties the captured attributes to preset and/or user-configurable settings. These attributes and settings are then correlated to an extensive library of profiles. For example, these profiles can be of devices like mobile phones (iPhones and Android phones), tablets (iPads and Android tablets), laptops, Chromebooks, and underlying operating systems (such as Windows, macOS, Linux, iOS, Android, and so on). Cisco ISE can also profile other systems such as printers, cameras, IP Phones, Internet of Things (IoT) devices, and so on.

After the endpoints are classified, they can be authorized to the network and granted access based on their profile. Let's take a look at the example in Figure 4-25. Endpoints that match the IP phone profile can be placed into a voice VLAN (VLAN 10) using MAC Authentication Bypass (MAB) as the authentication method.

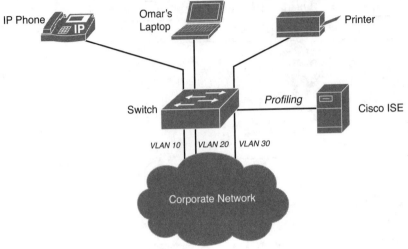

Figure 4-25 *Cisco ISE Profiling*

Users (based on their authentication and authorization) can also be assigned to different VLANs. In Figure 4-25, Omar's laptop is profiled and the user is authenticated and subsequently assigned to VLAN 20. A printer is also assigned to a separate VLAN (VLAN 30) using MAB as the authentication method.

Figure 4-26 provides another example. In this example, Cisco ISE can provide differentiated network access to users based on the device used. For instance, a user can get full access to the network resources when accessing the network from their corporate laptop. However, if they connect with their personal device (an iPhone in this example), they are granted limited network access.

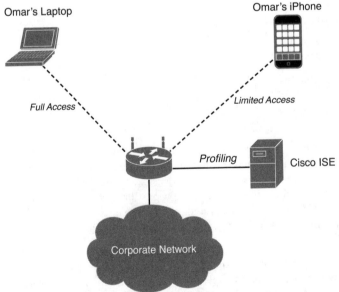

Figure 4-26 *Differentiated Network Access to a User Based on the Device Used*

Figure 4-27 shows the Cisco ISE Profiling Policies screen where numerous profile policies are listed (including different Apple devices and dozens of other devices from different manufacturers).

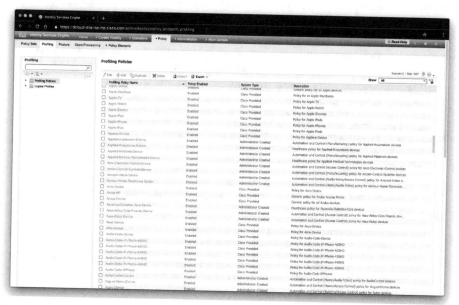

Figure 4-27 *Cisco ISE Profiling Policies*

> **TIP**   The following website provides additional detailed information about the profiling capabilities of Cisco ISE and related design and configuration: https://community.cisco.com/t5/security-documents/ise-profiling-design-guide/ta-p/3739456.

Cisco ISE Identity Services

Identity and authentication information can be gathered in several ways:

- **802.1X:** You learned that 802.1X is an industry standard for authentication and identity management. Cisco ISE supports 802.1X authentication in many different types of network access implementations (wired and wireless access).

- **VPN access with RADIUS authentication:** An example is a Cisco Firepower Threat Defense (FTD) device or Cisco ASA sending user credentials from the VPN client via RADIUS to Cisco ISE for VPN authentication.

- **Cisco ASA identity firewall:** Cisco ASA supports identity firewalling (IDFW) and can use Cisco ISE as the authentication server. IDFW is used to authenticate users before passing traffic through the firewall.

- **Web authentication:** Usually done via a URL redirect of the user's browser. The Cisco ISE has a built-in guest server that provides this web portal service. For instance, let's suppose the user (Omar) in Figure 4-26 connects to the wireless network infrastructure device and that device is configured in open mode. Then Omar's browser is then redirected to the login page hosted by the Cisco ISE, and, subsequently, the Cisco ISE server performs the authentication.

- **MAC Authentication Bypass (MAB):** You already learned that MAB relies on a MAC address for authentication. A MAC address is a globally unique identifier that is assigned to all network-attached devices. Consequently it can be used in authentication. However, since you can spoof a MAC address, MAB is not a strong form of authentication. Cisco ISE Profiling functionality combined with MAB provides you with a better alternative to just using MAB.

> **NOTE**   MAB implemented by itself is not an authentication mechanism. MAB bypasses authentication, which is why it is less secure if not used with profiling.

- **TrustSec Security Group Tags:** Cisco TrustSec is a solution for identity and policy enforcement. ISE can use security group tags (SGTs) for authentication and authorization. SGTs are values that are inserted into the client's data frames by a network device (for example, a switch, firewall, or wireless AP). This tag is then processed by another network device receiving the data frame and used to apply a security policy. For instance, data frames with a *finance_user* tag are allowed to communicate only with devices that have a *finance_net* tag. An IP address can be statically mapped to an SGT. Cisco ISE can gather and distribute all of the IP-to-SGT mapping tables to the network infrastructure devices to enforce policies.

- **Unauthenticated or authenticated guest access:** The Cisco ISE Guest Server functionality provides a guest user a captive portal (splash page) and, optionally, a user agreement page. This captive portal can be configured to ask for information from the user such as their email address, name, company, and other information. Guests that are allowed access to the network without providing identity information are classified as *unauthenticated guest access* users. Typically, you see this type of access in airports, coffee shops, and other places that offer free Internet access. Even though these guest users are not authenticated by Cisco ISE, all their actions and the information they provide are cataloged. In the case of authenticated guest access, users log in using temporary credentials that expire after a set time period. Guests can receive these credentials through text messages (SMS), a printed document, or other means.

Cisco ISE Authorization Rules

Cisco ISE can enforce policies (also known as authorization) after performing authentication. Cisco ISE supports dozens of policy attributes to each policy rule. These policy rules are maintained in a consolidated policy rule table for authorization. The following are examples of some of the more popular policy attributes supported by Cisco ISE:

- Posture assessment results (posture based on attributes collected from the endpoint, such as the version of the operating system, patches installed, applications installed, and more).

- Profiler match for device type.

- Active Directory group membership and Active Directory user-based attributes (such as company name, department, job title, physical address, and so on).

- Time and date.

- Location of the user.

- Access method (MAB, 802.1X, wired, wireless, and so on).

- Mobile Device Management (MDM) registration and enrollment (Cisco ISE supports the integration with third-party MDM solutions, as well).

- Information from digital certificates (digital certificates can be used to determine if the device that is trying to connect to the network is a corporate asset or a personal device).

- Hundreds of RADIUS attributes and values that can be used for both authentication and authorization.

Figure 4-28 shows an example of the Cisco ISE policy sets.

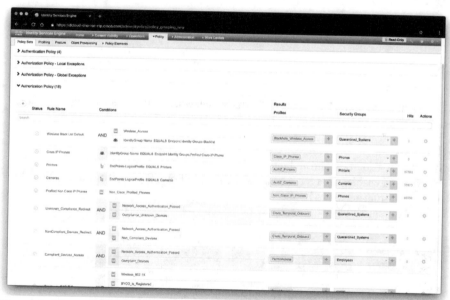

Figure 4-28   *Cisco ISE Policy Sets*

In the example illustrated in this figure, 18 authorization policies have been configured. The Cisco ISE policies are evaluated on a first-match basis (most common) or multiple-match basis. If there are no matches to any of the configured policies, a default "catch-all rule" is applied and enforced.

Figure 4-29 shows a few examples of other authorization policies configured in Cisco ISE. A policy with the rule name "Employees" is configured to allow network access to users connecting over wired and wireless connections using 802.1X authentication. After the

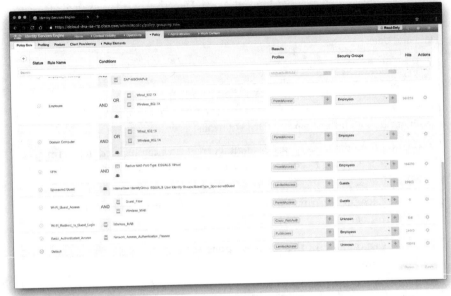

Figure 4-29   *Examples of Cisco ISE Authorization Policies*

"employees" are authenticated, they are granted access to the network. There are 560,219 hits to this policy.

Another example in Figure 4-29 is the VPN rule name, where VPN users are granted access to the network after successful authentication via RADIUS.

Cisco TrustSec

Cisco TrustSec is a solution and architecture that provides the ability to perform network segmentation and enables access controls primarily based on the role of the user (and other attributes) requesting access to the network. Figure 4-30 shows the key components of the Cisco TrustSec architecture.

Figure 4-30   *The Key Components of the Cisco TrustSec Architecture*

Cisco TrustSec uses the roles of supplicant, authentication server, and authenticator, just like 802.1X. All supplicants must join the TrustSec domain prior to sending packets to the network. Figure 4-31 shows the high-level steps of the Cisco TrustSec authentication process.

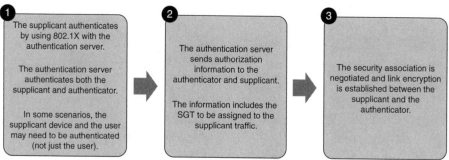

Figure 4-31 *Cisco TrustSec Authentication Process*

Security group tags (SGTs) are embedded within a Layer 2 frame. Figure 4-32 demonstrates how an SGT is embedded within a Layer 2 frame.

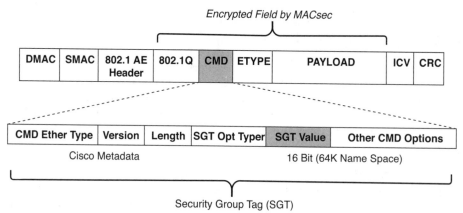

Figure 4-32 *SGT Embedded Within a Layer 2 Frame*

The access control in Cisco TrustSec is postulated by ingress tagging and egress enforcement. In other words, this means that packets are tagged based on their source once they enter the Cisco TrustSec domain. Then access control occurs at the egress point based on the destination.

NOTE   The access decision is based on SGACL implemented at the egress point.

Figure 4-33 demonstrates the TrustSec ingress tagging and egress enforcement.

The following are the steps demonstrated in Figure 4-33.

1. A user (Derek) sends HTTP packets to a web server.

2. The ingress switch to the TrustSec domain (TrustSec authenticator) modifies the packet and adds a source SGT. The SGT ID is 4 and corresponds to the "Engineering" group.

3. The packet is sent through the TrustSec domain and reaches the egress device (or egress point). The egress point checks the SGACL to verify if the Engineering group (4) is authorized to access the web server.

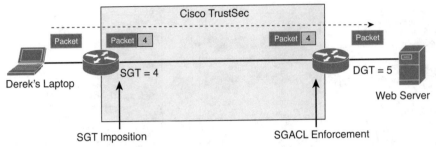

Figure 4-33   *TrustSec Ingress Tagging and Egress Enforcement*

  4. The packet also receives a destination SGT (DGT) with ID 5.

  5. The egress point removes the SGT if the Engineering group (4) is allowed to communicate with the web server and transmits the packet to the destination.

> **NOTE**   Adding the SGT requires the ingress point to be enabled for TrustSec. Most of the latest Cisco network infrastructure devices support TrustSec.

The protocol that allows software-enabled devices to participate in the TrustSec architecture is called the SGT Exchange Protocol (SXP). SXP uses an IP-address-to-SGT process to forward information about the SGT to the first TrustSec-enabled device that is in the path to the final destination. The first TrustSec-enabled device is the device that will insert a "tag" in the packet, which subsequently will reach to the destination (depending on the configured policies).

Posture Assessment

Posture assessment includes a set of rules in a security policy that define a series of checks before an endpoint is granted access to the network. Posture assessment checks include the installation of operating system patches, host-based firewalls, antivirus and anti-malware software, disk encryption, and more. The components illustrated in Figure 4-34 need to be taken into consideration when configuring posture assessment in Cisco ISE deployments.

The elements shown in Figure 4-34 ensure that each required section to configuring Cisco ISE for posture assessment is addressed.

You can also configure posture remediations. These posture remediations are the different methods that AnyConnect clients will use to handle non-compliant endpoints. The AnyConnect Posture Module is required to perform posture checks and remediation.

> **NOTE**   Some posture remediations can be achieved by AnyConnect. On the other hand, other remediations might need to be resolved manually by the end user. An endpoint is deemed compliant if it satisfies all the posture conditions.

You can deploy three types of agents to be used for posture assessment:

  ■ **Temporal Agent:** No permanent software is installed on the endpoint. This is ideal for guest or contractor endpoints. However, a disadvantage of the temporal agent is that it supports a limited number of posture conditions.

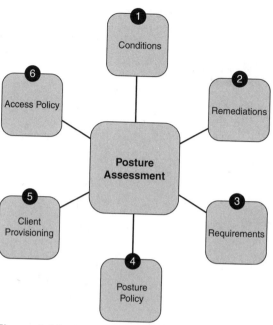

Figure 4-34 *Posture Assessment High-Level Configuration Components*

- **Stealth AnyConnect:** Provides a "headless" configuration and supports most of the posture conditions as the full AnyConnect agent. The Stealth AnyConnect agent will run as a background process, and it does not provide any GUI (unless it is included with other AnyConnect modules such as AMP Enabler and VPN).

- **AnyConnect:** Provides support for most posture conditions, automatic remediation, and passive reassessment.

Change of Authorization (CoA)

RADIUS Change of Authorization (CoA) is a feature that allows a RADIUS server to adjust an active client session. For instance, ISE can issue the CoA RADIUS attribute to an access device to force the session to be reauthenticated.

An example is the use of CoA when the Threat-Centric Network Access Control (TC-NAC) detects a vulnerability. The TC-NAC is a feature that enables ISE to collect threat and vulnerability data from many third-party threat and vulnerability scanners and software. The purpose of this feature is to allow ISE to have a threat and risk view into the hosts it is controlling access rights for.

> **NOTE**   The TC-NAC feature enables you to have visibility into any vulnerable hosts on your network and to take dynamic network quarantine actions when required. ISE can create authorization policies based on vulnerability attributes, such as Common Vulnerability Scoring System (CVSS) scores, received from your third-party threat and vulnerability assessment software. Threat severity levels and vulnerability assessment results can be used to dynamically control the access level of an endpoint or a user.

When a vulnerability event is received for an endpoint, Cisco ISE can automatically trigger a Change of Authorization (CoA) for that endpoint, as demonstrated in Figure 4-35.

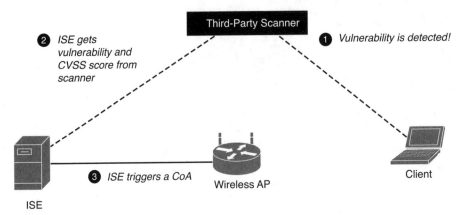

Figure 4-35   *TC-NAC Vulnerability Triggered CoA*

Cisco ISE supports the following the solutions for TC-NAC:

- Cisco Advanced Malware Protection (AMP) for Endpoints

- Cisco Cognitive Threat Analytics (CTA)

- Qualys

- Rapid7 Nexpose

- Tenable Security Center

NOTE   CoA is not triggered automatically in the Cisco ISE when a threat event (not a vulnerability event) is received and must be done manually.

CoA is defined in RFC 5176. There are a few RADIUS-related terms you need to be familiar with before we look at the details of the CoA messages:

- The device providing access to the network is the **Network Access Server (NAS).**

- The entity originating the CoA Requests or Disconnect Requests is the **Dynamic Authorization Client (DAC).** The DAC can be a "co-resident" element within a RADIUS server; however, in some cases that is not the case.

- The entity receiving CoA-Request or Disconnect-Request packets is the **Dynamic Authorization Server (DAS).** The DAS may be a NAS or a RADIUS proxy.

Figure 4-36 shows that the DAC sends a CoA-Request over UDP port 3799 to a NAS (wireless device in this example). The wireless AP (NAS) responds to this with a CoA-ACK if it's able to successfully change the authorizations for the user session. On the contrary, the NAS replies with a CoA-NAK if the CoA-Request is unsuccessful. In this example, the DAC is a "co-resident" element within the RADIUS server.

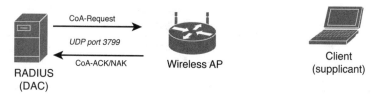

Figure 4-36   *CoA-Request from DAC to NAS*

Figure 4-37 shows the CoA packet format. A CoA-Request can include the Filter-ID or the NAS-Filter-Rule attributes. The Filter-ID is the name of a data filter list for the session that the identification attributes are mapped to. The NAS-Filter-Rule is the actual filter list to be applied to the session.

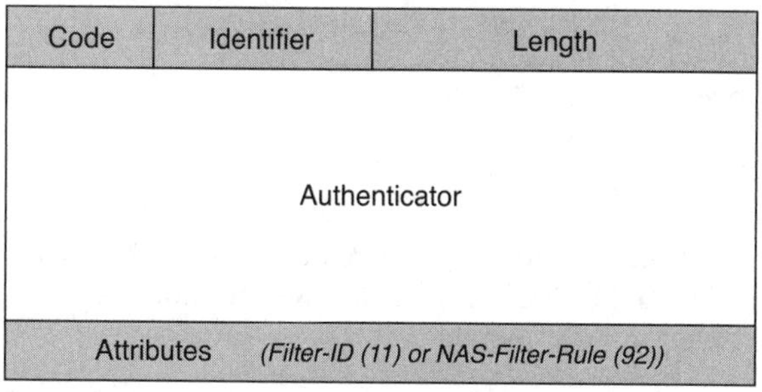

Code	Identifier	Length
	Authenticator	
Attributes	*(Filter-ID (11) or NAS-Filter-Rule (92))*	

Figure 4-37   *The CoA-Request Packet Format*

A DAC can also send a Disconnect-Request packet in order to terminate a user session on a NAS and discard all associated session context. Disconnect-Request packets are also sent over UDP port 3799.

If all of the session's context is discarded and the user sessions are no longer connected, the NAS replies with a Disconnect-ACK, as shown in Figure 4-38.

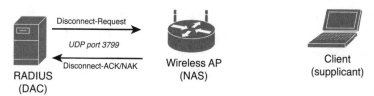

Figure 4-38   *Disconnect-Request Sent from a DAC to a NAS*

If the NAS is not able to disconnect one or more sessions or discard the associated session context, it will reply with a Disconnect-NAK packet.

> **TIP**   A CoA-Request is a RADIUS code 43 packet. A Disconnect-Request is a RADIUS code 40 packet.

Let's take a look at another example of a CoA implementation. In the example illustrated in Figure 4-39, a Cisco Next-Generation Firewall is configured for remote access VPN termination. CoA is also configured in this deployment. CoA allows the RADIUS server (Cisco ISE)

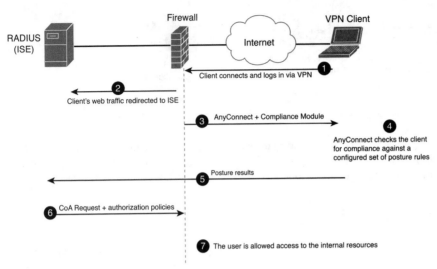

Figure 4-39   *CoA in VPN implementations*

to verify the posture of VPN users without the need for an Inline Posture Node (IPN). This posture assessment is natively performed with the Cisco AnyConnect Secure Mobility Client with the Compliance Module enabled.

The following steps are illustrated in Figure 4-39:

1. The VPN client connects to the firewall and logs in.

2. The firewall redirects web traffic to Cisco ISE.

3. The user is provisioned with AnyConnect and its Compliance Module.

4. AnyConnect checks the client for compliance against a configured set of posture rules (for example, operating system patches, antivirus and anti-malware software installed, services, application, registry entries, and so on).

5. After the posture validation is completed, the results are sent to the Cisco ISE.

6. If the client system is complaint, Cisco ISE sends a RADIUS CoA to the ASA with the new set of authorization policies.

7. After successful posture validation and CoA, the user is allowed access to the internal resources.

Configuring TACACS+ Access

Each of the concentration exams will focus on configuration and troubleshooting; however, in this section you will learn the concepts of TACACS+ access configurations in infrastructure devices such as routers and switches running Cisco IOS and Cisco IOS-XE software.

Let's suppose you are hired by a fictitious company called SecretCorp (secretcorp.org) to configure administrative access of infrastructure devices using TACACS+ authentication. You are being tasked to configure the routers (R1, R2, and R3) shown in Figure 4-40.

The goal is to configure TACACS+ authentication for Secure Shell (SSH) sessions. Let's start with Router 1 (R1); the rest will have basically the same configuration. The router will be

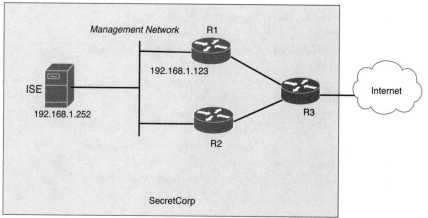

Figure 4-40 *SecretCorp's TACACS+ Authentication*

configured to communicate with Cisco ISE configured as a TACACS+ server. Authenticated users need to be authorized to have access to a command-line interface (EXEC) session, including the privilege level they should be placed into. The authorization check should be done by the router referring to the TACACS+ server. Example 4-1 shows the configuration of R1. The details about the commands entered are included in comments, which are preceded by exclamation marks (!).

Example 4-1 *AAA Router Configuration for TACACS+ Authentication*

```
! This command enables the configuration of the rest of the AAA
! If it is in the configuration, it doesn't need to be put in again.
! On most IOS systems, the default has aaa new-model disabled.


R1(config)# aaa new-model


! This authentication method list, when applied to a line such as the VTY
! lines will tell the router to prompt the user who is accessing that line
! for a username and password in order for that user to login.
! When the user supplies the username and password at the login prompt
! the router will send the credentials to a configured TACACS+ server
! and then the server can reply with a pass or fail message.
! This command indicates "group tacacs+" as the first method
! as there could be more than one server configured. If the AAA server does
! not respond after a short timeout the router will then try the
! second method in the method list which is "local" which means the
! router will then check the running configuration to see if there
! is a username and a matching password.


R1(config)# aaa authentication login AUTHEN_via_TACACS group tacacs+ local


! This next authorization method list, when applied to a line, will cause
! the router to check with the AAA server to verify that the user
```

```
! is authorized to gain access to the CLI. The CLI represents an
! Exec Shell. Not only can the ISE indicate to the router whether
! or not the user is authorized, but it can also indicate what privilege
! level the user is placed into.
! Both the username and password will need to be created on the ISE server
! for the previous authentication method, and the authorization
! for a CLI will also need to be configured on that same ISE server.
! This authorization list will use one or more configured ISE servers
! via TACACS+, and if there are no servers that respond, then the router
! will check locally regarding whether the command is authorized for this
! user based on privilege level of the user, and privilege level of the
! command being attempted.

R1(config)# aaa authorization exec Author-Exec_via_TACACS group tacacs+ local

! It is important to note that before we apply either of these method lists
! to the VTY lines, we should create at least one local user as a backup
! in the event the ISE server is unreachable, or not yet configured.
! In the example below it will create a user on the local database of the
! router including a username, password as well as a privilege level
! for that user. It is highly recommended that you use strong passwords
! when configuring any user or device credentials.

R1(config)# username admin privilege 15 secret supersecretpassword

! Next we need to create a least one ISE server that the router should try
! to use via TACACS+. This is the equivalent of creating a server group
! of one.
! The password is used as part of the encryption of the packets, and
! whatever password we configure here, we also need to configure on
! the ISE server.

R1(config)# tacacs-server host 192.168.1.252 key thisisapassword

! Verifying that the IP addresses reachable is a test that can be done
! even before the full ISE configuration is complete on the AAA server

R1(config)# do ping 192.168.1.252

Type escape sequence to abort.
Sending 5, 100-byte ICMP Echos to 192.168.1.252, timeout is 2 seconds:
!!!!!
Success rate is 100 percent (5/5), round-trip min/avg/max = 8/13/28 ms
```

```
! Next, for the authentication method list and authorization method list
! to be used we would need to apply them. In the example below
! we are applying both method lists to the first five VTY lines.


R1(config)# line vty 0 4
R1(config-line)# authorization exec Author-Exec_via_TACACS
R1(config-line)# login authentication AUTHEN_via_TACACS
! users connecting to these vty lines will now be subject to both
! authentication and authorization, based on the lists that are
! applied to these lines
```

With the authentication and authorization method lists created and applied, you could attempt to log in through one of the five vty lines, and here is what you would expect: You should be prompted for a username and password, the router should not be able to successfully contact the ISE server (because you have not configured the ISE part of it yet on that server), and then, after a short timeout, the router would use the second method in each of its lists, which indicates to use the local database for the authentication and the authorization. Because you do have a local user with a password and a privilege level assigned to that user, it should work.

By enabling a debug, and attempting to log in, you can see exactly what is happening, as shown in Example 4-2. If you are not connected to the device via the serial console, use the **terminal monitor** command to be able to see the debug messages in your screen.

Example 4-2   *Debugging TACACS+ in the Router*

```
R1# debug tacacs
TACACS access control debugging is on
TPLUS: Queuing AAA Authentication request 102 for processing
TPLUS: processing authentication start request id 102
TPLUS: Authentication start packet created for 102()
TPLUS: Using server 192.168.1.252
TPLUS(00000066)/0/NB_WAIT/6812DC64: Started 5 sec timeout


User Access Verification

! Timing out on TACACS+ regarding authentication because no server is responding
TPLUS(00000066)/0/NB_WAIT/6812DC64: timed out
TPLUS(00000066)/0/NB_WAIT/6812DC64: timed out, clean up
TPLUS(00000066)/0/6812DC64: Processing the reply packet


! Now moving to the local database on the router
Username: admin
Password: supersecretpassword
```

```
! Timing out on TACACS+ regarding authorization due to no server responding.
TPLUS: Queuing AAA Authorization request 102 for processing
TPLUS: processing authorization request id 102
TPLUS: Protocol set to None .....Skipping
TPLUS: Sending AV service=shell
TPLUS: Sending AV cmd*
TPLUS: Authorization request created for 102(admin)
TPLUS: Using server 192.168.1.252
TPLUS(00000066)/0/NB_WAIT/6812DC64: Started 5 sec timeout
TPLUS(00000066)/0/NB_WAIT/6812DC64: timed out
TPLUS(00000066)/0/NB_WAIT/6812DC64: timed out, clean up
TPLUS(00000066)/0/6812DC64: Processing the reply packet
! After timing out, the router again uses its local database for
! authorization and appropriate privilege level for the user.

! If we exit, and change the debugs slightly, and do it again, it will give
! us yet another perspective.

R1# debug aaa authentication
AAA Authentication debugging is on
R1# debug aaa authorization
AAA Authorization debugging is on
AAA/BIND(00000067): Bind i/f
! Notice it shows using the authentication list we assigned to the vty
! lines
AAA/AUTHEN/LOGIN (00000067): Pick method list 'AUTHEN_via_TACACS'

! Not shown here, but indeed the ISE server is timing out, due to not yet
! being configured, which causes the second entry in the list "local" to
! be used.

User Access Verification
Username: admin
Password: supersecretpassword

! Now the authorization begins, using the method list we configured for
! the vty lines
AAA/AUTHOR (0x67): Pick method list 'Author-Exec_via_TACACS'
AAA/AUTHOR/EXEC(00000067): processing AV cmd=
AAA/AUTHOR/EXEC(00000067): processing AV priv-lvl=15
AAA/AUTHOR/EXEC(00000067): Authorization successful
R1#
```

NOTE The 300-715 SISE exam (Implementing and Configuring Cisco Identity Services Engine [SISE]) focuses on the configuration and troubleshooting of Cisco ISE. However, the following are a few examples of the Cisco ISE configuration for TACACS+ access.

To configure TACACS+ in Cisco ISE, navigate to **Work Centers > Device Administration > Network Resources** and add a network device. You will see the screen in Figure 4-41.

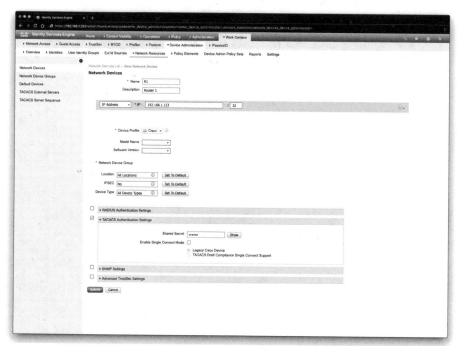

Figure 4-41 *Adding a Network Device for TACACS+ Authentication in ISE*

In Figure 4-41, the Router (R1) details are entered. The **TACACS Authentication Settings** checkbox is selected, and a shared secret used to authenticate the TACACS+ session between the router and ISE is entered.

> **NOTE**   The share secret (password) must match the password entered in the router's configuration.

You can create different policies to support different groups of people who require access to the organization's infrastructure devices, as shown in Figure 4-42 (network administrators, network operators, security administrators, help desk support, and so on).

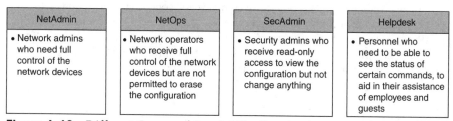

NetAdmin	NetOps	SecAdmin	Helpdesk
• Network admins who need full control of the network devices	• Network operators who receive full control of the network devices but are not permitted to erase the configuration	• Security admins who receive read-only access to view the configuration but not change anything	• Personnel who need to be able to see the status of certain commands, to aid in their assistance of employees and guests

Figure 4-42 *Different Groups of People Who Require Access to Infrastructure Devices*

To configure these groups and policies, navigate to **Work Centers > Device Administration > Policy Elements > Results > TACACS Profiles.** The screen in Figure 4-43 shows the TACACS+ profile of a NetAdmin.

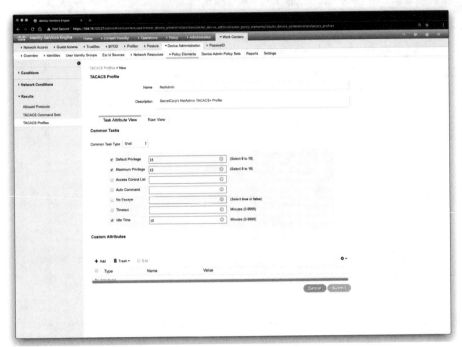

Figure 4-43   *NetAdmin Profile*

Configuring RADIUS Authentication

You can configure RADIUS authentication in multiple scenarios, including Remote Access VPN, Secure Network Access, 802.1X, and more. Chapter 8, "Virtual Private Networks (VPNs)," provides examples of remote access VPN configurations using RADIUS authentication. In the following section, you will learn how RADIUS can be configured in network switches and Cisco ISE for secure access with 802.1X authentication.

Cisco IOS 15.2.x and Cisco IOS-XE 3.6.x switches follow the Cisco Common Classification Policy Language (C3PL) style of configuration. C3PL is a structured replacement for the configuration commands of various features in Cisco IOS and Cisco IOS-XE. C3PL allows administrators to configure traffic policies based on events, conditions, and actions. This provides some intriguing and advanced authentication features, as well as a very different configuration style that has powerful options, but it can be confusing when learning how to use it. However, many administrators who start to use this configuration style end up loving it and rarely want to go back to the classic methods of configuration.

TIP   The default behavior of 802.1X is to deny access to the network when an authentication fails. In many of the early 802.1X deployments, this was a problem because it does not allow for guest access and does not allow employees to remediate their computer systems and gain full network access. The next phase in handling 802.1X authentication failures was to provide an "Auth-Fail VLAN" to allow a device/user that failed authentication to be granted access to a VLAN that provided limited resources. This was a step in the right direction, but it was still missing some practicality, especially in environments that must use MAC Authentication Bypass (MAB) for all the printers and other non-authenticating devices. With the default behavior of 802.1X, an administrator has to configure ports for printers and other devices that do not have supplicants differently from the ports where they plan to do authentication. In response to these issues, Cisco created Flexible Authentication (Flex-Auth). Flex-Auth enables a network administrator to set an authentication order and priority on the switch port, thereby allowing the port to attempt, in order, 802.1X, MAB, and then WebAuth. All of these functions are provided while maintaining the same configuration on all access ports, thereby providing a much simpler operational model for customers than is provided by traditional 802.1X deployments.

There are multiple methods of authentication on a switch port:

- 802.1X (dot1x)

- MAB

- WebAuth

With 802.1X authentication, the switch sends an identity request (EAP-Identity-Request) periodically after the link state has changed to up. Additionally, the endpoint supplicant should send a periodic EAP over LAN Start (EAPoL-Start) message into the switch port to speed up authentication. If a device is not able to authenticate, it merely waits until the dot1x timeout occurs, and then MAB occurs. Assuming the device MAC address is in the correct database, it is then authorized to access the network.

The default behavior of an 802.1X-enabled port is to authorize only a single MAC address per port. There are other options, most notably Multi-Domain Authentication (MDA) and Multiple Authentication (Multi-Auth) modes. During the initial phases of any Cisco TrustSec deployment, it is best practice to use Multi-Auth mode to ensure that there is no denial of service while deploying 802.1X.

TIP   Port Security is not compatible with 802.1X, because 802.1X handles this function natively. You will learn more about Port Security in Chapter 6, "Infrastructure Security."

Multi-Auth mode allows virtually unlimited MAC addresses per switch port, and requires an authenticated session for every MAC address. When the deployment moves into the late stages of the authenticated phase, or into the enforcement phase, it is then recommended that you use MDA mode, which allows a single MAC address in the Data domain and a single MAC address in the Voice domain per port.

802.1X is designed to clearly differentiate a successful authentication from an unsuccessful authentication. Successful authentication means the user is authorized to access the network.

Unsuccessful authentication means the user has no access to the network. This is problematic in a lot of environments. Most modern environments need to do workstation imaging with Preboot Execution Environments (PXEs), or they don't have any way to run a supplicant because they may have some thin clients that do not support it. When early adopters of 802.1X deployed authentication companywide, there were repercussions. Many issues arose. For instance, supplicants were misconfigured; there were unknown devices that could not authenticate because of a lack of supplicant, and other reasons.

> **TIP** Cisco created Open Authentication to aid with deployments. Open Authentication allows all traffic to flow through the switch port, even without the port being authorized. This feature permits authentication to be configured across the entire organization, but does not deny access to any device.

Several devices that support the Cisco Common Classification Policy Language (C3PL) style of configuration still accept the old style of commands. The legacy style of commands is the default in most of those platforms, and you must enable the C3PL style of commands with the global configuration command **authentication display new-style**. Even if the name of the command includes the word "display," the command changes much more than just the display of the commands. It also changes the way the network administrator interacts with the switch and configures the device. You can change back to the classic model using the **authentication display legacy** command.

> **TIP** After you start configuring the C3PL policies, you cannot revert to the legacy mode. You can only switch back if you haven't configured C3PL yet. That is, unless you erase the switch configuration and reload or restore an older backup configuration.

The C3PL syntax offers many benefits, most of which are transparent to the end user. For instance, C3PL allows the network device configuration to exist in memory once and be invoked multiple times. This is a resource efficient enhancement.

There are several additional benefits from the C3PL model. For example, 802.1X and MAB can run simultaneously without having to sequence the two distinctive authentication processes, whereas 802.1X authentication has to be failed for MAB to start when not using the C3PL model. You can also use service templates to control preconfigured access control lists on given interfaces in the event of RADIUS not being available.

In legacy devices, the sequencing of 802.1X and MAB can result in certain MAB endpoints not being able to obtain IP addresses via DHCP in a timely manner. Newer devices can process 802.1X and MAB simultaneously, allowing endpoints to obtain a DHCP-assigned IP address in a timely manner. Additionally, legacy devices require a static ACL often applied to interfaces in order to restrict network access for devices that have not yet authenticated. Consequently the ACL remains applied to devices attempting to connect while the RADIUS server is unavailable. This condition results in a denial of service until the RADIUS server is reachable. This may seem desirable in theory, but it is not recommended. This behavior actually makes life more difficult for the policy server administrator.

The ability to create service templates is a good enhancement of C3PL. A separate ACL that would permit network access can be applied to the interface using service templates. These rules can be configured to perform an action under a certain condition, such as when the RADIUS server is not reachable. This feature is known as the "Critical ACL functionality."

In addition, C3PL provides differentiated authentication. The differentiated authentication feature enables you to authenticate different methods with different servers. For instance, you can send MAB to one server and 802.1X authentications to another. Another interesting feature in C3PL is Critical MAB. Critical MAB allows the switch to use a locally defined list of MAC addresses when the centralized RADIUS server is unavailable.

Configuring 802.1X Authentication

Let's take a look at the topology shown in Figure 4-44. You are being hired to configure 802.1X in SecretCorp. The goal is to deploy 802.1X authentication in all of SecretCorp's switches and use ISE. SecretCorp's switch 1 (sc-sw1) is used in this example.

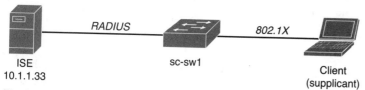

Figure 4-44 *SecretCorp 802.1X Deployment*

First, you need to configure certificates for URL redirection. To configure certificates for URL redirection, perform the following steps from global configuration mode on the switch (sc-sw1):

Step 1.   Configure the DNS domain name on the switch. The domain name is **secretcorp.org.**

```
sc-sw1(config)# ip domain-name secretcorp.org
```

NOTE   Cisco IOS does not allow for certificates, or even self-generated keys, to be created and installed without first defining a DNS domain name on the device.

Step 2.   Generate self-signed keys to be used for HTTPS. The following command generates a general-usage 2048-bit RSA key pair:

```
sc-sw1(config)# crypto key generate rsa general-keys mod
2048
```

Step 3.   Enable the HTTP server and configure HTTP Secure server in global configuration mode. Always use HTTPS instead of HTTP.

```
sc-sw1(config)# ip http server
sc-sw1(config)# ip http secure-server
```

TIP   In many cases, organizations require that this redirection process using the switch's internal HTTP server is decoupled from the management of the switch itself. If you are not using HTTP for management, then decoupling the HTTP server is highly recommended. This is done by following the next two commands:

```
sc-sw1(config)# ip http active-session-modules none
sc-sw1(config)# ip http secure-active-session-modules none
```

Step 4.   Enable the C3PL configuration style within privileged EXEC mode:

```
sc-sw1# authentication display new-style
```

Step 5.   After you have changed and enabled the C3PL configuration style, enable the AAA subsystem:

```
sc-sw1(config)# aaa new-model
```

Step 6.   Make sure you use a common **session-id** for the AAA network security services:

```
sc-sw1(config)# aaa session-id common
```

Step 7.   An authentication method is required to make sure the switch uses a particular group of RADIUS servers for 802.1X authentication requests. Create an authentication method for 802.1X:

```
sc-sw1(config)# aaa authentication dot1x default group
radius
```

Step 8.   Previously in this chapter you learned that authorization is what defines that the user or device is actually allowed to access the network, and what level of access is actually permitted. Create an authorization method for 802.1X:

```
sc-sw1(config)# aaa authorization network default group
radius
```

Step 9.   Earlier in this chapter you also learned about RADIUS accounting. Accounting packets ensure that the RADIUS server knows the exact state of the switch port and endpoint. Accounting provides information on when a session has been terminated, as well as local decisions made by the switch, such as authentication failure (AuthFail) VLAN assignments. Create an accounting method for 802.1X:

```
sc-sw1(config)# aaa accounting dot1x default start-stop
group radius
```

Step 10.   You can configure periodic RADIUS accounting packets to allow the RADIUS server (Cisco ISE) to track which sessions are still active on the network. In the following example, periodic updates are configured to be sent whenever there is new information, as well as a periodic update once every 24 hours (1440 minutes):

```
sc-sw1(config)# aaa accounting update newinfo periodic 1440
```

Step 11.   You can configure a proactive method to check the availability of the RADIUS server. When configured this way, the switch sends periodic test authentication messages to Cisco ISE (the RADIUS server in this case) and waits for a RADIUS response from the server. Within global configuration mode, you can add a username and password for the RADIUS keepalive. These credentials will also need to be added to the local user database in the RADIUS server (that is, Cisco ISE).

```
sc-sw1(config)# username radius-test password
this_is_a_password
```

Step 12.   Add the Cisco ISE as a RADIUS server. First, you create an object for the RADIUS server and then apply configuration to that object:

```
sc-sw1(config)# radius server 10.1.1.33

sc-sw1(config-radius-server)# address ipv4 address
auth-port 1812

acct-port 1813

sc-sw1(config-radius-server)# key this_is_a_key

sc-sw1(config-radius-server)# automate-tester username
radius-test probe-on
```

Step 13.   Now that you have configured the switch to proactively interrogate the Cisco ISE server for RADIUS responses, configure the counters on the switch to determine if the server is alive or dead. In the following example, the switch will wait 5 seconds for a response from the RADIUS server and then attempt the test three times before marking the server dead:

```
sc-sw1(config)# radius-server dead-criteria time 5
tries 3

sc-sw1(config)# radius-server deadtime 15
```

Step 14.   Earlier in this chapter you learned the details about Change of Authorization (CoA). Also in previous steps you defined the IP address of a RADIUS server that the switch will send RADIUS messages to. On the other hand, you need to also define the servers that are allowed to perform Change of Authorization (RFC 3576) operations within global configuration mode, as shown here:

```
sc-sw1(config)# aaa server radius dynamic-author

sc-sw1(config-locsvr-da-radius)# client 10.1.1.33
server-key this_is_a_shared_secret
```

Step 15.   You can configure the switch to send any defined vendor-specific attributes (VSA) to Cisco ISE during authentication requests and accounting updates:

```
sc-sw1(config)# radius-server vsa send authentication

sc-sw1(config)# radius-server vsa send accounting
```

Step 16.   Enable the VSAs:

```
sc-sw1(config)# radius-server attribute 6
on-for-login-auth

sc-sw1(config)# radius-server attribute 8
include-in-access-req

sc-sw1(config)# radius-server attribute 25 access-request
include

sc-sw1(config)# radius-server attribute 31 mac format
ietf upper-case

sc-sw1(config)# radius-server attribute 31 send nas-
port-detail mac-only
```

Step 17. Make sure that you configure the switch to always send traffic from the correct interface. Switches may often have multiple IP addresses associated to them. You should always force any management communications to occur through a specific interface. It is important also to know that the IP address of this interface must match the IP address defined in the Cisco ISE Network Device object. In the following example, the source interface is GigabitEthernet1/8:

```
sc-sw1 (config)# ip radius source-interface
GigabitEthernet1/8

sc-sw1 (config)# snmp-server trap-source
GigabitEthernet1/8

sc-sw1 (config)# snmp-server source-interface informs
GigabitEthernet1/8
```

Step 18. When you configure 802.1X, you configure the ports that will be accepting connections from any device trying to connect to the network in "monitor mode." Add the following ACL to be used on switch ports in monitor mode:

```
sc-sw1(config)# ip access-list extended MONITOR_MODE
sc-sw1(config-ext-nacl)# permit ip any any
```

Step 19. You typically need to allow DHCP, DNS, and PXE/TFTP communication in many different environments. These are classified as "low-impact." Add the following ACL to be used on switch ports in Low-Impact Mode:

```
sc-sw1(config)# ip access-list extended LOW_IMPACT
sc-sw1(config-ext-nacl)# remark DHCP
sc-sw1(config-ext-nacl)# permit udp any eq bootpc any eq
bootps
sc-sw1(config-ext-nacl)# remark DNS
sc-sw1(config-ext-nacl)# permit udp any any eq domain
sc-sw1(config-ext-nacl)# remark PXE / TFTP
sc-sw1(config-ext-nacl)# permit udp any any eq tftp
```

TIP   Keep in mind that there is a "default deny" at the end of any ACL.

Step 20. The following ACL can be used for URL redirection with web authentication:

```
sc-sw1(config)# ip access-list extended WEBAUTH-REDIRECT
sc-sw1(config-ext-nacl)# deny udp any any eq 53
sc-sw1(config-ext-nacl)# permit tcp any any eq 80
sc-sw1(config-ext-nacl)# permit tcp any any eq 443
```

Step 21. The Posture Agent used in the Cisco TrustSec solution is the Cisco AnyConnect Mobility Client. Add the following ACL to be used for URL redirection with the Posture Agent. This ACL only allows HTTP and HTTPS packets to be redirected.

```
sc-sw1(config)# ip access-list extended AGENT-REDIRECT
sc-sw1(config-ext-nacl)# permit tcp any any eq 80
sc-sw1(config-ext-nacl)# permit tcp any any eq 443
```

Step 22. Cisco ISE supports authorization profiles for devices joining the network. Cisco switches support something similar to an ISE authorization profile, but they are configured locally on the switch. Service templates are basically a list of VLANs, Timer, Named ACL, and URL Redirect strings that can be applied based on different criteria. These are also similar to downloadable ACLs (dACLs).

> **TIP** Per-user downloadable ACLs can be configured to provide different levels of network access and service to an 802.1X-authenticated user. When Cisco ISE (RADIUS server) authenticates a user connected to an 802.1X port, it retrieves the ACL attributes based on the user identity and sends them to the switch. Subsequently, the switch uses those attributes and enforces the policy on the 802.1X-enabled port for the duration of the user session. The downloadable (per-user) ACL is removed when the session is over. This per-user ACL is also removed if authentication fails or if a link-down condition occurs. The switch does not save RADIUS-specified ACLs in the running configuration. When the port is unauthorized, the switch removes the ACL from the port.

Similar to per-user ACLs, service templates can be centrally located on ISE and downloaded during authorization. On the other hand, you can configure a service template local to the switch to apply when none of the configured RADIUS servers (ISE PSNs) are reachable to process 802.1X or MAB requests (known as the critical-auth state). You can add the following service template, named CRITICAL_AUTH, to be used when no RADIUS servers are available, also known as the critical-auth state:

```
sc-sw1(config)# service-template CRITICAL_AUTH
sc-sw1(config-service-template)# access-group ACL-ALLOW
```

Step 23. Now let's configure 802.1X from global configuration mode:

```
sc-sw1(config)# dot1x system-auth-control
```

Please note that when you enable 802.1X globally, the switch does not actually enable authentication on any of the switch ports. You will enable that on each port at a later time.

Step 24. Next, enable the dACLs. You need to enable IP device tracking globally in order for dACLs to function properly:

```
sc-sw1(config)# ip device tracking
```

Step 25. Configure a control class. A control class defines the conditions under which the actions of a control policy are executed when none of the RADIUS servers are available (otherwise known as the critical-auth state).

```
sc-sw1(config)# class-map type control subscriber match-any RADIUS-SERVER-DOWN
sc-sw1(config-filter-control-classmap)# match result-type aaa-timeout
```

Step 26. The following command can be used to configure a control class for when 802.1X authentication fails for the session:

```
sc-sw1(config)# class-map type control subscriber match-
all DOT1X-FAILED
sc-sw1(config-filter-control-classmap)# match method
dot1x
sc-sw1(config-filter-control-classmap)# match result-type
method dot1x authoritative
```

Step 27. Control policies can be implemented to specify which actions should be taken in response to the specified events. Control policies include one or more rules that associate a control class with one or more actions. Configure a control policy that will be applied to all interfaces that are configured for 802.1X or MAB.

```
sc-sw1(config)# policy-map type control subscriber
DOT1X-DEFAULT
```

NOTE   You can enable 802.1X and MAB to run at the same time. You can do that by assigning a higher priority for 802.1X over MAB. On the other hand, this configuration is not recommended for production environments.

Step 28. Configure actions for policy violations:

```
sc-sw1(config-action-control-policymap)# event violation
match-all
sc-sw1(config-class-control-policymap)# 10 class always
do-all
sc-sw1(config-action-control-policymap)# 10 restrict
```

Step 29. Configure the switch to authenticate the endpoint trying to join the network using 802.1X. When 802.1X is enabled, authentication will be performed when a supplicant is detected on the endpoint:

```
sc-sw1(config-action-control-policymap)# event agent-
found match-all
sc-sw1(config-class-control-policymap)# 10 class always
do-all
sc-sw1(config-action-control-policymap)# 10 authenticate
using dot1x
```

Step 30. You can also configure the action the switch will take when 802.1X authentication fails, or when the RADIUS server is down:

```
sc-sw1(config-action-control-policymap)# event
authentication-failure match-all
sc-sw1(config-class-control-policymap)# 10 class RADIUS-
SERVER-DOWN do-all
sc-sw1(config-action-control-policymap)# 10 authorize
```

```
sc-sw1(config-action-control-policymap)# 20 activate
service-template CRITICAL_AUTH

sc-sw1(config-action-control-policymap)# 30 terminate dot1x

sc-sw1(config-action-control-policymap)# 40 terminate mab

sc-sw1(config-action-control-policymap)# 20 class DOT1X-
FAILED do-all

sc-sw1(config-action-control-policymap)# 10 authenticate
using mab
```

Step 31.   Now you can apply the control policy to the access-layer interfaces with the **service-policy** command. Please note that not all aspects of the 802.1X configuration are completed in C3PL. Consequently, this is why some configuration items will occur at the interfaces separately, as shown here:

```
sc-sw1(config)# interface range GigabitEthernet 1/0/1 - 10

sc-sw1(config-if-range)# description 802.x1 Enabled Ports

sc-sw1(config-if-range)# switchport host

sc-sw1(config-if-range)# service-policy type control
subscriber DOT1X-DEFAULT
```

The preceding configuration is applied to a range of ports (from GigabitEthernet 1/0/1 through GigabitEthernet 1/0/10).

Step 32.   Finally, you can complete the interface configuration parameters:

```
sc-sw1(config-if-range)# authentication periodic

sc-sw1(config-if-range)# authentication timer
reauthenticate server

sc-sw1(config-if-range)# mab

sc-sw1(config-if-range)# ip access-group DEFAULT-ACL in

sc-sw1(config-if-range)# access-session host-mode
multi-auth

sc-sw1(config-if-range)# no access-session closed

sc-sw1(config-if-range)# dot1x timeout tx-period 10

sc-sw1(config-if-range)# access-session port-control auto

sc-sw1(config-if-range)# no shutdown
```

Once again, the SCOR exam does not focus on the detailed configuration and implementation. The CCNP concentration exams and the CCIE Security lab require you to have hands-on experience with these configurations.

Additional Cisco ISE Design Tips

Many organizations deploy Cisco ISE in a distributed fashion. Deployment models can be distributed, standalone, or high availability in standalone mode.

Figure 4-45 illustrates a Cisco ISE deployed in standalone mode.

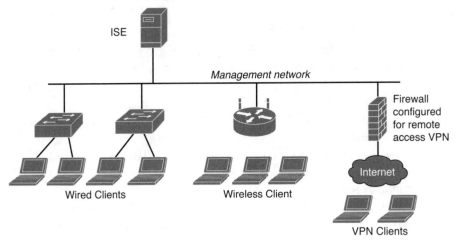

Figure 4-45 *Cisco ISE Standalone Deployment*

Figure 4-46 shows a topology where two Cisco ISE nodes are deployed in high-availability standalone mode. High-availability standalone mode is basically two Cisco ISE nodes configured in an active-standby backup mode.

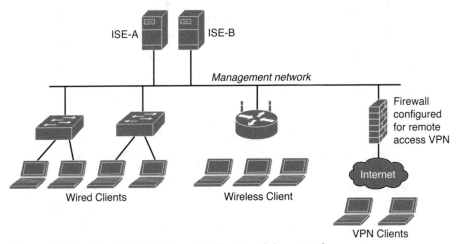

Figure 4-46 *Cisco ISE High-Availability Standalone Deployment*

When a Cisco ISE secondary node is registered, a data replication channel is created from the primary to the secondary node and the replication process starts automatically. Replication helps provide consistency among the configurations in all Cisco ISE nodes by sharing configuration data from the primary to the secondary nodes.

You can also deploy Cisco ISE in a scalable and distributed mode. Figure 4-47 explains the different types of Cisco ISE nodes that can be present in a distributed configuration model.

Figure 4-48 demonstrates that a Primary Administration Node (PAN) and a Secondary Administration Node (SAN) are deployed to control several Policy Service Nodes (PSNs)

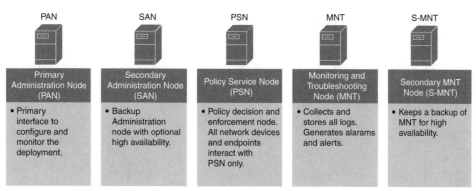

Figure 4-47   *Cisco ISE Distributed Mode Node Types*

across two different PSN Node Groups (PSN Node Group 1 and PSN Node Group 2). These groups can be deployed in different geographical locations within an organization. Monitoring and Troubleshooting (MNT) and Secondary MNT (S-MNT) nodes are also deployed to collect and store logs, as well as generating any alarms and alerts.

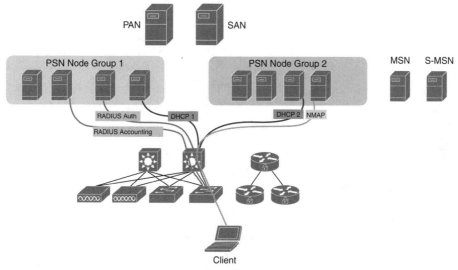

Figure 4-48   *Cisco ISE Distributed Mode Example Topology*

When you first register a Cisco ISE node as a secondary node, full replication starts automatically. Then incremental replication is performed on a periodic basis to ensure that any new changes, such as additions, modifications, or deletions to the configuration data in the PAN are synchronized to the secondary nodes.

TIP   As a best practice, you should always configure the Network Time Protocol (NTP) and you should also use the same NTP server for all the nodes. This is done to make sure that the logs, alarms, and reports from all nodes in your deployment are always synchronized with timestamps.

Advice on Sizing a Cisco ISE Distributed Deployment

Deployment sizing traditionally has been performed using the "maximum concurrent sessions" in Cisco ISE as the baseline. In addition, when you deploy Cisco ISE in a distributed way, you should consider the authentication attempts per second (auth/sec).

You can also configure a load balancer in front of the Cisco ISE PSNs to scale your deployment, as shown in Figure 4-49.

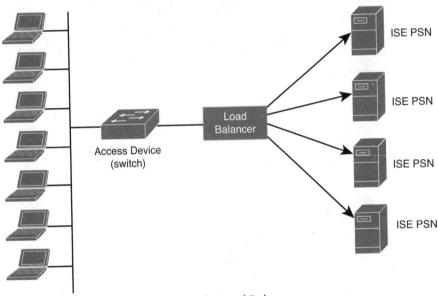

Figure 4-49 *Cisco ISE PSNs Behind a Load Balancer*

Another question that you may ask is, what happens during peak times or when a network device is rebooted? There is a feature called Anomalous Suppression Detection that can be enabled to suppress misbehaving clients as well as repeated successful authentications.

TIP The following whitepaper discusses different recommendations when deploying Cisco ISE in a distributed fashion: https://community.cisco.com/t5/security-documents/ise-performance-amp-scale/ta-p/3642148.

Exam Preparation Tasks

As mentioned in the section "How to Use This Book" in the Introduction, you have a couple of choices for exam preparation: the exercises here, Chapter 12, "Final Preparation," and the exam simulation questions in the Pearson Test Prep Software Online.

Review All Key Topics

Review the most important topics in this chapter, noted with the Key Topic icon in the outer margin of the page. Table 4-3 lists these key topics and the page numbers on which each is found.

Table 4-3   Key Topics for Chapter 4

Key Topic Element	Description	Page Number
Section	The Principle of Least Privilege and Separation of Duties	155
Section	Authentication	155
Section	Authentication by Knowledge	156
Section	Authentication by Ownership or Possession	157
Section	Authentication by Characteristic	158
Section	Multifactor Authentication	159
Section	Duo Security	159
Section	Zero Trust and BeyondCorp	161
Section	Single Sign-On	164
List	Recognize SSO and federated identity elements	165
Section	Authorization	167
List	Understand the importance of implicit deny and need-to-know authorization concepts	168
Section	Mandatory Access Control (MAC)	168
Section	Discretionary Access Control (DAC)	168
Section	Role-Based Access Control (RBAC)	168
Section	Rule-Based Access Control	169
Section	Attribute-Based Access Control	169
Section	Accounting	169
Section	Access Control Mechanisms	170
Section	RADIUS	173
Section	TACACS+	174
Table 4-2	The differences between the RADIUS and TACACS+ protocols	175
Section	802.1X	178
List	Explore the protocols used in 802.1X implementations	179
Section	Downloadable ACL	181
Section	Cisco Identity Services Engine (ISE)	181
Section	Cisco ISE Context and Identity Services	184
Section	Cisco ISE Profiling Services	184
Section	Cisco ISE Identity Services	187
Section	Cisco ISE Authorization Rules	188
Section	Cisco TrustSec	190
Section	Change of Authorization (CoA)	193
Paragraph	Explore the Cisco Common Classification Policy Language (C3PL) style of configuration	204
Section	Additional Cisco ISE Design Tips	211

4

Define Key Terms

Define the following key terms from this chapter and check your answers in the glossary:

AAA, identification, one-time passcode (OTP), out-of-band authentication, multifactor authentication, multilayer authentication, Security Assertion Markup Language (SAML), zero trust, BeyondCorp, Federated Identity Management, federation provider, forest, identity provider (IdP), Kerberos, multitenancy, OAuth, OpenID (or OpenID Connect), social identity provider (Social IdP), Web identity, Windows identity, WS-Federation, authorization, implicit deny, need to know, mandatory access control (MAC), discretionary access control (DAC), role-based access control (RBAC), attribute-based access control (ABAC), accounting, access control list (ACL), capability table, access control matrix (ACM), RADIUS, TACACS+, Diameter, 802.1X, authentication server, supplicant, authenticator, EAP over LAN (EAPoL), Extensible Authentication Protocol (EAP), VLAN ACLs, security group–based ACL (SGACL), downloadable ACL (dACL), pxGrid, MAC Authentication Bypass (MAB), TrustSec security group tag (SGT), posture assessment

Review Questions

1. Which of the following is a security model created by Google that is similar to the zero-trust concept?

   a. BeyondCorp

   b. TrustSec

   c. pxGrid

   d. Duo

2. Which of the following are technologies used in SSO implementations?

   a. SAML

   b. OpenID Connect

   c. Microsoft Account

   d. All of these options are correct.

3. Which of the following is true about delegation in SSO implementations? (Select all that apply.)

   a. SSO implementations use delegation to call external APIs to authenticate and authorize users.

   b. Delegation is used to make sure that applications and services do not store passwords and user information on-premises.

   c. Delegation uses multifactor authentication to provide identity services to other servers in the environment.

   d. pxGrid can be used for delegation between a PSN and PAN.

4. Which of the following statements are true about discretionary access controls (DACs)?

   a. Discretionary access controls (DACs) are defined by the owner of the object.

   b. DACs are used in commercial operating systems.

   c. The object owner builds an ACL that allows or denies access to the object based on the user's unique identity.

   d. All of these options are correct.

5. RADIUS accounting runs over what protocol and port?

    a. UDP port 1812

    b. UDP port 1813

    c. UDP port 1645

    d. None of these options is correct.

6. Which of the following is one primary difference between a malicious hacker and an ethical hacker?

    a. Malicious hackers use different tools and techniques than ethical hackers use.

    b. Malicious hackers are more advanced than ethical hackers because they can use any technique to attack a system or network.

    c. Ethical hackers obtain permission before bringing down servers or stealing credit card databases.

    d. Ethical hackers use the same methods but strive to do no harm.

7. You were hired to configure RADIUS authentication in a VPN implementation. You start RADIUS debugs in the VPN device and notice ACCESS-CHALLENGE messages. What do those messages mean?

    a. ACCESS-CHALLENGE messages are sent if additional information is needed. The RADIUS server needs to send an additional challenge to the access server before authenticating the user. The ACCESS-CHALLENGE will be followed by a new ACCESS-REQUEST message.

    b. ACCESS-CHALLENGE messages are sent if additional information is needed. The RADIUS server needs to send an additional challenge to the access server before authenticating the user. The ACCESS-CHALLENGE will be followed by a new ACCESS-REJECT message.

    c. ACCESS-CHALLENGE messages are sent if the client is using multifactor authentication with a mobile device. The ACCESS-CHALLENGE will be followed by a new ACCESS-REQUEST message.

    d. None of these options is correct.

8. Which of the following are TACACS+ exchange packets used during the authentication process?

    a. START

    b. REPLY

    c. CONTINUE

    d. All of these options are correct.

    e. None of these options is correct.

9. Which of the following is an entity that seeks to be authenticated by an authenticator (switch, wireless access point, and so on)? This entity could use software such as the Cisco AnyConnect Secure Mobility Client.

    a. PAN

    b. PSN

    c. Supplicant

    d. None of these options is correct.

10. 802.1X uses which of the following protocols?

   a. EAPoL

   b. EAP

   c. RADIUS

   d. All of these options are correct.

11. Which of the following statements is true about CoA?

   a. RADIUS CoA is a feature that allows a RADIUS server to adjust the authentication and authorization state of an active client session.

   b. RADIUS CoA is a feature that allows a RADIUS server to detect a change of configuration from other RADIUS servers and, subsequently, deny access to a client trying to connect to the network.

   c. RADIUS CoA is a feature that allows a RADIUS server to perform profiling and posture assessment simultaneously.

   d. None of these options is correct.

12. The _____ is a structured replacement for feature-specific configuration commands. This concept allows you to create traffic policies based on events, conditions, and actions.

   a. Cisco Common Classification Policy Language (C3PL)

   b. Cisco Policy Mapping

   c. Cisco TrustSec

   d. None of these options is correct.

CHAPTER 5

Network Visibility and Segmentation

This chapter covers the following topics:

Introduction to Network Visibility

NetFlow

IP Flow Information Export (IPFIX)

NetFlow Deployment Scenarios

Cisco Stealthwatch

Cisco Cognitive Threat Analytics (CTA) and Encrypted Traffic Analytics (ETA)

NetFlow Collection Considerations and Best Practices

Configuring NetFlow in Cisco IOS and Cisco IOS-XE

Configuring NetFlow in NX-OS

Introduction to Network Segmentation

Micro-segmentation with Cisco ACI

Segmentation with Cisco ISE

The following SCOR 350-701 exam objectives are covered in this chapter:

- **Domain 6: Secure Network Access, Visibility, and Enforcement**
  - 6.4 Describe the benefits of device compliance and application control
  - 6.5 Explain exfiltration techniques (DNS tunneling, HTTPS, email, FTP/SSH/SCP/SFTP, ICMP, Messenger, IRC, NTP)
  - 6.6 Describe the benefits of network telemetry
  - 6.7 Describe the components, capabilities, and benefits of these security products and solutions:
    - 6.7.a Cisco Stealthwatch
    - 6.7.b Cisco Stealthwatch Cloud
    - 6.7.a Cisco Stealthwatch

- 6.7.c Cisco pxGrid

- 6.7.d Cisco Umbrella Investigate

- 6.7.e Cisco Cognitive Threat Analytics

- 6.7.f Cisco Encrypted Traffic Analytics

- 6.7.g Cisco AnyConnect Network Visibility Module (NVM)

"Do I Know This Already?" Quiz

The "Do I Know This Already?" quiz allows you to assess whether you should read this entire chapter thoroughly or jump to the "Exam Preparation Tasks" section. If you are in doubt about your answers to these questions or your own assessment of your knowledge of the topics, read the entire chapter. Table 5-1 lists the major headings in this chapter and their corresponding "Do I Know This Already?" quiz questions. You can find the answers in Appendix A, "Answers to the 'Do I Know This Already?' Quizzes and Q&A Sections."

Table 5-1   "Do I Know This Already?" Section-to-Question Mapping

Foundation Topics Section	Questions
Introduction to Network Visibility	1
NetFlow	2–3
IP Flow Information Export (IPFIX)	4–5
NetFlow Deployment Scenarios	6
Cisco Stealthwatch	7–8
Cisco Cognitive Threat Analytics (CTA) and Encrypted Traffic Analytics (ETA)	9
NetFlow Collection Considerations and Best Practices	10
Configuring NetFlow in Cisco IOS and Cisco IOS-XE	11
Configuring NetFlow in NX-OS	12
Introduction to Network Segmentation	13
Micro-segmentation with Cisco ACI	14
Segmentation with Cisco ISE	15

CAUTION   The goal of self-assessment is to gauge your mastery of the topics in this chapter. If you do not know the answer to a question or are only partially sure of the answer, you should mark that question as wrong for the purposes of the self-assessment. Giving yourself credit for an answer you incorrectly guess skews your self-assessment results and might provide you with a false sense of security.

1. Which of the following technologies can be deployed to gain network visibility and awareness of security threats?

   a.   NetFlow

   b.   IPFIX

   c.   Cisco Stealthwatch

   d.   All of these answers are correct.

2. Which of the statements is true about NetFlow?

   a.   NetFlow supports IPv4 and IPv6.

   b.   NetFlow supports IPv4 and IPv6 was introduced with IPFIX.

   c.   IPFIX supports only IPv4.

   d.   None of these answers is correct.

3. A flow is a unidirectional series of packets between a given source and destination. In a *flow*, the same source and destination IP addresses, source and destination ports, and IP protocol are shared. This is often referred to as the _____.

   a   five-tuple

   b.   five elements

   c.   NetFlow intelligence

   d.   IPFIX

4. IPFIX was originally created based on which of the following?

   a.   NetFlow v5

   b.   NetFlow v9

   c.   Flexible NetFlow

   d.   None of the above

5. IPFIX is considered what type of protocol?

   a.   IPFIX is considered to be an active protocol.

   b.   IPFIX is considered to be a pull protocol.

   c.   IPFIX is considered to be a passive protocol.

   d.   IPFIX is considered to be a push protocol.

6. Which of the following is a NetFlow deployment best practice?

   a.   NetFlow should be enabled as close to the access layer as possible (user access layer, data center access layer, in VPN termination points, and so on).

   b.   All NetFlow records belonging to a flow should be sent to the same collector.

   c.   To gain network visibility, Test Access Ports (TAPs) or Switched Port Analyzer (SPAN) ports must be configured when the Cisco Stealthwatch FlowSensors are deployed.

   d.   All of these answers are correct.

7. Which of the following is a physical or virtual appliance that can generate NetFlow data when legacy Cisco network infrastructure components are not capable of producing line-rate, unsampled NetFlow data?

   a. Stealthwatch FlowSensor

   b. Stealthwatch FlowCollector

   c. Stealthwatch FlowReplicator

   d. Stealthwatch FlowGenerator

8. Which of the following statements is not true?

   a. In Amazon AWS, the equivalent of NetFlow is called VPC Flow Logs.

   b. Google Cloud Platform supports VPC Flow Logs (or Google-branded GPC Flow Logs).

   c. In Microsoft's Azure, traffic flows are collected in network security group (NSG) flow logs.

   d. In Microsoft's Azure, the equivalent of NetFlow is called VPC Flow Logs.

9. Which of the following are components of the Cisco ETA solution to identify malicious (malware) communications in encrypted traffic through passive monitoring, the extraction of relevant data elements, and a combination of behavioral modeling and machine learning?

   a. NetFlow

   b. Cisco Stealthwatch

   c. Cisco Cognitive Threat Analytics

   d. All of these answers are correct.

10. Which type of the following deployment models has the advantage of limiting the overhead introduced by NetFlow?

   a. FlowCollectors deployed at multiple sites and placed close to the source producing the highest number of NetFlow records.

   b. FlowCollectors deployed in a centralized area and placed to handle the highest number of NetFlow records.

   c. Using asymmetric routing to send NetFlow records to the same SMC, not to different collectors.

   d. None of the above.

11. Which of the following are the main Flexible NetFlow components?

   a. Records

   b. Flow monitors

   c. Flow exporters

   d. Flow samplers

   e. All of the options are correct.

12. In NX-OS, NetFlow CLI commands are not available until you enable which of the following commands?

   a. netflow collection enable

   b. feature netflow

   c. ip netflow enable

   d. ip netflow run

13. Which of the following are Layer 2 technologies that security professionals have used for policy enforcement and segmentation? (Select two.)

   a. VLANs

   b. Routing protocols

   c. VRFs

   d. Route reflectors

14. A micro-segment in ACI is also often referred to as _____.

   a. uSeg SGT

   b. uSeg EPGs

   c. SCVMM

   d. None of these answers is correct.

15. Cisco ISE scales by deploying service instances called "_____" in a distributed architecture.

   a. personas

   b. SGTs

   c. uSeg EPGs

   d. pxGrid

Foundation Topics

Introduction to Network Visibility

Network visibility is one of the most important pillars within any cybersecurity program and framework. In fact, two of the most important components of a cybersecurity program that go together are visibility and control. Total network visibility and complete control of all elements in your network is easier said than done (especially when organizations have their applications and data hosted in multi-cloud environments). However, at least maintaining a good level of visibility among all these environments is crucial to maintain services and business continuity. You must design an architecture that should be flexible while improving security without relying on a single technology or product. Multiple technologies and features are used throughout the network to obtain visibility into network behavior and to maintain control during abnormal or malicious behavior.

How good is your network if you cannot manage it when an outbreak or attack is underway? Visibility is twofold. Network administrators and cybersecurity professionals should always have complete visibility of networking devices and the traffic within their infrastructure. At the same time, intruders must not have visibility to unnecessary services or vulnerable systems that can be exploited within an organization.

The following are the most common technologies that can be used to obtain and maintain complete network visibility:

- NetFlow

- IPFIX

- Cisco Stealthwatch

- Intrusion detection system/intrusion prevention system (IDS/IPS)

- Cisco Advanced Malware Protection (AMP) for Endpoints and Networks

NOTE NetFlow, IPFIX, and Cisco Stealthwatch are covered in this chapter. IDS and IPS are covered in detailed in Chapter 7, "Cisco Next-Generation Firewalls and Cisco Next-Generation Intrusion Prevention Systems." Cisco AMP for Endpoints is covered in Chapter 11, "Endpoint Protection and Detection." Cisco AMP for Networks is also covered in Chapter 7.

NetFlow

NetFlow is a technology originally created by Cisco that provides comprehensive visibility into all network traffic that traverses a Cisco-supported device. NetFlow was initially created for billing and accounting of network traffic and to measure other IP traffic characteristics such as bandwidth utilization and application performance. NetFlow has also been used as a network-capacity planning tool and to monitor network availability. Nowadays, NetFlow is used as a network security tool because its reporting capabilities provide nonrepudiation, anomaly detection, and investigative capabilities. As network traffic traverses a NetFlow-enabled device, the device collects traffic flow information and provides a network administrator or security professional with detailed information about such flows.

NetFlow provides detailed network telemetry that allows the administrator to perform the tasks:

- See what is actually happening across the entire network.

- Identify DoS attacks.

- Quickly identify compromised endpoints and network infrastructure devices.

- Monitor network usage of employees, contractors, or partners.

- Obtain network telemetry during security incident response and forensics.

- Detect firewall misconfigurations and inappropriate access to corporate resources.

TIP NetFlow supports both IP Version 4 (IPv4) and IP Version 6 (IPv6).

Defending against cybersecurity attacks is becoming more challenging every day, and it is not going to get any easier. The threat landscape is evolving to a faster, more effective, and more efficient criminal economy profiting from attacks against users, enterprises, services providers, and governments. Organized cybercrime, with its exchange of exploits, is booming and fueling a very lucrative economy. Threat actors nowadays have a clear under-standing of the underlying security technologies and their vulnerabilities. Hacker groups now follow software development life cycles, just like enterprises follow their own. These bad actors perform quality-assurance testing against security products before releasing them into the underground economy. They continue to find ways to evade common security defenses. Attackers follow techniques such as the following:

- Port and protocol hopping

- Tunneling over many different protocols

- Encryption

- Utilization of droppers

- Social engineering

- Exploitation of zero-day vulnerabilities

Security technologies and processes should not focus only on defending against Internet threats, but should also provide the ability to detect and mitigate the impact after a successful attack.

Security professionals must maintain visibility and control across the extended network during the full attack continuum:

- Before the attack takes place

- During an active attack

- After an attacker starts to damage systems or steal information

Cisco next-generation security products provide protection throughout the attack continuum. Devices such as the Cisco Firepower Threat Defense (FTD) and Cisco AMP provide a security solution that helps discover threats and enforce and harden policies before an attack takes place. In addition, you can detect attacks before, during, and after they have already taken place with NetFlow. These solutions provide the capabilities to contain and remediate an attack to minimize data loss and additional network degradation.

The Network as a Sensor and as an Enforcer

Many organizations fail to use one of the strongest tools that can help protect against today's security threats: the network itself. For example, Cisco Catalyst switches, data center switches, Aggregation Services Routers (ASRs), Integrated Services Routers (ISRs), next-generation firewalls (NGFWs), next-generation intrusion prevention systems (NGIPs), NetFlow generation appliances (Cisco Stealthwatch FlowSensors), Advanced Malware Protection (AMP), and wireless products, in conjunction with the Cisco Application Centric Infrastructure, can protect before, during, and after an attack.

The network can be used in security in two different, fundamental ways:

- **The network as a sensor:** NetFlow allows you to use the network as a sensor, giving you deep and broad visibility into unknown and unusual traffic patterns, in addition to compromised devices.

- **The network as an enforcer:** You can use Cisco TrustSec to contain attacks by enforcing segmentation and user access control. Even when bad actors successfully breach your network defenses, you thus limit their access to only one segment of the network.

What Is a Flow?

A *flow* is a unidirectional series of packets between a given source and destination. In a flow, the same source and destination IP addresses, source and destination ports, and IP protocol are shared. This is often referred to as the *five-tuple*. Figure 5-1 illustrates a five-tuple example.

Five (5) Tuple

Source IP	Source Port	Destination IP	Destination Port	Protocol
10.1.1.2	1872	10.2.2.3	443	TCP

Figure 5-1 *Five-Tuple Example*

Figure 5-2 shows an example of a flow between a client and a server.

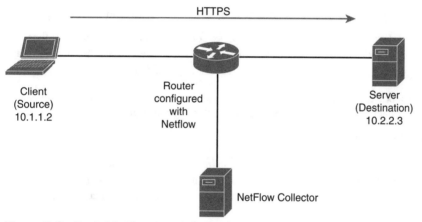

Figure 5-2 *Basic NetFlow Example Deployment*

In Figure 5-2, the client (source) establishes a connection to the server (destination). When the traffic traverses the router (configured for NetFlow), it generates a flow record. At the very minimum, the five-tuple is used to identify the flow in the NetFlow database of flows kept on the device.

> **TIP** The NetFlow database is often called the NetFlow cache.

Depending on the version of NetFlow, the router can also gather additional information, such as type of service (ToS) byte, differentiated services code point (DSCP), the device's input interface, TCP flags, byte counters, and start and end times.

Flexible NetFlow, Cisco's next-generation NetFlow, can track a wide range of Layer 2, IPv4, and IPv6 flow information, such as the following:

- Source and destination MAC addresses

- Source and destination IPv4 or IPv6 addresses

- Source and destination ports

- ToS

- DSCP

- Packet and byte counts

- Flow timestamps

- Input and output interface numbers

- TCP flags and encapsulated protocol (TCP/UDP) and individual TCP flags

- Sections of packet for deep packet inspection

- All fields in IPv4 header, including IP-ID, TTL, and others

- All fields in IPv6 header, including Flow Label, Option Header, and others

- Routing information such as next-hop address, source autonomous system number (ASN), destination ASN, source prefix mask, destination prefix mask, Border Gateway Protocol (BGP) next hop, and BGP policy accounting traffic index

NetFlow protocol data units (PDUs), also referred to as flow records, are generated and sent to a NetFlow collector after the flow concludes or expires (times out).

There are three types of NetFlow cache:

- **Normal cache:** This is the default cache type in many infrastructure devices enabled with NetFlow and Flexible NetFlow. The entries in the flow cache are removed (aged out) based on the configured **timeout active** *seconds* and **timeout inactive** *seconds* settings.

- **Immediate cache:**

  - Flow accounts for a single packet

  - Desirable for real-time traffic monitoring and distributed DoS (DDoS) detection

  - Used when only very small flows are expected (for example, sampling)

NOTE The use of the immediate cache may result in a large amount of export data. This subsequently increases the CPU and memory utilization of the network infrastructure device.

- **Permanent cache:**

  - Used to track a set of flows without expiring the flows from the cache.

  - The entire cache is periodically exported (update timer).

  - The cache is a configurable value.

  - After the cache is full, new flows will not be monitored.

  - Uses update counters rather than delta counters.

Many people often confuse a *flow* with a *session*. All traffic in a flow is going in the same direction; however, when the client establishes the HTTP connection (session) to the server and accesses a web page, it represents two separate flows. The first flow is the traffic from the client to the server, and the other flow is from the server to the client.

NetFlow was originally created for IP accounting and billing purposes; however, it plays a crucial role for the following:

- Network security

- Traffic engineering

- Network planning

- Network troubleshooting

> **TIP**   Do not confuse the feature in Cisco IOS and Cisco IOS-XE software called *IP Accounting* with NetFlow. IP Accounting is a great Cisco IOS and Cisco IOS-XE tool, but it is not as robust or as well-known as NetFlow.

NetFlow for Network Security and Visibility

NetFlow is a tremendous security tool. It provides nonrepudiation, anomaly detection, and investigative capabilities. Complete visibility is one of the key requirements when identifying and classifying security threats.

The first step in the process of preparing your network and staff to successfully identify security threats is achieving complete network visibility. You cannot protect against or mitigate what you cannot view/detect. You can achieve this level of network visibility through existing features on network devices you already have and on devices whose potential you do not even realize. In addition, you should create strategic network diagrams to clearly illustrate your packet flows and where, within the network, you may enable security mechanisms to identify, classify, and mitigate the threat. Remember that network security is a constant war. When defending against the enemy, you must know your own territory and implement defense mechanisms in place.

NetFlow for Anomaly Detection and DDoS Attack Mitigation

You can use NetFlow as an anomaly-detection tool. Anomaly-based analysis keeps track of network traffic that diverges from "normal" behavioral patterns. Of course, you must first define what is considered to be normal behavior. You can use anomaly-based detection to mitigate DDoS attacks and zero-day outbreaks. DDoS attacks are often used maliciously to consume the resources of your hosts and network that would otherwise be used to serve legitimate users. The goal with these types of attacks is to overwhelm the victim network resources, or a system's resources such as CPU and memory. In most cases, this is done by sending numerous IP packets or forged requests.

A particularly dangerous attack is when an attacker builds up a more powerful attack with a more sophisticated and effective method of compromising multiple hosts and installing small attack daemons. This is what many call *zombies* or *bot hosts/nets*. Subsequently, an attacker can launch a coordinated attack from thousands of zombies onto a single victim. This daemon typically contains both the code for sourcing a variety of attacks and some basic communications infrastructure to allow for remote control.

Typically, an anomaly-detection system monitors network traffic and alerts and then reacts to any sudden increase in traffic and any other anomalies.

> **TIP** NetFlow, along with other mechanisms such as syslog and SNMP, can be enabled within your infrastructure to provide the necessary data used for identifying and classifying threats and anomalies. Before implementing these anomaly-detection capabilities, you should perform traffic analysis to gain an understanding of general traffic rates and patterns. In anomaly detection, learning is generally performed over a significant interval, including both the peaks and valleys of network activity.

Figure 5-3 shows a dashboard from the Cisco Stealthwatch Cloud, which is displaying network traffic during a DDoS attack.

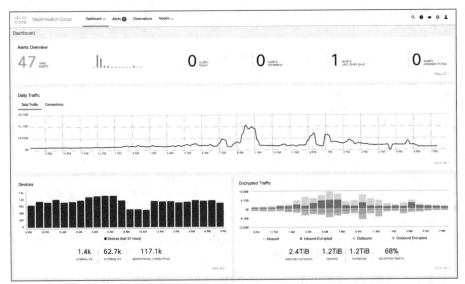

Figure 5-3 *Anomaly Detection with NetFlow and Stealthwatch Cloud Example*

As you can see in the graph illustrated in Figure 5-3, the amount of traffic (in gigabits per second) increases after 8:00 a.m. until after 9 a.m. If this spike in traffic is not a normal occurrence on other days, this may be an anomaly. In some cases, misconfigured hosts and servers can send traffic that consumes network resources unnecessarily. Having the necessary tools and mechanisms to identify and classify security threats and anomalies in the network is crucial.

NOTE   Cisco Stealthwatch will be covered later in this chapter.

Data Leak Detection and Prevention

Many network administrators, security professionals, and business leaders struggle in the effort to prevent data loss within their organizations. The ability to identify anomalous behavior in data flows is crucial to detect and prevent data loss. The application of analytics to data collected via NetFlow can aid security professionals in detecting anomalous large amounts of data leaving the organization and abnormal traffic patterns inside of the organization.

Using NetFlow along with identity management systems, an administrator can detect who initiated the data transfer, the hosts (IP addresses) involved, the amount of data transferred, and the services used.

In addition, the administrator can measure how long the communications lasted and the frequency of the same connection attempts.

TIP   Often, tuning is necessary because certain traffic behavior could cause false positives. For instance, your organization may be legitimately sharing large amounts of data or streaming training to business partners and customers. In addition, analytics software that examines baseline behavior may be able to detect typical file transfers and incorporate them into existing baselines.

Incident Response, Threat Hunting, and Network Security Forensics

NetFlow is often compared to a phone bill. When police want to investigate criminals, for instance, they often collect and investigate their phone records. NetFlow provides information about all network activity that can be very useful for incident response and network forensics. This information can help you discover indicators of compromise (IOCs).

The National Institute of Standards and Technology (NIST) created the following methodology on security incident handling, which has been adopted by many organizations, including service providers, enterprises, and government organizations:

Step 1.   Preparation

Step 2.   Detection and analysis

Step 3.   Containment, eradication, and recovery

Step 4.   Post-incident activity (postmortem and lessons learned)

NOTE   The NIST Computer Security Incident Handling Guide is Special Publication 800-61 Revision 2. The NIST SP 800-61 R2 publication can be downloaded from https://nvlpubs. nist.gov/nistpubs/SpecialPublications/NIST.SP.800-61r2.pdf.

NetFlow plays a crucial role in the preparation phase and the detection and analysis phase. Information collected in NetFlow records can be used to identify, categorize, and scope suspected incidents as part of the identification. NetFlow data also provides great benefits for attack traceback and attribution. In addition, NetFlow provides visibility of what is getting into your network and what information is being exfiltrated out of your network.

Figure 5-4 shows an example of how a botnet is performing a DDoS attack against the corporate network. NetFlow in this case can be used as an anomaly-detection tool for the DDoS attack and also as a forensics tool to potentially find other IOCs of more sophisticated attacks that may be carried out incognito.

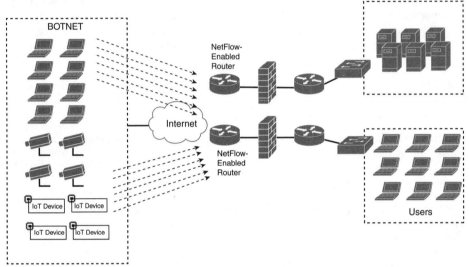

Figure 5-4 *Detecting What Is Getting into Your Network*

Figure 5-5 shows how a "stepping-stone" attack is carried out in the corporate network. A compromised host in the call center department is exfiltrating large amounts of sensitive data to an attacker on the Internet from a server in the data center.

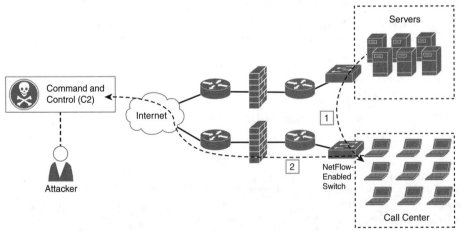

Figure 5-5 *Detecting What Is Getting Out of Your Network*

You can also use NetFlow in combination with DNS records to help you detect suspicious and malicious traffic, such as the following:

- Suspicious requests to .gov, .mil, and .edu sites when you do not even do business with any of those entities

- Large amount of traffic leaving the organization late at night to suspicious sites

- Traffic to embargoed countries that should not have any business partners or transactions

- Suspicious virtual private network (VPN) requests and VPN traffic

- Requests and transactions to sites without any content

- Pornography sites or any other corporate policy violations

- Illegal file-sharing sites

- Crypto mining informational sites

Syslog and packet captures are also often used in network forensics; however, an area where these traditional network forensics tools fall short is in coverage. For instance, it is very difficult to deploy hundreds of sniffers (packet-capture devices) in the networks of large organizations. In addition, the cost will be extremely high. When a security incident or breach is detected, the incident responders need answers fast! They do not have time to go over terabytes of packet captures, and they can definitely not analyze every computer on the network to find the root cause, miscreant, and source of the breach. You can use NetFlow to obtain a high-level view of what is happening in the network, and then the incident responder can perform a deep-dive investigation with packet captures and other tools later in the investigation. Sniffers can be then deployed as needed in key locations where suspicious activity is suspected. The beauty of NetFlow is that you can deploy it anywhere you have a supported router, switch, Cisco ASA, or Cisco FTD; alternatively, you can use Cisco Stealthwatch FlowSensor.

TIP The Cisco Stealthwatch FlowSensor is a network appliance that functions similarly to a traditional packet capture appliance or IDS in that it connects into a Switch Port Analyzer (SPAN), mirror port, or a Test Access Port (TAP). The Cisco Stealthwatch FlowSensor augments visibility where NetFlow is not available in the infrastructure device (router, switch, and so on) or where NetFlow is available but you want deeper visibility into performance metrics and packet data. You typically configure the Cisco Stealthwatch FlowSensor with the Cisco Stealthwatch Flow Collector. You will learn more about the Cisco Stealthwatch solution later in this chapter.

NetFlow can fill in some of the gaps and challenges regarding the collection of packet captures everywhere in the network. It is easier to store large amounts of NetFlow data because it is only a transactional record. Therefore, administrators can keep a longer history of events that occurred on their networks. Historical records can prove very valuable when investigating a breach. Network transactions can show you where an initial infection came from, what command-and-control channel was initiated by the malware, what other computers on the internal network were accessed by that infected host, and whether other hosts in the network reached out to the same attacker or command-and-control system, as demonstrated earlier at a high level.

The logging facility on Cisco routers, switches, Cisco ASA, Cisco FTD, and other infrastructure devices allows you to save syslog messages locally or to a remote host. By default, routers send logging messages to a logging process. The logging process controls the delivery of logging messages to various destinations, such as the logging buffer, terminal lines, a syslog server, or a monitoring event correlation system such as Elastic Search, Logstash and Kibana (known as the ELK stack), Graylog, Splunk, and others. You can set the severity level of the messages to control the type of messages displayed, in addition to a timestamp to successfully track the reported information. Every security professional and incident responder knows how important it is to have good logs. There is no better way to find out what was happening in a router, switch, and firewall at the time that an attack occurred. However, like all things, syslog has limitations. You have to enable the collection of logs from each endpoint; so in many environments, syslog coverage is incomplete, and after a computer has been compromised, it is not possible to trust the logs coming from that device anymore. Syslog is extremely important, but it cannot tell you everything.

Many network telemetry sources can also be correlated with NetFlow while responding to security incidents and performing network forensics, including the following:

- Dynamic Host Configuration Protocol (DHCP) logs

- VPN logs

- Network address translation (NAT) information

- 802.1X authentication logs

- Server logs (syslog)

- Web proxy logs

- Spam filters from e-mail security appliances such as the Cisco Email Security Appliance (ESA)

Figures 5-6, 5-7, and 5-8 list different event types, their source, and respective events that can be combined with NetFlow while responding to security incidents and performing network forensics.

Figure 5-6 shows event types and respective sources that can be used for attribution.

Figure 5-7 shows event types and respective sources of underlying system activity.

Figure 5-8 shows event types of web proxy, spam filters, and firewall logs.

It is extremely important that your syslog and other messages are timestamped with the correct date and time. This is why the use of Network Time Protocol (NTP) is strongly recommended.

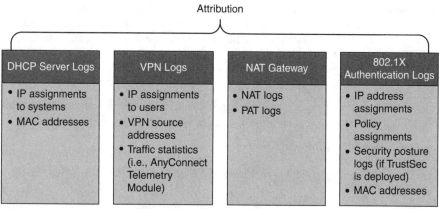

Figure 5-6 *Event Types and Sources for Attribution*

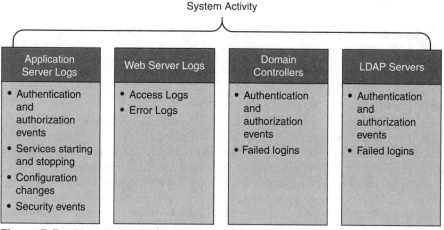

Figure 5-7 *Event Types and Sources to Understand Underlying System Activity*

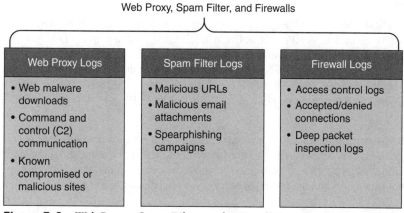

Figure 5-8 *Web Proxy, Spam Filter, and Firewall Logs for Incident Response*

Network forensics can be an intimidating topic for many security professionals. Everyone knows that forensic investigation may entail many other sources of information from end hosts, servers, and any affected systems. Each forensics team needs to be focused on many different areas, such as the following:

- Having a thorough knowledge of assets, risks, impact, and likelihood of events.

- Practicing incident response policies and procedures in mock events and collecting things like NetFlow on a regular basis to analyze what is happening in the network.

- Awareness of current vulnerabilities and threats.

- Understanding evidence handling and chain of custody. (Even NetFlow events can be used as evidence.)

- Enacting mitigation based on evidence collected.

- Knowing the documentation requirements for evidence, depending on your country and local laws.

- Understanding the analysis process during and after the incident.

- Having a framework for communications, both within the team and external to the team.

Traffic Engineering and Network Planning

NetFlow can also be used for traffic engineering and network planning. For example, you can use NetFlow to understand many things that can be tuned and further engineered in the network, such as understanding what are the non-mission-critical traffic-taxing network resources (for example, end users downloading audio or video files, or visiting Facebook, Twitter, and other social networking sites). Network configuration problems in infrastructure devices can be inferred by observing NetFlow. In the case of service providers, NetFlow correlation with BGP benefits peering and IP transit analysis (profitability, costs, violations, and so on). In enterprise networks, correlation with internal gateway protocols, such as Open Shortest Path First (OSPF), Enhanced Interior Gateway Routing Protocol (EIGRP), and Routing Information Protocol (RIP), can provide similar benefits.

You can also use NetFlow as a troubleshooting tool. For instance, you might have users complaining that Voice over IP (VoIP) calls or Cisco TelePresence video calls are failing or that the quality is very poor. Reviewing NetFlow data might indicate that many users are inappropriately streaming YouTube, Netflix, or Hulu videos. Using this data, the network administrator can create access lists to block traffic to some of these sites (or all) and/or create Quality of Service (QoS) policies to prioritize the VoIP and Cisco TelePresence traffic.

An example of capacity planning may be a planned merger with another company that will add hundreds or thousands of new users to internal corporate resources. Another example is a new application that may have numerous users where the network administrator might need to provision additional bandwidth or *capacity* to different areas of the network.

NetFlow Versions

There have been several versions of NetFlow throughout the years. Table 5-2 lists all versions of NetFlow and a brief description of the features supported.

Table 5-2   NetFlow Versions

NetFlow Version	Description
Version 1 (v1)	(Obsolete) The first implementation of NetFlow. NetFlow v1 was limited to IPv4 without network masks and autonomous system numbers (ASNs).
Version 2 (v2)	Never released to the public.
Version 3 (v3)	Never released to the public.
Version 4 (v4)	Never released to the public.
Version 5 (v5)	One of the most popular versions of NetFlow in earlier implementations. Although it is still being used in some environments, NetFlow v5 is now obsolete.
Version 6 (v6)	(Obsolete) No longer supported.
Version 7 (v7)	(Obsolete) No longer supported. Introduced a source router field.
Version 8 (v8)	(Obsolete) No longer supported.
Version 9 (v9)	Template-based implementation of NetFlow. Introduced support for IPv6, Multiprotocol Label Switching (MPLS), and Border Gateway Protocol (BGP) next hop.
Flexible NetFlow	Template-based and modeled after NetFlow v9.
IPFIX	Although IPFIX is not a version of NetFlow, it is the IETF standard based on NetFlow v9 that also introduced several extensions.

IP Flow Information Export (IPFIX)

The Internet Protocol Flow Information Export (IPFIX) is a network flow standard led by the Internet Engineering Task Force (IETF). IPFIX was created for a common, universal standard of export for the flow information from routers, switches, firewalls, and other infrastructure devices. IPFIX defines how flow information should be formatted and transferred from an exporter to a collector. IPFIX is documented in RFC 7011 through RFC 7015 and RFC 5103. Cisco NetFlow Version 9 is the basis and main point of reference for IPFIX. IPFIX changes some of the terminologies of NetFlow, but in essence they are the same principles as NetFlow Version 9.

IPFIX defines different elements that are grouped into the following 12 categories according to their applicability:

1. Identifiers
2. Metering and exporting process configuration
3. Metering and exporting process statistics
4. IP header fields
5. Transport header fields
6. Sub-IP header fields
7. Derived-packet properties

8. Min/max flow properties

9. Flow timestamps

10. Per-flow counters

11. Miscellaneous flow properties

12. Padding

IPFIX is considered to be a *push protocol*. Each IPFIX-enabled device regularly sends IPFIX messages to configured collectors (receivers) without any interaction by the receiver. The sender controls most of the orchestration of the IPFIX data messages. IPFIX introduces the concept of *templates*, which make up these flow data messages to the receiver. IPFIX also allows the sender to use user-defined data types in its messages. IPFIX prefers the Stream Control Transmission Protocol (SCTP) as its transport layer protocol; however, it also supports the use of the Transmission Control Protocol (TCP) or User Datagram Protocol (UDP) messages.

Traditional Cisco NetFlow records are usually exported via UDP messages. The IP address of the NetFlow collector and the destination UDP port must be configured on the sending device. The NetFlow standard (RFC 3954) does not specify a specific NetFlow listening port. The standard or most common UDP port used by NetFlow is UDP port 2055, but other ports like 9555, 9995, 9025, and 9026 can also be used. UDP port 4739 is the default port used by IPFIX.

IPFIX Architecture

IPFIX uses the following architecture terminology:

- **Metering process (MP):** Generates flow records from packets at an observation point. The metering process timestamps, samples, and classifies flows. The MP also maintains flows in an internal data structure and passes complete flow information to an exporting process (EP).

- **EP:** Sends flow records via IPFIX from one or more MPs to one or more collecting processes (CPs).

- **CP:** Receives records via IPFIX from one or more EPs.

Figure 5-9 illustrates these concepts and the architecture.

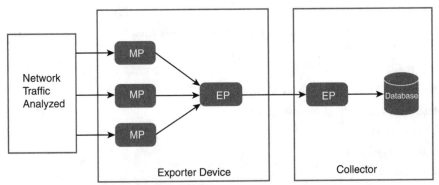

Figure 5-9 *IPFIX High-Level Architecture*

Understanding IPFIX Mediators

IPFIX introduces the concept of *mediators*. Mediators collect, transform, and re-export IPFIX streams to one or more collectors. Their main purpose is to allow federation of IPFIX messages. Mediators include an intermediate process (ImP) that allows for the following:

- NetFlow data to be kept anonymously

- NetFlow data to be aggregated

- Filtering of NetFlow data

- Proxying of web traffic

- IP translation

Figure 5-10 shows a sample architecture that includes an IPFIX mediator.

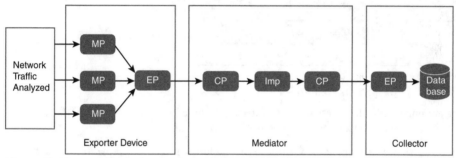

Figure 5-10   *IPFIX Mediator Example*

IPFIX Templates

An IPFIX template describes the structure of flow data records within a data set. Templates are identified by a template ID, which corresponds to a set ID in the set header of the data set. Templates are composed of (information element [IE] and length) pairs. IEs provide field type information for each template. Figure 5-11 illustrates these concepts.

IPFIX covers nearly all common flow collection use cases, such as the following:

- The traditional five-tuple (source IP address, destination IP address, source port, destination port, and IP protocol)

- Packet treatment such as IP next-hop IPv4 addresses, BGP destination ASN, and others

- Timestamps to nanosecond resolution

- IPv4, IPv6, ICMP, UDP, and TCP header fields

- Sub-IP header fields such as source MAC address and wireless local area network (WLAN) service set identifier (SSID)

- Various counters (packet delta counts, total connection counts, top talkers, and so on)

- Flow metadata information such as ingress and egress interfaces, flow direction, virtual routing and forwarding (VRF) information

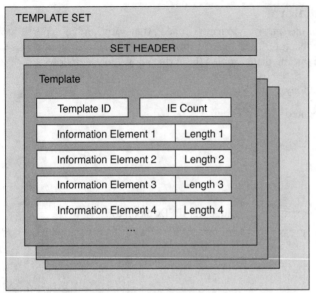

Figure 5-11 *IPFIX Template Structure*

There are numerous others defined at the Internet Assigned Numbers Authority (IANA) website: http://www.iana.org/assignments/ipfix/ipfix.xhtml.

Figure 5-12 shows an example of a template that includes different information element lengths and the association with the respective data set of flow records.

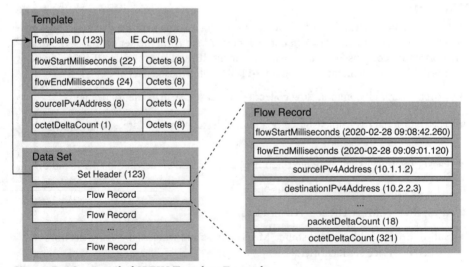

Figure 5-12 *Detailed IPFIX Template Example*

As shown in Figure 5-12, the template ID matches the set header of the related data set (123) in this example. Then the data set includes a series of flow records. An example of the flow record is shown in the block diagram on the right.

Option Templates

Option templates are a different type of IPFIX templates used to define records referred to as *options* that are associated with a specified scope. A scope may define an entity in the IPFIX architecture, including the exporting process, other templates, or a property of a collection of flows. Flow records describe flows, and option records define things other than flows, such as the following:

- Information about the collection infrastructure

- Metadata about flows or a set of flows

- Other properties of a set of flows

Understanding the Stream Control Transmission Protocol (SCTP)

IPFIX uses SCTP, which provides a packet transport service designed to support several features beyond TCP or UDP capabilities. These features include the following:

- Packet streams

- Partial reliability (PR) extension

- Unordered delivery of packets or records

- Transport layer multihoming

Many refer to SCTP as a simpler state machine than features provided by TCP with an "a la carte" selection of features. PR-SCTP provides a reliable transport with a mechanism to skip packet retransmissions. It allows multiple applications with different reliability requirements to run on the same flow association. In other words, it combines the best effort reliability of UDP while still providing TCP-like congestion control.

SCTP ensures that IPFIX templates are sent reliably by improving end-to-end delay. RFC 6526 introduces additional features such as per-template drop counting with partial reliability and fast template reuse.

Exploring Application Visibility and Control and NetFlow

The Cisco Application Visibility and Control (AVC) solution is a collection of services available in several Cisco network infrastructure devices to provide application-level classification, monitoring, and traffic control. The Cisco AVC solution is supported by the Cisco Integrated Services Routers (ISR), Cisco ASR 1000 Series Aggregation Service Routers (ASR 1000s), and Cisco Wireless LAN Controllers (WLCs). The following are the capabilities Cisco AVC provides:

- Application recognition

- Metrics collection and exporting

- Management and reporting systems

- Network traffic control

Application Recognition

Cisco AVC uses existing Cisco Network-Based Application Recognition Version 2 (NBAR2) to provide deep packet inspection (DPI) technology to identify a wide variety of applications within the network traffic flow, using Layer 3 to Layer 7 data.

NBAR works with QoS features to help ensure that the network bandwidth is best used to fulfill its main primary objectives. The benefits of combining these features include the ability to guarantee bandwidth to critical applications, limit bandwidth to other applications, drop selective packets to avoid congestion, and mark packets appropriately so that the network and the service provider's network can provide QoS from end to end.

Metrics Collection and Exporting

Cisco AVC includes an embedded monitoring agent that is combined with NetFlow to provide a wide variety of network metrics data. The types of metrics the monitoring agent collects include the following:

- TCP performance metrics such as bandwidth usage, response time, and latency

- VoIP performance metrics such as packet loss and jitter

These metrics are collected and exported in NetFlow v9 or IPFIX format to a management and reporting system.

In Cisco IOS and Cisco IOS-XE routers, metrics records are sent out directly from the data plane when possible, to maximize system performance. However, if more complex processing is required on the Cisco AVC–enabled device, such as if the user requests that a router keep a history of exported records, the records may be exported from the route processor at a lower speed.

You can use QoS capabilities to control application prioritization. Protocol discovery features in Cisco AVC show you the mix of applications currently running on the network. This helps you define QoS classes and polices, such as how much bandwidth to provide to mission-critical applications and how to determine which protocols should be policed. Per-protocol bidirectional statistics are available, such as packet and byte counts as well as bit rates.

After administrators classify the network traffic, they can apply the following QoS features:

- Class-based weighted fair queuing (CBWFQ) for guaranteed bandwidth

- Enforcing bandwidth limits using policing

- Marking for differentiated service downstream or from the service provider using type of service (ToS) bits or DSCPs in the IP header

- Drop policy to avoid congestion using weighted random early detection (WRED)

 ## NetFlow Deployment Scenarios

You can enable NetFlow on network devices at all layers of the network to record and analyze all network traffic and identify threats such as malware that could be spreading laterally through the internal network; in other words, malware that spreads between adjacent hosts in the network. The following sections describe different types of NetFlow deployment scenarios within an enterprise network. These deployment scenarios include the following:

- User access layer

- Wireless LANs

- Internet edge

- Data center

- NetFlow in site-to-site and remote-access VPNs

- NetFlow in cloud environments

> **TIP**   As a best practice, NetFlow should be enabled as close to the access layer as possible (user access layer, data center access layer, in VPN termination points, and so on). Another best practice is that all NetFlow records belonging to a flow should be sent to the same collector.

NetFlow Deployment Scenario: User Access Layer

Figure 5-13 shows an enterprise corporate network where NetFlow is enabled on the access layer switches on different sections of the network.

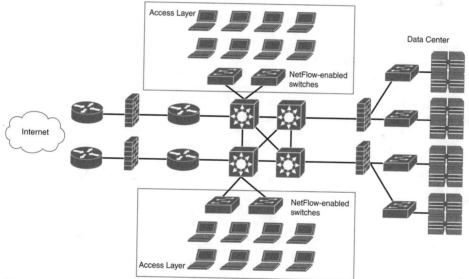

Figure 5-13   *NetFlow Enabled at the User Access Layer*

If the access switches do not support NetFlow, you can deploy a Cisco Stealthwatch FlowSensor to generate the NetFlow data. The Cisco Stealthwatch FlowSensor must be placed in a Layer 1 or Layer 2 adjacent manner to the monitoring area of the network.

> **TIP**   To gain network visibility, Test Access Ports (TAPs) or Switched Port Analyzer (SPAN) ports must be configured when the Cisco Stealthwatch FlowSensors are deployed.

NetFlow Deployment Scenario: Wireless LAN

NetFlow can be enabled in wireless LAN (WLAN) deployments. An administrator can configure NetFlow in combination with Cisco AVC in Cisco WLCs. The following are a few facts about the hardware and software requirements for AVC and NetFlow support in Cisco WLCs:

- Cisco AVC works on traffic from Cisco wireless access points (APs) configured in local mode or enhanced local mode. APs configured in local mode provide wireless service to clients in addition to limited time-sliced attacker scanning. There is an enhanced local mode feature (just called *enhanced local mode*) that, like local mode, provides wireless service to clients, but when scanning off-channel, the radio dwells on the channel for an extended period of time, allowing enhanced attack detection.

- Cisco AVC also works with FlexConnect central switching and OfficeExtend Access Point (OEAP) traffic.

- Cisco AVC is based on port, destination, and heuristics, which allows reliable packet classification with deep visibility.

- Cisco AVC looks into the initial setup of the client flow (first 10 to 20 packets), so loading on the controller system is minimal.

- Cisco AVC and NetFlow support is available for all Cisco controllers supporting Cisco WLC Software Version 7.4 or later.

Figure 5-14 shows a simple topology where an AP is providing wireless connectivity to clients in the corporate network while managed by a Cisco WLC. NetFlow is enabled in the Cisco WLC, as well as other Cisco AVC features.

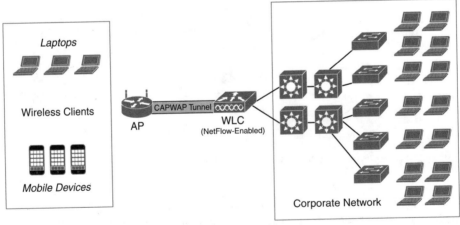

Figure 5-14 *NetFlow-Enabled Wireless Network Devices*

Cisco WLC and APs establish a Control And Provisioning of Wireless Access Points (CAPWAP) tunnel between them for communication. CAPWAP is an IETF standard that is based on its predecessor, the Lightweight Access Point Protocol (LWAPP). CAPWAP provides an upgrade path from Cisco products that use LWAPP to next-generation Cisco wireless products to interoperate with third-party APs and to manage radio frequency identification (RFID) readers and similar devices.

NetFlow Deployment Scenario: Internet Edge

You can also deploy NetFlow in the Internet edge to see what traffic is knocking on your door and what is leaving your network. The Cisco ASR 1000 series routers provide multigigabit performance to meet the requirements for Internet gateway functions for medium and large organizations. The architecture of the Cisco ASRs include the Cisco QuantumFlow Processor, which provides a lot of high-performance features such as application layer gateways (ALGs), all Layer 4 to Layer 7 zone-based firewall session processing, high-speed NAT and firewall translation logging, as well as NetFlow Event Logging (NEL). NEL uses NetFlow v9 templates to log binary syslog to NEL collectors, allowing not only the use of NAT at multigigabit rates but also the ability to record NAT and firewall session creation and teardown records at very high speeds.

TIP   You can use Cisco Feature Navigator to get details of all NAT- and firewall-related features for the Cisco ASR 1000 series routers at https://cfn.cloudapps.cisco.com/ITDIT/ CFN/jsp/by-feature-technology.jsp.

Figure 5-15 shows how two ASR Cisco ASR 1000 series routers are connected to two different Internet service providers (ISPs) and enabled for NetFlow. The ISP edge is a critical part of the Internet edge because it provides the interface to the public Internet infrastructure.

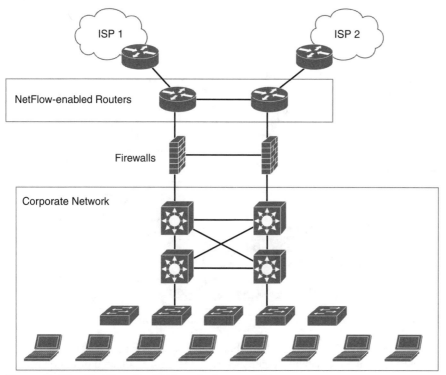

Figure 5-15   *NetFlow Enabled at the Internet Edge*

The Internet edge can be built out of many platforms and components that might fail or that might be subject to attack. Designing a redundant architecture helps eliminate single points of failure, thereby improving the availability and resiliency of the network.

TIP   The Internet edge should be designed with several layers of redundancy, including redundant interfaces and standby devices. Redundant interfaces at various points of the architecture provide alternative paths. Dynamic routing protocols should be used for path selection. The design allows for the use of multiple ISP connections, each served by different interfaces.

NetFlow Deployment Scenario: Data Center

The data center can be a very complex world. It not only provides a rich set of services and architectures, but it also hosts the crown jewels of your organization. It is extremely important to maintain visibility of everything that is happening in the data center. The concepts of "north-to-south" and "east-to-west" are often used when trying to describe the types of communication (or flow) within and to the outside of the data center:

- North-to-south IP traffic is the communication with end users and external entities.

- East-to-west is the communication between entities in the data center.

Figure 5-16 demonstrates the concepts of north-to-south and east-to-west communication.

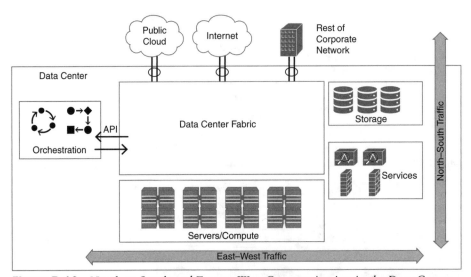

Figure 5-16   *North-to-South and East-to-West Communication in the Data Center*

The data center has many different high-throughput and low-latency requirements, in addition to increasing high-availability requirements. In addition, automated provisioning and control with orchestration, monitoring, and management tools are crucial.

The data center architecture consists of three primary modular layers with hierarchical interdependencies:

- **Data center foundation:** Primary building block of the data center on which all other services rely. Despite the size of the data center, the foundation must be resilient, scalable, and flexible to support data center services that add value, performance, and reliability. The data center foundation provides the computing necessary to support the applications that process information and the seamless transport between servers, storage, and the end users who access the applications.

- **Data center services:** Includes infrastructure components to enhance the security of the applications and access to critical data. It also includes virtual switching services to extend the network control in a seamless manner from the foundation network into the hypervisor systems on servers to increase control and lower operational costs (as well as other application resilience services).

- **User services:** Includes email, order processing, and file sharing or any other applications in the data center that rely on the data center foundation and services like database applications, modeling, and transaction processing.

Examples of the data center service insertion components include the following:

- Firewalls

- Intrusion prevention systems (IPSs)

- Application delivery features

- Server load balancing

- Network analysis tools (such as NetFlow)

- Virtualized services deployed in a distributed manner along with virtual machines

- Application Centric Infrastructure (ACI) automated framework components for service insertion

NetFlow collection in large data centers can be very challenging because of the large amount of data being generated at very high rates. In this case, the Cisco Stealthwatch FlowSensor provides a high-performance solution for flow visibility in multigigabit data centers.

NOTE The Cisco Stealthwatch FlowSensors come in different form factors and support millions of active flows. For the most up-to-date models and information, visit https://www.cisco.com/c/en/us/support/security/stealthwatch-flow-sensor-series/tsd-products-support-series-home.html.

Figure 5-17 illustrates how two Cisco Stealthwatch FlowSensors are deployed in the data center services architecture.

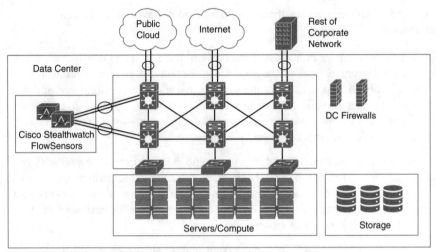

Figure 5-17 *Cisco Stealthwatch FlowSensors in the Data Center*

The Cisco Stealthwatch FlowSensor can be placed to receive data from the physical access, aggregation, and core layers to maintain complete visibility of all traffic within the data center, in addition to traffic that is leaving the data center (north-to-south and east-to-west traffic). Traffic within the virtual environment (VM-to-VM traffic) can be monitored using the Nexus 1000V, and traffic entering and leaving the data center can be monitored using edge devices such as the Cisco ASA, Cisco FTD, and Nexus 7000 and Nexus 9000 switches. Strategically placing the Stealthwatch FlowSensors in the aggregation and core layers ensures effective monitoring of traffic within the data center and provides additional statistics for traffic leaving the data center.

If the deployment of the Cisco Stealthwatch FlowSensor is not possible, you can enable Net-Flow in the Cisco ASA or enabled in any other strategic areas of the data center such as data center distribution switches and access switches.

NetFlow Deployment Scenario: NetFlow in Site-to-Site and Remote VPNs

Many organizations deploy VPNs to provide data integrity, authentication, and data encryption to ensure confidentiality of the packets sent over an unprotected network or the Internet. VPNs are designed to avoid the cost of unnecessary leased lines. You can also strategically enable NetFlow in VPN termination points for both site-to-site and remote-access VPN scenarios.

Remote-access VPNs enable users to work from remote locations such as their homes, hotels, and other premises as if they were directly connected to their corporate network.

Figure 5-18 shows how two remote-access VPN clients connect to a Cisco ASA in the corporate network.

The remote-access clients use the Cisco AnyConnect Secure Mobility Client. One is creating a Secure Sockets Layer (SSL) VPN tunnel, and the other is creating an IPsec tunnel.

NetFlow Secure Event Logging (NSEL) can be enabled at the Cisco ASA to monitor traffic from any type of remote-access client VPN termination.

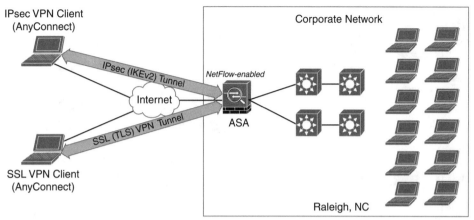

Figure 5-18 *NetFlow in Remote-Access VPNs*

You can also configure the Cisco AnyConnect client with the Network Visibility Module (NVM). NVM collects network flow information from an endpoint on or off premises. This allows you to maintain visibility into network-connected devices and user behaviors by sending this flow information to Cisco Stealthwatch. You can also take advantage of this feature to perform capacity and service planning, auditing, compliance, and security analytics. NVM monitors application use by leveraging expanded IPFIX elements. You can also use NVM to classify logical groups of applications, users, or endpoints.

TIP   The Cisco AnyConnect NVM sends flow information only when it is on the trusted network. Visibility (flow) data is collected only when configured in the AnyConnect NVM profile. If collection is done on an untrusted network, it is cached and sent when the endpoint is on a trusted network. This is possible by leveraging another feature called trusted network detection (TDN). You will learn more about AnyConnect and different VPN deployment models in Chapter 8, "Virtual Private Networks (VPNs)."

Site-to-site (otherwise known as LAN-to-LAN) VPNs enable organizations to establish VPN tunnels between two or more network infrastructure devices in different sites so that they can communicate over a shared medium such as the Internet. Many organizations use IPsec, generic routing encryption (GRE), Dynamic Multipoint VPN (DMVPN), FlexVPN, and Multiprotocol Label Switching (MPLS) VPN as site-to-site VPN protocols. Typically, site-to-site VPN tunnels are terminated between two or more network infrastructure devices, whereas remote-access VPN tunnels are formed between a VPN headend device and an end-user workstation or hardware VPN client. Figure 5-19 shows a site-to-site VPN example.

In Figure 5-19, two site-to-site VPNs are terminated in the router (R1) at the corporate headquarters. One of the tunnels connects to a router (R2) at a business partner in Boston, Massachusetts, and the second tunnel connects the corporate headquarters to a remote branch office router (R3) in Raleigh, North Carolina. In this example, NetFlow is enabled in R1 to monitor traffic coming from both of the remote locations.

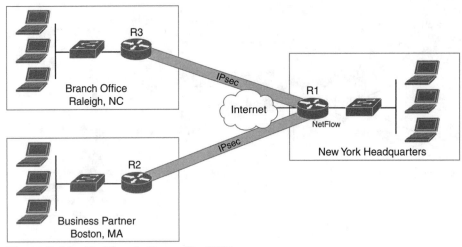

Figure 5-19 *NetFlow in Site-to-Site VPNs*

Cisco Stealthwatch

Cisco acquired Lancope several years ago and further developed their Stealthwatch solution. The Cisco Stealthwatch solution uses NetFlow telemetry and contextual information from the Cisco network infrastructure. This solution allows network administrators and cybersecurity professionals to analyze network telemetry in a timely manner to defend against advanced cyber threats, including the following:

- Network reconnaissance

- Malware proliferation across the network for the purpose of stealing sensitive data or creating back doors to the network

- Communications between the attacker (or command-and-control servers) and the compromised internal hosts

- Data exfiltration

Cisco Stealthwatch aggregates and normalizes considerable amounts of NetFlow data to apply security analytics to detect malicious and suspicious activity. The Cisco Stealthwatch Management Console (SMC) provides a rich graphical unit interface (GUI) with many visualizations and telemetry information.

The following are the primary components of the Cisco Stealthwatch solution:

- **FlowCollector:** A physical or virtual appliance that collects NetFlow data from infrastructure devices.

- **Stealthwatch Management Console (SMC):** The main management application that provides detailed dashboards and the ability to correlate network flow and events.

- **Flow licenses:** Required to aggregate flows at the Stealthwatch Management Console. (Flow licenses define the volume of flows that may be collected.)

The following are optional components of the Cisco Stealthwatch System:

- **FlowSensor:** A physical or virtual appliance that can generate NetFlow data when legacy Cisco network infrastructure components are not capable of producing line-rate, unsampled NetFlow data.

- **FlowReplicator:** A physical appliance used to forward NetFlow data as a single data stream to other devices. The FlowReplicator is also known as the UDP Director.

Figure 5-20 illustrates the components of the Cisco Stealthwatch solution and its integration with Cisco ISE for identity services and Cisco Talos for threat intelligence ingest.

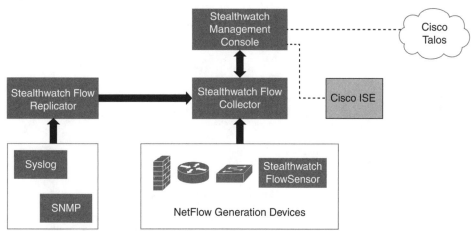

Figure 5-20   *The Cisco Stealthwatch Solution Components*

TIP   Cisco Stealthwatch components can be deployed as physical or virtual appliances. The two minimum required components are the SMC and the Stealthwatch Flow Collector.

Stealthwatch Cloud

Stealthwatch Cloud is a Software as a Service (SaaS) cloud solution. You can use Stealthwatch Cloud to monitor many different public cloud environments, such as Amazon's AWS, Google Cloud Platform, and Microsoft Azure. All of these cloud providers support their own implementation of NetFlow:

- In Amazon AWS, the equivalent of NetFlow is called VPC Flow Logs. You can obtain detailed information about VPC Flow Logs in AWS at https://docs.aws.amazon.com/vpc/latest/userguide/flow-logs.html.

- Google Cloud Platform also supports VPC Flow Logs (or Google-branded GPC Flow Logs). You can obtain detailed information about VPC Flow Logs in Google Cloud Platform at https://cloud.google.com/vpc/docs/using-flow-logs.

- In Microsoft's Azure, traffic flows are collected in Network Security Group (NSG) flow logs. NSG flow logs are a feature of Network Watcher. You can obtain additional information about Azure's NSG flow logs and Network Watcher at https://docs.microsoft.com/en-us/azure/network-watcher/network-watcher-nsg-flow-logging-overview.

Figure 5-21 shows the Cisco Stealthwatch Cloud AWS Visualizations Network Graph. The AWS Visualizations Network Graph allows you to explore the nodes you have deployed in AWS, and when you mouse over each node, additional information is provided.

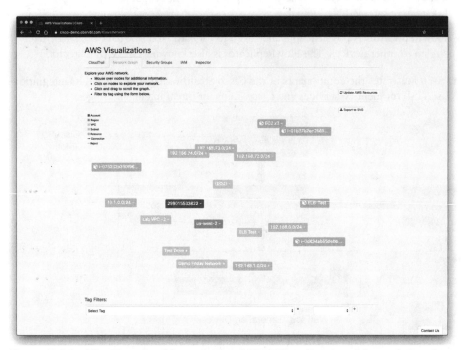

Figure 5-21   *The Cisco Stealthwatch Cloud AWS Visualizations Network Graph*

Figure 5-22 shows the Cisco Stealthwatch Cloud AWS Visualizations Security Groups screen, which allows you to explore the AWS security groups that are deployed.

Figure 5-23 shows the Cisco Stealthwatch Cloud AWS Visualizations Identity and Access Management (IAM) screen. The AWS Visualizations IAM screen allows you to explore the AWS IAM permissions. You can obtain detailed information about AWS IAM at https://aws.amazon.com/iam/.

If you navigate to the **Alerts** tab in the Cisco Stealthwatch Cloud portal, you can drill down to any of the alerts being generated. Figure 5-24 shows the details about an alert (Excessive Access Attempts).

You can click any of the "supporting observations" shown at the bottom of Figure 5-24 to get additional details. Figure 5-25 shows the details of one of the supporting observations. In Figure 5-25, you can see a visualization of the history of the device metrics, the different IP addresses associated with such an alert, as well as other traffic statistics.

You can perform profiling of the applications and associated connections and then obtain the results under the **Profiling** tab, as shown Figure 5-26.

If you navigate to **Dashboard > Network Dashboard**, you can obtain very detailed visualizations of everything that is happening in the observed environment. Figure 5-27 shows the Cisco Stealthwatch Cloud Network Dashboard.

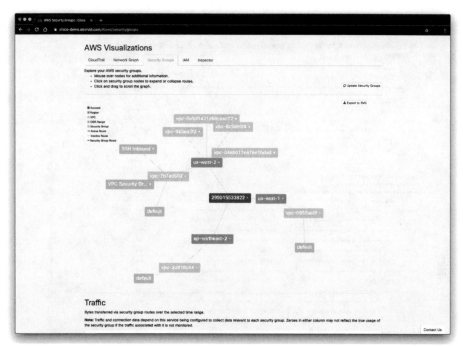

Figure 5-22 *The Cisco Stealthwatch Cloud AWS Visualizations Security Groups*

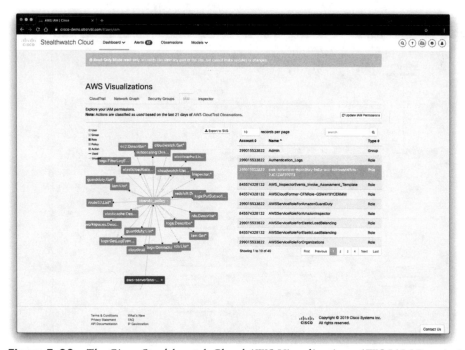

Figure 5-23 *The Cisco Stealthwatch Cloud AWS Visualizations AWS IAM Permissions*

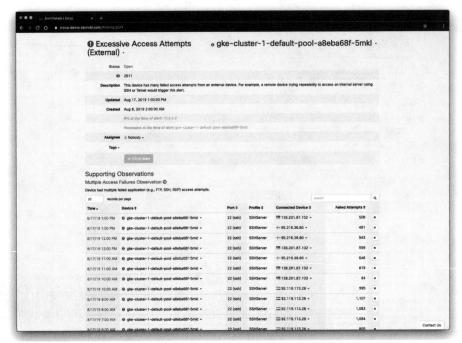

Figure 5-24 *Excessive Access Attempts Alert in Cisco Stealthwatch Cloud*

Figure 5-25 *Details of the Supporting Observations in Cisco Stealthwatch Cloud*

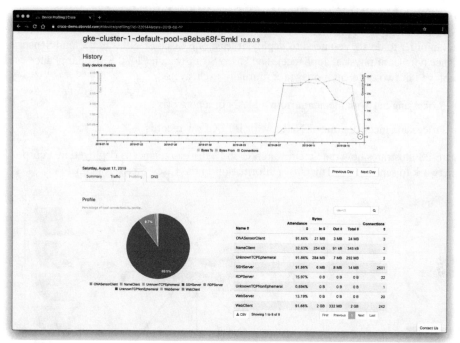

Figure 5-26 *Connection Profiling in Cisco Stealthwatch Cloud*

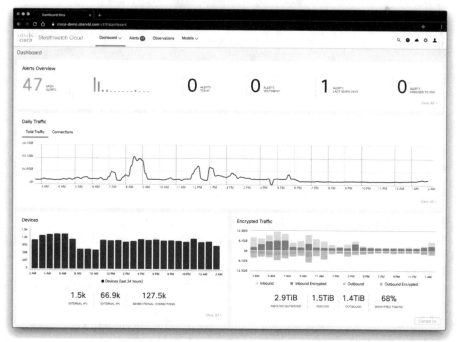

Figure 5-27 *The Cisco Stealthwatch Cloud Network Dashboard*

On-Premises Monitoring with Cisco Stealthwatch Cloud

You can also monitor on-premises networks in your organizations using Cisco Stealthwatch Cloud. In order to do so, you need to deploy at least one Cisco Stealthwatch Cloud Sensor appliance (virtual or physical appliance). The Cisco Stealthwatch Cloud Sensor appliance can be deployed in two different modes (not mutually exclusive):

- Processing network metadata from a SPAN or a network TAP

- Processing metadata out of NetFlow or IPFIX flow records

Figure 5-28 illustrates how the Stealthwatch Cloud Sensor appliance is deployed in a corporate network to send network metadata information to the Cisco Stealthwatch Cloud.

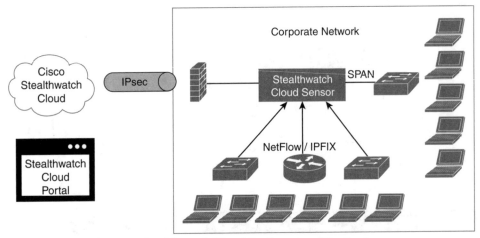

Figure 5-28 *The Cisco Stealthwatch Cloud On-Premises Monitoring*

Cisco Stealthwatch Cloud Integration with Meraki and Cisco Umbrella

Cisco Stealthwatch Cloud can also be integrated with Meraki and Cisco Umbrella. The following document details the integration with Meraki: https://www.cisco.com/c/dam/en/us/td/docs/security/stealthwatch/cloud/portal/SWC_Meraki_DV_1_1.pdf.

The following document outlines how to integrate Stealthwatch Cloud with the Cisco Umbrella Investigate REST API in order to provide additional information in the Stealthwatch Cloud environment for external entity domain information: https://www.cisco.com/c/dam/en/us/td/docs/security/stealthwatch/cloud/portal/SWC_Umbrella_DV_1_0.pdf.

Exploring the Cisco Stealthwatch On-Premises Appliances

The Cisco Stealthwatch Security Insight Dashboard displayed in Figure 5-29 can be used to quickly see the events that have triggered alarms within your premises. Some of the graphics provide data for the past 7 days (such as the Alarms by Type), and some information concerns data for the last 24 hours (such as the Today's Alarms statistics, the Flow Collection Trend, and the Top Applications visualization).

If you click any of the alarming hosts, you can drill down to the specifics of each host and associated traffic, as demonstrated in Figure 5-30.

Figure 5-29 *The Cisco Stealthwatch On-Premises Security Insight Dashboard*

Figure 5-30 *The Cisco Stealthwatch Host Report*

In the Host Report page shown in Figure 5-30, you can view information about a single host's activity as far back as the last 7 days to determine if there is a security risk associated with that host. The information shown pertains to an individual host. You can also use this page to assign a host to one or more host groups.

You can also perform very detailed flow searches by navigating to **Analyze > Flow Search**, as demonstrated in Figure 5-31.

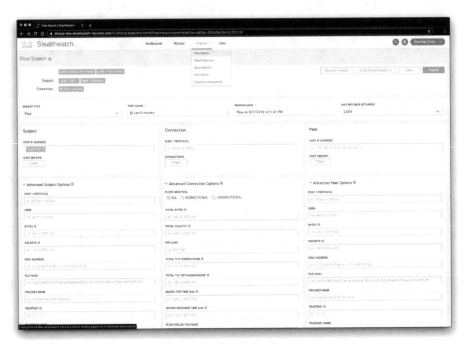

Figure 5-31   *The Cisco Stealthwatch Flow Search*

The Flow Search page shown in Figure 5-31 is useful as a way of monitoring the behavior of a host or range of hosts at a high level. If you need more detail or want to perform an in-depth forensic investigation, you can then use the **Advanced Options** section. The **Advanced Options** section allows you to define a search using more specific settings than those in the main section. You can use this capability to perform in-depth forensic investigation of suspicious behavior involving a specific host or for a specific time interval. You can use the **Advanced Options** section to fine tune settings, so the search returns the data that is most relevant for your cybersecurity forensics investigation.

Figure 5-32 shows the results of the flow search.

Threat Hunting with Cisco Stealthwatch

Threat hunting is the concept of "proactively" or "actively" searching for advanced threats that may evade your security products and capabilities. There are different methodologies for threat hunting:

- Threat intelligence-driven hunting using threat intelligence feeds and reports, malware analysis, and other vulnerability assessment methods.

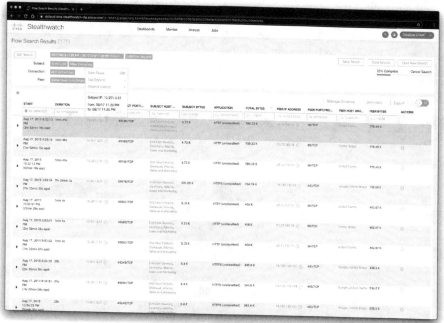

Figure 5-32 *The Cisco Stealthwatch Flow Search Results*

- Machine learning-based or analytics-driven methods.

- Situational or "Crown Jewel" analysis, where you start with a critical system or network in mind and base your "hypothesis" of what an attacker can do to those resources and what indicators should be collected.

Threat hunting is all about a hypothesis of what an advanced threat actor could do to compromise your assets, systems and/or underlying network.

TIP One of the best resources to learn about the techniques and tactics used by advanced threat actors (attackers) is the MITRE ATT&CK. I strongly suggest that you become familiar with the techniques outlined in MITRE's ATT&CK, not solely for preparing for this exam, but also to keep up with the new techniques used by threat actors and to better protect your infrastructure. You can access the MITRE ATT&CK resources at https://attack.mitre.org.

You can use many of the capabilities of Cisco Stealthwatch to perform threat hunting. For instance, the Host Report shown in Figure 5-30 can be used to quickly identify high-level, important information about a specific host, its location, the groups to which it belongs, and the policy assigned to it. You can then start answering questions like the following: Who is this host contacting? Based on the Traffic by Peer Host Group, are there any host groups that raise concerns or questions? Is the volume of traffic reasonable? What is this host doing that I should be concerned about? Are any of the alarm categories at the top of the page showing a volume of alarms that causes concern? If so, you can click the number of active alarms for the category you want to investigate.

You can also use the Application Traffic visualizations to determine which applications this host is using the most. Which ones are transferring the most data? How many are being used to contact inside hosts, and how many are being used to contact outside hosts? Do the results represent normal trends for your system and this host?

You can also navigate to **Monitor > Hosts** to access the **Inside Hosts** dashboard shown in Figure 5-33.

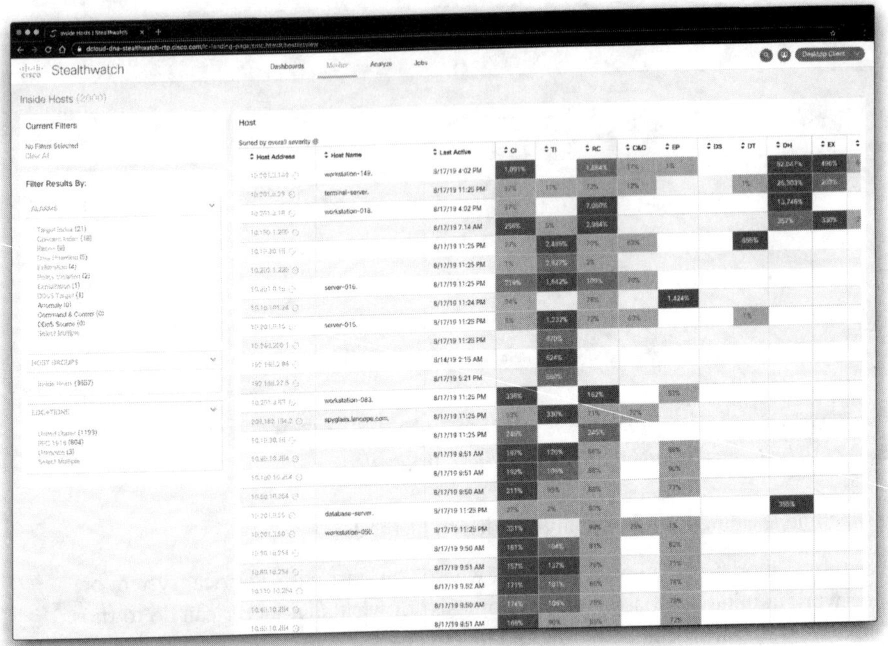

Figure 5-33   *The Cisco Stealthwatch Flow Inside Hosts Dashboard*

In the dashboard shown in Figure 5-33, you can view the inside hosts that have been active on your network in the last hour. You can also see the host group to which they belong, their location, and the categories of alarms the hosts are triggering.

TIP   If you reboot the Cisco Stealthwatch Flow Collector, it deletes all alarm history; however, if you replace your Flow Collector, the new Flow Collector retains the alarm history from the old Flow Collector instead of deleting it.

You can perform a similar analysis for users by navigating to **Monitor > Users**, as shown in Figure 5-34.

In the dashboard shown in Figure 5-34, you can see which users have been active in the last hour, as well as information about their sessions, location, and current alarms.

NOTE   In order to see user data, you must integrate the Cisco Identity Services Engine (ISE) with the Cisco Stealthwatch deployment.

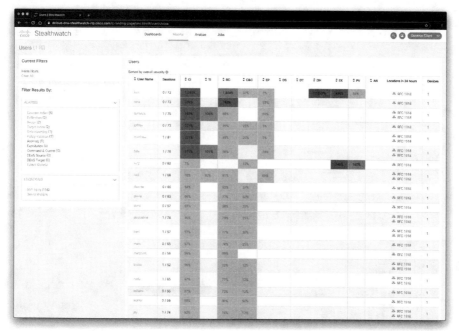

Figure 5-34   *The Cisco Stealthwatch Flow Users Dashboard*

To investigate the activity of a specific user, click that user's name to open the User Report. To investigate an alarm category for a specific user, click the alarm category column in that row.

Another very useful dashboard that can help you perform threat hunting is the Host Group Report. The Top Host Groups by Traffic screen allows you to visualize the top 10 inside and outside host groups with which the current host group has communicated within the last 12 hours. When you hover your cursor over a line connecting two host groups, the total amount of traffic transferred between these two host groups is displayed. The Host Group Report screen is shown in Figure 5-35.

If you click a host group, a column, or the line between two host groups, you can access the context menu from which you can run a flow search or top report to view the traffic associated with the two host groups to conduct a deeper analysis into specific areas of interest. To view flows associated with a particular host, host group, or application, run a flow search. To view top reports that will give more in-depth information about specific areas of interest, run a top report search.

In Figure 5-35, the Concern Index (CI) indicates that the host's concern index has either exceeded the CI threshold or rapidly increased. The Target Index indicates that the target IP (or host) has been the recipient of more than an acceptable number of scans or other malicious attacks.

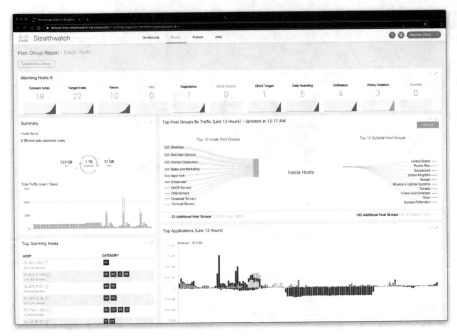

Figure 5-35   *The Cisco Stealthwatch Flow Host Group Report*

Cisco Cognitive Threat Analytics (CTA) and Encrypted Traffic Analytics (ETA)

This section includes information about two solutions that integrate with Cisco Stealthwatch: Cisco Cognitive Threat Analytics (CTA) and Cisco Encrypted Traffic Analytics (ETA).

What Is Cisco ETA?

Cisco ETA can identify malicious (malware) communications in encrypted traffic through passive monitoring, the extraction of relevant data elements, and a combination of behavioral modeling and machine learning. Cisco ETA is able to do this without decrypting the packets traversing the network.

The Cisco ETA components are as follows:

- NetFlow
- Cisco Stealthwatch
- Cisco Cognitive Threat Analytics

What Is Cisco Cognitive Threat Analytics?

Cisco Cognitive Threat Analytics (CTA) is a cloud-based Cisco solution that uses machine learning and statistical modeling of networks. Cisco CTA creates a baseline of the traffic in your network and identifies anomalies. Cisco CTA can also analyze user and device behavior, as well as web traffic to uncover malicious command-and-control communications and data exfiltration.

You can combine Cisco CTA, Cisco Stealthwatch, and Cisco ETA to provide a very comprehensive visibility solution within your organization. In the following section you will learn how Cisco Stealthwatch integrates with Cisco CTA and Cisco ETA to provide visibility of malicious encrypted communications without actually decrypting any packets.

Let's start by analyzing potential ransomware encrypted traffic in the network. In Figure 5-29, you saw a user called "Dustin Hilton" that was flagged as an affected user under the Cognitive Threat Analytics widget in the Stealthwatch Security Insight dashboard. If you click that user, the screen shown in Figure 5-36 is displayed.

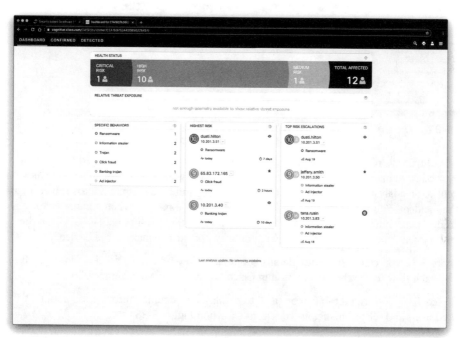

Figure 5-36 *The Cisco Cognitive Threat Analytics Dashboard*

You must have Cisco CTA configured on your network to display the information shown in Figure 5-36. Cognitive Threat Analytics quickly detects suspicious web traffic and/or Net-Flow and responds to attempts to establish a presence in your environment and to attacks that are already underway. The Cisco CTA main dashboard identifies the suspicious traffic and displays the number of affected users or hosts, sorted by incident risk, that Cisco CTA has identified as active on your network. Additionally, it shows the top affected hosts with the incident type. You can hover the incident risk number for more information about the incident risk type.

When you click the user in the Cisco CTA main dashboard, the screen shown in Figure 5-37 is displayed.

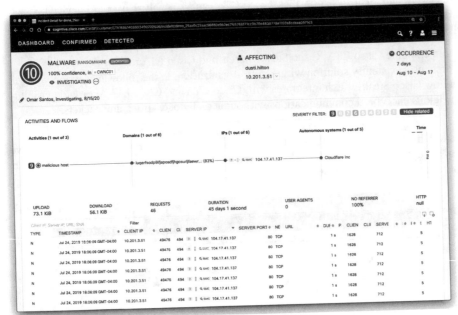

Figure 5-37 *The Cisco Cognitive Threat Analytics Malware Investigation*

You should always review the connected lines and check activities that are tied to the high-severity ones. In many cases, hosts may contain more malware traffic. You can assign an incident to an analyst to investigate. In the example in Figure 5-37, Omar Santos is investigating the incident. You can mark the incident after investigation. Even when communication is blocked, the machine is still infected. You should always aim to full remediation. Clean the infected system, and in case of reinfection, you may consider reimaging the device.

In Figure 5-37, you can also see the details about each flow. You can also click on the SMC icon by each flow to see the information in Cisco Stealthwatch Management Console.

The Cisco CTA Confirmed tab can be used to display very detailed information about affected users and related malicious traffic, as shown in Figure 5-38.

In the screen shown in Figure 5-38, you can add notes to the specific threat, see the affected users, and each of the malware occurrences. You can also see the infection history and the types of risks in the graphics displayed on the right of the screen. In addition, you can scroll down in the page and see statistics from Cisco's Global Intelligence (Threat Grid), as shown in Figure 5-39. Threat Grid is the name of a company that Cisco acquired several years ago. The Threat Grid solution provides a way to analyze malware in a sandbox. Threat Grid combines advanced sandboxing capabilities with threat intelligence, and it is now integrated in many Cisco solutions such as Cisco CTA and Cisco AMP for Endpoints.

Figure 5-38 *The Cisco Cognitive Threat Analytics Confirmed Dashboard*

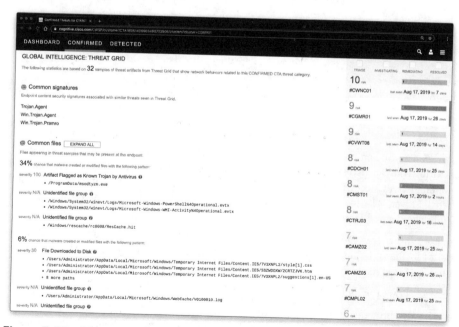

Figure 5-39 *The Cisco Cognitive Threat Analytics Global Threat Intelligence and Threat Grid Integration*

In Figure 5-39, you can see that the statistics are based on 32 samples of threat artifacts from Threat Grid that show network behaviors related to this "Confirmed" CTA threat category. You can also see the endpoint content security signatures associated with similar threats (Trojan.Agent, Win.Trojan.Agent, and Win.Trojan.Pramro) and common files appearing in threat samples that may be present at the endpoint.

If you keep scrolling down on the same page, the screen shown in Figure 5-40 is displayed.

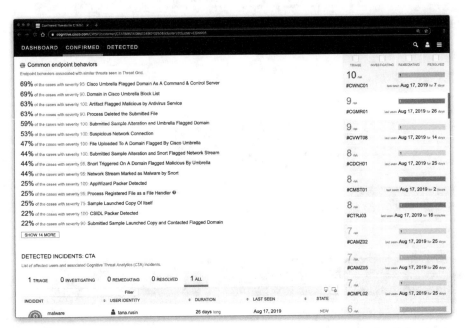

Figure 5-40   *The Cisco Cognitive Threat Analytics Common Endpoint Behaviors and Integration with Cisco Umbrella*

In Figure 5-40, you can see the endpoint behaviors associated with similar threats seen in Threat Grid. You can also see that some of the entries include threat intelligence information also coming from Cisco Umbrella, such as "Cisco Umbrella Flagged Domain As A Command & Control Server," "Domain in Cisco Umbrella Block List," "Submitted Sample Alteration and Umbrella Flagged Domain," and others.

Figure 5-41 shows the "Detected" dashboard of Cisco CTA. In there you can see the number of incidents being triaged, investigated, remediated, resolved, or all of the incidents. In Figure 5-41, the malware (ransomware) incident is displayed.

Figure 5-42 shows the incidents being remediated.

Cisco CTA also provides a high-level incident response guide describing how the CTA system assigns each incident a risk value based on expected damage. The incident response guide is shown in Figure 5-43.

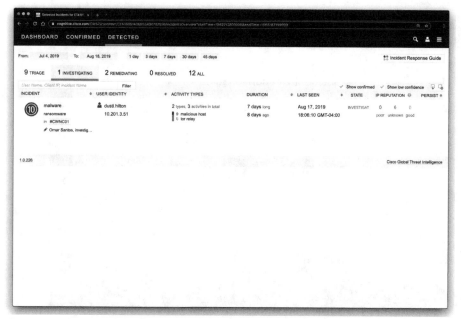

Figure 5-41   *The Cisco Cognitive Threat Analytics Detected Dashboard Incident Being Investigated*

Figure 5-42   *The Cisco Cognitive Threat Analytics Detected Dashboard Incidents Being Remediated*

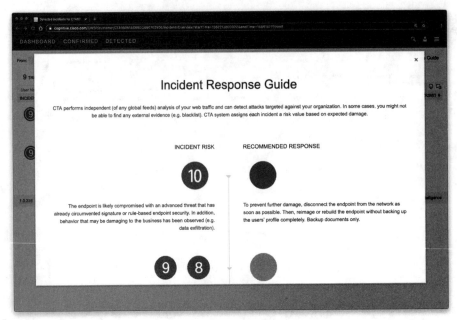

Figure 5-43   *The Cisco Cognitive Threat Analytics Incident Response Guide*

NetFlow Collection Considerations and Best Practices

Let's switch back to NetFlow. The following are several best practices and general recommendations when preparing and designing where to enable NetFlow in your organization:

- **Minimizing NetFlow overhead:** NetFlow collection should be done as close to the NetFlow generator as possible. For instance, in the data center, NetFlow can be enabled close to the servers or assets you want to monitor. Another example is in access switches of a network segment where the users you want to monitor reside.

- **Asymmetric routing considerations:** All devices in the asymmetric route should send NetFlow records to the same collector, not to different collectors.

- **Distributed deployment:** In a distributed deployment, FlowCollectors are deployed at multiple sites and are usually placed close to the source producing the highest number of NetFlow records. This deployment has the advantage of limiting the overhead introduced by NetFlow.

- **Centralized deployment:** In a centralized deployment, all NetFlow collectors are placed in a single location. In some cases, the collectors can be configured behind a load balancer. This provides the benefit of a single collection location and possibly a single IP address globally for NetFlow collection. This deployment offers advantages in environments where NetFlow generators are far apart.

- **Bandwidth consideration:** It is not recommended to collect NetFlow over wide area network (WAN) connections because there might be limitations in bandwidth between sites to consider, as well. You should plan ahead and identify the monitoring locations that make more sense for your environment.

Determining the Flows per Second and Scalability

One of the most important steps in the planning and design of NetFlow deployments is to determine and measure the flows per second (fps) volume that will be generated by the monitoring locations. The volume or fps indicates how many records the collectors must be able to receive and analyze.

Many different factors affect the volume of NetFlow records generated by network infrastructure devices. Forecasting an exact number can be difficult. As a general rule, a NetFlow-enabled device can generate between 1000 and 5000 fps per 1 gigabit per second (Gbps) of traffic passing through such a device. The number of fps fundamentally depends on the following:

- Number of unique flows passing through the NetFlow-enabled device.

- Number of new connections per second.

- The lifetime of each of the flows. Some flows can be short-lived and others may be long-lived.

The network overhead introduced by NetFlow is also influenced by the number of fps and the NetFlow record size.

Careful planning is required when enabling NetFlow in high-impact areas of your network. You can start by enabling NetFlow in certain areas of your network and becoming familiar with the impact it may have in the rest of your deployment. A few tools are available that you can use to forecast the impact of enabling NetFlow in your network.

> **TIP**   Using random-sampled NetFlow provides NetFlow data for a subset of traffic in a Cisco router by processing only one randomly selected packet out of a series of sequential packets. The number of packets or "randomness" is a user-configurable parameter. This type of statistical traffic sampling substantially reduces the overheard in CPU and other resources while providing NetFlow data. However, the main use of random-sampled NetFlow is for traffic engineering, capacity planning, and applications where full NetFlow is not needed for an accurate view of network traffic.

Configuring NetFlow in Cisco IOS and Cisco IOS-XE

In this section, you will learn how to configure NetFlow and specifically Flexible NetFlow. Flexible NetFlow provides enhanced optimization of the network infrastructure, reduces costs, and improves capacity planning and security detection beyond other flow-based technologies available today. Flexible NetFlow supports IPv6 and Network-Based Application Recognition (NBAR) 2 for IPv6 starting in Cisco IOS Software Version 15.2(1)T. It also supports IPv6 transition techniques (IPv6 inside IPv4). Flexible NetFlow can detect the following tunneling technologies that give full IPv6 connectivity for IPv6-capable hosts that are on the IPv4 Internet but that have no direct native connection to an IPv6 network:

- Teredo

- Intra-Site Automatic Tunnel Addressing Protocol (ISATAP)

- 6to4

- 6rd

Simultaneous Application Tracking

Flexible NetFlow tracks different applications simultaneously. For instance, security monitoring, traffic analysis, and billing can be tracked separately, and the information customized per application.

Flexible NetFlow allows the network administrator or security professional to create multiple flow caches or information databases to track. Conventionally, NetFlow has a single cache, and all applications use the same cache information. Flexible NetFlow supports the collection of specific security information in one flow cache and traffic analysis in another. Consequently, each NetFlow cache serves a different purpose. For instance, multicast and security information can be tracked separately and the results sent to two different collectors. Figure 5-44 shows the Flexible NetFlow model and how three different monitors are used. Monitor 1 exports Flexible NetFlow data to Exporter 1, Monitor 2 exports Flexible NetFlow data to Exporter 2, and Monitor 3 exports Flexible NetFlow data to Exporter 1 and Exporter 3.

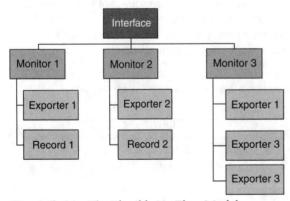

Figure 5-44 *The Flexible NetFlow Model*

The following are the Flexible NetFlow components:

- Records

- Flow monitors

- Flow exporters

- Flow samplers

In Flexible NetFlow, the administrator can specify what to track, resulting in fewer flows. This helps to scale in busy networks and use fewer resources that are already taxed by other features and services.

Flexible NetFlow Records

Records are a combination of key and non-key fields. In Flexible NetFlow, records are appointed to flow monitors to define the cache that is used for storing flow data. There are seven default attributes in the IP packet identity or "key fields" for a flow and for a device to determine whether the packet information is unique or similar to other packets sent over the network. Fields such as TCP flags, subnet masks, packets, and number of bytes are *non-key fields*. However, they are often collected and exported in NetFlow or in IPFIX.

Flexible NetFlow Key Fields

There are several Flexible NetFlow key fields in each packet that is forwarded within a NetFlow-enabled device. The device looks for a set of IP packet attributes for the flow and determines whether the packet information is unique or similar to other packets. In Flexible NetFlow, key fields are configurable, which enables the administrator to conduct a more granular traffic analysis.

Figure 5-45 lists the key fields related to the actual flow, device interface, and Layer 2 services.

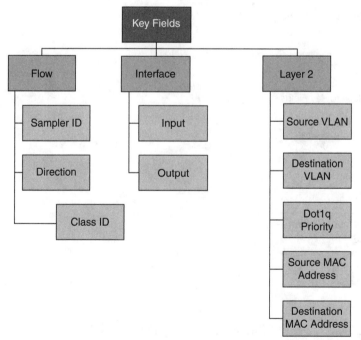

Figure 5-45 *Flexible NetFlow Key Fields Related to Flow, Interface, and Layer 2*

Figure 5-46 lists the IPv4- and IPv6-related key fields.

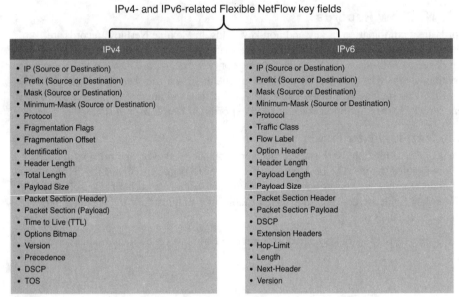

IPv4- and IPv6-related Flexible NetFlow key fields

IPv4	IPv6
• IP (Source or Destination)	• IP (Source or Destination)
• Prefix (Source or Destination)	• Prefix (Source or Destination)
• Mask (Source or Destination)	• Mask (Source or Destination)
• Minimum-Mask (Source or Destination)	• Minimum-Mask (Source or Destination)
• Protocol	• Protocol
• Fragmentation Flags	• Traffic Class
• Fragmentation Offset	• Flow Label
• Identification	• Option Header
• Header Length	• Header Length
• Total Length	• Payload Length
• Payload Size	• Payload Size
• Packet Section (Header)	• Packet Section Header
• Packet Section (Payload)	• Packet Section Payload
• Time to Live (TTL)	• DSCP
• Options Bitmap	• Extension Headers
• Version	• Hop-Limit
• Precedence	• Length
• DSCP	• Next-Header
• TOS	• Version

Figure 5-46 *Flexible IPv4 and IPv6 Key Fields*

Figure 5-47 lists the Layer 3 routing protocol-related key fields.

Layer 3 Routing Protocol–Related Key Fields

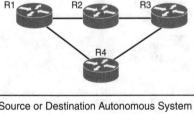

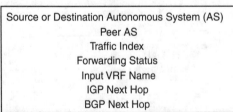

Source or Destination Autonomous System (AS)
Peer AS
Traffic Index
Forwarding Status
Input VRF Name
IGP Next Hop
BGP Next Hop

Figure 5-47 *Flexible NetFlow Layer 3 Routing Protocol Key Fields*

Figure 5-48 lists the transport-related key fields.

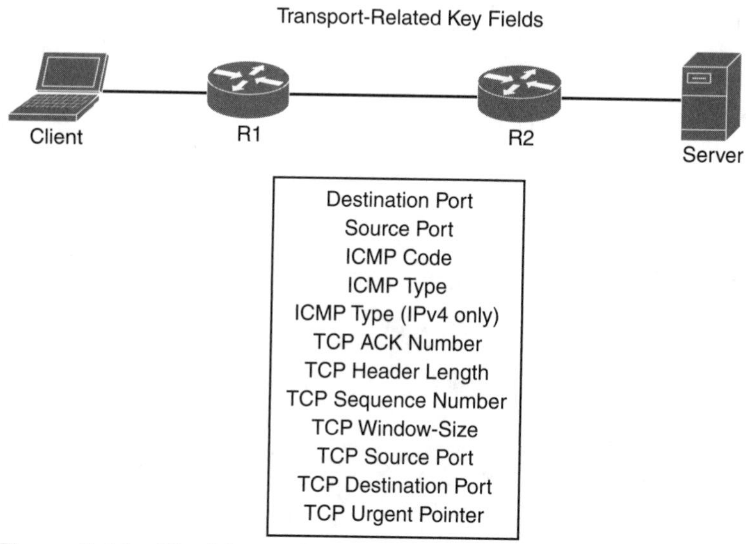
Transport-Related Key Fields
Client
R1
R2
Server
Destination Port
Source Port
ICMP Code
ICMP Type
ICMP Type (IPv4 only)
TCP ACK Number
TCP Header Length
TCP Sequence Number
TCP Window-Size
TCP Source Port
TCP Destination Port
TCP Urgent Pointer

Figure 5-48 *Flexible NetFlow Transport Key Fields*

Figure 5-49 lists the multicast-related key fields.

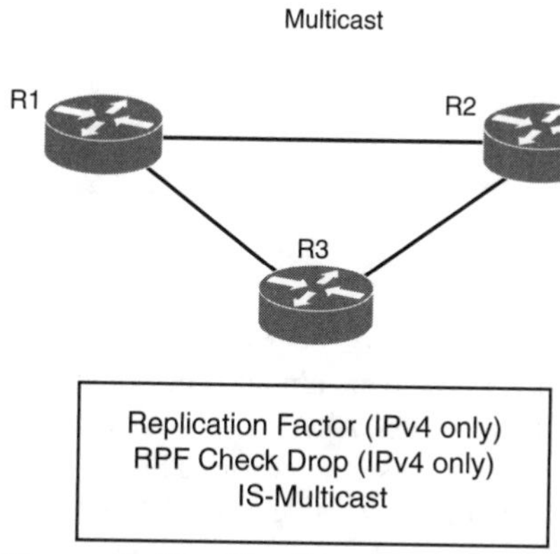
Multicast
R1
R2
R3
Replication Factor (IPv4 only)
RPF Check Drop (IPv4 only)
IS-Multicast

Figure 5-49 *Flexible NetFlow Multicast Key Fields*

Flexible NetFlow Non-Key Fields

There are several non-key Flexible NetFlow fields. Network administrators can use non-key fields for different purposes. For instance, the number of packets and amount of data (bytes) can be used for capacity planning and also to identify denial-of-service (DoS) attacks, in addition to other anomalies in the network.

The following are the non-key fields related to counters:

- byte counts
- bytes long
- bytes square sum
- bytes square sum long

- number of packets

- packets long

- bytes replicated

- bytes replicated long

The following are the timestamp-related non-key fields:

- sysUpTime first packet

- sysUpTime last packet

- absolute first packet

- absolute last packet

The following are the IPv4-only non-key fields:

- total length minimum

- total length maximum

- TTL minimum

- TTL maximum

The following are the IPv4 and IPv6 non-key fields:

- total length minimum

- total length maximum

NetFlow Predefined Records

Flexible NetFlow includes several predefined records that can help an administrator or security professional start deploying NetFlow within their organization. Alternatively, they can create their own customized records for more granular analysis. As Cisco evolves Flexible NetFlow, many popular user-defined flow records could be made available as predefined records to make them easier to implement.

The predefined records guarantee backward compatibility with legacy NetFlow collectors. Predefined records have a unique blend of key and non-key fields that allows network administrators and security professionals to monitor different types of traffic in their environment without any customization.

NOTE Flexible NetFlow predefined records that are based on the aggregation cache schemes in legacy NetFlow do not perform aggregation. Alternatively, the predefined records track each flow separately.

User-Defined Records

As the name indicates, Flexible NetFlow gives network administrators and security professionals the flexibility to create their own records (user-defined records) by specifying key and non-key fields to customize the data collection. The values in non-key fields are added to flows to provide additional information about the traffic in the flows. A change in the value of a non-key field does not create a new flow. In most cases, the values for non-key fields are taken from only the first packet in the flow. Flexible NetFlow enables you to capture counter values such as the number of bytes and packets in a flow as non-key fields.

Flexible NetFlow adds a new NetFlow v9 export format field type for the header and packet section types. A device configured for Flexible NetFlow communicates with the collector what the configured section sizes are in the corresponding NetFlow v9 export template fields.

Flow Monitors

In Flexible NetFlow, *flow monitors* are applied to the network device interfaces to perform network traffic monitoring. Flow data is collected from the network traffic and added to the flow monitor cache during the monitoring process based on the key and non-key fields in the flow record.

Flow Exporters

The entities that export the data in the flow monitor cache to a remote system are called *flow exporters*. Flow exporters are configured as separate entities. Flow exporters are assigned to flow monitors. An administrator can create several flow exporters and assign them to one or more flow monitors. A flow exporter includes the destination address of the reporting server, the type of transport (User Datagram Protocol [UDP] or Stream Control Transmission Protocol [SCTP]), and the export format corresponding to the NetFlow version or IPFIX. You can configure up to eight flow exporters per flow monitor.

Flow Samplers

Flow samplers are created as separate components in a router's configuration. Flow samplers are used to reduce the load on the device running Flexible NetFlow by limiting the number of packets selected for analysis.

Flow sampling exchanges monitoring accuracy for router performance. When you apply a sampler to a flow monitor, the overhead load on the router from running the flow monitor is reduced because the number of packets the flow monitor must analyze is reduced. The reduction in the number of packets that are analyzed by the flow monitor causes a corresponding reduction in the accuracy of the information stored in the flow monitor's cache.

Flexible NetFlow Configuration

The sections that follow provide step-by-step configuration guidance on how to enable and configure Flexible NetFlow in a Cisco IOS or Cisco IOS-XE device. Figure 5-50 lists the Flexible NetFlow configuration steps in a sequential graphical representation.

The topology shown in Figure 5-51 is used in the following examples.

A Cisco ASR 1009-X at the SecretCorp New York headquarters is configured for Flexible NetFlow.

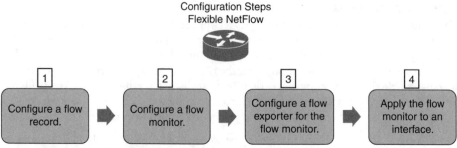

Configuration Steps
Flexible NetFlow

1		2		3		4
Configure a flow record.	➡	Configure a flow monitor.	➡	Configure a flow exporter for the flow monitor.	➡	Apply the flow monitor to an interface.

Figure 5-50 *Flexible NetFlow Configuration Steps*

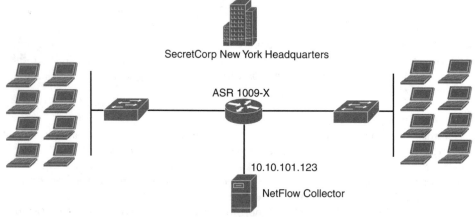

SecretCorp New York Headquarters

ASR 1009-X

10.10.101.123

NetFlow Collector

Figure 5-51 *NetFlow Configuration Example Topology*

Configure a Flow Record

The following are the steps required to configure a customized flow record.

> **TIP** There are hundreds of possible ways to configure customized flow records. The following steps can be followed to create one of the possible variations. You can create a customized flow record depending on your organization's requirements.

Step 1.   Log in to your router and enter into enable mode with the **enable** command:

```
NY-ASR1009-X>enable
```

Step 2.   Enter into configuration mode with the **configure terminal** command:

```
NY-ASR1009-X#configure terminal
Enter configuration commands, one per line.  End with
CNTL/Z.
```

Step 3.   Create a flow record with the **flow record** command. In this example, the record name is NY-ASR-FLOW-RECORD-1. After you enter the **flow record** command, the router enters flow record configuration mode. You can also use the **flow record** command to edit an existing flow record:

```
NY-ASR1009-X(config)# flow record NY-ASR-FLOW-RECORD-1
```

Step 4.   (Optional) Enter a description for the new flow record:

```
NY-ASR1009-X(config-flow-record)# description FLOW
RECORD 1 for basic traffic analysis
```

Step 5.   Configure a key field for the flow record using the **match** command. In this example, the IPv4 destination address is configured as a key field for the record:

```
NY-ASR1009-X(config-flow-record)# match ipv4 destination
address
```

The output of the **match ?** command shows all the primary options for the key field categories that you learned earlier in this chapter:

```
NY-ASR1009-X(config-flow-record)# match ?
    application   Application fields
    flow          Flow identifying fields
    interface     Interface fields
    ipv4          IPv4 fields
    ipv6          IPv6 fields
    routing       Routing attributes
    transport     Transport layer fields
```

Step 6.   Configure a non-key field with the **collect** command. In this example, the input interface is configured as a non-key field for the record:

```
NY-ASR1009-X(config-flow-record)# collect interface input
```

The output of the **collect ?** command shows all the options for the non-key field categories that you learned earlier in this chapter:

```
NY-ASR1009-X(config-flow-record)# collect ?
    application   Application fields
    counter       Counter fields
    flow          Flow identifying fields
    interface     Interface fields
    ipv4          IPv4 fields
    ipv6          IPv6 fields
    routing       Routing attributes
    timestamp     Timestamp fields
    transport     Transport layer fields
```

Step 7.   Exit configuration mode with the **end** command and return to privileged EXEC mode:

```
NY-ASR1009-X(config-flow-record)# end
```

NOTE You can configure Flexible NetFlow to support NBAR with the **match application** *name* command under Flexible NetFlow flow record configuration mode.

You can use **show flow record** to show the status and fields for the flow record. If multiple flow records are configured in the router, you can use the **show flow record** *name* command to show the output of a specific flow record, as shown here:

```
NY-ASR1009-X# show flow record NY-ASR-FLOW-RECORD-1

flow record NY-ASR-FLOW-RECORD-1:
   Description:         Used for basic traffic analysis
   No. of users:        0
   Total field space:   8 bytes
   Fields:
      match ipv4 destination address
      collect interface input
```

Use the **show running-config flow record** command to show the flow record configuration in the running configuration, as shown next:

```
NY-ASR1009-X# show running-config flow record

Current configuration:

!

flow record NY-ASR-FLOW-RECORD-1
 description Used for basic traffic analysis
 match ipv4 destination address
 collect interface input
```

Configure a Flow Monitor for IPv4 or IPv6

The following are the steps required to configure a flow monitor for IPv4 or IPv6 implementations. In the following examples, a flow monitor is configured for the previously configured flow record.

Step 1. Log in to your router and enter into enable mode with the **enable** command:

```
NY-ASR1009-X>enable
```

Step 2. Enter into configuration mode with the **configure terminal** command:

```
NY-ASR1009-X# configure terminal
Enter configuration commands, one per line.  End with
CNTL/Z.
```

Step 3. Create a flow monitor with the **flow monitor** command. In this example, the flow monitor is called **NY-ASR-FLOW-MON-1**:

```
NY-ASR1009-X(config)# flow monitor NY-ASR-FLOW-MON-1
```

Step 4.   (Optional) Enter a description for the new flow monitor:

```
NY-ASR1009-X(config-flow-monitor)# description monitor
for IPv4 traffic in NY
```

Step 5.   Identify the record for the flow monitor:

```
NY-ASR1009-X(config-flow-monitor)# record netflow
NY-ASR-FLOW-RECORD-1
```

In the following example, the **record ?** command is used to see all the flow monitor **record** options:

```
NY-ASR1009-X(config-flow-monitor)# record ?
  NY-ASR-FLOW-RECORD-1   Used for basic traffic analysis
  netflow                Traditional NetFlow collection
                         schemes
  netflow-original       Traditional IPv4 input NetFlow
                         with origin ASs
```

Step 6.   Exit configuration mode with the **end** command and return to privileged EXEC mode:

```
NY-ASR1009-X(config-flow-record)# end
```

You can use **show flow monitor** to show the status and configured parameters for the flow monitor, as shown next:

```
NY-ASR1009-X# show flow monitor

Flow Monitor NY-ASR-FLOW-MON-1:

  Description:        monitor for IPv4 traffic in NY

  Flow Record:        NY-ASR-FLOW-RECORD-1

  Cache:

    Type:             normal (Platform cache)

    Status:           not allocated

    Size:             200000 entries

    Inactive Timeout: 15 secs

    Active Timeout:   1800 secs

    Update Timeout:   1800 secs
```

Use the **show running-config flow monitor** command to display the flow monitor configuration in the running configuration, as shown here:

```
NY-ASR1009-X# show running-config flow monitor

Current configuration:

!
```

```
flow monitor NY-ASR-FLOW-MON-1

 description monitor for IPv4 traffic in NY

 record NY-ASR-FLOW-RECORD-1

 cache entries 200000
```

Configure a Flow Exporter for the Flow Monitor

Complete the following steps to configure a flow exporter for the flow monitor to export the data that is collected by NetFlow to a remote system for further analysis and storage. This is an optional step. IPv4 and IPv6 are supported for flow exporters.

> **NOTE**   Flow exporters use UDP as the transport protocol and use the NetFlow v9 export format. Each flow exporter supports only one destination. If you want to export the data to multiple destinations, you must configure multiple flow exporters and assign them to the flow monitor.

Step 1.   Log in to the router and enter into enable and configuration mode, as you learned in previous steps.

Step 2.   Create a flow exporter with the **flow exporter** command. In this example, the exporter name is NY-EXPORTER-1:

```
NY-ASR1009-X(config)# flow exporter NY-EXPORTER-1
```

Step 3.   (Optional) Enter a description for the exporter:

```
NY-ASR1009-X(config-flow-exporter)# description exports
to New York Collector
```

Step 4.   Configure the export protocol using the **export-protocol** command. In this example, NetFlow v9 is used. You can also configure legacy NetFlow v5 with the **netflow-v5** keyword or IPFIX with the **ipfix** keyword. IPFIX support was added in Cisco IOS Software Release 15.2(4)M and Cisco IOS XE Release 3.7S.

```
NY-ASR1009-X(config-flow-exporter)# export-protocol
netflow-v9
```

Step 5.   Enter the IP address of the destination host with the **destination** command. In this example, the destination host is 10.10.10.123:

```
NY-ASR1009-X(config-flow-exporter)# destination
10.10.10.123
```

Step 6.   You can configure the UDP port used by the flow exporter with the **transport udp** command. The default is UDP port 9995.

Step 7.   Exit the Flexible NetFlow flow monitor configuration mode with the **exit** command and specify the name of the exporter in the flow monitor:

```
NY-ASR1009-X(config)# flow monitor NY-ASR-FLOW-MON-1

NY-ASR1009-X(config-flow-monitor)# exporter NY-EXPORTER-1
```

You can use the **show flow exporter** command to view the configured options for the Flexible NetFlow exporter, as demonstrated here:

```
NY-ASR1009-X# show flow exporter

Flow Exporter NY-EXPORTER-1:

  Description:                 exports to New York Collector

  Export protocol:            NetFlow Version 9

  Transport Configuration:

    Destination IP address:   10.10.10.123

    Source IP address:        209.165.200.225

    Transport Protocol:       UDP

    Destination Port:         9995

    Source Port:              55939

    DSCP:                     0x0

    TTL:                      255

    Output Features:          Used
```

You can use the **show running-config flow exporter** command to view the flow exporter configuration in the command-line interface (CLI):

```
NY-ASR1009-X# show running-config flow exporter

Current configuration:

!

flow exporter NY-EXPORTER-1

 description exports to New York Collector

 destination 10.10.10.123
```

You can use the **show flow monitor name NY-ASR-FLOW-MON-1 cache format record** command to display the status and flow data in the NetFlow cache for the flow monitor, as demonstrated here:

```
NY-ASR1009-X# show flow monitor name NY-ASR-FLOW-MON-1 cache
format record

  Cache type:                    Normal  (Platform cache)

  Cache size:                    200000

  Current entries:               4

  High Watermark:                4

  Flows added:                   132
```

```
   Flows aged:                                           42

        - Active timeout    (  3600 secs)               3

        - Inactive timeout  (    15 secs)              94

        - Event aged                                    0

        - Watermark aged                                0

        - Emergency aged                                0

   IPV4 DESTINATION ADDRESS:   10.10.20.5

   ipv4 source address:        10.10.10.42

   trns source port:           25

   trns destination port:      25

   counter bytes:              34320

   counter packets:            1112

   IPV4 DESTINATION ADDRESS:   10.10.1.2

   ipv4 source address:        10.10.10.2

   trns source port:           20

   trns destination port:      20

   counter bytes:              3914221

   counter packets:            5124

   IPV4 DESTINATION ADDRESS:   10.10.10.200

   ipv4 source address:        10.20.10.6

   trns source port:           32

   trns destination port:      3073

   counter bytes:              82723

   counter packets:            8232
```

Apply a Flow Monitor to an Interface

A flow monitor must be applied to at least one interface. To apply the flow monitor to an interface, use the **ip flow monitor** *name* **input** command in interface configuration mode, as demonstrated in Example 5-1.

Example 5-1 *Applying a Flow Monitor to an Interface*

```
NY-ASR1009-X(config)# interface  GigabitEthernet0/1/1
NY-ASR1009-X(config-if)# ip flow monitor NY-ASR-FLOW-MON-1 input
```

The flow monitor NY-ASR-FLOW-MON-1 is applied to interface GigabitEthernet0/1/1.

Example 5-2 shows the complete configuration.

Example 5-2 *Flexible NetFlow Configuration*

```
flow record NY-ASR-FLOW-RECORD-1
 description Used for basic traffic analysis
 match ipv4 destination address
 collect interface input
!
!
flow exporter NY-EXPORTER-1
 description exports to New York Collector
 destination 10.10.10.123
!
!
flow monitor NY-ASR-FLOW-MON-1
 description monitor for IPv4 traffic in NY
 record NY-ASR-FLOW-RECORD-1
 exporter NY-EXPORTER-1
 cache entries 200000
!
interface GigabitEthernet0/1/1
 ip address 209.165.200.233 255.255.255.248

 ip flow monitor NY-ASR-FLOW-MON-1 input
```

Flexible NetFlow IPFIX Export Format

You can also export Flexible NetFlow packets using the IPFIX export protocol. This feature is enabled with the **export-protocol ipfix** subcommand under the flow exporter. Example 5-3 shows how the Flexible NetFlow IPFIX Export Format feature is enabled in the flow exporter configured in the previous example.

Example 5-3 *Flexible NetFlow Configuration*

```
flow exporter NY-EXPORTER-1
 description exports to New York Collector
 destination 10.10.10.123
  export-protocol ipfix
```

Configuring NetFlow in NX-OS

The NetFlow configuration in the Cisco NX-OS software is very similar to the configuration in Cisco IOS and Cisco IOS-XE. The steps are the same:

Step 1.   Define a flow record.

Step 2.   Define a flow exporter.

Step 3.   Define a flow monitor.

Step 4.   Apply the flow monitor to an interface.

NetFlow CLI commands are not available until you enable the NetFlow feature with the **feature netflow** command.

Example 5-4 demonstrates how to define a flow record in Cisco NX-OS software.

Example 5-4 *Defining a Flow Record in Cisco NX-OS*

```
nx-os-sw1(config)# feature netflow
nx-os-sw1(config)# flow record myRecord
nx-os-sw1(config-flow-record)# description Custom-Flow-Record
nx-os-sw1(config-flow-record)# match ipv4 source address
nx-os-sw1(config-flow-record)# match ipv4 destination address
nx-os-sw1(config-flow-record)# match transport destination-port
nx-os-sw1(config-flow-record)# collect counter bytes
nx-os-sw1(config-flow-record)# collect counter packets
```

Example 5-5 demonstrates how to define a flow exporter in the Cisco NX-OS software.

Example 5-5 *Defining a Flow Exporter in Cisco NX-OS software*

```
nx-os-sw1(Config)# flow exporter myExporter
nx-os-sw1(Config-flow-exporter)# destination 192.168.11.2
nx-os-sw1(Config-flow-exporter)# source Ethernet2/2
nx-os-sw1(Config-flow-exporter)# version 9
```

Example 5-6 demonstrates how to define a flow monitor with a custom record in Cisco NX-OS software.

Example 5-6 *Defining a Flow Monitor with a Custom Record in Cisco NX-OS software*

```
nx-os-sw1(config)# flow monitor myMonitor
nx-os-sw1(config-flow-monitor)# description Applied Inbound on Ethernet 2/1
nx-os-sw1(config-flow-monitor)# record myRecord
nx-os-sw1(config-flow-monitor)# exporter myExporter
```

Example 5-7 demonstrates how to define a flow monitor with an original record in Cisco NX-OS software.

Example 5-7 *Defining a Flow Monitor with an Original Record in Cisco NX-OS software*

```
nx-os-sw1(config)# flow monitor myMonitor2
nx-os-sw1(config-Netflow-Monitor)# description monitor using predefined record
nx-os-sw1(config-Netflow-Monitor)# record netflow-original
nx-os-sw1(config-Netflow-Monitor)# exporter myExporter
```

Example 5-8 demonstrates how to adjust the NetFlow timers in Cisco NX-OS software.

Example 5-8   *Adjusting the NetFlow Timers in Cisco NX-OS software*

```
nx-os-sw1(config)# flow timeout active 120
nx-os-sw1(config)# flow timeout inactive 32
nx-os-sw1(config)# flow timeout fast 32 threshold 100
nx-os-sw1(config)# flow timeout session
nx-os-sw1(config)# flow timeout aggressive threshold 75
```

Example 5-9 demonstrates how to configure sampled NetFlow in Cisco NX-OS software.

Example 5-9   *Configuring Sampled NetFlow in Cisco NX-OS software*

```
nx-os-sw1(config)# sampler mySampler
nx-os-sw1(config-flow-sampler)# mode 1 out-of 1000
```

Example 5-10 demonstrates how to apply a NetFlow monitor and sampler to an interface in Cisco NX-OS software.

Example 5-10   *Applying a NetFlow Monitor and Sampler*

```
nx-os-sw1(config)# interface GigabitEthernet2/1
nx-os-sw1(config-if)# ip flow monitor myMonitor input sampler mySampler
```

Introduction to Network Segmentation

Network segmentation is the process of logically grouping network assets, resources, and applications. Segmentation provides the flexibility to implement a variety of services, authentication requirements, and security controls. Working from the inside out, network segments include the following types:

- **Enclave network:** A segment of an internal network that requires a higher degree of protection. Internal accessibility is further restricted through the use of firewalls, VPNs, VLANs, and network access control (NAC) devices.

- **Trusted network (wired or wireless):** The internal network that is accessible to authorized users. External accessibility is restricted through the use of firewalls, VPNs, and IDS/IPS devices. Internal accessibility may be restricted through the use of VLANs and NAC devices.

- **Semi-trusted network, perimeter network, or DMZ:** A network that is designed to be Internet accessible. Hosts such as web servers and email gateways are generally located in the DMZ. Internal and external accessibility is restricted through the use of firewalls, VPNs, and IDS/IPS devices.

- **Guest network (wired or wireless):** A network that is specifically designed for use by visitors to connect to the Internet. There is no access from the guest network to the internal trusted network.

- **Untrusted network:** A network outside your security controls. The Internet is an untrusted network.

Data-Driven Segmentation

In the past, the consolidation and centralization of network infrastructure has been a key driver for segmentation. Previously isolated application infrastructures were migrated to common shared physical and virtual networks that require separation to maintain some level of isolation. Furthermore, numerous organizations host applications in the cloud (or multiple cloud providers) and use Software as a Service (SaaS) offerings. Because of this, network infrastructures have gone through a dramatic shift over the past few years with the introduction of network function virtualization, containers, and the Internet of Things (IoT) dilemma.

Numerous security professionals have used policy enforcement through Layer 2 technologies such as VLANs, virtual routing and forwarding (VRF), and virtual firewalls to provide network segmentation. You may ask yourself, if organizations have been already segmenting their network for many years, why do we still have so many data breaches where the root cause was "the lack of segmentation"? Before answering this question, let's go over a few topics.

Traditionally, network architectures were built by placing the most important data in a well-guarded fortress (the data center). By doing this, you think that all your critical assets are protected by "the perimeter" and nothing can pass through your defenses if not explicitly allowed. What about insiders? What if an unauthorized entity is already inside your fortress? What if the unauthorized individual, malware, or bot has found a way to move your critical data out of your "perimeter"?

If you have limited segmentation and hundreds or thousands of users and applications, you may experience the "N×R" problem, where N is the number of user groups and R is the number of critical resources. In short, in most cases, every user group has access to pretty much every application in the organization. This concept is illustrated in Figure 5-52.

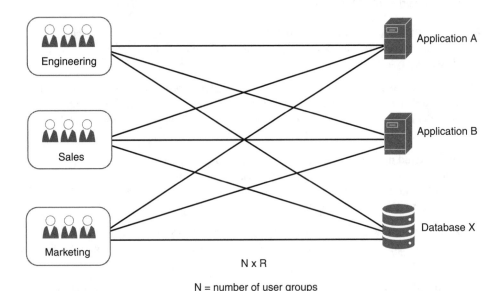

Figure 5-52   *User Group to Resource (N×R) Relationship*

This problem gets worse if access is provided at an individual user level without grouping users by a set of common characteristics. If you use the principle of least privilege, you can try to simplify this problem by explicitly allowing user groups to only access authorized resources they need to carry out their jobs. If you group the authorized user groups and related resources, the magnitude of this issue is reduced significantly. Figure 5-53 demonstrates a one-way segmentation policy allowing user groups to have appropriate access to the authorized resources.

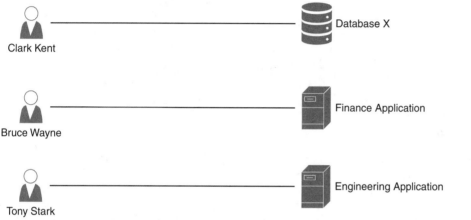

Figure 5-53   *User Group to Resource (N+R) Relationship (one-way segmentation policy)*

Traditionally, applications and services were simpler and not as diverse as they are nowadays. Only a few applications were used throughout an enterprise by a select group of users. These applications were hosted in a data center protected by a set of security and monitoring solutions. Nowadays, new applications are being created at a very fast pace, and those applications integrate with numerous others via APIs, streaming services, and so on. In addition, users are not confined to the physical perimeter of an office. Conventional data protection models are practically obsolete. Data itself could be stored anywhere from within your data center, the cloud (public, private, hybrid), or a business partner's network.

Once-acceptable security measures such as segmenting the network, configuring VLANs, and deploying firewalls are no longer enough. Placing users and apps into VLANs and filtering traffic through access control lists achieves limited traffic separation. With network virtualization, cloud adoption, and the proliferation of devices, it is imperative to look at the entire context of the connection before allowing access to critical data. With cyber threats evolving, providing segmentation strictly at the network layer is not enough to ensure complete data protection.

Segmentation policies should be built based on data gathered about what users, applications, systems, and resources need to communicate. The segmentation policy should start at a high level, outlining the different zones within traditional network boundaries, such as the demilitarized zone (DMZ), data center, Internet edge, campus, then gradually drilling into each zone. This process should continue all the way to the application itself (more on this topic later in this chapter). Once all objects have been discovered, you should develop the segmentation policy based on the type and location of those objects and on the users who are requesting access to various resources hosting or containing data. Figure 5-54 illustrates this concept.

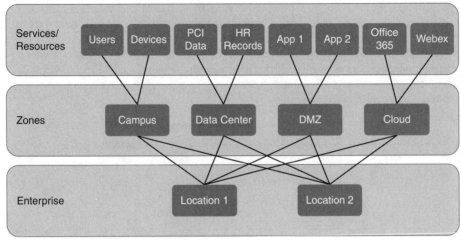

Figure 5-54 *High-level segmentation policy*

In Chapter 1, "Cybersecurity Fundamentals," you learned that there are multiple access control models to choose from. The access control model used depends on the environment. Traditional network engineers are mostly familiar with network-based access control lists (ACLs). While they do provide a way to control access between the larger zones, they are difficult to be implemented in a very large environment and to be able to scale—mainly because they are mostly static and become difficult to manage over time. Many organizations nowadays are using a hybrid model, one that does not rely entirely on Open Systems Interconnection (OSI) Layer 3 and Layer 4 but also takes into consideration multiple access control models and that can provide application segmentation in a granular way.

Application-Based Segmentation

Another dilemma is the machine-to-machine communication between different systems and applications. How do you also segment and protect that in an effective manner?

In today's virtualized and containerized environments, traffic between applications may never leave a physical device or server, as illustrated in Figure 5-55.

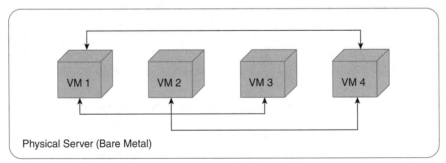

Figure 5-55 *VM Traffic Never Leaving the Physical Server*

This is why micro-segmentation is so popular nowadays. A solution of the past is to include virtual firewalls between VMs, as shown in Figure 5-56.

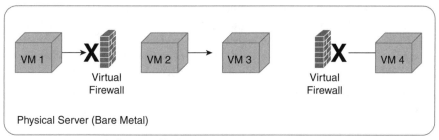

Figure 5-56 *Virtual Firewalls for Segmentation*

Machine-to-machine (or application-to-application) communication also needs to be segmented within an organization. For instance, let's take a look at Figure 5-57. Do your Active Directory (AD) servers need to communicate to Network Time Protocol (NTP) servers? What is their relationship and data interaction?

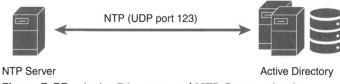

Figure 5-57 *Active Directory and NTP Communication*

NTP implementations send and receive timestamps using UDP port 123. However, NTP servers can also use broadcasting or multicasting, where clients passively listen to time updates after an initial round-trip calibrating exchange. Which of the two need to be allowed or protected? Of course, this is a simple example, and in this case UDP port 123 can be used. However, what if an attacker can tunnel traffic over NTP and exfiltrate information from the organization? All these questions need to be answered when you are creating your segmentation strategy.

Furthermore, do you know what applications are running in your environment? What applications are talking to the cloud? Which are hosted in the cloud? How do you do application enumeration and mapping?

One of the most elegant solutions for micro-segmentations is the one provided by Cisco Application Centric Infrastructure (ACI).

Micro-Segmentation with Cisco ACI

Cisco ACI allows organizations to automatically assign endpoints to logical security zones called endpoint groups (EPGs). EPGs are used to group VMs within a tenant and apply filtering and forwarding policies to them. These EPGs are based on various network-based or VM-based attributes.

A micro-segment in ACI is also often referred to as a uSeg EPG. You can group endpoints in existing application EPGs into new micro-segment (uSeg) EPGs and configure network or VM-based attributes for those uSeg EPGs. With these uSeg EPGs, you can apply dynamic policies. You can also apply policies to any endpoints within the tenant. For instance, let's say that you want to assign web servers to an EPG and then apply similar policies. By default, all endpoints within an EPG can communicate with each other. On the other hand, you can restrict access if this web EPG contains a mix of production and development web servers. To accomplish this, you can create a new EPG and automatically assign endpoints based on their VM name attribute, such as "prod-xxxx" or "dev-xxx."

Micro-segmentation in Cisco ACI can be accomplished by integrating with vCenter or Microsoft System Center Virtual Machine Manager (SCVMM), Cisco ACI API (controller), and leaf switches.

Applying attributes to uSeg EPGs enables you to apply forwarding and security policies with greater granularity than you can to EPGs without attributes. Attributes are unique within the tenant.

You can apply uSeg EPGs using two types of attributes:

- **Network-based attributes:** IP (IP address filter) and MAC (MAC Address Filter). You can apply one or more MAC or IP addresses to a uSeg EPG.

- **VM-based attributes:** You can apply multiple VM-based attributes to a uSeg EPG. The VM-based attributes include Operating System Type, VMM Domain, Hypervisor Identifier, Datacenter, VM Identifier, VM Name, vNIC Domain Name, custom attributes, and tags.

Let's take a look at the example illustrated in Figure 5-58.

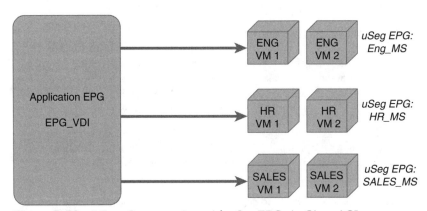

Figure 5-58   *Micro-Segmentation with uSeg EPGs in Cisco ACI*

In Figure 5-58, all Virtual Desktop Infrastructure (VDI) environments from the Engineering, Human Resources (HR), and Sales groups are moved to new uSeg EPGs (Eng_MS, HR_MS, and SALES_MS, respectively) to provide granular policies.

Segmentation with Cisco ISE

In Chapter 4, "Authentication, Authorization, Accounting (AAA) and Identity Management," you learned about Cisco ISE, 802.1X, and the Cisco TrustSec solution. One of the benefits of these technologies is the ability to enforce granular policies and perform network segmentation. The first step when deploying Cisco ISE for segmentation is to know which assets need protecting. After that is accomplished, classification groups and mechanisms can be designed.

You typically create groups of assets, applications, and devices. These group names should be meaningful to the organization and to the protected asset (for instance, production servers belong to the "prod_servers" group, development servers belong to the "dev_servers" group, and so on). Subsequently, the security policy will reference these group names, so using meaningful names makes policy management very simple and straightforward.

You can perform classification in a static or dynamic way. For example, employees, guests, and contractors can be dynamically classified utilizing AAA services from Cisco ISE. If a supplicant is available on the clients, then 802.1X can be utilized, and if no supplicant is available, then MAC Authentication Bypass (MAB) or PassiveID/Easy Connect can be used. You can also configure Cisco ISE to display a web portal for guests to register and access the network (Web-Auth). Cisco ISE can use these methods to dynamically assign the users into the respective groups.

TIP   During the classification process, Security Group Tags (SGTs) are assigned to IP addresses (IPv4 or IPv6). You can obtain the latest platform capability matrix for classification support at the following website: https://www.cisco.com/c/en/us/solutions/enterprise-networks/trustsec/solution-overview-listing.html.

When selecting your classification choices, using the classification precedence can really help when rolling out services. A strict order of precedence for classification methods is documented at the following site: https://community.cisco.com/t5/security-documents/trustsec-troubleshooting-guide/ta-p/3647576#toc-hId-891661665. An example of using the precedence to your advantage is when rolling out an 802.1X supplicant to employees. Plan to deploy VLAN-to-SGT or Subnet-to-SGT mapping initially, to provide a coarse-grained classification, and then slowly roll out 802.1X supplicants to employees where dynamic SGT assignment will take precedence.

You can carry SGTs over VXLAN environments. If VXLAN is configured in your environment, the source SGT can be carried in every packet towards the enforcement point and enforced, as demonstrated in Figure 5-59. VXLAN is the protocol used when propagating the SGT across a Software-Defined Access (SDA) fabric, as shown in the following figure.

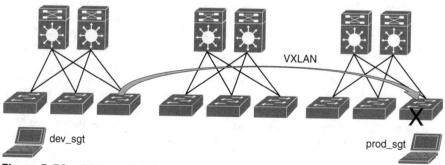

Figure 5-59   *SGTs and VXLANs*

The Scalable Group Tag Exchange Protocol (SXP)

The Scalable Group Tag Exchange Protocol (SXP) is a control plane protocol used to convey IP-to-SGT mappings to network devices when you cannot perform inline tagging. SXP provides capabilities to identify and classify IP packets to corresponding SGTs tracked in the mapping table within network devices. SPX uses peer-to-peer TCP connections over TCP port 64999. IP-to-SGT mappings are sent from the SXP speaker end of the connection to the SXP listener end. Cisco ISE supports both the SXP speaker and listener functionality. You can simplify propagation in your design by using Cisco ISE as an SXP speaker.

Cisco ISE learns the IP address of the user's system when they authenticate onto the network. Then Cisco ISE assigns an SGT via the authorization table. Cisco ISE is the first platform in the network that learns of the IP-to-SGT mapping for dynamically authenticated endpoints. When Cisco ISE is configured for SXP and there is an SXP connection provisioned from ISE to the enforcement point, then the IP-to-SGT mappings for the users will be directly sent to the enforcement point, negating the need for inline tagging or multiple SXP connections from different zones in your network (campus, DMZ, data center, and so on).

The method of sending out IP-to-SGT mappings from ISE is particularly useful if the access switch does not support TrustSec. This process works with any vendor switch or wireless AP that is able to create and teardown sessions on Cisco ISE using RADIUS Authentication and Accounting.

TIP   In Chapter 3, "Software-Defined Networking Security and Network Programmability," and Chapter 4, "Authentication, Authorization, Accounting (AAA) and Identity Management," you learned about the Cisco Platform Exchange Grid (pxGrid). You can use pxGrid to send SGTs and group membership information to Cisco security products like Cisco FTD, Cisco Web Security Appliance (WSA), and Cisco Stealthwatch, as well as third-party platforms. To learn more about the products that support pxGrid, refer to the following website:

https://www.cisco.com/c/en/us/products/security/pxgrid.html

Cisco ISE is a critical component to a segmentation design for many organizations. Cisco ISE interacts with network infrastructure devices and transfers SGTs, IP-to-SGT mapping propagation, policy enforcement, and SGACL download.

TIP   Cisco ISE scales by deploying service instances (called "personas") in a distributed architecture. You learned about the different elements in a distributed Cisco ISE deployment in Chapter 4. As a refresher, a Cisco ISE node can assume the Administration, Policy Service, or Monitoring persona. Such a node can provide different services based on the persona it assumes. Each Cisco ISE node in a deployment can assume the Administration, Policy Service, or Monitoring persona. In a distributed deployment, you can have a primary and secondary Administration node for high availability. You can also have a single or a pair of non-Administration nodes for health check of Administration nodes for automatic failover. Also, you can have a pair of health check nodes or a single health check node for Primary Administration Node (PAN) automatic failover, and you can have one or more Policy Service Nodes (PSNs) for session failover.

Even though all services can be enabled within a single instance, it is recommended to separate the services for increased scale and performance.

Figure 5-60 illustrates a distributed architecture, although for high availability, a PAN, three PSNs, a Monitoring and Troubleshooting (MnT) node, and an optional pxGrid node are deployed.

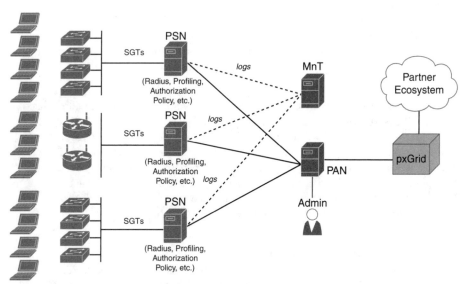

Figure 5-60 *Cisco ISE Distributed Deployment for Network Segmentation*

NOTE When you first deploy a Cisco ISE node, all the default services provided by the Administration, Policy Service, and Monitoring personas will be enabled. The node will be in a standalone mode. You have to connect and log in to the administrator portal to configure it. On the other hand, you cannot edit the personas or services of a standalone Cisco ISE node. Instead, you can edit the personas and services of the primary and secondary Cisco ISE nodes. You must first configure a primary ISE node and then register secondary ISE nodes to the primary ISE node. You should always refer to the latest Cisco ISE performance and scalability guide posted at the following link: https://community.cisco.com/t5/security-documents/ise-performance-amp-scale/ta-p/3642148.

During the assignment of scalable groups in Cisco ISE deployments, most administrators initially think about user groups or server groups. On the other hand, you should also think about "network-device-to-network-device" and "network-device-to-network-services" communications. For instance, network-device-to-network-device communications include things like routing protocols (for example, EIGRP, OSPF, BGP), Cisco Discovery Protocol (CDP), Link Aggregation Control Protocol (LACP), and Link Layer Discovery Protocol (LLDP). Network-device-to-network-services communications include things like DHCP, DNS, RADIUS, SYSLOG, and SNMP. Cisco ISE provides an SGT for these types of communications and assigns it to network devices themselves. By default, this is SGT number 2 (SGT 2) and is named TrustSec_Devices. The recommended way to assign the TrustSec_Devices SGT to network devices is by deploying it within the *environment-data download* when the network device authenticates with Cisco ISE.

After the PAN and PSN nodes are installed and network devices are configured to send authentication to the PSN (or Cisco DNA Center in DNA-based deployments), the identity and location of the users are learned. MAC Authentication Bypass (MAB) or PassiveID can be used if devices do not support 802.1X. In addition, guest users can connect to the network using a configured captive portal and web authentication (WebAuth).

Cisco ISE can determine the type of device or endpoint connecting to the network by performing "profiling." Profiling is done by using DHCP, SNMP, Span, NetFlow, HTTP, RADIUS, DNS, or NMAP scans to collect as much metadata as possible to learn the device fingerprint. The metadata or attributes collected are used to automate the categorization or "classification" of the device and to enforce appropriate access control policies.

TIP   You can also try to identify hosts based on "what they are" and "what they do." Cisco Stealthwatch using NetFlow is typically used to identify flow behavior per asset. Cisco Stealthwatch performs flow de-duplication, flow stitching, and flow anomaly detection for this classification and for additional visibility. For instance, Cisco Stealthwatch can discover all hosts communicating with web servers—who, what, when, where, and how. This information can be used to characterize the "normal behavior" of a web server; then, policies can be built around this information. Furthermore, different alerts and logs can be generated for endpoints or servers behaving outside the baseline behavior and thresholds.

SGT Assignment and Deployment

When SGTs are provisioned by Cisco ISE, they are downloaded to network devices within the environment data. Here are a few things to remember about classification and SGT provisioning:

- Typically, servers are classified into groups using static classification.

- IP- and Subnet-to-SGT mappings can be centrally managed from Cisco ISE and deployed to networking devices using SSH or SXP.

- Additional static classification configurations can also be provisioned via CLI on the network devices.

- Dynamic classification is typically used for user, endpoint, or guest authentications by using 802.1X, MAB, WebAuth, or PassiveID, or they can also be learned from a Cisco ACI APIC (in the case of a Cisco ACI deployment).

- The network access switch or wireless AP needs to have AAA, RADIUS, and interface configuration deployed. In addition, Cisco ISE needs to have the network devices and authentication and authorization rules added. Within SDA, the total switch configuration is orchestrated by Cisco DNA Center, as are the network devices in ISE.

Initially Deploying 802.1X and/or TrustSec in Monitor Mode

Many organizations initially deploy 802.1X in monitor mode to scope the deployment and prevent user productivity from being impacted while changes are being implemented. This is because 802.1X Monitor Mode allows network connectivity even if authentication fails. Users will still be able to log in to the network in the case of any deployment problems, and the administrators will be able to gain visibility into any misconfigurations or connectivity challenges.

You can configure the Cisco ISE with the "log" keyword in each "permit statement" (policies allowing IP traffic) in order to obtain visibility of the deployment and user behavior. The "log" keyword on an access control entry generates a syslog message when the entry is matched. Policies will only be enforced if the **cts role-based enforcement** configuration command is applied to network devices (Cisco DNA Center controls this configuration within an SDA fabric). The **cts role-based enforcement** command can be used to enable enforcement on Layer 3 interfaces, and the **cts role-based enforcement vlan-list** *vlan id* command can be used to enable enforcement on Layer 2 interfaces.

> **NOTE**   You can obtain detailed information about Monitor Mode at the following website:
> https://www.cisco.com/c/dam/en/us/solutions/collateral/enterprise/design-zone-security/
> howto_21_monitor_mode_deployment_guide.pdf.

Active Policy Enforcement

After you learn and tweak your configuration by deploying 802.1X and TrustSec in monitoring mode, the policies can be provisioned and enforced. Once you configure your deployment in "enforcement mode," if there are IP-to-SGT mappings, then the associated policy protecting those SGTs will be downloaded from ISE. If you have an SDA deployment, contracts and policies are added in Cisco DNA Center and pushed to Cisco ISE via the Cisco ISE API.

To start the deployment of TrustSec and segmentation policies in Cisco ISE, you can navigate to **Work Centers > TrustSec > Overview**, as shown in Figure 5-61.

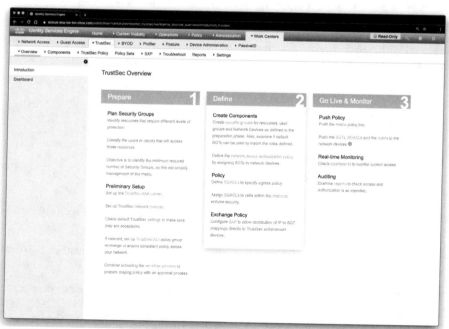

Figure 5-61   *Cisco ISE TrustSec Overview*

The screen shown in Figure 5-61 provides an overview of the TrustSec configuration and deployment steps in Cisco ISE.

Figure 5-62 shows the Cisco ISE TrustSec dashboard.

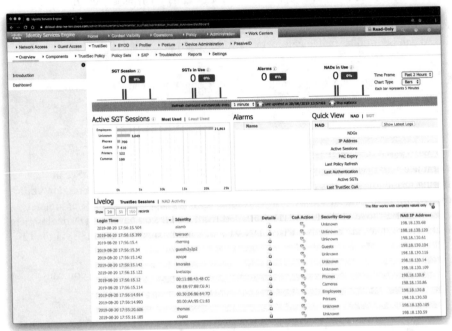

Figure 5-62 *Cisco ISE TrustSec Dashboard*

The Cisco ISE TrustSec dashboard is a centralized monitoring tool for the TrustSec deployment. The Cisco ISE TrustSec dashboard contains the following information:

- General Metrics

- Active SGT Sessions

- Alarms

- NAD / SGT Quick View

- TrustSec Sessions / Live Log

Figure 5-63 shows information about the different TrustSec components and security groups configured in Cisco ISE.

To view the configured policies in Cisco ISE, navigate to **Work Centers > TrustSec > Policy Sets**, as shown in Figure 5-64.

You learned that the Source Group Tag (SGT) Exchange Protocol (SXP) is used to propagate the SGTs across network devices that do not have hardware support for TrustSec. SXP is used to transport an endpoint's SGT, along with the IP address, from one SGT-aware network device to another. The data that SXP transports is called an IP-SGT mapping. The SGT to which an endpoint belongs can be assigned statically or dynamically, and the SGT

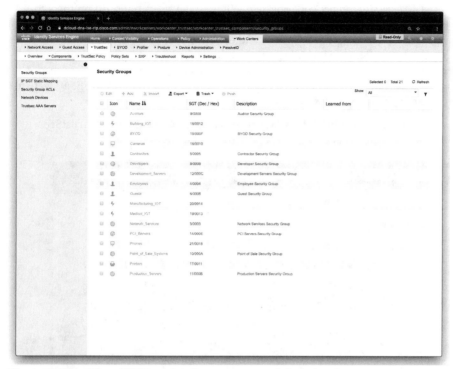

Figure 5-63 *Cisco ISE TrustSec Components and Security Groups*

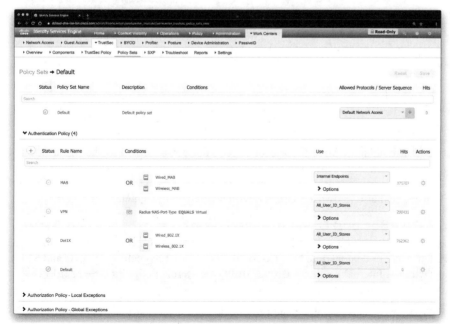

Figure 5-64 *Cisco ISE TrustSec Policy Sets*

can be used as a classifier in network policies. SXP uses TCP as its transport protocol to set up an SXP connection between two separate network devices. Each SXP connection has one peer designated as the SXP speaker and the other peer as the SXP listener. The peers can also be configured in a bidirectional mode, where each of them acts as both speaker and listener. Connections can be initiated by either peer, but mapping information is always propagated from a speaker to a listener. To configure SPX in Cisco ISE, navigate to **Work Centers > TrustSec > SPX**.

To view the TrustSec Policy Production Matrix, navigate to **Work Centers > TrustSec > TrustSec Policy > Egress Policy > Matrix**, as shown in Figure 5-65.

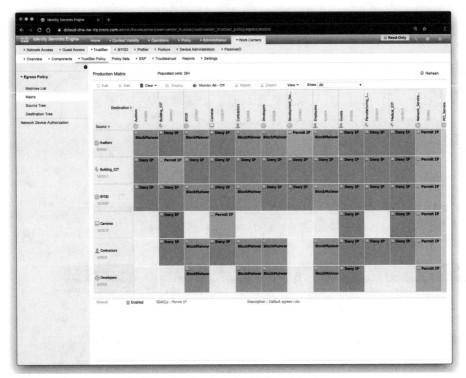

Figure 5-65 *Cisco ISE TrustSec Production Policy Matrix*

The screen shown in Figure 5-65 displays source-scalable groups down the left-hand side and destination-scalable groups across the top. In the intersection of the rows and columns, you find the actual enforcement instructions, otherwise known as Scalable Group Access Control Lists (SGACLs), and also known as contracts within Cisco DNA Center.

Cisco ISE TrustSec and Cisco ACI Integration

Although the SCOR exam does not cover ACI configuration or integration, this section is for your reference and to demonstrate how Cisco ISE allows you to synchronize SGTs and SXP mappings with the Internal Endpoint Groups (IEPGs), External Endpoint Groups (EEPGs), and endpoint (EP) configuration of Cisco ACI.

Cisco ISE supports packets coming from the ACI domain to the TrustSec domain by synchronizing the IEPGs and creating correlating read-only SGTs in ISE. These SGTs are used to map the endpoints configured in ACI and create correlating SXP mappings in ISE. These SGTs are displayed on the Security Groups page (with the value "ACI" in the Learned From field). You can view the SXP mappings on the All SXP Mappings page. These mappings are propagated to ACI only if the SXP device belongs to an SXP domain that is selected in the ACI Settings page. You can select the SGTs that are propagated to ACI. While adding a security group, you can specify whether the SGT should be propagated to ACI or not by using the Propagate to ACI option. When this option is enabled, the SXP mappings that are related to this SGT are propagated to ACI if the SXP device belongs to a VPN that is selected in the ACI Settings page.

You must enable the SXP service to use the ACI integration feature. ACI supports the packets coming from the TrustSec domain to the ACI domain by synchronizing the SGTs and creating correlating EEPGs. ACI creates subnets under EEPG based on the SXP mappings that are propagated from Cisco ISE. These subnets are not deleted from ACI, when the corresponding SXP mappings are deleted in Cisco ISE.

When an IEPG is updated in ACI, the corresponding SGT configuration is updated in Cisco ISE. A new EEPG is created in ACI when an SGT is added in Cisco ISE. When an SGT is deleted, the corresponding EEPG will be deleted in ACI. When an endpoint is updated in ACI, the corresponding SXP mapping is updated in Cisco ISE. If the connection with the ACI server is lost, Cisco ISE resynchronizes the data again when the connection is reestablished.

To integrate Cisco ISE with Cisco ACI, you must first import the Cisco ACI certificate into the Trusted Certificates Store under **Administration > System > Certificates > Trusted Certificates > Import**. Then you navigate to **Work Centers > TrustSec > Settings > ACI Settings**, as shown in Figure 5-66.

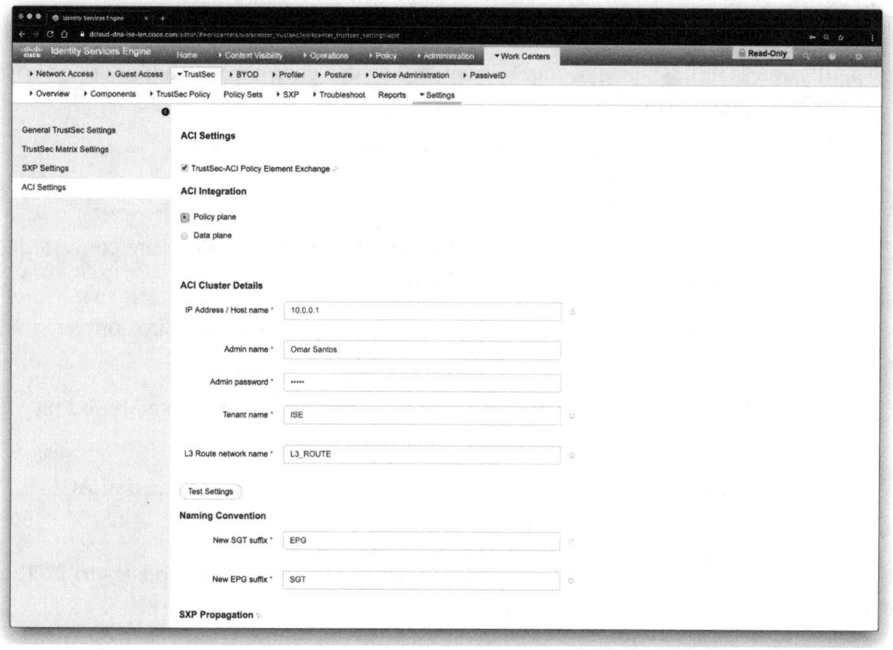

Figure 5-66 *Integration of Cisco ISE TrustSec and Cisco ACI*

After navigating to the screen shown in Figure 5-66, check the TrustSec-ACI Policy Element Exchange check box to synchronize SGTs and SXP mappings with IEPGs, EEPGs, and endpoint configuration of ACI. You can select **Policy Plane** if you want Cisco ISE to interact only with APIC data center to interchange SGT, EPG, and SXP information. Alternatively, if you select the **Data Plane** option, in addition to SGT, EPG, and SXP information, additional information is provided to the ASR devices that are connected between the TrustSec network and the APIC-controlled network. These ASR devices must contain the translation tables for SGT-to-EPG and EPG-to-SGT conversion.

Enter the following details if you have selected the Policy Plane option:

- IP address/hostname

- Admin name/password

- The name of the tenant that is configured on the ACI

- The name of the Layer 3 route network that is configured on the ACI for synchronizing the policy elements

- The new SGT suffix that will be added to the SGTs that are newly created based on the EPGs learned from ACI

- The new EPG suffix that will be added to the EPGs that are newly created in ACI based on the SGTs learned from Cisco ISE.

> **NOTE**   You can click **Test Settings** to check the connectivity with the ACI server.

Enter the following details if you have selected the Data Plane option:

- The IP address or hostname of the ACI server. You can enter three IP addresses or host names separated by commas.

- The username and password of the ACI admin user.

- The name of the tenant that is configured on the ACI.

- The maximum number of IEPGs that will be converted to SGTs. IEPGs are converted in alphabetical order. The default value is 1000.

- The maximum number of SGTs that will be converted to IEPGs. SGTs are converted in alphabetical order. The default value is 500.

- The new SGT suffix that will be added to the SGTs that are newly created based on the EPGs learned from ACI.

- The new EPG suffix that will be added to the EPGs that are newly created in ACI based on the SGTs learned from Cisco ISE.

Check the **Propagate using SXP** check box if you want to propagate the mappings to ACI using SXP.

TIP One of the most comprehensive lists of resources for Cisco ISE deployments can be obtained from the Cisco ISE Community Resources site at https://community.cisco.com/t5/security-documents/ise-community-resources/ta-p/3621621. Not only you can obtain detailed documentation and sample configuration, but you can also interact with other engineers and Cisco experts.

Exam Preparation Tasks

As mentioned in the section "How to Use This Book" in the Introduction, you have a couple of choices for exam preparation: the exercises here, Chapter 12, "Final Preparation," and the exam simulation questions in the Pearson Test Prep Software Online.

Review All Key Topics

Review the most important topics in this chapter, noted with the Key Topic icon in the outer margin of the page. Table 5-3 lists these key topics and the page numbers on which each is found.

Table 5-3 Key Topics for Chapter 5

Key Topic Element	Description	Page Number
Section	What Is a Flow?	227
Section	NetFlow for Network Security and Visibility	229
List	Listing and recognizing the methodology on security incident handling created by NIST that has been adopted by many organizations, including service providers, enterprises, and government organizations	231
Tip	Understanding how Cisco Stealthwatch FlowSensor can be deployed to a SPAN or network TAP	233
List	Describing the network telemetry sources that can also be correlated with NetFlow while responding to security incidents and performing network forensics	234
Paragraph	Describing the importance of time synchronization for network telemetry	234
Section	IP Flow Information Export (IPFIX)	237
Section	IPFIX Templates	239
Section	Understanding the Stream Control Transmission Protocol (SCTP)	241
Section	Exploring Application Visibility and Control and NetFlow	241
Section	NetFlow Deployment Scenarios	242
Section	Cisco Stealthwatch	250
Section	Threat Hunting with Cisco Stealthwatch	258

Key Topic Element	Description	Page Number
Section	What Is Cisco ETA?	262
Section	What Is Cisco Cognitive Threat Analytics?	262
Section	NetFlow Collection Considerations and Best Practices	268
Section	Configuring NetFlow in Cisco IOS and Cisco IOS-XE	269
Section	Data-Driven Segmentation	286
Section	Application-Based Segmentation	288
Section	Micro-Segmentation with Cisco ACI	289
Section	Segmentation with Cisco ISE	290

Define Key Terms

Define the following key terms from this chapter and check your answers in the glossary:

NetFlow, the "five-tuple," IPFIX, Stream Control Transmission Protocol (SCTP), FlowCollector, Stealthwatch Management Console (SMC), FlowSensor, FlowReplicator, endpoint groups (EPGs), uSeg EPG, Scalable Group Tag Exchange Protocol (SXP)

Review Questions

The answers to these questions appear in Appendix A. For more practice with exam format questions, use the Pearson Test Prep Software Online.

1. You have been asked to provide a segmentation strategy for applications residing in Docker containers and in virtual machines in a large data center. Which of the following technologies will you choose for such deployment?

   a. Cisco ETA

   b. Cisco ACI

   c. VLANs and firewalls

   d. None of these answers is correct.

2. When you first deploy a Cisco ISE node, all the default services provided by the Administration, Policy Service, and Monitoring personas will be enabled. Which of the following statements is not true?

   a. The Cisco ISE node will be in a standalone mode.

   b. You must first configure a primary ISE node and then register secondary ISE nodes to the primary ISE node.

   c. You must first configure the secondary ISE node and then the primary ISE node to avoid network disruption.

   d. You cannot edit the personas or services of a standalone Cisco ISE node.

3. When SGTs are provisioned by Cisco ISE, they are downloaded to network devices within the environment data. Which of the following are things to take into consideration about classification and SGT provisioning?

    a. Typically, servers are classified into groups using static classification.

    b. IP and Subnet-to-SGT mappings can be centrally managed from Cisco ISE and deployed to networking devices using SSH or SXP.

    c. Dynamic classification is typically used for user, endpoint, or guest authentications by using 802.1X, MAB, WebAuth, or PassiveID, or they can also be learned from a Cisco ACI APIC (in the case of a Cisco ACI deployment).

    d. All of the options are correct.

4. Many organizations initially deploy 802.1X in _____ mode to scope the deployment and prevent user productivity from being impacted while changes are being implemented.

    a. monitor

    b. active

    c. standby

    d. high-availability

5. Depending on the version of NetFlow, the router can also gather additional information, such as which of the following?

    a. Type of service (ToS) byte

    b. Differentiated services code point (DSCP)

    c. The device's input interface

    d. TCP flags

    e. All of the options are correct.

6. Which of the following statements is not true?

    a. The Cisco Stealthwatch FlowSensor is a network appliance that functions similarly to a traditional packet capture appliance or IDS in that it connects into a Switch Port Analyzer (SPAN), mirror port, or a Test Access Port (TAP).

    b. The Cisco Stealthwatch FlowSensor augments visibility where NetFlow is not available in the infrastructure device (router, switch, and so on) or where NetFlow is available but you want deeper visibility into performance metrics and packet data.

    c. You typically configure the Cisco Stealthwatch FlowSensor in combination with a Cisco Stealthwatch FlowCollector.

    d. A Cisco Stealthwatch FlowSensor replaces a Cisco Stealthwatch FlowCollector in several deployment models.

7. Which of the following network telemetry sources can also be correlated with Net-Flow while responding to security incidents and performing network forensics?

    a. Syslog

    b. 802.1X authentication logs

    c. VPN logs

    d. All of these options are correct.

5

8. Which of the following NetFlow versions support templates? (Select all that apply.)

   a. Flexible NetFlow

   b. NetFlow v2

   c. NetFlow v9

   d. NetFlow v5

   e. NetFlow v8

9. IPFIX uses which of the following protocols to provide a packet transport service designed to support several features beyond TCP or UDP capabilities?

   a. SCP

   b. SCTP

   c. pxGrid

   d. EPG

10. Cisco Stealthwatch components can be deployed as physical or virtual appliances. The two minimum required components are _____.

   a. SMC and FlowSensor

   b. SMC and FlowCollector

   c. FlowSensor and FlowCollector

   d. None of these options is correct.

CHAPTER 6

Infrastructure Security

This chapter covers the following topics:

Securing Layer 2 Technologies

Common Layer 2 Threats and How to Mitigate Them

Network Foundation Protection

Understanding and Securing the Management Plane

Understanding the Control Plane

Understanding and Securing the Data Plane

Securing Management Traffic

Implementing Logging Features

Configuring NTP

Securing the Network Infrastructure Device Image and Configuration Files

Securing the Data Plane in IPv6

Securing Routing Protocols and the Control Plane

In this chapter, you will learn how to secure Layer 2 technologies as well as what are the common Layer 2 threats and how to mitigate them. You will learn what is Network Foundation Protection. This chapter helps you gain an understanding of the management, control, and data planes and how to secure them. You will learn how to implement logging features and how to configure infrastructure devices to synchronize the time with Network Time Protocol (NTP) servers. You will also learn how to secure the network infrastructure device image and configuration files, secure the data plane in IPv6, and how to secure routing protocol implementations.

The following SCOR 350-701 exam objectives are covered in this chapter:

- Domain 2: Network Security
    - 2.4 Configure and verify network infrastructure security methods (router, switch, wireless)
        - 2.4.a Layer 2 methods (network segmentation using VLANs and VRF-lite; Layer 2 and port security; DHCP snooping; Dynamic ARP inspection; storm control; PVLANs to segregate network traffic; and defenses against MAC, ARP, VLAN hopping, STP, and DHCP rogue attacks

- 2.4.b Device hardening of network infrastructure security devices (control plane, data plane, management plane, and routing protocol security)

- 2.8 Configure secure network management of perimeter security and infrastructure devices (secure device management, SNMPv3, views, groups, users, authentication, and encryption, secure logging, and NTP with authentication)

"Do I Know This Already?" Quiz

The "Do I Know This Already?" quiz allows you to assess whether you should read this entire chapter thoroughly or jump to the "Exam Preparation Tasks" section. If you are in doubt about your answers to these questions or your own assessment of your knowledge of the topics, read the entire chapter. Table 6-1 lists the major headings in this chapter and their corresponding "Do I Know This Already?" quiz questions. You can find the answers in Appendix A, "Answers to the 'Do I Know This Already?' Quizzes and Q&A Sections."

Table 6-1 "Do I Know This Already?" Section-to-Question Mapping

Foundation Topics Section	Questions
Securing Layer 2 Technologies	1
Common Layer 2 Threats and How to Mitigate Them	2
Network Foundation Protection	3
Understanding and Securing the Management Plane	4
Understanding the Control Plane	5
Understanding and Securing the Data Plane	6
Securing Management Traffic	7
Implementing Logging Features	8
Configuring NTP	9
Securing the Network Infrastructure Device Image and Configuration Files	10
Securing the Data Plane in IPv6	11
Securing Routing Protocols and the Control Plane	12

CAUTION The goal of self-assessment is to gauge your mastery of the topics in this chapter. If you do not know the answer to a question or are only partially sure of the answer, you should mark that question as wrong for purposes of the self-assessment. Giving yourself credit for an answer you incorrectly guess skews your self-assessment results and might provide you with a false sense of security.

1. Which of the following are different STP port states?

    a. Root port

    b. Designated

    c. Nondesignated

    d. All of these answers are correct.

2. Which of the following are Layer 2 best practices?

   a. Avoid using VLAN 1 anywhere, because it is a default.

   b. Administratively configure access ports as access ports so that users cannot negotiate a trunk and disable the negotiation of trunking (no Dynamic Trunking Protocol [DTP]).

   c. Turn off Cisco Discovery Protocol (CDP) on ports facing untrusted or unknown networks that do not require CDP for anything positive. (CDP operates at Layer 2 and may provide attackers information we would rather not disclose.)

   d. On a new switch, shut down all ports and assign them to a VLAN that is not used for anything else other than a parking lot. Then bring up the ports and assign correct VLANs as the ports are allocated and needed.

   e. All of these answers are correct.

3. The Network Foundation Protection (NFP) framework is broken down into which of the following three basic planes (also called sections/areas)?

   a. Controller plane, administrative plane, management plane

   b. Management plane, control plane, administrative plane

   c. Management plane, control plane, data plane

   d. None of these answers is correct.

4. Which of the following are best practices for securing the management plane?

   a. Enforce password policy, including features such as maximum number of login attempts and minimum password length.

   b. Implement role-based access control (RBAC).

   c. Use AAA services, and centrally manage those services on an authentication server (such as Cisco ISE).

   d. Keep accurate time across all network devices using secure Network Time Protocol (NTP).

   e. All of these answers are correct.

5. Which of the following statements is not true?

   a. Control Plane Protection, or CPPr, allows for a more detailed classification of traffic (more than CoPP) that is going to use the CPU for handling.

   b. The benefit of CPPr is that you can rate-limit and filter this type of traffic with a more fine-toothed comb than CoPP.

   c. Using CoPP or CPPr, you can specify which types of management traffic are acceptable at which levels. For example, you could decide and configure the router to believe that SSH is acceptable at 100 packets per second, syslog is acceptable at 200 packets per second, and so on.

   d. Routing protocol authentication is not a best practice for securing the control plane; it is a best practice to protect the management plane.

6. You were hired to help increase the security of a new company that just deployed network devices in two locations. You are tasked to deploy best practices to protect the data plane. Which of the following techniques and features should you consider deploying to protect the data plane? (Select all that apply.)

    a. Use TCP Intercept and firewall services to reduce the risk of SYN-flood attacks.

    b. Filter (deny) packets trying to enter your network (from the outside) that claim to have a source IP address that is from your internal network.

    c. Deploy CoPP and CPPr in firewalls and IPS systems, as well as routing protocol authentication.

    d. Configure NetFlow and NETCONF for Control Plane Protection.

7. Which of the following are best practices to protect the management plane and management traffic?

    a. Deploy Login Password Retry Lockout to lock out a local AAA user account after a configured number of unsuccessful attempts by the user to log in using the username that corresponds to the AAA user account.

    b. Enable role-based access control (RBAC).

    c. Use NTP to synchronize the clocks on network devices so that any logging that includes timestamps may be easily correlated. Preferably, use NTP Version 3 to leverage its ability to provide authentication for time updates.

    d. All of these answers are correct.

8. Which of the following commands enable timestamps in syslog messages?

    a. **service syslog timestamps log datetime**

    b. **logging timestamps log datetime**

    c. **service timestamps log datetime**

    d. None of these answers is correct.

9. You were hired to configure all networking devices (routers, switches, firewalls, and so on) to generate syslog messages to a security information and event management (SIEM) system. Which of the following is recommended that you do on each of the infrastructure devices to make sure that the SIEM is able to correctly correlate all syslog messages ?

    a. Enable OSPF.

    b. Configure the network infrastructure devices to send syslog messages in batches (at a scheduled interval).

    c. Configure the SIEM to process the syslog messages at a scheduled interval.

    d. Enable NTP.

10. Which of the following is a feature that's intended to improve recovery time by making a secure working copy of a router or switch image and the startup configuration files (which are referred to as the primary bootset) so that they cannot be deleted by a remote user?

    a. Cisco Resilient Configuration

    b. Cisco Secure Firmware Configuration

    c. Address Space Layout Randomization (ASLR)

    d. None of these answers is correct.

6

11. If an attacker attempts to spoof many IPv6 destinations in a short time, the router can get overwhelmed while trying to store temporary cache entries for each destination. The _____ feature blocks data traffic from an unknown source and filters IPv6 traffic based on the destination address. It populates all active destinations into the IPv6 first-hop security binding table, and it blocks data traffic when the destination is not identified.

   a. IPv6 Destination Guard

   b. IPv6 Neighbor Cache Guard

   c. IPv6 Hop-by-hop Extension Header

   d. IPv6 Neighbor Cache Resource Starvation

12. BGP keychains enable keychain authentication between two BGP peers. The BGP endpoints must both comply with a keychain on one router and a password on the other router. Which of the following statements is not true regarding BGP keychains?

   a. BGP is able to use the keychain feature to implement hitless key rollover for authentication.

   b. Key rollover specification is time based, and in the event of clock skew between the peers, the rollover process is impacted.

   c. The configurable tolerance specification allows for the accept window to be extended (before and after) by that margin. This accept window facilitates a hitless key rollover for applications (for example, routing and management protocols).

   d. Routing protocols all support a different set of cryptographic algorithms. BGP supports only HMAC-SHA1-12.

Foundation Topics

Securing Layer 2 Technologies

We often take for granted Layer 2 in the network because it just works. Address Resolution Protocol (ARP) and Layer 2 forwarding on Ethernet are both proven technologies that work very well. The CCNP Security and CCIE Security certifications are built with the presumption that candidates understand some of the fundamentals of routing and switching. With this knowledge, your understanding of the details about VLANs, trunking, and inter-VLAN routing is presumed. However, so that you absolutely understand these fundamental concepts, this section begins with a review.

It is important to make sure that the basics are in place so that you can fully understand the discussion about protecting Layer 2 in the last section of this chapter, which covers the really important "stuff." That section focuses on just a few Layer 2–related security vulnerabilities and explains exactly how to mitigate threats at Layer 2. If you are currently comfortable with VLANs, trunking, and routing between VLANs, you might want to jump right to the last section.

VLAN and Trunking Fundamentals

In Chapter 5, "Network Visibility and Segmentation," you already learned that VLANs can be used to segment your network and are assigned to switch ports as well as wireless clients to enforce policy. In this chapter, you will also learn about several security challenges when protecting Layer 2 networks, VLAN assignment, and trunking protocols. However, you must understand the basics of how VLANs and trunking operate before you can learn to secure those features. This section reviews how VLANs and trunking are configured and how they operate.

Figure 6-1 serves as a reference for the discussion going forward. You might want to bookmark this page or take a moment to make a simple drawing of the topology. You will want to refer to this illustration often during the discussion.

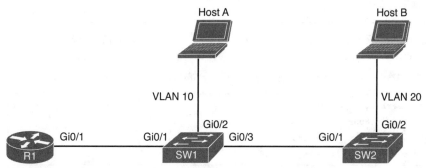

Figure 6-1   *VLAN Example*

What Is a VLAN?

One way to identify a local area network is to say that all the devices in the same LAN have a common Layer 3 IP network address and that they also are all located in the same Layer 2 broadcast domain. A virtual LAN (VLAN) is another name for a Layer 2 broadcast domain. VLANs are controlled by the switch. The switch also controls which ports are associated with which VLANs. In Figure 6-1, if the switches are in their default configuration, all ports by default are assigned to VLAN 1, and that means all the devices, including the two users and the router, are all in the same broadcast domain, or VLAN.

As you start adding hundreds of users, you might want to separate groups of users into individual subnets and associated individual VLANs. To do this, you assign the switch ports to the VLAN, and then any device that connects to that specific switch port is a member of that VLAN. Hopefully, all the devices that connect to switch ports that are assigned to a given VLAN also have a common IP network address configured so that they can communicate with other devices in the same VLAN. Often, Dynamic Host Configuration Protocol (DHCP) is used to assign IP addresses from a common subnet range to the devices in a given VLAN.

If you want to move the two users in Figure 6-1 to a new common VLAN, you create the VLAN on the switches and then assign the individual access ports that connect the users to the network to that new VLAN, as shown in Example 6-1.

Example 6-1   *Creating New VLANs*

```
sw1(config)# vlan 10
sw1(config-vlan)# name VLAN10
sw1(config-vlan)# state active
sw1(config)# vlan 20
sw1(config-vlan)# name VLAN20
sw1(config-vlan)# state active
```

In Example 6-2, interface GigabitEthernet0/2 of switch 1 (sw1) is configured as an access port (interface) and assigned to VLAN 10.

Example 6-2 *Assigning VLAN 10 to an Interface in Switch 1*

```
sw1(config)# interface GigabitEthernet0/2
sw1(config-if)# switchport
sw1(config-if)# switchport mode access
sw1(config-if)# switchport access vlan 10
```

In Example 6-3, interface GigabitEthernet0/2 of switch 2 (sw2) is configured as an access port (interface) and assigned to VLAN 20.

Example 6-3 *Assigning VLAN 20 to an Interface in Switch 2*

```
sw2(config)# interface GigabitEthernet0/2
sw2(config-if)# switchport
sw2(config-if)# switchport mode access
sw2(config-if)# switchport access vlan 20
```

You can do a quick verification of the newly created VLAN and associated ports with the **show vlan brief** command, as demonstrated in Example 6-4.

Example 6-4 *Output of the show vlan brief Command in Switch 1*

```
sw1# show vlan brief

VLAN Name                             Status     Ports
---- -------------------------------- ---------  ------------------------------
1    default                          active
10   VLAN10                           active     Gi0/2
1002 fddi-default                     act/unsup
1003 token-ring-default               act/unsup
1004 fddinet-default                  act/unsup
1005 trnet-default                    act/unsup
sw1#
```

Example 6-5 demonstrates another way to verify that the port is assigned to the VLAN using the **show vlan id** command.

Example 6-5 *Output of the show vlan id 10 Command in Switch 1*

```
sw1# show vlan id 10
VLAN Name                             Status     Ports
---- -------------------------------- ---------  ------------------------------
10   VLAN10                           active     Gi0/2


VLAN Type  SAID       MTU   Parent RingNo BridgeNo Stp  BrdgMode Trans1 Trans2
---- ----- ---------- ----- ------ ------ -------- ---- -------- ------ -----
10   enet  100010     1500  -      -      -        -    -        0      0
```

```
Remote SPAN VLAN

----------------

Disabled


Primary Secondary Type              Ports
------- --------- ---------------- -----------------------------------------
sw1#
```

Example 6-6 demonstrates yet another way to verify the port VLAN assignment using the **show interfaces Gi0/2 switchport** command.

Example 6-6 *Output of the show interfaces Gi0/2 switchport Command in Switch 1*

```
sw1# show interfaces Gi0/2 switchport
Name: Gi0/2
Switchport: Enabled
Administrative Mode: static access
Operational Mode: static access
Administrative Trunking Encapsulation: negotiate
Operational Trunking Encapsulation: native
Negotiation of Trunking: Off
Access Mode VLAN: 10 (VLAN10)
Trunking Native Mode VLAN: 1 (default)
Administrative Native VLAN tagging: enabled
Voice VLAN: none
Administrative private-vlan host-association: none
Administrative private-vlan mapping: none
Administrative private-vlan trunk native VLAN: none
Administrative private-vlan trunk Native VLAN tagging: enabled
Administrative private-vlan trunk encapsulation: dot1q
Administrative private-vlan trunk normal VLANs: none
Administrative private-vlan trunk associations: none
Administrative private-vlan trunk mappings: none
Operational private-vlan: none
Trunking VLANs Enabled: ALL
Pruning VLANs Enabled: 2-1001
Capture Mode Disabled
Capture VLANs Allowed: ALL
Protected: false
Appliance trust: none
```

Trunking with 802.1Q

One problem with having two users in the same VLAN but not on the same physical switch is how SW1 tells SW2 that a broadcast or unicast frame is supposed to be for VLAN 10. The answer is simple. For connections between two switches that contain ports in VLANs that exist in both switches, you configure specific trunk ports instead of configuring access ports. If the two switch ports are configured as trunks, they include additional information called a tag that identifies which VLAN each frame belongs to. 802.1Q is the standard protocol for this tagging. The most critical piece of information (for this discussion) in this tag is the VLAN ID.

Currently, the two users cannot communicate because they are in the same VLAN (VLAN 10), but the inter-switch links (between the two switches) are not configured as trunks. To configure both sets of interfaces as trunks, you would specify the trunk method of 802.1Q and then turn on the feature, as shown in Example 6-7.

Example 6-7 *Configuring Interfaces as Trunk Ports*

```
sw1(config-if)# interface GigabitEthernet0/3
sw1(config-if)# description to sw2
sw1(config-if)# switchport trunk encapsulation dot1q
sw1(config-if)# switchport mode trunk
```

You will repeat the same procedure on switch 2 (sw2). To verify the trunk configuration, you can use the **show interface trunk** command, as demonstrated in Example 6-8.

Example 6-8 *Output of the show interface trunk Command in Switch 1*

```
sw1# show interface trunk
Port            Mode            Encapsulation   Status          Native vlan
Gi0/3           on              802.1q          trunking        1
Port            Vlans allowed on trunk
Gi0/3           1-4094
Port            Vlans allowed and active in management domain
Gi0/3           1,10
Port            Vlans in spanning tree forwarding state and not pruned
Gi0/3           1,10
sw1#
```

Example 6-9 shows the output of the **show interface trunk** command in switch 2.

Example 6-9 *Output of the show interface trunk Command in Switch 2*

```
sw2# show interface trunk
Port            Mode            Encapsulation   Status          Native vlan
Gi0/1           on              802.1q          trunking        1
Port            Vlans allowed on trunk
Gi0/1           1-4094
Port            Vlans allowed and active in management domain
Gi0/1           1,10,20
Port            Vlans in spanning tree forwarding state and not pruned
Gi0/1           1,10,20
sw2#
```

Another way to verify the trunk configuration is with the **show interfaces** *interface* **switchport** command, as demonstrated in Example 6-10.

Example 6-10 *Output of the show interfaces Gi0/1 switchport Command in Switch 2*

```
sw2# show interfaces Gi0/1 switchport
Name: Gi0/1
Switchport: Enabled
Administrative Mode: trunk
Operational Mode: trunk
Administrative Trunking Encapsulation: dot1q
Operational Trunking Encapsulation: dot1q
Negotiation of Trunking: On
Access Mode VLAN: 1 (default)
Trunking Native Mode VLAN: 1 (default)
Administrative Native VLAN tagging: enabled
Voice VLAN: none
Administrative private-vlan host-association: none
Administrative private-vlan mapping: none
Administrative private-vlan trunk native VLAN: none
Administrative private-vlan trunk Native VLAN tagging: enabled
Administrative private-vlan trunk encapsulation: dot1q
Administrative private-vlan trunk normal VLANs: none
Administrative private-vlan trunk associations: none
Administrative private-vlan trunk mappings: none
Operational private-vlan: none
Trunking VLANs Enabled: ALL
Pruning VLANs Enabled: 2-1001
Capture Mode Disabled
Capture VLANs Allowed: ALL
Protected: false
Appliance trust: none
sw2#
```

Let's Follow the Frame, Step by Step

A broadcast frame sent from HOST A and received by SW1 would forward the frame over the trunk tagged as belonging to VLAN 10 to SW2. SW2 would see the tag, know it was a broadcast associated with VLAN 20, remove the tag, and forward the broadcast to all other interfaces associated with VLAN 20, including the switch port that is connected to HOST B. These two core components (access ports being assigned to a single VLAN, and trunk ports that tag the traffic so that a receiving switch knows which VLAN a frame belongs to) are the core building blocks for Layer 2 switching, where a VLAN can extend beyond a single switch.

What Is the Native VLAN on a Trunk?

From the output in the earlier example, we verified our trunk interfaces between the two switches. One option shown in the output was a native VLAN. By default, the native VLAN is VLAN 1. So, what does this mean, and why do we care? If a user is connected to an access port that is assigned to VLAN 1 on SW1, and that user sends a broadcast frame, when SW1 forwards that broadcast to SW2, because the frame belongs to the native VLAN (and both switches agree to using the same native VLAN), the 802.1Q tagging is simply left off.

This works because when the receiving switch receives a frame on a trunk port, if that frame is missing the 802.1Q tag completely, the receiving switch assumes that the frame belongs to the native VLAN (in this case, VLAN 1).

This is not a huge problem until somebody tries to take advantage of this, as discussed later in this chapter. In the meantime, just know that using a specific VLAN as the native VLAN (different from the default of VLAN 1) and never using that same VLAN for user traffic is a prudent idea.

So, What Do You Want to Be? (Asks the Port)

Trunks can be automatically negotiated between two switches, or between a switch and a device that can support trunking. Automatic negotiation to determine whether a port will be an access port or a trunk port is risky because an attacker could potentially negotiate a trunk with a switch; then the attacker could directly access any available VLANs simply by illegally tagging the traffic directly from his PC.

Understanding Inter-VLAN Routing

Suppose that there are two hosts communicating with each other in VLAN 10 (which is also the same IP subnet), but they cannot communicate with devices outside their local VLAN without the assistance of a default gateway. A router could be implemented with two physical interfaces: one connecting to an access port on the switch that is been assigned to VLAN 10, and another physical interface connected to yet a different access port that has been configured for yet a different VLAN. With two physical interfaces and a different IP address on each, the router could perform routing between the two VLANs.

What Is the Challenge of Only Using Physical Interfaces?

So here is the problem: What if you have 50 VLANs? Purchasing 50 physical interfaces for the router would be pricey, let alone the fact that you would also be using 50 physical interfaces on the switch. One solution is to use a technique called *router-on-a-stick*. Consider Figure 6-1. R1 has one physical interface physically connected to the switch. So, from a physical topology perspective, it looks like the router is a lollipop or a "router on a stick," and that is where it gets its name.

Using Virtual "Sub" Interfaces

To use one physical interface, we have to play a little game, where we tell the switch that we are going to do trunking out to the router, which from the switch perspective looks exactly like trunking to another switch. And on the router, we tell the router to pay attention to the 802.1Q tags and assign frames from specific VLANs, based on the tags, to logical sub-interfaces. Each sub-interface has an IP address from different subnets, as shown in Example 6-11.

Example 6-11 *Configuring Router-on-a-Stick and Switch Support for the Router*

```
! Enabling trunking on the switchport connected to the router


SW1(config)# interface Gi0/1
SW1(config-if)# switchport trunk encapsulation dot1q
SW1(config-if)# switchport mode trunk


! Move to R1:
! Make sure the physical interface isn't shutdown
```

```
R1(config)# interface Gi0/1
R1(config-if)# no shutdown


! Create a logical sub interface, using any number following the "0/1"
! The sub interface is configured as "0/1.10" to correspond to VLAN 10)
R1(config-if)# interface Gi0/1.10


! Tell the router to process any dot1q frames tagged with VLAN ID 10 with
! this logical interface
R1(config-subif)# encapsulation dot1q 10


! Provide an IP address in the same subnet as other devices in VLAN 10
R1(config-subif)# ip address 10.0.0.1 255.255.255.0


! Verify that this router can ping devices in VLAN 10
R1(config-subif)# do ping 10.0.0.11


Type escape sequence to abort.
Sending 5, 100-byte ICMP Echos to 10.0.0.11, timeout is 2 seconds:
!!!!!
Success rate is 100 percent (5/5), round-trip min/avg/max = 1/2/4 ms
R1(config-subif)# do ping 10.0.0.22


Type escape sequence to abort.
Sending 5, 100-byte ICMP Echos to 10.0.0.22, timeout is 2 seconds:
!!!!!
Success rate is 100 percent (5/5), round-trip min/avg/max = 1/2/4 ms
R1(config-subif)#
```

The HOST A computer needs to configure R1's address of 10.0.0.1 as its default gateway when trying to reach nonlocal networks (that is, VLANs other than VLAN 10).

You can repeat the process of creating additional sub-interfaces on the router to support more VLANs until the router has a sub-interface in every VLAN you want.

Spanning Tree Fundamentals

This section discusses the basics of how the Spanning Tree Protocol (STP) can avoid loops at Layer 2 of the OSI model. It is important to understand how it works so that you can fully understand correct mitigation techniques.

NOTE   Refer to Figure 6-1 again for this section.

Loops in networks are bad. Without STP, whenever we have parallel connections between Layer 2 devices, such as the connections between SW1 and SW2, we would have Layer 2 loops. Let's take a look at this using the network configured in the previous section. STP is on by default on most Cisco switches, but for the purposes of this discussion, assume that STP is not running, at least for now.

Let's discuss the "life of a loop." If HOST A sends an ARP request into the network, SW1 receives it and knows that this frame belongs to VLAN 10, because of the access port it came in on, and forwards it out all other ports that are also assigned to VLAN 10, in addition to any trunk ports that are allowing VLAN 10. By default, trunk ports allow all VLAN traffic. Therefore, this broadcast is tagged as belonging to VLAN 10.

Just for a moment, let's follow just one of those ports. The traffic is being sent down port 23, and SW2 sees it and decides it needs to forward this traffic out all other ports assigned to VLAN 10, including port number 2, which is an access port assigned to VLAN 10, and also trunk port 24. Now, SW2 sends the same broadcast to SW1 on port 24. SW1 repeats the process and sends it out port 23, and there would be a loop. The loop happens in the other direction, as well. Besides having a loop, both switches become confused about which port is associated with the source MAC address of HOST A. Because a looping frame is seen inbound on both ports 23 and 24, due to the loop going both directions, MAC address flapping occurs in the dynamically learned MAC address table of the switch. Not only does this lead to excessive and unnecessary forwarding of the ARP request to switch ports, it also could present a denial-of-service condition if the switch is unable to perform all of its functions because it is wasting resources due to this loop in the network.

The Solution to the Layer 2 Loop

STP, or 802.1D, was developed to identify parallel Layer 2 paths and block on one of the redundant paths so that a Layer 2 loop would not occur. A single switch with the lowest bridge ID becomes the root bridge, and then all the other non-root switches determine whether they have parallel paths to the root and block on all but one of those paths. STP communicates using bridge protocol data units (BPDUs), and that is how negotiation and loop detection are accomplished.

Example 6-12, which contains comments and annotations, allows you to both review how STP operates and see the commands to verify it at the same time; it uses the topology from the beginning of this chapter.

Example 6-12 *STP Verification and Annotations*

```
SW1# show spanning-tree vlan 10
VLAN0010

! This top part indicates the Bridge ID of the root bridge, which is a
! combination of the Bridge Priority and Base MAC address.
! The switch with the lowest overall Bridge ID of all switches in the
! network becomes the Root Bridge.
! NOTE: If all switches in a network are enabled with default
! spanning-tree settings (default bridge priority is 32768), the switch
! with the lowest MAC address becomes the Root Bridge.
! This switch is claiming victory over the other switch (SW2)
! This is due to this switch having a lower bridge ID
```

```
 Spanning tree enabled protocol ieee
   Root ID    Priority    32778
              Address     0019.060c.9080
              This bridge is the root
              Hello Time   2 sec  Max Age 20 sec  Forward Delay 15 sec

! This is the output about the local switch.  Because this is the root
! switch,
! this information will be identical to the information above regarding the
! bridge ID, which is a combination of the Priority and Base MAC address
   Bridge ID  Priority    32778  (priority 32768 sys-id-ext 10)
              Address     0019.060c.9080
              Hello Time   2 sec  Max Age 20 sec  Forward Delay 15 sec
              Aging Time   300 sec

! This specifies the state of each interface, and the default costs
! associated with each interface if trying to reach the root switch.
! Because this switch is the root bridge/switch, the local costs are
! not relevant.
! This also shows the forwarding or blocking state.
! All ports on the root switch will be forwarding, every time, for the VLAN
! for which it is the root bridge.
Interface           Role Sts Cost      Prio.Nbr Type
------------------- ---- --- --------- -------- ----------------------------
Gi0/1               Desg FWD 19         128.3    P2p
Gi0/3               Desg FWD 19         128.5    P2p
Gi0/23              Desg FWD 19         128.25   P2p
Gi0/24              Desg FWD 19         128.26   P2p

! Road trip over to SW2, who didn't win the STP election
SW2# show spanning-tree vlan 10

! This first part identifies who the root is, and the cost for this switch
! to get to the root switch.
! SW1 advertised BPDUs that said the cost to reach me (SW1)
! is 0, and then this switch SW2 added that advertised cost to its
! only local interface cost to get to 19 as the cost for this switch to reach
! the root bridge.

VLAN0010
  Spanning tree enabled protocol ieee
   Root ID    Priority    32778
              Address     0019.060c.9080
              Cost        19
              Port        25 (GigabitEthernet0/23)
              Hello Time   2 sec  Max Age 20 sec  Forward Delay 15 sec
```

```
! This part identifies the local switch STP information. If you compare the
! bridge ID of this switch, to the bridge ID of SW1 (the root switch), you
! will notice that the priority values are the same, but SW1's MAC address
! is slightly lower (".060c" is lower than ".0617"), and as a result has
! a lower Bridge ID, which caused SW1 to win the election for root bridge
! of the spanning tree for VLAN 10.

  Bridge ID  Priority    32778  (priority 32768 sys-id-ext 10)
             Address     0019.0617.6600
             Hello Time   2 sec  Max Age 20 sec  Forward Delay 15 sec
             Aging Time  300 sec

! This is the port forwarding/blocking information for SW2.  SW2 received
! BPDUs from root bridge on both 23 and 24, and so it knew there was a
! loop. It decided to block on port 24.   The cost was the same on both
! ports, and STP used the advertised port priority of the sending switch,
! and chose the lower value. In STP lower is always preferred.   By
! default, lower numbered physical ports have lower numbered port
! priorities.


Interface          Role Sts Cost      Prio.Nbr Type
------------------ ---- --- --------- -------- ----------------------------
Gi0/2              Desg FWD 19        128.4    P2p
Gi0/23             Root FWD 19        128.25   P2p
Gi0/24             Altn BLK 19        128.26   P2p

! The blocking on port 24 is also reflected in the output of the show
! commands for trunking.  Notice that port 23 is forwarding for both
! VLAN 1 and 10, while port 24 is not forwarding for either VLAN.
SW2# show interfaces trunk

Port      Mode         Encapsulation  Status        Native vlan
Gi0/23    on           802.1Q         trunking      1
Gi0/24    on           802.1Q         trunking      1


Port      Vlans allowed on trunk
Gi0/23    1-4094
Gi0/24    1-4094


Port      Vlans allowed and active in management domain
Gi0/23    1,10
Gi0/24    1,10


Port      Vlans in spanning tree forwarding state and not pruned
Gi0/23    1,10
Gi0/24    none
```

STP is on by default and will have a separate instance for each VLAN. So, if you have five VLANs, you have five instances of STP. Cisco calls this default implementation Per-VLAN Spanning Tree Plus (PVST+).

STP consists of the following port states:

- **Root Port:** The switch port that is closest to the root bridge in terms of STP path cost (that is, it receives the best BPDU on a switch) is considered the root port. All switches, other than the root bridge, contain one root port.

- **Designated:** The switch port that can send the best BPDU for a particular VLAN on a switch is considered the designated port.

- **Nondesignated:** These are switch ports that do not forward packets, so as to prevent the existence of loops within the networks.

STP Is Wary of New Ports

When an interface is first brought up and receives a link signal from a connected device, such as a PC or router that is connected, STP is cautious before allowing frames in on the interface. If another switch is attached, there is a possible loop. STP cautiously waits for a total of 30 seconds (by default) on a recently brought-up port before letting frames go through that interface; the first 15 seconds of that is the listening state, where STP is seeing whether any BPDUs are coming in. During this time, it does not record source MAC addresses in its dynamic table. During the next 15 seconds (out of the total of 30 seconds), STP is still looking for BPDUs, but STP also begins to populate the MAC address table with the source MAC addresses it sees in frames. This is called the learning state. After listening and learning have completed (full 30 seconds), the switch can begin forwarding frames. If a port is in a blocking state at first, an additional 20-second delay might occur as the port determines that the parallel path is gone, before moving to listening and learning.

For most administrators and users, this delay is too long. When configured, enhancements to STP, including the PortFast feature, can tell the switch to bypass the listening and learning stage and go right to forwarding. This leaves a small window for a loop if a parallel path is injected in the network.

Improving the Time Until Forwarding

Cisco had some proprietary improvements to the 802.1D (traditional STP) that allowed faster convergence in the event of a topology change and included many features, such as the Port-Fast, UplinkFast, and BackboneFast. Many of these features were used in a newer version of STP called Rapid Spanning Tree (also known as 802.1w). Enabling PortFast for traditional STP and configuring Rapid Spanning Tree globally are shown in Example 6-13.

Example 6-13 *Configuring PortFast and Then Rapid Spanning Tree*

```
! PortFast can be enabled locally per interface

SW2(config)# interface Gi0/2
SW2(config-if)# spanning-tree portfast
%Warning: portfast should only be enabled on ports connected to a single
 host. Connecting hubs, concentrators, switches, bridges, etc... to this
 interface  when portfast is enabled, can cause temporary bridging loops.
```

```
Use with CAUTION

%Portfast has been configured on GigabitEthernet0/2 but will only
 have effect when the interface is in a non-trunking mode.
SW2(config-if)#

! or PortFast could be enabled globally
SW2(config)# spanning-tree portfast default
%Warning: this command enables portfast by default on all interfaces. You
 should now disable portfast explicitly on switched ports leading to hubs,
 switches and bridges as they may create temporary bridging loops.

! To change the STP from 802.1D to 802.1w, it requires just this one command
SW2(config)# spanning-tree mode rapid-pvst

! The show command will display RSTP, instead of the original IEEE for the
! mode
SW2# show spanning-tree vlan 10

VLAN0010
  Spanning tree enabled protocol rstp

<snip>
```

Common Layer 2 Threats and How to Mitigate Them

This section discusses many security threats that focus on Layer 2 technologies and addresses how to implement countermeasures against those risks.

Think about the saying: *"Disrupt the Bottom of the Wall, and the Top Is Disrupted, Too"*. Everything at Layer 3 and higher is encapsulated into some type of Layer 2 frame. If the attacker can interrupt, copy, redirect, or confuse the Layer 2 forwarding of data, that same attacker can also disrupt any type of upper-layer protocols that are being used.

 Let's begin with best practices for securing your switches and then discuss in more detail which best practice mitigates which type of attack.

Best practices for securing your infrastructure, including Layer 2, include the following:

- Select an unused VLAN (other than VLAN 1) and use that for the native VLAN for all your trunks. Do not use this native VLAN for any of your enabled access ports.

- Avoid using VLAN 1 anywhere, because it is a default.

- Administratively configure access ports as access ports so that users cannot negotiate a trunk and disable the negotiation of trunking (no Dynamic Trunking Protocol [DTP]).

- Limit the number of MAC addresses learned on a given port with the port security feature.

- Control spanning tree to stop users or unknown devices from manipulating spanning tree. You can do so by using the BPDU Guard and Root Guard features.

- Turn off Cisco Discovery Protocol (CDP) on ports facing untrusted or unknown networks that do not require CDP for anything positive. (CDP operates at Layer 2 and may provide attackers information we would rather not disclose.)

- On a new switch, shut down all ports and assign them to a VLAN that is not used for anything else other than a parking lot. Then bring up the ports and assign correct VLANs as the ports are allocated and needed.

To control whether a port is an access port or a trunk port, you can revisit the commands used earlier in this chapter, including the ones shown in Example 6-14.

Example 6-14 *Administratively Locking Down Switch Ports*

```
SW2(config)# interface Gi0/2


! Specifies that this is an access port, not a trunk, and specifies VLAN
! association
SW2(config-if)# switchport mode access
SW2(config-if)# switchport access VLAN 10


! Disables the ability to negotiate, even though we hard coded the port as
! an access port.  This command disables DTP, which otherwise is still on
! by default, even for an interface configured as an access port.
SW2(config-if)# switchport nonegotiate


SW2(config-if)# interface Gi 0/23
! Specifies the port as a trunk, using dot1q
SW2(config-if)# switchport trunk encapsulation dot1q
SW2(config-if)# switchport mode trunk


! Specify a VLAN that exists, but isn't used anywhere else, as the native
! VLAN
SW2(config-if)# switchport trunk native vlan 3


! Disables the ability to negotiate, even though we hard coded the port as
! a trunk
SW2(config-if)# switchport nonegotiate


! Note, negotiation packets  are still being sent unless we disable -
! negotiation
```

Do Not Allow Negotiations

The preceding example prevents a user from negotiating a trunk with the switch, maliciously, and then having full access to each of the VLANs by using custom software on the computer that can both send and receive dot1q-tagged frames. A user with a trunk established could perform "VLAN hopping" to any VLAN he desires by just tagging frames with the VLAN of choice. Other malicious tricks could be done, as well, but forcing the port to an access port with no negotiation removes this risk.

Layer 2 Security Toolkit

Cisco has many tools for protecting Layer 2, including the following:

- **BPDU Guard:** If BPDUs show up where they should not, the switch protects itself.

- **Root Guard:** Controls which ports are not allowed to become root ports to remote root switches.

- **Port security:** Limits the number of MAC addresses to be learned on an access switch port.

- **DHCP snooping:** Prevents rogue DHCP servers from impacting the network.

- **Dynamic ARP inspection:** Prevents spoofing of Layer 2 information by hosts.

- **IP Source Guard:** Prevents spoofing of Layer 3 information by hosts.

- **802.1X:** You learned about 802.1X in Chapter 4, "Authentication, Authorization, Accounting (AAA) and Identity Management." With 802.1X, you can authenticate users before allowing their data frames into the network.

- **Storm Control:** Limits the amount of broadcast or multicast traffic flowing through the switch.

- **Access control lists:** Used for traffic control and to enforce policy.

The key Layer 2 security technologies we focus on in the following sections include BPDU Guard, Root Guard, port security, CDP and LLDP, and DHCP snooping. With a review of the switching technologies and how they operate now in mind, let's take a specific look at implementing security features on our switches.

BPDU Guard

When you enable BPDU Guard, a switch port that was forwarding now stops and disables the port if a BPDU is seen inbound on the port. A user should never be generating legitimate BPDUs. This configuration, applied to ports that should only be access ports to end stations, helps to prevent another switch (that is sending BPDUs) from being connected to the network. This could prevent manipulation of your current STP topology. Example 6-15 shows the implementation of BPDU Guard.

Example 6-15 *Implementing BPDU Guard on a Switch Port*

```
SW2(config-if)# interface Gi0/2
SW2(config-if)# spanning-tree bpduguard enable

! Verify the status of the switchport
SW2# show interface Gi0/2 status

Port      Name         Status       Vlan      Duplex  Speed  Type
Gi0/2                  connected    10        a-full  a-100  10/100BaseTX
SW2#
```

A port that has been disabled because of a violation shows a status of "err-disabled." To bring the interface back up, issue a **shutdown** and then a **no shutdown** in interface configuration mode.

You can also configure the switch to automatically bring an interface out of the err-disabled state, based on the reason it was placed there and how much time has passed before bringing the interface back up. To enable this for a specific feature, follow Example 6-16.

Example 6-16 *Configuring the Switch to Automatically Restore Err-Disabled Ports*

```
SW2(config)# errdisable recovery cause bpduguard

! err-disabled ports will be brought back up after 30 seconds of no bpdu
! violations
SW2(config)# errdisable recovery interval 30

! You can also see the timeouts for the recovery

SW2# show errdisable recovery
ErrDisable Reason          Timer Status
-----------------          -------------
arp-inspection             Disabled
bpduguard                  Enabled
<snip>

Timer interval: 30 seconds
Interfaces that will be enabled at the next timeout:

SW2#
```

Root Guard

Your switch might be connected to other switches that you do not manage. If you want to prevent your local switch from learning about a new root switch through one of its local ports, you can configure Root Guard on that port, as shown in Example 6-17. This will also help in preventing tampering with your existing STP topology.

Example 6-17 *Controlling Which Ports Face the Root of the Spanning Tree*

```
SW1(config)# interface Gi0/24

SW1(config-if)# spanning-tree guard root
%SPANTREE-2-ROOTGUARD_CONFIG_CHANGE: Root guard enabled on port GigabitEthernet0/24.
```

Port Security

How many MAC addresses should legitimately show up inbound on an access port?

Port security controls how many MAC addresses can be learned on a single switch port. This feature is implemented on a port-by-port basis. A typical user uses just a single MAC address. Exceptions to this may be a virtual machine or two that might use different MAC addresses than their host, or if there is an IP phone with a built-in switch, which may also account for additional MAC addresses. In any case, to avoid a user connecting dozens of devices to a rogue switch that is then connected to their access port, you can use port security to limit the number of devices (MAC addresses) on each port.

This also protects against malicious applications that may be sending thousands of frames into the network, with a different bogus MAC address for each frame, as the user tries to exhaust the limits of the dynamic MAC address table on the switch, which might cause the switch to forward all frames to all ports within a VLAN so that the attacker can begin to sniff all packets. This is referred to as a *CAM table overflow attack*. Content-addressable memory (CAM) is a fancy way to refer to the MAC address table on the switch.

Port security also prevents the client from depleting DHCP server resources, which could have been done by sending thousands of DHCP requests, each using a different source MAC address. DHCP spoofing attacks take place when devices purposely attempt to generate enough DHCP requests to exhaust the number of IP addresses allocated to a DHCP pool.

With the port security feature, the default violation action is to shut down the port. Alternatively, we can configure the violation response to be to "protect," which will not shut down the port but will deny any frames from new MAC addresses over the set limit. The "restrict" action does the same as protect but generates a syslog message as well.

To implement port security, follow Example 6-18.

Example 6-18 *Implementing Port Security*

```
SW2(config-if)# interface Gi0/2

! Enable the feature per interface
SW2(config-if)# switchport port-security

! Set the maximum to desired number.  Default is 1. If we administratively
! set the maximum to 1, the command won't show in the running configuration
! because the configuration matches the default value. It is handy to know
! this behavior, so you won't be surprised by what may seem to be a missing
! part of your configuration.
SW2(config-if)# switchport port-security maximum 5

! Set the violation action.  Default is err-disable. Protect will simply
! not allow
! frames from MAC addresses above the maximum.
SW2(config-if)# switchport port-security violation protect

! This will cause the dynamic mac addresses to be placed into running
! -config to save them to startup config, use copy run start
SW2(config-if)# switchport port-security mac-address sticky

! To verify settings, use this command
SW2# show port-security
Secure Port  MaxSecureAddr  CurrentAddr  SecurityViolation  Security Action
             (Count)        (Count)      (Count)
----------------------------------------------------------------------------
   Gi0/2          5             1             0              Protect
----------------------------------------------------------------------------
```

```
Total Addresses in System (excluding one mac per port)    : 0
Max Addresses limit in System (excluding one mac per port) : 6144
! This can also provide additional information about port security:

SW2# show port-security interface Gi0/2
Port Security              : Enabled
Port Status                : Secure-up
Violation Mode             : Protect
Aging Time                 : 0 mins
Aging Type                 : Absolute
SecureStatic Address Aging : Disabled
Maximum MAC Addresses      : 5
Total MAC Addresses        : 1
Configured MAC Addresses   : 0
Sticky MAC Addresses       : 1
Last Source Address: Vlan  : 0000.2222.2222:10
Security Violation Count   : 0
```

CDP and LLDP

Cisco Systems introduced the Cisco Discovery Protocol (CDP) in 1994 to provide a mechanism for the management system to automatically learn about devices connected to the network. CDP runs on Cisco devices (routers, switches, phones, and so on) and is also licensed to run on some network devices from other vendors. Using CDP, network devices periodically advertise their own information to a multicast address on the network, making it available to any device or application that wishes to listen and collect it.

Over time, enhancements have been made to discovery protocols to provide greater capabilities. Applications (such as voice) have become dependent on these capabilities to operate properly, leading to interoperability problems between vendors. Therefore, to allow interworking between vendor equipment, it has become necessary to have a single, standardized discovery protocol. Cisco has been working with other leaders in the Internet and IEEE community to develop a new, standardized discovery protocol, 802.1AB (Station and Media Access Control Connectivity Discovery, or Link Layer Discovery Protocol [LLDP]). LLDP, which defines basic discovery capabilities, was enhanced to specifically address the voice application; this extension to LLDP is called LLDP-MED, or LLDP for Media Endpoint Devices.

As mentioned previously, a recommended best practice is to disable CDP on any ports facing untrusted or unknown networks that do not require CDP. CDP operates at Layer 2 and can provide attackers with information (for example, device types, hardware and software versions, and VLAN and IP address details) that you would rather not disclose. Example 6-18 details the configuration steps necessary to disable CDP on a global and per-interface basis.

In the same way it is recommended to disable CDP, it is also a best practice to disable LLDP in areas of the network that it is not needed. Example 6-19 also includes the configuration steps necessary to disable LLDP on a global basis.

Example 6-19 *Disabling CDP*

```
! Disable CDP on Interface Gi0/24
sw2(config)# interface Gi0/24
sw2(config-if)# no cdp ?
  enable  Enable CDP on interface
sw2(config-if)# no cdp enable
! Disable CDP Globally on switch
sw2(config)# no cdp run
sw2(config)# exit
sw2#
! Verify CDP has been disabled
sw2(config)# do show cdp
% CDP is not enabled
! Confirm LLDP is enabled on switch
sw2# show lldp

Global LLDP Information:
    Status: ACTIVE
    LLDP advertisements are sent every 30 seconds
    LLDP hold time advertised is 120 seconds
    LLDP interface reinitialisation delay is 2 seconds
sw2#
! Disable LLDP on a global basis
sw2# conf t
Enter configuration commands, one per line.  End with CNTL/Z.
sw2(config)# no lldp run
sw2(config)# exit
! Confirm LLDP has been disabled
sw2# show llpd
% LLDP is not enabled
sw2#
```

DHCP Snooping

DHCP snooping is a security feature that acts like a firewall between untrusted hosts and trusted DHCP servers. The DHCP snooping feature performs the following activities:

■ Validates DHCP messages received from untrusted sources and filters out invalid messages

■ Rate-limits DHCP traffic from trusted and untrusted sources

■ Builds and maintains the DHCP snooping binding database, which contains information about untrusted hosts with leased IP addresses

■ Utilizes the DHCP snooping binding database to validate subsequent requests from untrusted hosts

Other security features, such as dynamic ARP inspection (DAI), which is described in the next section, also use information stored in the DHCP snooping binding database.

DHCP snooping is enabled on a per-VLAN basis. By default, the feature is inactive on all VLANs. You can enable the feature on a single VLAN or a range of VLANs.

As mentioned previously, DHCP spoofing attacks take place when devices purposely attempt to generate enough DHCP requests to exhaust the number of IP addresses allocated to a DHCP pool.

The DHCP snooping feature determines whether traffic sources are trusted or untrusted. An untrusted source may initiate traffic attacks or other hostile actions. To prevent such attacks, the DHCP snooping feature filters messages and rate-limits traffic from untrusted sources.

The following steps are required to implement DHCP snooping on your network:

Step 1. Define and configure the DHCP server. Configuration of this step does not take place on the switch or router and is beyond the scope of this book.

Step 2. Enable DHCP snooping globally.

Step 3. Enable DHCP snooping on at least one VLAN. By default, DHCP snooping is inactive on all VLANs.

Step 4. Ensure that the DHCP server is connected through a trusted interface.

By default, the trust state of all interfaces is untrusted.

Step 5. Configure the DHCP snooping database agent. This step ensures that database entries are restored after a restart or switchover.

The DHCP snooping feature is not active until you complete this step. Example 6-20 provides the configuration details necessary to implement DHCP snooping to mitigate the effects of DHCP spoofing attacks.

Example 6-20   *Configuring DHCP Snooping*

```
! Enable DHCP Snooping Globally

sw2(config)# ip dhcp snooping
! Enable DHCP Snooping on VLAN 10
sw2(config)# ip dhcp snooping vlan 10
! Configure Interface Gi1/0/24 as a Trusted interface
sw2(config)# interface Gi1/0/24
sw2(config-if)# ip dhcp snooping trust
! Configure the DHCP snooping database agent to store the bindings at a
! given location
sw2(config)# ip dhcp snooping database tftp://10.1.1.1/directory/file
sw2(config)# exit
sw2#
! Verify DHCP Snooping Configuration
```

```
sw2# show ip dhcp snooping
Switch DHCP snooping is enabled
DHCP snooping is configured on following VLANs:
10
DHCP snooping is operational on following VLANs:
none
DHCP snooping is configured on the following L3 Interfaces:

Insertion of option 82 is enabled
    circuit-id default format: vlan-mod-port
    remote-id: 000f.90df.3400 (MAC)
Option 82 on untrusted port is not allowed
Verification of hwaddr field is enabled
Verification of giaddr field is enabled
DHCP snooping trust/rate is configured on the following Interfaces:


Interface               Trusted    Allow option    Rate limit (pps)
----------------------  -------    ------------    ----------------
GigabitEthernet1/0/24      yes         yes            unlimited
  Custom circuit-ids:
```

Dynamic ARP Inspection

ARP provides IP communication within a Layer 2 broadcast domain by mapping an IP address to a MAC address. For example, Host B wants to send information to Host A but does not have the MAC address of Host A in its Address Resolution Protocol (ARP) cache. Host B generates a broadcast message for all hosts within the broadcast domain to obtain the MAC address associated with the IP address of Host A. All hosts within the broadcast domain receive the ARP request, and Host A responds with its MAC address.

ARP spoofing attacks and ARP cache poisoning can occur because ARP allows a gratuitous reply from a host even if an ARP request was not received. After the attack, all traffic from the device under attack flows through the attacker's computer and then to the router, switch, or host.

An ARP spoofing attack can target hosts, switches, and routers connected to your Layer 2 network by poisoning the ARP caches of systems connected to the subnet and by intercepting traffic intended for other hosts on the subnet. Figure 6-2 shows an example of ARP cache poisoning.

Hosts A, B, and C are connected to the switch on interfaces A, B, and C, all of which are on the same subnet. Their IP and MAC addresses are shown in parentheses; for example, Host A uses IP address IA and MAC address MA. When Host A needs to communicate to Host B at the IP layer, it broadcasts an ARP request for the MAC address associated with IP address IB. When the switch and Host B receive the ARP request, they populate their ARP caches with an ARP binding for a host with the IP address IA and a MAC address MA; for example, IP address IA is bound to MAC address MA. When Host B responds, the switch and Host A populate their ARP caches with a binding for a host with the IP address IB and the MAC address MB.

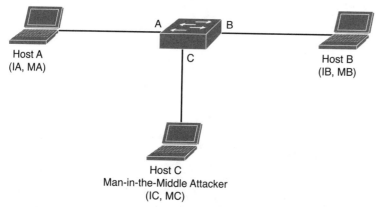

Figure 6-2 *ARP Cache Poisoning*

Host C can poison the ARP caches of the switch for Host A and Host B by broadcasting forged ARP responses with bindings for a host with an IP address of IA (or IB) and a MAC address of MC. Hosts with poisoned ARP caches use the MAC address MC as the destination MAC address for traffic intended for IA (or IB). This means that Host C intercepts that traffic. Because Host C knows the true MAC addresses associated with IA and IB, it can forward the intercepted traffic to those hosts by using the correct MAC address as the destination. Host C has inserted itself into the traffic stream from Host A to Host B, which is the topology of the classic man-in-the middle attack.

Dynamic ARP inspection (DAI) is a security feature that validates ARP packets in a network. DAI intercepts, logs, and discards ARP packets with invalid IP-to-MAC address bindings. This capability protects the network from some man-in-the-middle attacks.

DAI determines the validity of an ARP packet based on valid IP-to-MAC address bindings stored in a trusted database, the DHCP snooping binding database. As described in the previous section, this database is built by DHCP snooping if DHCP snooping is enabled on the VLANs and on the switch. If the ARP packet is received on a trusted interface, the switch forwards the packet without any checks. On untrusted interfaces, the switch forwards the packet only if it is valid.

You can configure DAI to drop ARP packets when the IP addresses in the packets are invalid or when the MAC addresses in the body of the ARP packets do not match the addresses specified in the Ethernet header.

Example 6-21 provides the configuration details necessary to implement DAI to mitigate the effects of ARP spoofing attacks.

Example 6-21 *Configuring Dynamic ARP Inspection (DAI)*

```
! Enable DAI on VLAN 10
sw2(config)# ip arp inspection vlan 10
sw2(config)# exit
! Verify DAI Configuration for VLAN 10
sw2# show ip arp inspection vlan 10
```

```
Source Mac Validation      : Disabled
Destination Mac Validation : Disabled
IP Address Validation      : Disabled


 Vlan      Configuration    Operation   ACL Match        Static ACL
 ----      -------------    ---------   ---------        ----------
   10      Enabled          Inactive


 Vlan      ACL Logging      DHCP Logging      Probe Logging
 ----      -----------      ------------      -------------
   10      Deny             Deny              Off


! Configure Interface Gi1/0/24 as a Trusted DAI Interface
sw2(config)# interface Gi1/0/24
sw2(config-if)# ip arp inspection trust
sw2(config-if)# exit
sw2(config)# exit
sw2# show ip arp inspection interfaces
 Interface        Trust State      Rate (pps)    Burst Interval
 ---------------  -----------      ----------    --------------
 Gi1/0/1          Untrusted            15             1
 Gi1/0/2          Untrusted            15             1
! output removed for brevity
 Gi1/0/23         Untrusted            15             1
 Gi1/0/24         Trusted            None            N/A
```

Network Foundation Protection

The network infrastructure primarily consists of routers and switches and their intercon-
necting cables. The infrastructure has to be healthy and functional if we want to be able to
deliver network services reliably.

If we break a big problem down into smaller pieces, such as security and what an attacker
might do, we can then focus on individual components and parts. By doing this, we make the
work of implementing security less daunting. That is what Network Foundation Protection
(NFP) is all about: breaking the infrastructure down into smaller components and then
systematically focusing on how to secure each of those components.

The Importance of the Network Infrastructure

Many pieces and parts make up a network, and even one simple component that is not
working can cause a failure of the network. If a network does not work, revenue and
productivity suffer. In a nutshell, if you have vulnerabilities such as weak passwords (or no
passwords), software vulnerabilities, or misconfigured devices, that leaves the door open
to attackers. The impact of a down network is huge; it normally affects the workforce and
other systems and customers that rely on that network. The NFP framework is designed to
assist you to logically group functions that occur on the network and then focus on specific
security measures you can take with each of these functions.

The Network Foundation Protection Framework

For Cisco IOS routers and switches, the Network Foundation Protection (NFP) framework is broken down into three basic planes (also called sections/areas):

- **Management plane:** This includes the protocols and traffic that an administrator uses between his workstation and the router or switch itself. An example is using a remote management protocol such as Secure Shell (SSH) to monitor or configure the router or switch. The management plane is listed here first because until the device is configured (which occurs in the management plane), the device will not be very functional in a network. If a failure occurs in the management plane, it may result in losing the ability to manage the network device altogether.

- **Control plane:** This includes protocols and traffic that the network devices use on their own without direct interaction from an administrator. An example is a routing protocol. A routing protocol can dynamically learn and share routing information that the router can then use to maintain an updated routing table. If a failure occurs in the control plane, a router might lose the capability to share or correctly learn dynamic routing information, and as a result might not have the routing intelligence to be able to route for the network.

- **Data plane:** This includes traffic that is being forwarded through the network (sometimes called transit traffic). An example is a user sending traffic from one part of the network to access a server in another part of the network; the data plane represents the traffic that is either being switched or forwarded by the network devices between clients and servers. A failure of some component in the data plane results in the customer's traffic not being able to be forwarded. Other times, based on policy, you might want to deny specific types of traffic traversing the data plane.

Interdependence

Some interdependence exists between these three planes. For example, if the control plane fails, and the router does not know how to forward traffic, this scenario impacts the data plane because user traffic cannot be forwarded. Another example is a failure in the management plane that might allow an attacker to configure devices and, as a result, could cause both a control plane and data plane failure.

Implementing NFP

You learn more about each of these three planes later in this chapter as well as in subsequent chapters dedicated to each of the three planes individually. Before that, however, Table 6-2 describes security measures you can use to protect each of the three planes.

Table 6-2   Management, Control, and Data Plane Security

Management Plane	Control Plane	Data Plane
Authentication, authorization, accounting (AAA)	Control Plane Policing (CoPP)	Access control lists (ACLs)
Authenticated Network Time Protocol (NTP)	Control Plane Protection (CPPr)	Layer 2 controls, such as private VLANs and Spanning Tree Protocol (STP) Guards

Management Plane	Control Plane	Data Plane
Secure Shell (SSH)	Authentication Routing Protocol Updates	Cisco IOS and Cisco IOS-XE Zone-based Firewall
Transport Layer Security (TLS)		SD-WAN Security
Protected syslog		
Simple Network Management Protocol Version 3 (SNMPv3)		
NETCONF/RESTCONF		

As you might have noticed, NFP is not a single feature but rather is a holistic approach that covers the three components (that is, planes) of the infrastructure, with recommendations about protecting each one using a suite of features you can implement across your network.

A command-line utility called **auto secure** implements security measures (several in each category) across all three of the planes.

When implementing the best practices described by NFP, does that mean your network is going to be up forever and not have any problems? Of course not. If the network is designed poorly, with no fault tolerance, for example, and a device fails (because of a mechanical or software failure or a physical problem or because cables were removed), if you do not have the failovers in place to continue to move traffic, your data plane is going to suffer. Other factors, such as lack of change control or an administrator accidentally putting in the incorrect configuration, are, of course, ongoing potential opportunities for the network to stop functioning.

Understanding and Securing the Management Plane

This section examines what you can do to protect management access and management protocols used on the network.

As mentioned earlier, the management plane is covered first in this discussion. After all, without a configured router (whether configured through the console port or through an IP address with a secure remote-access tool such as SSH), the network device is not much good without a working configuration that either an administrator or some other type of management system such as Cisco DNA Center has put in place. (A basic Layer 2 switch with all ports in the same VLAN would be functional, but this is unlikely to be the desired configuration for that device.)

 ## Best Practices for Securing the Management Plane

To secure the management plane, adhere to these best practices:

- Enforce password policy, including features such as maximum number of login attempts and minimum password length.

- Implement role-based access control (RBAC). This concept has been around for a long time in relation to groups. By creating a group that has specific rights and then placing users in that group, you can more easily manage and allocate administrators. With RBAC, we can create a role (like a group) and assign that role to the users who will be acting in it. With the role comes the permissions and access. Ways to implement RBACs include using Access Control Server (ACS) and CLI parser views (which restrict the commands that can be issued in the specific view the administrator is in). Custom privilege level assignments are also an option to restrict what a specific user may do while operating at that custom privilege level.

- Use AAA services, and centrally manage those services on an authentication server (such as the Cisco ISE). With AAA, a network router or switch can interact with a centralized server before allowing any access, before allowing any command to be entered, and while keeping an audit trail that identifies who has logged in and what commands they executed while there. Your policies about who can do what can be configured on the central server, and then you can configure the routers and switches to act as clients to the server as they make their requests, asking whether it is okay for a specific user to log in or if it is okay for a specific user to issue a specific command.

- Keep accurate time across all network devices using secure Network Time Protocol (NTP).

- Use encrypted and authenticated versions of SNMP, which includes Version 3 (and some features from Version 2).

- Control which IP addresses are allowed to initiate management sessions with the network device.

- Lock down syslog. Syslog is sent in plain text. On the infrastructure of your network, only permit this type of traffic between the network device's IP address and the destinations that the network device is configured to send the syslog messages to. In practice, not too many people are going to encrypt syslog data, although it is better to do so. Short of doing encryption, we could use an out-of-band (OOB) method to communicate management traffic between our network devices and the management stations. An example is a separate VLAN that user traffic never goes on to, and using that separate VLAN just for the management traffic. If management traffic is sent in-band, which means the management traffic is using the same networks (the same VLANs, for instance), all management traffic needs to have encryption, either built in or protected by encryption (such as using IPsec).

- Disable any unnecessary services, especially those that use User Datagram Protocol (UDP). These are infrequently used for legitimate purposes, but can be used to launch denial-of-service (DoS) attacks. Following are some services that should be disabled if they are not needed:

  - TCP and UDP small services

  - Finger

  - BOOTP

  - DHCP

  - Maintenance Operation Protocol (MOP)

  - DNS

  - Packet assembler/disassembler (PAD)

- HTTP server and Secure HTTP (HTTPS) server

- CDP

- LLDP

Understanding the Control Plane

This section reviews what you can do to protect network devices in the event of attacks involving traffic directed to (nontransit traffic) the network device itself.

The route processor, the CPU on a router, can only do so much. So, whenever possible, the router is going to cache information about how to forward packets (transit packets going from one device on the network to some other device). By using cached information when a packet shows up that needs to be forwarded, the CPU has to expend little effort. Forwarding of traffic is function of the data plane, and that is what really benefits from using cached information.

So, what has that got to do with the control plane? If a packet, such as an Open Shortest Path First (OSPF) or Enhanced Interior Gateway Routing Protocol (EIGRP) routing advertisement packet, is being sent to an IP address on the router, it is no longer a transit packet that can be simply forwarded by looking up information in a route cache of some type. Instead, because the packet is addressed to the router itself, the router has to spend some CPU cycles to interpret the packet, look at the application layer information, and then potentially respond. If an attacker sends thousands of packets like these to the router, or if there is a botnet of hundreds of thousands of devices, each configured to send these types of packets to the router, the router could be so busy just processing all these requests that it might not have enough resources to do its normal work. Control plane security is primarily guarding against attacks that might otherwise negatively impact the CPU, including routing updates (which are also processed by the CPU).

Best Practices for Securing the Control Plane

You can deploy CoPP and CPPr to protect the control plane. Control Plane Policing, or CoPP, can be configured as a filter for any traffic destined to an IP address on the router itself. For instance, you can specify that management traffic, such as SSH/HTTPS/SSL and so on, can be rate-limited (policed) down to a specific level or dropped completely. This way, if an attack occurs that involves an excessive amount of this traffic, the excess traffic above the threshold set could simply be ignored and not have to be processed directly by the CPU. Another way to think of this is as applying Quality of Service (QoS) to the valid management traffic and policing to the bogus management traffic.

> **TIP** CoPP is applied to a logical control plan interface (not directly to any Layer 3 interface) so that the policy can be applied globally to the router.

Control Plane Protection, or CPPr, allows for a more detailed classification of traffic (more than CoPP) that is going to use the CPU for handling. The specific sub-interfaces can be classified as follows:

- **Host sub-interface:** Handles traffic to one of the physical or logical interfaces of the router.

- **Transit sub-interface:** Handles certain data plane traffic that requires CPU intervention before forwarding (such as IP options).

- **Cisco Express Forwarding (CEF):** Exception traffic (related to network operations, such as keepalives or packets with Time-to-Live [TTL] mechanisms that are expiring) that has to involve the CPU.

The benefit of CPPr is that you can rate-limit and filter this type of traffic with a more fine-toothed comb than CoPP.

> **TIP**   CPPr is also applied to a logical control plane interface so that regardless of the logical or physical interface on which the packets arrive, the router processor can still be protected.

Using CoPP or CPPr, you can specify which types of management traffic are acceptable at which levels. For example, you could decide and configure the router to believe that SSH is acceptable at 100 packets per second, syslog is acceptable at 200 packets per second, and so on. Traffic that exceeds the thresholds can be safely dropped if it is not from one of your specific management stations. You can specify all those details in the policy.

Routing protocol authentication is another best practice for securing the control plane. If you use authentication, a rogue router on the network will not be believed by the authorized network devices (routers). The attacker may have intended to route all the traffic through his device, or perhaps at least learn detailed information about the routing tables and networks.

Although not necessarily a security feature, Selective Packet Discard (SPD) provides the ability to prioritize certain types of packets (for example, routing protocol packets and Layer 2 keepalive messages, which are received by the route processor [RP]). SPD provides priority of critical control plane traffic over traffic that is less important or, worse yet, is being sent maliciously to starve the CPU of resources required for the RP.

Understanding and Securing the Data Plane

This section covers the methods available for implementing policy related to traffic allowed through (transit traffic) network devices.

For the data plane, this discussion concerns traffic that is going through your network device rather than to a network device. This is traffic from a user going to a server, and the router is just acting as a forwarding device. This is the data plane. The following are some of the prevalent ways to control the data plane (which may be implemented on an IOS router).

Best Practices for Protecting the Data Plane

To secure the data plane, adhere to these best practices:

- Block unwanted traffic at the router. If your corporate policy does not allow TFTP traffic, just implement ACLs that deny traffic that is not allowed. You can implement ACLs inbound or outbound on any Layer 3 interface on the router. With extended ACLs, which can match based on the source and/or destination address, placing the ACL closer to the source saves resources because it denies the packet before it consumes network bandwidth and before route lookups are done on a router that is filtering inbound rather than outbound. Filtering on protocols or traffic types known to be malicious is a good idea.

■ Reduce the chance of DoS attacks. Techniques such as TCP Intercept and firewall services can reduce the risk of SYN-flood attacks.

■ Reduce spoofing attacks. For example, you can filter (deny) packets trying to enter your network (from the outside) that claim to have a source IP address that is from your internal network.

■ Provide bandwidth management. Implementing rate-limiting on certain types of traffic can also reduce the risk of an attack (Internet Control Message Protocol [ICMP], for example, which would normally be used in small quantities for legitimate traffic).

■ When possible, use an IPS to inhibit the entry of malicious traffic into the network.

Additional Data Plane Protection Mechanisms

Normally, for data plane protection we think of Layer 3 and routers. Obviously, if traffic is going through a switch, a Layer 2 function is involved, as well. Layer 2 mechanisms that you can use to help protect the data plane include the following:

■ Port security to protect against MAC address flooding and CAM (content-addressable memory) overflow attacks. When a switch has no more room in its tables for dynamically learned MAC addresses, there is the possibility of the switch not knowing the destination Layer 2 address (for the user's frames) and forwarding a frame to all devices in the same VLAN. This might give the attacker the opportunity to eavesdrop.

■ Dynamic Host Configuration Protocol (DHCP) snooping to prevent a rogue DHCP server from handing out incorrect default gateway information and to protect a DHCP server from a starvation attack (where an attacker requests all the IP addresses available from the DHCP server so that none are available for clients who really need them).

■ Dynamic ARP inspection (DAI) can protect against Address Resolution Protocol (ARP) spoofing, ARP poisoning (which is advertising incorrect IP-to-MAC-address mapping information), and resulting Layer 2 man-in-the-middle attacks.

■ IP Source Guard, when implemented on a switch, verifies that IP spoofing is not occurring by devices on that switch.

Securing Management Traffic

Fixing a problem is tricky if you are unaware of the problem. So, this first section starts by classifying and describing management traffic and identifying some of the vulnerabilities that exist. It also identifies some concepts that can help you to protect that traffic. This chapter then provides implementation examples of the concepts discussed earlier.

What Is Management Traffic and the Management Plane?

When you first get a new router or switch, you connect to it for management using a blue rollover cable that connects from your computer to the console port of that router or switch. This is your first exposure to the concept of management traffic. By default, when you connect to a console port, you are not prompted for a username or any kind of password. By requiring a username or password, you are taking the first steps toward improving what is called the management plane on this router or switch.

The management plane includes not only configuration of a system, but also who may access a system and what they are allowed to do while they are logged in to the system. The management plane also includes messages to or from a Cisco router or switch that are used to maintain or report on the current status of the device, such as a management protocol like Simple Network Management Protocol (SNMP).

Beyond the Console Cable

Using the console cable directly connected to the console port is fairly safe. Unfortunately, it is not very convenient to require the use of a console port when you are trying to manage several devices that are located in different buildings, or even on different floors of the same building. A common solution to this problem is to configure the device with an IP address that you can then use to connect to that device remotely. It is at this moment that the security risk goes up. Because you are connecting over IP, it might be possible for an unauthorized person to also connect remotely. The management plane, if it were secure, would enable you to control who may connect to manage the box, when they may connect, what they may do, and report on anything that they did. At the same time, you want to ensure that all the packets that go between the device being managed and the computer where the administrator is sitting are encrypted so that anyone who potentially might capture the individual packets while going through the network cannot interpret the contents of the packets (which might contain sensitive information about the configuration or passwords used for access).

Management Plane Best Practices

When implementing a network, remember the following best practices. Each one, when implemented, improves the security posture of the management plane for your network. In other words, each additional best practice, when put in place, raises the level of difficulty required on behalf of the attackers to compromise your device:

- **Strong passwords:** Whenever and wherever you use passwords, make them complex and difficult to guess. An attacker can break a password in several ways, including a dictionary and/or a brute-force attack. A dictionary attack automates the process of attempting to log in as the user, running through a long list of words (potential passwords); when one attempt fails, the attack just tries the next one (and so on). A brute-force attack does not use a list of words, but rather tries thousands or millions of possible character strings trying to find a password match (modifying its guesses progressively if it incorrectly guesses the password or stops before it reaches the boundary set by the attacker regarding how many characters to guess, with every possible character combination being tried). A tough password takes longer to break than a simple password.

- **User authentication and AAA:** Require administrators to authenticate using usernames and passwords. This is much better than just requiring a password and not knowing exactly who the user is. To require authentication using usernames and passwords, you can use a method for authentication, authorization, and accounting (AAA). Using this, you can control which administrators are allowed to connect to which devices and what they can do while they are there, and you can create an audit trail (accounting records) to document what they actually did while they were logged in.

- **Login Password Retry Lockout:** The Login Password Retry Lockout feature allows system administrators to lock out a local AAA user account after a configured number of unsuccessful attempts by the user to log in using the username that corresponds to the AAA user account. A locked-out user cannot successfully log in again until the user account is unlocked by the administrator.

■ **Role-based access control (RBAC):** Not every administrator needs full access to every device, and you can control this through AAA and custom privilege levels/parser views. For example, if there are junior administrators, you might want to create a group that has limited permissions. You could assign users who are junior administrators to that group; they then inherit just those permissions. This is one example of using RBAC. Another example of RBAC is creating a custom privilege level and assigning user accounts to that level. Regardless of how much access an administrator has, a change management plan for approving, communicating, and tracking configuration changes should be in place and used before changes are made.

■ **Encrypted management protocols:** When using either in-band or out-of-band management, encrypted communications should be used, such as Secure Shell (SSH) or Hypertext Transfer Protocol Secure (HTTPS). Out-of-band (OOB) management implies that there is a completely separate network just for management protocols and a different network for end users and their traffic. In-band management is when the packets used by your management protocols may intermingle with the user packets (considered less secure than OOB). Whether in-band or OOB, if a plaintext management protocol must be used, such as Telnet or HTTP, use it in combination with a virtual private network (VPN) tunnel that can encrypt and protect the contents of the packets being used for management.

■ **Logging and monitoring:** Logging is a way to create an audit trail. Logging includes not only what configuration changes have been made by administrators, but also system events that are generated by the router or switch because of some problem that has occurred or some threshold that has been reached. Determine the most important information to log and identify the appropriate logging levels to use. A logging level simply specifies how much detail to include in logging messages, and may also indicate that some less-serious logging messages do not need to be logged. Because the log messages may include sensitive information, the storage of the logs and the transmission of the logs should be protected to prevent tampering or damage. Allocate sufficient storage capacity for anticipated logging demands. Logging may be done in many different ways, and your logging information may originate from many different sources, including messages that are automatically generated by the router or switch and sent to a syslog server. A syslog server is a computer that is set up to receive and store syslog messages generated from network devices. If SNMP is used, Version 3 is recommended because of its authentication and encryption capabilities. You can use SNMP to change information on a router or switch, and you can also use it to retrieve information from the router or switch. An SNMP trap is a message generated by the router or switch to alert the manager or management station of some event.

■ **Network Time Protocol (NTP):** Use NTP to synchronize the clocks on network devices so that any logging that includes timestamps may be easily correlated. Preferably, use NTP Version 3 to leverage its ability to provide authentication for time updates. This becomes very important to correlate logs between devices in case there is ever a breach and you need to reconstruct (or prove in a court of law) what occurred.

■ **Secure system files:** Make it difficult to delete, whether accidentally or on purpose, the startup configuration files and the IOS images that are on the file systems of the local routers and switches. You can do so by using built-in IOS features discussed later in this chapter.

> **NOTE**   Even though OOB management is usually preferred over in-band management, some management applications benefit from in-band management. For example, consider a network management application that checks the reachability of various hosts and subnets. To check this reachability, an application might send a series of pings to a remote IP address, or check the availability of various Layer 4 services on a remote host. To perform these "availability" checks, the network management application needs to send traffic across a production data network. Also, in-band network management often offers a more economical solution for smaller networks. Even if you're using in-band management, it should be a separate subnet/VLAN, and one that only a select few people/devices have access to get to. This reduces your footprint for possible attack vectors.

Password Recommendations

Using passwords is one way to restrict access to only authorized users. Using passwords alone is not as good as requiring a user ID or login name associated with the password for a user.

Here are some guidelines for password creation:

■ It is best to have a minimum of eight characters for a password; bigger is better. This rule can be enforced by the local router if you are storing usernames and passwords on the router in the running config. The command **security passwords** *min-length* followed by the minimum password length enforces this rule on new passwords that are created, including the enable secret and line passwords on the vty, AUX, and console 0. Preexisting passwords will still operate even if they are less than the new minimum specified by the command.

■ Passwords can include any alphanumeric character, a mix of uppercase and lowercase characters, and symbols and spaces. As a general security rule, passwords should not use words that may be found in a dictionary, because they are easier to break. Leading spaces in a password are ignored, but any subsequent spaces, including in the middle or at the end of a password, literally become part of that password and are generally a good idea. Another good practice is using special characters or even two different words (that are not usually associated with each other) as a passphrase when combined together. Caution should be used to not require such a complex password that the user must write it down to remember it, which increases the chance of it becoming compromised.

Passwords in a perfect environment should be fairly complex and should be changed periodically. The frequency of requiring a change in passwords depends on your security policy. Passwords changed often are less likely to be compromised.

■ From a mathematical perspective, consider how many possibilities someone would need to try to guess a password. If only capital letters are used, you have 26 possibilities for each character. If your password is one character long, that is 26^1, or 26 possible variants. If you have a two-character password, that is 26^2, or 676 possible variants.

If you start using uppercase (26) and lowercase (26), numerals (10), and basic special characters (32), your starting set becomes 94 possible variants per character. Even if we look at using an eight-character password, that is 94^8, or 6,095,689,385,410,816 (6.1 quadrillion), possibilities.

■ Multifactor authentication and Duo. In Chapter 4, you learned that the process of authentication requires the subject to supply verifiable credentials. The credentials are often referred to as "factors." *Single-factor* authentication is when only one factor is presented. The most common method of single-factor authentication is the use of passwords. *Multifactor authentication* is when two or more factors are presented. *Multilayer authentication* is when two or more of the same type of factors are presented. Data classification, regulatory requirements, the impact of unauthorized access, and the likelihood of a threat being exercised should all be considered when you're deciding on the level of authentication required. The more factors, the more robust the authentication process. Duo Security was a company acquired by Cisco that develops a very popular multifactor authentication solution used by many small, medium, and large organizations. Duo provides protection of on-premises and cloud-based applications. This is done by both preconfigured solutions and generic configurations via RADIUS, Security Assertion Markup Language (SAML), LDAP, and more. Duo integrates with many different third-party applications, cloud services, and other solutions. Duo allows administrators to create policy rules around who can access applications and under what conditions. You can customize policies globally or per user group or application. User enrollment strategy will also inform your policy configuration. Duo Beyond subscribers can benefit from additional management within their environment by configuring a Trusted Endpoints policy to check the posture of the device that is trying to connect to the network, application, or cloud resource.

Using AAA to Verify Users

Unauthorized user access to a network creates the potential for network intruders to gain information or cause harm, or both. Authorized users need access to their network resources, and network administrators need access to the network devices to configure and manage them. In Chapter 4, you learned that AAA offers a solution for both. In a nutshell, the goal of AAA is to identify who users are before giving them any kind of access to the network, and once they are identified, only give them access to the part they are authorized to use, see, or manage. AAA can create an audit trail that identifies exactly who did what and when they did it. That is the spirit of AAA. User accounts may be kept on the local AAA database or on a remote AAA server. The local database is a fancy way of referring to user accounts that are created on the local router and are part of the running configuration.

Router Access Authentication

Note that we must choose authentication first if we want to also use authorization for a user or administrator. We cannot choose authorization for a user without knowing who that user is through authentication first.

Typically, if we authenticate an administrator, we also authorize that administrator for what we want to allow him to do. Administrators traditionally are going to need access to the CLI. When an administrator is at the CLI, that interface is provided by something called an EXEC shell. If we want to authorize the router to provide this CLI, that is a perfect example of using AAA to first authenticate the user (in this case, the administrator) and then authorize that user to get a CLI prompt (the EXEC shell) and even place the administrator at the correct privilege level. This type of access (CLI) could also be referred to as character mode. Simply think of an administrator at a CLI typing in characters to assist you in remembering that this is "character" mode. With the administrator, we would very likely authenticate his login request and authorize that administrator to use an EXEC shell. If we were using a remote ISE server for this authentication and authorization of an administrator, we would very likely use TACACS+ (between the router and the ISE server) because it has the most granular control, compared with RADIUS, which is the alternative. TACACS+ and RADIUS were both discussed in Chapter 4.

The AAA Method List

To make implementing AAA modular, we can specify individual lists of ways we want to authenticate, authorize, and account for the users. To do this, we create a method list that defines what resource will be used (such as the local database, an ISE server via TACACS+ protocol or an ISE server via RADIUS protocol, and so forth). To save time, we can create a default list or custom lists. We can create authentication method lists that define the authentication methods to use, authorization method lists that define which authorization methods to use, and accounting method lists that specify which accounting methods to use. A default list, if created, applies to the entire router or switch. A custom list, to be applied, must be first created and then specifically referenced in line or interface configuration mode. You can apply a custom list over and over again in multiple lines or interfaces. The type of the method list may be authentication, authorization, or accounting.

The syntax for a method list is as follows:

aaa type {**default** | *list-name*} *method-1* [*method-2 method-3 method-4*]

Here, *list-name* is used to create a custom method list. This is the name of the list, and it is used when this list is applied to a line (such as vty lines 0 through 4). The methods (*method-1*, *method-2*, and so on) are a single list that can contain up to four methods, which are tried in order from left to right. At least one method must be specified. To use the local user database, use the **local** keyword. In the case of an authentication method list, methods include the following:

- **enable:** The enable password is used for authentication. This could be a good choice as the last method in a method list. This way, if the previous methods are not available (for example, if the AAA server is down or not configured), the router times out on the first methods and eventually prompts the user for the enable secret as a last resort.

- **krb5:** Kerberos version 5 is used for authentication.

- **krb5-telnet:** Kerberos version 5 Telnet authentication protocol is used when using Telnet to connect to the router. This is not a secure method. Using Telnet is not recommended and should be avoided.

- **line:** The line password (the one configured with the password command, on the individual line) is used for authentication.

- **local:** The local username database (in the router's running configuration) is used for authentication.

6

Role-Based Access Control

The concept of role-based access control (RBAC) is to create a set of permissions or limited access and assign that set of permissions to users or groups. Those permissions are used by individuals for their given roles, such as a role of administrator, a role of a help desk person, and so on. There are different ways to implement RBAC, including creating custom privilege levels and creating parser views (coming up later in this section). In either case, the custom level or view can be assigned the permissions needed for a specific function or role, and then users can use those custom privilege levels or parser views to carry out their job responsibilities on the network, without being given full access to all configuration options.

Custom Privilege Levels

When you first connect to a console port on the router, you are placed into user mode. User mode is really defined as privilege level 1. This is represented by a prompt that ends with **>**. When you move into privileged mode by typing the **enable** command, you are really moving into privilege level 15. A user at privilege level 15 has access and can issue all the commands that are attached to or associated with level 15 and below. Nearly all the configuration commands, and the commands that get us into configuration mode, are associated by default with privilege level 15.

By creating custom privilege levels (somewhere between levels 2 and 14, inclusive), and assigning commands that are normally associated with privilege level 15 to this new level, you can give this subset of new commands to the individual who either logs in at this custom level or to the user who logs in with a user account that has been assigned to that level.

Limiting the Administrator by Assigning a View

Working with individual commands and assigning them to custom privilege levels is tedious at best, and it is for that reason that this method is not used very often. So, what can be done if we need users to have a subset of commands available to them, but not all of them? You can use parser views, which are also referred to as simply "a view." You can create a view and associate it with a subset of commands. When the user logs in using this view, that same user is restricted to only being able to use the commands that are part of his current view. You can also associate multiple users with a single view.

Encrypted Management Protocols

It is not always practical to have console access to the Cisco devices you manage. There are several options for remote access via IP connectivity, and the most common is an application called Telnet. The problem with Telnet is that it uses plain text, and anyone who gets a copy of those packets can identify our usernames and passwords used for access and any other information that goes between the administrator and the router being managed (over the management plane). One solution to this is to not use Telnet. If Telnet must be used, it should only be used out of band, or placed within a VPN tunnel for privacy, or both.

Secure Shell (SSH) provides the same functionality as Telnet, in that it gives you a CLI to a router or switch; unlike Telnet, however, SSH encrypts all the packets that are used in the session. So, with SSH, if a packet is captured and viewed by an unauthorized individual, it will not have any meaning because the contents of each packet are encrypted, and the attacker or unauthorized person will not have the keys or means to decrypt the information. The encryption provides the feature of confidentiality.

With security, bigger really is better. With SSH, Version 2 is bigger and better than Version 1. Either version, however, is better than the unencrypted Telnet protocol. When you type in **ip ssh version 2** (to enable Version 2), the device may respond with "Version 1.99 is active." This is a function of a server that runs 2.0 but also supports backward compatibility with older versions. For more information, see RFC 4253, section 5.1. You should use SSH rather than Telnet whenever possible.

For graphical user interface (GUI) management tools, use HTTPS rather than HTTP because, like SSH, it encrypts the session, which provides confidentiality for the packets in that session.

Using Logging Files

I still recall an incident on a customer site when a database server had a failed disk and was running on its backup. It was like that for weeks until they noticed a log message. If a second failure had occurred, the results would have been catastrophic. Administrators should, on a regular basis, analyze logs, especially from their routers, in addition to logs from other network devices. Logging information can provide insight into the nature of an attack. Log information can be used for troubleshooting purposes. Viewing logs from multiple devices can provide event correlation information (that is, the relationship between events occurring on different systems). For proper correlation of events, accurate time stamps on those events are important. Accurate time can be implemented through Network Time Protocol (NTP).

Cisco network infrastructure devices can send log output to a variety of destinations, including the following:

- **Console:** A router's console port can send log messages to an attached terminal (such as your connected computer, running a terminal emulation program).

- **vty lines:** Virtual tty (vty) connections (used by SSH and Telnet connections) can also receive log information at a remote terminal (such as an SSH or Telnet client). However, the **terminal monitor** command should be issued to cause log messages to be seen by the user on that vty line.

- **Buffer:** When log messages are sent to a console or a vty line, those messages are not later available for detailed analysis. However, log messages can be stored in router memory. This "buffer" area can store messages up to the configured memory size, and then the messages are rotated out, with the first in being the first to be removed (otherwise known as first in, first out [FIFO]). When the router is rebooted, these messages in the buffer memory are lost.

- **SNMP server:** When configured as an SNMP device, a router or switch can generate log messages in the form of SNMP traps and send them to an SNMP manager (server).

- **Syslog server:** A popular choice for storing log information is a syslog server, which is easily configured and can store a large volume of logs. Syslog messages can be directed to one or more syslog servers from the router or switch.

A syslog logging solution consists of two primary components: syslog servers and syslog clients. A syslog server receives and stores log messages sent from syslog clients such as routers and switches.

Not all syslog messages are created equal. Specifically, they have different levels of severity. Table 6-3 lists the eight levels of syslog messages. The higher the syslog level, the more detailed the logs. Keep in mind that more-detailed logs require a bit more storage space, that syslog messages are transmitted in clear text, and that the higher levels of syslog logging consume higher amounts of CPU processing time. For these reasons, you should take care when logging to the console at the debugging level.

Table 6-3   Syslog Severity Levels

Level	Name	Description
0	Emergencies	System is unusable.
1	Alerts	Immediate action needed.
2	Critical	Critical conditions.
3	Error	Error conditions.
4	Warnings	Warning conditions.
5	Notifications	Normal, but significant conditions.
6	Informational	Informational messages.
7	Debugging	Highly detailed information based on debugging that is turned on.

The syslog log entries contain timestamps, which are helpful in understanding how one log message relates to another. The log entries include severity level information in addition to the text of the syslog messages. Having synchronized time on the routers and including timestamps in the syslog messages make correlation of the syslog messages from multiple devices more meaningful.

Understanding NTP

Network Time Protocol (NTP) uses UDP port 123, and it allows network devices to synchronize their time. Ideally, they would synchronize their time to a trusted time server. You can configure a Cisco router to act as a trusted NTP server for the local network, and in the same way, that trusted NTP server (Cisco router) could turn around and be an NTP client to a trusted NTP server either on the Internet or reachable via network connectivity. NTP Version 3 supports cryptographic authentication between NTP devices, and for this reason its use is preferable over any earlier versions.

One benefit of having reliable synchronized time is that log files and messages generated by the router can be correlated. In fact, if we had 20 routers, and they were all reporting various messages and all had the same synchronized time, we could very easily correlate the events across all 20 routers if we looked at those messages on a common server such as a syslog server.

Protecting Cisco IOS, Cisco IOS-XE, Cisco IOS-XR, and Cisco NX-OS Files

Similar to the computers we use every day, a router also uses an operating system. The traditional Cisco operating system on the router is called IOS, or sometimes classic IOS. However, the newer operating system for enterprise routers is Cisco IOS-XE, for service provider routers it is Cisco IOS-XR, and for data center infrastructure devices it is NX-OS. For the purpose of this section, we will refer to all of them as "IOS."

When a router first boots, it performs a power-on self-test and then looks for an image of IOS on the flash. After loading the IOS into RAM, the router then looks for its startup configuration. If for whatever reason an IOS image or the startup configuration cannot be found or loaded properly, the router will effectively be nonfunctional as far as the network is concerned.

To help protect a router from accidental or malicious tampering of the IOS or startup configuration, Cisco offers a resilient configuration feature. This feature maintains a secure working copy of the router IOS image and the startup configuration files at all times. Once the feature is enabled, the administrator cannot disable it remotely (but can if connected directly on the console). The secure files are referred to as a "secure bootset."

Implementing Security Measures to Protect the Management Plane

Let's look at some practical examples of implementing infrastructure security best practices. It requires both the understanding and implementation of these best practices to secure your networks.

Implementing Strong Passwords

The privileged EXEC secret (the one used to move from user mode to privileged mode) should not match any other password that is used on the system. Many of the other passwords are stored in plain text (such as passwords on the vty lines). If an attacker discovers these other passwords, he might try to use them to get into privileged mode, and that is why the enable secret should be unique. Service password encryption scrambles any plaintext passwords as they are stored in the configuration. This is useful for preventing someone who is looking over your shoulder from reading a plaintext password that is displayed in the configuration on the screen. Any new plaintext passwords are also scrambled as they are stored in the router's configuration. Example 6-22 shows the use of strong passwords.

Example 6-22 *Using Strong Passwords*

```
! Use the "secret" keyword instead of the "password" for users

! This will create a secured password in the configuration by default
! The secret is hashed using the MD5 algorithm as it is stored in the
! configuration
R1(config)# username admin secret CeyeSc01$24


! At a minimum, require a login and password for access to the console port
! Passwords on lines, including the console, are stored as plain text, by
! default, in the configuration
R1(config)# line console 0
R1(config-line)# password k4(1fmMsS1#
R1(config-line)# login
R1(config-line)# exit


! At a minimum, require a login and password for access to the vty lines which
! is where remote users connect when using Telnet
! Passwords on lines, including the vty lines, are stored as plain text, by
```

```
! default, in the configuration
R1(config)# line vty 0 4
R1(config-line)# password 8wT1*eGP5@
R1(config-line)# login


! At a minimum, require a login and password for access to the AUX line
! and disable the EXEC shell if it will not be used
R1(config-line)# line aux 0
R1(config-line)# no exec
R1(config-line)# password 1wT1@ecP27
R1(config-line)# login
R1(config-line)# exit


! Before doing anything else, look at the information entered.
R1(config)# do show run | include username
username admin secret 5 $1$XJdX$9hqvG53z3lesP5BLOqggO.
R1(config)#
R1(config)# do show run | include password
no service password-encryption
 password k4(1fmMsS1#
 password 8wT1*eGP5@
 password 1wT1@ecP27
R1(config)#


! Notice that we cannot determine the admin user's password, since
! it is automatically hashed using the MD5 algorithm because of using
! the secret command, however, we can still see all the other plain text
! passwords.


! Encrypt the plain text passwords so that someone reading the configuration
! won't know what the passwords are by simply looking at the configuration.
R1(config)# service password-encryption


! Verify that the plain text passwords configured are now scrambled due to the
! command "service password-encryption"
R1(config)# do show run | begin line
line con 0
 password 7 04505F4E5E2741631A2A5454
 login
line aux 0
 no exec
 login
 password 7 075E36781F291C0627405C
line vty 0 4
 password 7 065E18151D040C3E354232
 login
!
end
```

User Authentication with AAA

In Chapter 4, you learned details about AAA and identity management. Example 6-23 shows the use of AAA method lists, both named and default.

Example 6-23 *Enabling AAA Services and Working with Method Lists*

```
! Enable aaa features, if not already present in the running configuration

R1(config)# aaa new-model

! Identify a AAA server to be used, and the password it is expecting with
! requests from this router. This server would need to be reachable and
! configured as a TACACS+ server to support R1's requests

R1(config)# tacacs-server host 10.8.4.1
R1(config)# tacacs-server key ToUgHPaSsW0rD-1#7

! configure the default method list for the authentication of character
! mode login (where the user will have access to the CLI)
! This default method list, created below has two methods listed "local"
! and "enable"
! This list is specifying that the local database (running-config) will
! be used first to look for the username.  If the username isn't in the
! running-config, then it will go to the second method in the list.
! The second method of "enable" says that if the user account isn't found
! in the running config, then to use the enable secret to login.
! This default list will apply to all SSH, Telnet, vty, AUX and Console
! sessions unless there is another (different) custom method list that is
! created and directly applied to one of those lines.

R1(config)# aaa authentication login default local enable

! The next authentication method list is a custom authentication
! method list named MY-LIST-1.This method list says that the first attempt
! to verify the user's name and password should be done through one of the
! tacacs servers (we have only configured one so far), and then if the server
! doesn't respond, use the local database (running-config), and if the
! username isn't in the running configuration to then use the enable secret
! for access to the device.  Note: this method list is not used until
! applied to a line elsewhere in the configuration, i.e. the default list
! configured previously is used unless MY-LIST-1 is specifically configured.

R1(config)# aaa authentication login MY-LIST-1 group tacacs local enable

! These next method lists are authorization method lists.
! We could create a default one as well, using the key
! word "default" instead of a name.   These custom method lists for
```

6

```
! authorization won't be used until we apply them
! elsewhere in the configuration, such as on a VTY line.
! The first method list called TAC1 is an authorization
! method list for all commands at user mode (called privilege level 1).
! The second method list called TAC15 is an
! authorization method list for commands at level 15 (privileged exec mode).
! If these method lists are applied to a line, such as the
! console or VTY lines, then before any commands
! are executed at user or privileged mode, the router will check
! with an ISE server that is one of the "tacacs+" servers, to see if the user
! is authorized to execute the command. If a tacacs+ server isn't
! reachable, then the router will use its own database of users (the local
! database) to determine if the user trying to issue the command
! is at a high enough privilege level to execute the command.

R1(config)# aaa authorization commands 1 TAC1 group tacacs+ local
R1(config)# aaa authorization commands 15 TAC15 group tacacs+ local

! The next 2 method lists are accounting method lists that will record the
! commands issued at level 1 and 15 if the lists are applied to a line, and
! if an administrator connects to this device via that line.
! Accounting method lists can have multiple methods but can't log to the
! local router.

R1(config)# aaa accounting commands 1 TAC-act1 start-stop group tacacs+
R1(config)# aaa accounting commands 15 TAC-act15 start-stop group tacacs+

! Creating a user with level 15 access on the local router is a good idea,
! in the event the ISE server can't be
! reached, and a backup method has been specified as the local database.

R1(config)# username admin privilege 15 secret 4Je7*1swEsf

! Applying the named method lists is what puts them in motion.
! By applying the method lists to the vty lines
! any users connecting to these lines will be authenticated by the
! methods specified by the lists that are applied
! and also accounting will occur, based on the lists that are applied.
R1(config)# line vty 0 4
R1(config-line)# login authentication MY-LIST-1
R1(config-line)# authorization commands 1 TAC1
R1(config-line)# authorization commands 15 TAC15
R1(config-line)# accounting commands 1 TAC-act1
R1(config-line)# accounting commands 15 TAC-act15

! Note: on the console and AUX ports, the default list will be applied,
! due to no custom method list being applied
! directly to the console or AUX ports.
```

Using debug as a tool to verify what you think is happening is a good idea. In Example 6-24 we review and apply AAA and perform a debug verification.

Example 6-24 *Another Example of Creating and Applying a Custom Method List to vty Lines*

```
! Creating the method list, which in this example has 3 methods.

! First the local database
! if the username exists in the configuration, and if not
! then the enable secret (if configured),  and if not then no
! authentication required
! (none)
R2(config)# aaa authentication login MY-AUTHEN-LIST-1 local enable none

! Applying the method list to the vty lines 0-4
R2(config)# line vty 0 4
R2(config-line)# login authentication MY-AUTHEN-LIST-1
R2(config-line)# exit

! Creating a local username in the local database (running-config)
R2(config)# username omar secret s0m3passwd!

! Setting the password required to move from user mode to privileged mode
R2(config)# enable secret S0mecisc0!enablePAssWD
R2(config)# interface loopback 0

! Applying an IP address to test a local telnet to this same local router
! Not needed if the device has another local IP address that is in use
R2(config-if)# ip address 10.2.2.2 255.255.255.0
R2(config-if)# exit

! Enable logging so we can see results of the upcoming debug
R2(config)# logging buffered 7
R2(config)# end

! Enabling debug of aaa authentication, so we can see what the router is
! thinking regarding aaa authentication
R2# debug aaa authentication
AAA Authentication debugging is on

R2# clear log
Clear logging buffer [confirm]
```

```
! Telnet to our own address
R2# telnet 10.2.2.2
Trying 10.2.2.2 ... Open


User Access Verification
Username: omar
AAA/BIND(00000063): Bind i/f
AAA/AUTHEN/LOGIN (00000063): Pick method list 'MY-AUTHEN-LIST-1'
Password: ***** password not shown when typing it in


R2>


! Below, after issuing the who command, we can see that bob is connected via line
! vty 0, and that from the debug messages above
! the correct authentication list was used.
R2>who
    Line       User      Host(s)           Idle        Location
   0 con 0               2.2.2.2           00:00:00
 * 2 vty 0     bob       idle              00:00:00 2.2.2.2
R2> exit


! If we exit back out, and remove all the users in the local database,
! (including omar) then the same login authentication will fail on the first
! method of the "local" database (no users there), and will go to the second
! method in the list, which is "enable", meaning use the enable secret if
! configured.


! As soon as I supply a username, the router discovers that there are no
! usernames configured in running configuration (at least none that match
! the user who is trying to login), and fails on the first method "local" in the list.
! It then tries the next method of just caring about the enable secret.


R2# telnet 10.2.2.2
Trying 10.2.2.2 ... Open
User Access Verification


AAA/BIND(00000067): Bind i/f
AAA/AUTHEN/LOGIN (00000067): Pick method list 'MY-AUTHEN-LIST-1'


! Note: omar is not a configured user in the local database on the router
Username: omar
Password: **** not shown while typing.
! This is the enable secret that was previously set.
AAA/AUTHEN/ENABLE(00000067): Processing request action LOGIN
AAA/AUTHEN/ENABLE(00000067): Done status GET_PASSWORD
```

```
R2>
AAA/AUTHEN/ENABLE(00000067): Processing request action LOGIN
AAA/AUTHEN/ENABLE(00000067): Done status PASS
R2> exit


! One more method exists in the method list we applied to the vty lines.
! If the local fails, and the enable secret fails (because neither of these
! is configured on the router), then the third method in the method list
! 'MY-AUTHEN-LIST-1' will be tried. The third method we specified is none,
! meaning no authentication required, come right in.  After removing the
! enable secret, we try once more.


R2# telnet 10.2.2.2
Trying 10.2.2.2 ... Open


User Access Verification


AAA/BIND(00000068): Bind i/f
AAA/AUTHEN/LOGIN (00000068): Pick method list 'MY-AUTHEN-LIST-1'
Username: doesn't matter
R2>
AAA/AUTHEN/ENABLE(00000068): Processing request action LOGIN
AAA/AUTHEN/ENABLE(00000068): Done status FAIL - secret not configured
R2>
! No password was required.  All three methods of the method lists were
! tried.
! The first two methods failed, and the third of "none" was accepted.
```

Using the CLI to Troubleshoot AAA for Cisco Routers

One tool you can use when troubleshooting AAA on Cisco routers is the **debug** command. You may use three separate **debug** commands to troubleshoot the various aspects of AAA:

- **debug aaa authentication:** Use this command to display debugging messages for the authentication functions of AAA.

- **debug aaa authorization:** Use this command to display debugging messages for the authorization functions of AAA.

- **debug aaa accounting:** Use this command to display debugging messages for the accounting functions of AAA.

Each of these commands is executed from privileged EXEC mode. To disable debugging for any of these functions, use the **no** form of the command, such as **no debug aaa authentication**. If you want to disable all debugging, issue the **undebug all** command.

Example 6-25 shows an example of debugging login authentication, EXEC authorization, and commands at level 15 authorization. As shown in the example, you can use **debug** not only for verification, as in the preceding example, but also as a troubleshooting method.

Example 6-25   *Using debug Commands*

```
! R4 will have a loopback, so we can telnet to ourselves to test


R4(config-if)# ip address 10.4.4.4 255.255.255.0
R4(config-if)# exit


! Local user in the database has a privilege level of 15
R4(config)# username admin privilege 15 secret cisco1sNotAGoodPasswd!


! This method list, if applied to a line, will specify local authentication
R4(config)# aaa authentication login AUTHEN_Loc local


! This next method list, if applied to a line, will require authorization
! before giving the administrator an exec shell.   If the user has a valid
! account in the running configuration, the exec shell will be created for
! the authenticated
! user, and it will place the user in their privilege level automatically


R4(config)# aaa authorization exec AUTHOR_Exec_Loc local


! This method list, if applied to a line, will require authorization for
! each and every level 15 command issued.   Because the user is at
! privilege level 15 the router will say "yes" to any level 15 commands
! that may be issued by the user
R4(config)# aaa authorization commands 15 AUTHOR_Com_15 local


! Next we will apply the 3 custom method lists to vty lines 0-4, so that
! when anyone connects via these vty lines, they will be subject to the
! login authentication, the exec authorization, and the level 15 command
! authorizations for the duration of their session.


R4(config)# line vty 0 4
R4(config-line)# login authentication AUTHEN_Loc
R4(config-line)# authorization exec AUTHOR_Exec_Loc
R4(config-line)# authorization commands 15 AUTHOR_Com_15
R4(config-line)# exit
R4(config)#
R4(config)# do debug aaa authentication
AAA Authentication debugging is on
R4(config)# do debug aaa authorization
AAA Authorization debugging is on
R4(config)# exit
```

```
! Now test to see it all in action.
R4# telnet 10.4.4.4
Trying 10.4.4.4 ... Open
User Access Verification


Username: admin
Password: ******  (password not displayed when entering)


! The router used the login authentication list we specified
AAA/BIND(00000071): Bind i/f
AAA/AUTHEN/LOGIN (00000071): Pick method list 'AUTHEN_Loc'


! The router picked the authorization list specified for the exec shell
AAA/AUTHOR (0x71): Pick method list 'AUTHOR_Exec_Loc'
AAA/AUTHOR/EXEC(00000071): processing AV cmd=
AAA/AUTHOR/EXEC(00000071): processing AV priv-lvl=15
AAA/AUTHOR/EXEC(00000071): Authorization successful


! It picked the command level 15 authorization list, when we issued the
! configure terminal command, which is a level 15 command.
R4# config t
Enter configuration commands, one per line.  End with CNTL/Z.
R4(config)#
AAA/AUTHOR: auth_need : user= 'admin' ruser= 'R4' rem_addr= '10.4.4.4' priv= 15 list=
'AUTHOR_Com_15' AUTHOR-TYPE= 'command'
AAA: parse name=tty2 idb type=-1 tty=-1
AAA: name=tty2 flags=0x11 type=5 shelf=0 slot=0 adapter=0 port=2 channel=0
AAA/MEMORY: create_user (0x6A761F34) user='admin' ruser='R4' ds0=0 port='tty2'
rem_addr='10.4.4.4' authen_type=ASCII service=NONE priv=15 initial_task_id='0',
vrf= (id=0)
tty2 AAA/AUTHOR/CMD(1643140100): Port='tty2' list='AUTHOR_Com_15' service=CMD
AAA/AUTHOR/CMD: tty2(1643140100) user='admin'
tty2 AAA/AUTHOR/CMD(1643140100): send AV service=shell
tty2 AAA/AUTHOR/CMD(1643140100): send AV cmd=configure
tty2 AAA/AUTHOR/CMD(1643140100): send AV cmd-arg=terminal
tty2 AAA/AUTHOR/CMD(1643140100): send AV cmd-arg=<cr>
tty2 AAA/AUTHOR/CMD(1643140100): found list "AUTHOR_Com_15"
tty2 AAA/AUTHOR/CMD(1643140100): Method=LOCAL
AAA/AUTHOR (1643140100): Post authorization status = PASS_ADD
AAA/MEMORY: free_user (0x6A761F34) user='admin' ruser='R4' port='tty2'
rem_addr='10.4.4.4' authen_type=ASCII service=NONE priv=15 vrf= (id=0)
R4(config)#
! It made a big splash, with lots of debug output, but when you boil it all
! down it means the user was authorized to issue the configure terminal
! command.
```

There is also a **test aaa** command that is very useful when verifying connectivity with a remote ISE server.

RBAC Privilege Level/Parser View

You may implement RBAC through AAA, with the rules configured on an ISE server, but you may implement it in other ways, too, including creating custom privilege levels and having users enter those custom levels where they have a limited set of permissions, or creating a parser view (also sometimes simply called a view), which also limits what the user can see or do on the Cisco device. Each option can be tied directly to a username so that once users authenticate, they may be placed at the custom privilege level or in the view that is assigned to them.

Let's implement a custom privilege level first, as shown in Example 6-26. The example includes explanations throughout.

Example 6-26 *Creating and Assigning Commands to a Custom Privilege Level*

```
! By default,  we use privilege level 1 (called user mode), and privilege
! level 15 (called privileged mode).  By creating custom levels, (between
! 1-15) and assigning commands to those levels, we are creating custom
! privilege levels.
! A user connected at level 8, would have any of the new commands -
! associated with level 8, as well as any commands that have been custom
! assigned or defaulted to levels 8 and below.  A user at level 15 has
! access to all commands at level 15 and below.
! This configuration assigns the command "configure terminal" to privilege
! level 8
R2(config)# privilege exec level 8 configure terminal

! This configuration command assigns the password for privilege level 8
! The keyword "password" could be used instead of secret, but is less secure
! as the "password" doesn't use the MD5 hash to protect the password.
! The "0" before the password implies that we are inputting a non-hashed
! (to begin with) password.  The system will hash this for us, because we
! used the enable "secret" keyword.
R2(config)# enable secret level 8 0 NewPa5s123&
R2(config)# end
R2#
%SYS-5-CONFIG_I: Configured from console by console

! To enter this level, use the enable command, followed by the level you want
! to enter.  If no level is specified, the default level is 15.

R2# disable

! Validate that user mode is really privilege level 1
R2> show privilege
```

```
Current privilege level is 1
! Context sensitive help shows that we can enter a level number after the
! word enable

R2> enable ?
  <0-15>  Enable level
  view    Set into the existing view
  <cr>

R2> enable 8
Password: [NewPa5s123&]
! note: password doesn't show when typing it in

R2# show privilege
Current privilege level is 8
! We can go into configuration mode, because "configure terminal" is at our
! level

R2# configure terminal

Enter configuration commands, one per line.  End with CNTL/Z.
! Notice we don't have further ability to configure the router, because
! level 8 doesn't include the interface configuration or other router -
! configuration commands.
R2(config)# ?
Configure commands:
  beep     Configure BEEP (Blocks Extensible Exchange Protocol)
  call     Configure Call parameters
  default  Set a command to its defaults
  end      Exit from configure mode
  exit     Exit from configure mode
  help     Description of the interactive help system
  license  Configure license features
  netconf  Configure NETCONF
  no       Negate a command or set its defaults
  oer      Optimized Exit Routing configuration submodes
  sasl     Configure SASL
  wsma     Configure Web Services Management Agents
```

If we are requiring login authentication, we can associate a privilege level with a given user account, and then when users authenticate with their username and password, they will automatically be placed into their appropriate privilege level. Example 6-27 shows an example of this.

Example 6-27   *Creating a Local User and Associating That User with Privilege Level 8 and Assigning Login Requirements on the vty Lines*

```
! Create the user account in the local database (running-config) and
! associate that user with the privilege level you want that user to use.
R2(config)# username Bob privilege 8 secret Cisco1231sAlsoNotAG00dPasswd
R2(config)# line vty 0 4

! "login local" will require a username and password for access if the "aaa
! new-model" command is not present.   If we have set the aaa new-model,
! then we would also want to create a default or named method list that
! specifies we want to use the local database for authentication.
R2(config-line)# login local

! Note:  Once bob logs in, he would have access to privilege level 8 and
! below, (including all the normal show commands at level 1)
```

Implementing Parser Views

To restrict users without having to create custom privilege levels, you can use a parser view, also referred to as simply "a view." A view can be created with a subset of privilege level 15 commands, and when the user logs in using this view, that same user is restricted to only being able to use the commands that are part of his current view.

To create a view, an enable secret password must first be configured on the router. AAA must also be enabled on the router (**aaa new-model** command).

Example 6-28 shows the creation of a view.

Example 6-28   *Creating and Working with Parser Views*

```
! Set the enable secret, and enable aaa new-model
! (unless already in place)
R2(config)# enable secret aBc!2#&iU
R2(config)# aaa new-model
R2(config)# end

! Begin the view creation process by entering the "default" view, using the
! enable secret
R2# enable view
Password: ***** note password not shown when typed

R2#
%PARSER-6-VIEW_SWITCH: successfully set to view 'root'.
R2# configure terminal

! As the administrator in the root view, create a new custom view
R2(config)# parser view New_VIEW
%PARSER-6-VIEW_CREATED: view 'New_VIEW' successfully created.
```

```
! Set the password required to enter this new view
R2(config-view)# secret New_VIEW_PW

! Specify which commands you want to include as part of this view.
! commands "exec" refer to commands issued from the command prompt
! commands "configure" refer to commands issued from privileged mode
R2(config-view)# commands exec include ping
R2(config-view)# commands exec include all show
R2(config-view)# commands exec include configure

! This next line adds the ability to configure "access-lists" but nothing
! else
R2(config-view)# commands configure include access-list
R2(config-view)# exit
R2(config)# exit

! Test the view, by going to user mode, and then back in using the new view
R2# disable

R2>enable view New_VIEW
Password: [New_VIEW_PW] Password not shown when typed in

! Console message tells us that we are using the view
%PARSER-6-VIEW_SWITCH: successfully set to view 'New_VIEW'.

! This command reports what view we are currently using
R2# show parser view
Current view is 'New_VIEW'

! We can verify that the commands assigned to the view work.
! Note: we only assigned configure, not configure terminal so we have to
! use the configure command, and then tell the router we are configuring
! from the terminal.   We could have assigned the view "configure terminal"
! to avoid this.
R2# configure terminal
Enter configuration commands, one per line.  End with CNTL/Z.

! Notice that the only configuration options we have are for access-list,
! per the view
R2(config)# ?
Configure commands:
  access-list  Add an access list entry
  do           To run exec commands in config mode
  exit         Exit from configure mode
```

We could also assign this view to a user account so that when users log in with their username and password, they are automatically placed into their view, as shown in Example 6-29.

Example 6-29 *Associating a User Account with a Parser View*

```
R2(config)# username Lois view New_VIEW secret Cisco1231sNotAG00dPasswd
```

TIP This creation of a username and assigning that user to a view needs to be done by someone who is at privilege level 15.

SSH and HTTPS

Because Telnet sends all of its packets as plain text, it is not secure. SSH allows remote management of a Cisco router or switch, but unlike Telnet, SSH encrypts the contents of the packets to protect them from being interpreted if they fall into the wrong hands.

To enable SSH on a router or switch, the following items need to be in place:

- Hostname other than the default name of router.

- Domain name.

- Generating a public/private key pair, used behind the scenes by SSH.

- Requiring user login via the vty lines, instead of just a password. Local authentication or authentication using an ISE server are both options.

- Having at least one user account to log in with, either locally on the router or on an ISE server.

Example 6-30 shows how to implement these components, along with annotations and examples of what happens when the required parts are not in place. If you have a nonproduction router or switch handy, you might want to follow along.

Example 6-30 *Preparing for SSH*

```
! To create the public/private key pair used by SSH, we would issue the
! following command.   Part of the key pair will be the hostname and the
! domain name.
! If these are not configured first, the crypto key generate command will
! tell you as shown in the next few lines.
Router(config)# crypto key generate rsa
% Please define a hostname other than Router.
Router(config)# hostname R1
R1(config)# crypto key generate rsa
% Please define a domain-name first.
R1(config)# ip domain-name cisco.com

! Now with the host and domain name set, we can generate the key pair
R1(config)# crypto key generate rsa
The name for the keys will be: R1.cisco.com
```

```
Choose the size of the key modulus in the range of 360 to 2048 for your
   General Purpose Keys. Choosing a key modulus greater than 512 may take
   a few minutes.


! Bigger is better with cryptography, and we get to choose the size for the
! modulus.
! The default is 512 on many systems, but you would want to choose 1024 or
! more to improve security. SSH has several flavors, with version 2 being
! more secure than version 1. To use version 2, you would need at least a
! 1024 size for the key pair.
How many bits in the modulus [512]: 1024
% Generating 1024 bit RSA keys, keys will be non-exportable...


R1(config)#
%SSH-5-ENABLED: SSH 1.99 has been enabled
! Note the "1.99" is based on the specifications for SSH from RFC 4253
! which indicate that an SSH server may identify its version as 1.99 to
! identify that it is compatible with current and older versions of SSH.


! Create a user in the local database
R1(config)# username omar secret Ci#kRk*ks


! Configure the vty lines to require user authentication
R1(config)# line vty 0 4
R1(config-line)# login local


! Alternatively, we could do the following for the requirement of user
! authentication
! This creates a method list which points to the local database, and then
! applies that list to the VTY lines
R1(config)# aaa new-model
R1(config)# aaa authentication login Omar-List-1 local
R1(config)# line vty 0 4
R1(config-line)# login authentication Omar-List-1


! To test this we could SSH to ourselves from the local machine, or from
! another router that has IP connectivity to this router.


R1# ssh ?
  -c    Select encryption algorithm
  -l    Log in using this user name
  -m    Select HMAC algorithm
  -o    Specify options
  -p    Connect to this port
  -v    Specify SSH Protocol Version
```

```
   -vrf   Specify vrf name
   WORD   IP address or hostname of a remote system


! Note: one of our local IP addresses is 10.1.0.1
R1# ssh -l omar 10.1.0.1


Password: <password for Omar goes here>


R1>
! to verify the current SSH session(s)
R1> show ssh
Connection Version Mode Encryption  Hmac       State            Username
0            2.0     IN  aes128-cbc hmac-sha1  Session started  omar
0            2.0     OUT aes128-cbc hmac-sha1  Session started  omar
%No SSHv1 server connections running.
R1>
```

Perhaps you want to manage a router via HTTPS. If so, you can implement HTTPS functionality, as shown in Example 6-31.

Example 6-31   *Preparing for HTTPS*

```
! Enable the SSL service on the local router.   If it needs to generate
! keys for this feature, it will do so on its own in the background.
R1(config)# ip http secure-server


! Specify how you want users who connect via HTTPS to be authenticated.
R1(config)# ip http authentication ?
   aaa     Use AAA access control methods
   enable  Use enable passwords
   local   Use local username and passwords


R1(config)# ip http authentication local


! If you are using the local database, make sure you have at least one user
! configured in the running-config so that you can login.  To test, open
! a browser to HTTPS://a.b.c.d where a.b.c.d is the IP address on the
! router.
```

Implementing Logging Features

Logging is important as a tool for discovering events that are happening in the network and for troubleshooting. Correctly configuring logging so that you can collect and correlate events across multiple network devices is a critical component for a secure network.

Configuring Syslog Support

Example 6-32 shows a typical syslog message and how to control what information is included with the message.

Example 6-32 *Using Service Timestamps with Syslog Events*

```
R4(config)# interface Gi0/0

R4(config-if)# shut
%LINK-5-CHANGED: Interface GigabitEthernet0/0, changed state to administratively
down
%LINEPROTO-5-UPDOWN: Line protocol on Interface GigabitEthernet0/0,
changed state to down
R4(config-if)#

! If we add time stamps to the syslog messages, those time stamps can assist
! in correlating events that occurred on multiple devices

R4(config)# service timestamps log datetime
R4(config)# int Gi0/0
R4(config-if)# no shutdown

! These syslog messages have the date of the event, the event (just after
! the %) a description, and also the level of the event (the first event in
! the example below is level 3 with the second event being level 5).
*Sep 22 12:08:13: %LINK-3-UPDOWN: Interface GigabitEthernet0/0,
changed state to up
*Sep 22 12:08:14: %LINEPROTO-5-UPDOWN: Line protocol on
Interface GigabitEthernet0/0, changed state to up
```

Configuring NTP

Because time is such an important factor, you should use Network Time Protocol (NTP) to synchronize the time in the network so that events that generate messages and timestamps can be correlated.

To configure the NTP, you first need to know what the IP address is of the NTP server you will be working with, and you also want to know what the authentication key is and the key ID. NTP authentication is not required to function, but it's a good idea to ensure that the time is not modified because of a rogue NTP server sending inaccurate NTP messages using a spoofed source IP address.

NTP supports authentication on a Cisco router because the router supports NTPv3. Example 6-33 demonstrates how to configure NTPv3 in a router. The NTP server's IP address is 192.168.78.96 and is reachable via the GigabitEthernet0/1 interface.

Example 6-33   *Using Authentication via Keys with NTPv3*

```
ntp authentication-key 1 md5 141411050D 7
ntp authenticate
ntp trusted-key 1
ntp update-calendar
ntp server 192.168.78.96 key 1 prefer source GigabitEthernet0/1
```

To verify the status on this router acting as an NTP client, you could use the commands from the CLI shown in Example 6-34.

Example 6-34   *Verifying Synchronization from the NTP Client*

```
LAB-Router# show ntp status

Clock is synchronized, stratum 4, reference is 192.168.78.96
nominal freq is 250.0000 Hz, actual freq is 249.9980 Hz, precision is
2**24 reference time is D8147295.4E6FD112 (13:11:49.306 UTC Mon Feb 3 2020)
clock offset is -0.3928 msec, root delay is 83.96 msec
root dispersion is 94.64 msec, peer dispersion is 2.22 msec
loopfilter state is 'CTRL' (Normal Controlled Loop),
drift is 0.000007749 s/s, system poll interval is 64,
last update was 126 sec ago.
LAB-Router#

LAB-Router# show ntp association

  address         ref clock        st   when   poll reach  delay offset    disp
*~192.168.78.96    208.75.89.4     3     49     64    377  1.341  -0.392  2.424
 * sys.peer, # selected, + candidate, - outlyer, x falseticker, ~ configured
LAB-Router#
```

NOTE   NTP uses UDP port 123. If NTP does not synchronize within 15 minutes, you may want to verify that connectivity exists between this router and the NTP server that it is communicating to. You also want to verify that the key ID and password for NTP authentication are correct.

Securing the Network Infrastructure Device Image and Configuration Files

If a router has been compromised, and the flash file system and NVRAM have been deleted, there could be significant downtime as the files are put back in place before restoring normal router functionality. The Cisco Resilient Configuration feature is intended to improve the recovery time by making a secure working copy of the IOS image and startup configuration files (which are referred to as the primary bootset) that cannot be deleted by a remote user.

To enable and save the primary bootset to a secure archive in persistent storage, follow Example 6-35.

Example 6-35 *Creating a Secure Bootset*

```
! Secure the IOS image

R6(config)# secure boot-image
%IOS_RESILIENCE-5-IMAGE_RESIL_ACTIVE: Successfully secured running image

! Secure the startup-config
R6(config)# secure boot-config
%IOS_RESILIENCE-5-CONFIG_RESIL_ACTIVE: Successfully secured config archive
   [flash:.runcfg-20111222-230018.ar]

! Verify the bootset
R6(config)# do show secure bootset
IOS resilience router id FTX1036A13J

IOS image resilience version 16.4 activated at 23:00:10 UTC Thu Dec 22 2011
Secure archive flash:router-164-24.bin type is image (elf) []
  file size is 60303612 bytes, run size is 60469256 bytes
  Runnable image, entry point 0x80010000, run from ram

IOS configuration resilience version 16.4 activated at
23:00:18 UTC Thu Dec 22 2019
Secure archive flash:runcfg-20111222-230018.ar type is config
configuration archive size 1740 bytes

! Note: to undo this feature, (using the "no" option in front of the command)
! you must be connected via the console.  This prevents remote users from
! disabling the feature.
```

Securing the Data Plane in IPv6

This book and the exam assume that you are familiar with IPv6 fundamentals. To make certain that you are, this chapter explains how IPv6 works and how to configure it, and then you learn how to develop a security plan for IPv6.

Understanding and Configuring IPv6

When compared to IPv4, both similarities and differences exist as to how IPv6 operates. The CCNP Security and CCIE Security certifications require that you know the fundamentals of IPv6, and that is the focus of this section, which first reviews IPv6 basics and then shows you how to configure it.

Why IPv6? Here are two good reasons to move to IPv6:

- IPv6 has more address space available.

- We are running out of public IPv4 addresses.

For more than a decade, the requirement to implement IPv6 has been threatened as imminent. The lifetime of its predecessor (IPv4) was extended more than a decade because of features such as network address translation (NAT), which enables you to hide thousands of users with private IP addresses behind a single public IP address.

With IPv6, upper-layer applications still work like you are used to with IPv4. The biggest change is that we are doing a forklift upgrade to Layer 3 of the OSI model. Along with this change, there are some modifications as to how IPv6 interacts with the rest of the protocol stack and some modifications to its procedures for participating on the network.

Table 6-4 describes a few of the notable differences and some similarities between IPv4 and IPv6.

Table 6-4 IPv4 Versus IPv6

IPv4	IPv6
32-bit (4-byte) address supports 2^{32} (or 4,294,967,296) addresses.	128-bit (16-byte) address supports 2^{128} (about 3.4 x 10^{38}) addresses. 438 quintillion addresses per square inch of land on Earth (so a LOT!).
You can use NAT to extend address space limitations.	Does not support NAT by design, since it has plenty of addresses available.
Administrators must use Dynamic Host Configuration Protocol (DHCP) or static configuration to assign IP addresses to hosts.	Hosts can use stateless address autoconfiguration to assign an IP address to themselves, but can also use DHCP features to learn more information, such as information about DNS servers.
IPsec support is an optional add-on concept to protect IP packets by using encryption, validating a peer, ensuring data integrity, and supporting anti-replay.	IPsec is fully supported in IPv6. In fact, IPsec was supposed to be mandatory when using IPv6; however, IPv6 does not actually require IPsec to be enabled in order for IPv6 to work.
An IPv4 header consists of multiple fields.	Simplified (but larger) IPv6 header, with options for header extension as needed.
Uses broadcasts for several functions, including Address Resolution Protocol (ARP).	Does not use any broadcasts and does not use ARP. Instead, it uses multicast addresses and Neighbor Discovery Protocol (NDP), also called ND. ND replaces ARP. Devices can automatically discover the IPv6 network address and many other housekeeping features such as discovering any routers on the network. ND uses IPv6's version of Internet Control Message Protocol (ICMP) as the workhorse behind most of its functions.
Both support common Layer 4 protocols such as UDP and TCP.	
Both support common application layer (Layer 7) protocols (HTTP, FTP, and so on) that are encapsulated in their respective Layer 4 protocols.	
Both support common Layer 2 technologies, such as Ethernet and WAN standards.	
Both contain two parts in the IP address: the network on the left side of the address and the host part on the right side of the address.	
Both use a network mask to identify which part of the address is the network (on the left), indicated by the mask bits that are on (or 1), and the rest of the bits to the right represent the host part of the IPv4 or IPv6 address.	

The Format of an IPv6 Address

Understanding the basic format of an IPv6 address is important for certification and for the actual implementation of IPv6. Here are a few key details about IPv6 and its format:

- **Length:** IPv6 addresses are 128 bits (16 bytes) long.

- **Groupings:** IPv6 addresses are segmented into eight groups of four hex characters.

- **Separation of groups:** Each group is separated by a colon (:).

- **Length of mask:** Usually 50 percent (64 bits long) for network ID, which leaves 50 percent (also 64 bits) for interface ID (using a 64-bit mask).

- **IPv6 Address Types:** There are three types of IPv6 addresses (global unicast addresses, link local addresses, and unique local addresses).

We can represent an IPv6 address the hard way or the easier way. The hard way is to type in every hexadecimal character for the IP address. In Example 6-36, we put in a 128-bit IPv6 address, typed in as 32 hexadecimal characters, and a 64-bit mask.

Example 6-36 *An IPv6 Address Configured the Hard Way*

```
R1(config-if)# ipv6 address 2001:0db8:0000:0000:1234:0000:0052:0001/64

! The output reflects how we could have used some shortcuts in representing
! the groups of zeros.
R1(config-if)# do show ipv6 interface brief
GigabitEthernet0/1            [up/up]
    FE80::C800:41FF:FE32:6
    2001:DB8::1234:0:52:1
R1(config-if)#
```

Understanding the Shortcuts

Example 6-36 shows the address being configured and the abbreviated address from the output of the **show** command. When inserting a group of four hexadecimal numbers, you can limit your typing a bit. For example, if any leading characters in the group are 0, you can omit them (just as the 0 in front of DB8 is in the second group from the left). In addition, if there are one or more consecutive groups of all 0s, you can input them as a double colon (::). The system knows that there should be eight groups separated by seven colons, and when it sees ::, it just looks at how many other groups are configured and assumes that the number of missing groups plus the existing groups that are configured totals eight. In the example, the first two groups of consecutive 0s are shortened in the final output. This shortcut may be done only once for any given IPv6 address. This example contains three groupings of 0s, and if you use the shortcut twice, the system will not know whether there should be four 0s after the DB8: and eight 0s (or two groups) after the 1234:, or vice versa.

Did We Get an Extra Address?

Besides the IPv6 global address configured in Example 6-36, the system automatically configured for itself a second IPv6 address known as a link-local address, which begins with FE80. A link-local address is an IPv6 address that you can use to communicate with other IPv6 devices on the same local network (local broadcast domain). If an IPv6 device wants to communicate with a device that is remote, it needs to use its global and routable IPv6 address for that (not the link-local one). To reach remote devices, you also need to have a route to that remote network or a default gateway to use to reach the remote network.

The following section covers the other types of addresses you will work with in IPv6 networks.

IPv6 Address Types

In IPv6, you must be familiar with several types of addresses. Some are created and used automatically; others you must configure manually. These address types include the following:

- **Link-local address:** Link-local addresses may be manually configured, but if they are not, they are dynamically configured by the local host or router itself. These always begin with the characters FE80. The last 64 bits are the host ID (also referred to as the interface ID), and the device uses the modified EUI-64 format (by default) to create that. The modified EUI-64 uses the MAC address (if on Ethernet; and if not on Ethernet, it borrows the MAC address of another interface) and realizes it is only 48 bits. To get to 64 bits for the host ID, it inserts four hexadecimal characters of FFFE (which is the 16 more bits we need) and injects those into the middle of the existing MAC address to use as the 64-bit host ID. It also looks at the seventh bit from the left (of the original MAC address) and inverts it. If it is a 0 in the MAC address, it is a 1 in the host ID, and vice versa. To see an example of this, look back at Example 6-35, at the output of the **show** command there, focusing on the address that begins with FE80.

- **Loopback address:** In IPv4, this was the 127 range of IP addresses. In IPv6, the address is ::1 (which is 127 0s followed by a 1).

- **All-nodes multicast address:** In IPv6, multicasts begin with FF*xx*: (where the *xx* is some other hex number). The number 02 happens to designate a multicast address that is link-local in scope. There are other preset scopes, but you do not have to worry about them here. The IPv6 multicast group that all IPv6 devices join is FF02::1. If any device needs to send a packet/frame to all other local IPv6 devices, it can send the packet to the multicast address of FF02::1, which translates to a specific multicast Layer 2 address, and then all the devices that receive those frames continue to de-encapsulate them. If a device receives a frame, and the receiving device determines that the Layer 2 destination in the frame is not destined for itself and not destined for any multicast groups that the local device has joined, it discards the frame instead of continuing to de-encapsulate it to find out what is inside (in the upper layers).

- **All-routers multicast address:** In addition to the multicast group address of FF02::1 that is joined by all devices configured for IPv6, routers that have had routing enabled for IPv6 also join the multicast group FF02::2. By doing so, any client looking for a router can send a request to this group address and get a response if there is a router on the local network. You might have noticed a pattern here: FF02 is just like 224.0.0.x in IPv4 multicast: 224.0.0.1 = all devices, and 224.0.0.2 = all routers.

- **Unicast and anycast addresses (configured automatically or manually):** A global IPv6 address, unlike a link-local address, is routable and can be reached through one or more routers that are running IP routing and that have a correct routing table. Global IPv6 unicast addresses have the first four characters in the range of 2000 to 3FFF, and may be manually configured, automatically discovered by issuing a router solicitation request to a local router, or be learned via IPv6 Dynamic Host Configuration Protocol (DHCP). An anycast address can be a route or an IP address that appears more than one time in a network, and then it is up to the network to decide the best way to reach that IP. Usually, two DNS servers, if they both use the same anycast address, are functional to the users so that regardless of which DNS server that packets are forwarded to, the client gets the DNS response it needs.

- **Solicited-node multicast address for each of its unicast and anycast addresses:** When a device has global and link-local addresses, it joins a multicast group of FF02::1:FF*xx:xxxx* The *x* characters represent the last 24 bits of the host ID being used for the addresses. If a device needs to learn the Layer 2 address of a peer on the same network, it can send out a neighbor solicitation (request) to the multicast group that the device that has that address should have joined. This is the way IPv6 avoids using broadcasts.

- **Multicast addresses of all other groups to which the host belongs:** If a router has enabled IPv6 routing, it joins the FF02::2 group (all routers), as mentioned earlier. If a router is running RIPng (the IPv6 flavor), it joins the multicast group for RIPng, which is FF02::9, so that it will process updates sent to that group from other RIP routers. Notice again some similarities. RIPv2 in IPv4 uses 224.0.0.9 as the multicast address.

Example 6-37 shows the output for a router that has been enabled for IPv6 routing, RIPng, and has an IPv6 global address.

Example 6-37   *IPv6 Interface Information*

```
! MAC address, for reference, that is currently used on the Gi0/1
! interface
R1# show interfaces Gi0/1 | include bia
  Hardware is i82543 , address is ca00.4132.0006 (bia ca00.4132.0006)


R1# show ipv6 interface Gi0/1
GigabitEthernet0/1 is up, line protocol is up



! Link-local address, beginning with FE80::
! and using modified EUI-64 for the host ID.
! Notice that CA from the MAC address is C8 in the host ID
! due to inverting the 7th bit for the modified EUI-64 formatting
  IPv6 is enabled, link-local address is FE80::C800:41FF:FE32:6

! Global addresses have the first group range of 2000-3fff
  Global unicast address(es):
    2001:DB8::1234:0:52:1, subnet is 2001:DB8::/64
```

```
! Multicast begins with FFxx:
  Joined group address(es):

! Because we are enabled for IPv6 on this interface
    FF02::1

! Because we are enabled for IPv6 routing
    FF02::2

! Because we are enabled for RIPng
    FF02::9

! Because our link-local address ends in 32:0006
! this is a solicited node multicast group
    FF02::1:FF32:6

! Because our global address ends in 52:0001
! this is a solicited node multicast group
    FF02::1:FF52:1

<snip>
```

Configuring IPv6 Routing

To support multiple IPv6 networks and allow devices to communicate between those networks, you need to tell the routers how to reach remote IPv6 networks. You can do so through static routes, IPv6 routing protocols, and default routes. For the router to route any customer's IPv6 traffic, you need to enable unicast routing for IPv6 from global configuration mode. If you fail to do this, the router can send and receive its own packets, but it will not forward packets for others, even if it has the routes in its IPv6 routing table.

IPv6 can use the new and improved flavors of these dynamic routing protocols with their versions that support IPv6:

- RIP, called RIP next generation (RIPng)

- OSPFv3

- EIGRP for IPv6

One difference with the interior gateway routing protocols for IPv6 is that none of them support network statements. To include interfaces of the routing process, you use interface commands. For EIGRP, you also need to issue the **no shutdown** command in EIGRP router configuration mode. Example 6-38 shows the enabling of unicast routing and the configuring of IPv6 routing protocols.

Example 6-38   *Enabling IPv6 Routing and Routing Protocols*

```
! Enables IPv6 routing of other devices' packets

R1(config)# ipv6 unicast-routing

! Enabling all 3 IGPs on interface Gi0/1
! Note: that in a production network, we would only need 1 routing protocol
! on a given interface.   If we did have multiple identical learned routes
! the Administrative Distance (same as in IPv4) would determine which
! routing protocols would be the "best" and be placed in the routing table.
R1(config)# interface Gi0/1

! Enabling RIPng on the interface
! Simply create a new "name" for the process.  I called this one "MYRIP"
! Use the same name on all the interfaces on the local router where you
! want RIPng to be used on that same router.

R1(config-if)# ipv6 rip MYRIP enable

! Enabling OSPFv3 on the interface
! Syntax is the keywords ipv6 ospf, followed by the process ID (the process ID
! is locally assigned and can be a positive integer from 1 to 65535), then the
! area information
R1(config-if)# ipv6 ospf 1 area 0

! Enabling IPv6 EIGRP on the interface
R1(config-if)# ipv6 eigrp 1
R1(config-if)# exit

! Bringing the EIGRP routing process out of its default shutdown state
! Note: This is done in global (not interface) configuration mode.
! This is not needed for RIPng or OSPFv3.
R1(config)# ipv6 router eigrp 1
R1(config-rtr)# no shutdown

! Verify which routing protocols are running
R1# show ipv6 protocol
IPv6 Routing Protocol is "connected"
IPv6 Routing Protocol is "rip MYRIP"
  Interfaces:
    GigabitEthernet0/1
  Redistribution:
    None
IPv6 Routing Protocol is "ospf 1"
  Interfaces (Area 0):
    GigabitEthernet0/1
```

6

```
   Redistribution:
     None
IPv6 Routing Protocol is "eigrp 1"
  EIGRP metric weight K1=1, K2=0, K3=1, K4=0, K5=0
  EIGRP maximum hopcount 100
  EIGRP maximum metric variance 1
  Interfaces:
     GigabitEthernet0/1
  Redistribution:
     None
  Maximum path: 16
  Distance: internal 90 external 170
```

The command **show ipv6 route** outputs the IPv6 routes the router knows how to reach, including the ones learned through dynamic routing protocols.

Moving to IPv6

Moving to IPv6 will be more of a transition or migration than a one-time event. As such, there are mechanisms in place to support coexistence between IPv4 and IPv6, including the ability for a router or network device to run both protocol stacks at the same time, and the ability to perform tunneling. A tunneling example would be when there are two isolated portions of the network that want to run IPv6, and between them, there is a big patch of IPv4 only. Tunneling would take the IPv6 packets and re-encapsulate them into IPv4 for transport across the IPv4 portion of the network. At the other end of the tunnel, the router would de-encapsulate the IPv6 from the IPv4 shell and then continue forwarding the IPv6 packet on toward its final destination.

Developing a Security Plan for IPv6

Most security risks associated with the older IPv4 are the same security risks associated with the newer IPv6. Now what does that mean to you? It means that you need to make sure you have considered and implemented security controls to address both protocol stacks. This section discusses many security threats common to both IPv4 and IPv6 (and some specific to IPv6) and how to address them.

Best Practices Common to Both IPv4 and IPv6

For both protocol stacks, here are some recommended best practices—a great place to start your network configuration:

- **Physical security:** Keep the room where the router is housed free (safe) from electrostatic and magnetic interference. It should also be temperature and humidity controlled. There should be controlled and logged access to that physical room. Redundant systems for electricity that feed into the routers are part of this, as well.

- **Device hardening:** Disable services that are not in use and features and interfaces that are not in use. A great reference for these best practices is the Cisco Guide to Harden Cisco IOS Devices (http://www.cisco.com/c/en/us/support/docs/ip/access-lists/13608-21.html).

- **Control access between zones:** Enforce a security policy that clearly identifies which packets are allowed between networks. You can use either simple access list controls or more advanced controls such as stateful inspection, which leverages firewall features on a router or a dedicated firewall appliance (all of which are covered extensively in other chapters in this book).

- **Routing protocol security:** Use authentication with routing protocols to help stop rogue devices from abusing the information being used in routing updates by your routers.

- **Authentication, authorization, and accounting (AAA):** Require AAA so that you know exactly who is accessing your systems, when they are accessing your systems, and what they are doing. You learned about AAA in earlier chapters. Network Time Protocol (NTP) is a critical part to ensure that timestamps reflect reality. Check log files periodically. All management protocols should be used with cryptographic services. Secure Shell (SSH) and Hypertext Transfer Protocol Secure (HTTPS) include these features. Place Telnet and HTTP inside of an encrypted virtual private network (VPN) tunnel to meet this requirement.

- **Mitigating DoS attacks:** Denial of service refers to willful attempts to disrupt legitimate users from getting access to the resources they intend to use. Although no complete solution exists, administrators can do specific things to protect the network from a DoS attack and to lessen its effects and prevent a would-be attacker from using a system as a source of an attack directed at other systems. These mitigation techniques include filtering based on bogus source IP addresses trying to come into the networks, and vice versa. Unicast reverse path verification is one way to assist with this, as are access lists. Unicast reverse path verification looks at the source IP address as it comes into an interface, and then looks at the routing table. If the source address seen would not be reachable out of the same interface it is coming in on, the packet is considered bad, potentially spoofed, and is dropped.

- **Have and update a security policy:** A security policy should be referenced and possibly updated whenever major changes occur to the administrative practices, procedures, or staff. If new technologies are implemented, such as a new VPN or a new application that is using unique protocols or different protocols than your current security policy allows, this is another reason to revisit the security policy. Another time a security policy might need to be updated is after a significant attack or compromise to the network has been discovered.

Threats Common to Both IPv4 and IPv6

The following threats and ways to mitigate them apply to both IPv4 and IPv6:

- **Application layer attacks:** An attacker is using a network service in an unexpected or malicious way. To protect against this, you can place filters to allow only the required protocols through the network. This will prevent services that aren't supposed to be available from being accessed through the network. You can also use application inspection, done through the Adaptive Security Appliance (ASA) or the IOS zone-based firewall, or intrusion prevention system (IPS) functionality to identify and filter protocols that are not being used in their intended way. Being current on system and application patches will also help mitigate an application layer attack.

■ **Unauthorized access:** Individuals not authorized for access are gaining access to network resources. To protect against this, use AAA services to challenge the user for credentials, and then authorize that user for only the access they need. This can be done for users forwarding traffic through the network and for administrators who want to connect directly for network management. Accounting records can create a detailed record of the network activity that has taken place.

■ **Man-in-the-middle attacks:** Someone or something is between the two devices who believe they are communicating directly with each other. The "man in the middle" may be eavesdropping or actively changing the data that is being sent between the two parties. You can prevent this by implementing Layer 2 dynamic ARP inspection (DAI) and Spanning Tree Protocol (STP) guards to protect spanning tree. You can implement it at Layer 3 by using routing protocol authentication. Authentication of peers in a VPN is also a method of preventing this type of attack.

■ **Sniffing or eavesdropping:** An attacker is listening in on the network traffic of others. This could be done in a switched environment, where the attacker has implemented a content-addressable memory (CAM) table overflow, causing the switch to forward all frames to all other ports in the same VLAN. To protect against this, you can use switch port security on the switches to limit the MAC addresses that could be injected on any single port. In general, if traffic is encrypted as it is transported across the network, either natively or by a VPN, that is a good countermeasure against eavesdropping.

■ **Denial-of-service (DoS) attacks:** Making services that should be available unavailable to the users who should normally have the access/services. Performing packet inspection and rate limiting of suspicious traffic, physical security, firewall inspection, and IPS can all be used to help mitigate a DoS attack.

■ **Spoofed packets:** Forged addressing or packet content. Filtering traffic that is attempting to enter the network is one of the best first steps to mitigating this type of traffic. Denying inbound traffic that is claiming to originate from inside the network will stop this traffic at the edge of the network. Reverse path checks can also help mitigate this type of traffic.

■ **Attacks against routers and other network devices:** Turning off unneeded services and hardening the routers will help the router be less susceptible to attacks against the router itself.

The Focus on IPv6 Security

With IPv6, you do have a few advantages related to security. If an attacker issues a ping sweep of your network, he will not likely find all the devices via a traditional ping sweep to every possible address, so reconnaissance will be tougher for the attacker using that method because there are potentially millions of addresses on each subnet—264 possibilities, or about 18 quintillion! Be aware, however, that this is a double-edged sword, because each device on an IPv6 network joins the multicast group of FF02::1. So, if the attacker has local access to that network, he could ping that local multicast group and get a response that lets him know about each device on the network. FF02::1 is local in scope, so the attacker cannot use this technique remotely; he would have to be on the local network.

The scanners and worms that used to operate in IPv4 will still very likely be able to operate in IPv6, but they will just use a different mechanism to do it. Customers unaware that IPv6 is even running on their workstations represent another security risk. They could be using IPv4 primarily but still have an active IPv6 protocol stack running. An attacker may leverage a newfound vulnerability in some aspect of IPv6 and then use that vulnerability to gain access to the victim's computer. Disabling an unused protocol stack (in this case, the unused IPv6 stack) would appropriately mitigate this risk.

New Potential Risks with IPv6

Any new feature or way of operating could be open to a new form of attack. Here is a list of features that are implemented differently or have slightly different methods than IPv4, and as a result, any manipulation of how these features work could result in a compromise of the network:

- **Network Discovery Protocol (NDP):** Clients discover routers using NDP, and if a rogue router is present, it could pretend to be a legitimate router and send incorrect information to the clients about the network, the default gateway, and other parameters. This could also lead to a man-in-the-middle attack, where the rogue router now has the opportunity to see all packets from the hosts that are being sent to remote networks.

- **Neighbor cache resource starvation:** If an attacker attempts to spoof many IPv6 destinations in a short time, the router can get overwhelmed while trying to store temporary cache entries for each destination. The IPv6 Destination Guard feature blocks data traffic from an unknown source and filters IPv6 traffic based on the destination address. It populates all active destinations into the IPv6 first-hop security binding table, and it blocks data traffic when the destination is not identified.

- **DHCPv6:** A rogue router that has fooled a client about being a router could also manipulate the client into using incorrect DHCP-learned information. This could cause a man-in-the-middle attack because the host could be using the address of the rogue router as the default gateway.

- **Hop-by-hop extension headers:** With IPv4, there were IP options that could be included in IP headers. Malicious use of these headers could cause excessive CPU utilization on the routers that receive or forward these packets, in addition to dictating the path the packet should take through the network. There are no IP options in IPv6; instead, there are IPv6 extensions, which can also be misused. One of the IPv6 extension headers is the Routing Header, type 0 (also referred to as RH0). RH0 can be used to identify a list of one or more intermediate nodes to be included on the path toward the final destination (think IP source routing). This can enable an attacker to dictate the path a packet can take through the network. By default, Cisco routers and switches disable the processing of RH type 0 headers on most of the current versions of Cisco IOS, Cisco IOS-XE, Cisco NX-OS, and Cisco IOS-XR. You can find more information on the use of IPv6 extension headers in the Cisco.com document "IPv6 Extension

Headers Review and Considerations" (http://tinyurl.com/ipv6ext-headers). As noted in this white paper, there is always the possibility that IPv6 traffic with a significant number of (or very long) extension headers is sent into the network maliciously to attempt to overwhelm the HW resources of the router. Regardless of the platform HW design, this provides for a distributed DoS (DDoS) attack vector, and security mechanisms should be put in place to reduce the risk of a DDoS attack. To protect the CPU from being overwhelmed by high rates of this type of traffic, Cisco routers implement rate limiting of packets that are diverted from the hardware to software path. This rate limiting reduces the chance that the CPU resources of the router will be depleted while trying to process the combination of extensions headers.

- **Packet amplification attacks:** Using multicast addresses rather than IPv4 broadcast addresses could allow an attacker to trick an entire network into responding to a request. An example is to send a neighbor solicitation request (which is part of the NDP) to the all-hosts multicast address of FF02::1, which would cause all devices to respond. Another example is if a packet is sent with the header extensions set so that a packet is just looped around the network until the Time-To-Live (TTL) mechanism expires, and perhaps injecting thousands of these to consume bandwidth and resources on the network devices forwarding them.

- **ICMPv6:** This protocol is used extensively by IPv6 as its NDP. A great deal of potential harm can result from manipulation of this protocol by an attacker.

- **Tunneling options:** Tunneling IPv6 through IPv4 parts of a network may mean that the details inside the IPv6 packet might not be inspected or filtered by the IPv4 network. Filtering needs to be done at the edges of the tunnel to ensure that only authorized IPv6 packets are successfully sent end to end.

- **Autoconfiguration:** Because an IPv6 host can automatically configure an IP address for itself, any trickery by a rogue router could also cause the host's autoconfiguration to be done incorrectly, which could cause a failure on the network or a man-in-the-middle attack as the client tries to route all traffic through the rogue router.

- **Dual stacks:** If a device is running both IPv4 and IPv6 at the same time, but is aware of only one (or is primarily only using one), the other protocol stack, if not secured, provides a potential vector for an attacker to remotely access the device. Once access is obtained this way, the attacker could then change IP settings or other configuration options based on the end goal the attacker is trying to achieve.

- **Bugs in code:** Any software has the potential to have bugs, including the software that is supporting the IPv6 features in the network or end-station devices.

IPv6 Best Practices

Implementing security measures at the beginning of a deployment improves the initial security posture instead of waiting until after an attack has occurred. IPv6 best practices include the following:

- **Filter bogus addresses:** Drop, at the edge of your network, any addresses that should never be valid source or destination addresses. These are also referred to as "bogon addresses."

- **Filter nonlocal multicast addresses:** If you are not running multicast applications, you should never need multicast to be forwarded beyond a specific VLAN. Local multicast is often used by IPv6 (for example, in routing updates and neighbor discovery).

- **Filter ICMPv6 traffic that is not needed on your specific networks:** Normal NDP uses ICMPv6 as its core protocol. A path's maximum transmission unit (MTU) is also determined by using ICMP. Outside of its normal functionality, you want to filter the unused parts of ICMP so that an attacker cannot use it against your network.

- **Drop Routing Header type 0 packets:** Routing Header 0, also known as RH0, may contain many intermediate next hops, and if followed an attacker could control the path of a packet through a network. The attacker could also use this to create an amplification attack that could loop until the TTL expires on the packet. Cisco routers, by default, drop packets with this type of header.

- **Use manual tunnels rather than automatic tunnels:** If tunneling, do not use automatic tunnel mechanisms such as automatic 6to4, because you cannot control all of them. (They are dynamic.) With the manual tunnels, avoid allowing the tunnels to go through the perimeter of your network, as you will not have tight controls on the contents of the tunneled packets.

- **Protect against rogue IPv6 devices:** There are a number of mechanisms available within IPv6 to help defend against the spoofing of IPv6 neighbors. These include the following:

  - **IPv6 first-hop security binding table:** This table is used to validate that the IPv6 neighbors are legitimate.

  - **IPv6 device tracking:** This feature provides the IPv6 neighbor table with the ability to immediately reflect changes when an IPv6 host becomes inactive.

  - **IPv6 port-based access list support:** Similar to IPv4 port access control lists (PACL), this feature provides access control on Layer 2 switch ports for IPv6 traffic.

  - **IPv6 RA Guard:** Provides the capability to block or reject rogue RA Guard messages that arrive at the network switch platform.

  - **IPv6 ND Inspection:** IPv6 ND inspection analyzes neighbor discovery messages to build a trusted binding table database, and IPv6 neighbor discovery messages that do not conform are dropped.

- **Secure Neighbor Discovery in IPv6 (SeND):** Although platform support of SeND still remains limited, this feature defines a set of new ND options, and two new ND messages, Certification Path Solicitation (CPS) and Certification Path Answer (CPA), to help mitigate the effects of the ND spoofing and redirection.

IPv6 Access Control Lists

As with IPv4, network administrators can use access control lists (ACLs) on IOS devices to filter and restrict the types of IPv6 traffic that enter the network at ingress points. The configuration in Example 6-39 prevents unauthorized IPv6 packets on UDP port 53 (DNS) from entering the network from interface Gigabit0/0. In this example, 2001:DB8:1:60::/64 represents the IP address space that is used by DNS servers that the network administrator is trying to protect, and 2001:DB8::100:1 is the IP address of the host that is allowed to access the DNS servers.

TIP   Be careful to ensure that all required traffic for routing and administrative access is allowed in the ACL before denying all unauthorized traffic.

Example 6-39   *IPv6 Access Control List*

```
LAB-Router-1# configure terminal

Enter configuration commands, one per line.  End with CNTL/Z.
LAB-Router-1(config)# ipv6  access-list IPv6-ACL
!
!Include explicit permit statements for trusted sources that
!require access on UDP port 53 (DNS)
!
LAB-Router-1(config-ipv6-acl)# permit udp host
2001:DB8::100:1 2001:DB8:1:60::/64 eq 53
LAB-Router-1(config-ipv6-acl)#  deny udp any 2001:DB8:1:60::/64 eq 53


! Allow IPv6 neighbor discovery packets, which
! include neighbor solicitation packets and neighbor
! advertisement packets

LAB-Router-1(config-ipv6-acl)#  permit icmp any any nd-ns
LAB-Router-1(config-ipv6-acl)#  permit icmp any any nd-na
!
!-- Explicit deny for all other IPv6 traffic
!
LAB-Router-1(config-ipv6-acl)#  deny ipv6 any any
! There is an explicit deny at the end of an ACL. The deny statement
! is configured as a demonstration, but it is not needed since the router
! will deny all remaining IPv6 traffic that does not match any of the
! permit statements.
!
! Apply ACL to interface in the ingress direction
!
LAB-Router-1(config-ipv6-acl)# interface GigabitEthernet0/0
LAB-Router-1(config-if)# ipv6 traffic-filter IPv6-ACL in
LAB-Router-1(config-if)# exit
LAB-Router-1(config)# exit
LAB-Router-1#
```

Securing Routing Protocols and the Control Plane

Protection of the control plane of a network device is critical because the control plane ensures that the management and data planes are maintained and operational. If the control plane were to become unstable during a security incident, it can be impossible for you to recover the stability of the network.

Control plane packets are network device–generated or received packets that are used for the creation and operation of the network itself. From the perspective of the network device, control plane packets always have a receive destination IP address and are handled by the CPU in the network device route processor. Some examples of control plane functions include routing protocols (for example, BGP, OSPF, and EIGRP) as well as protocols like Internet Control Message Protocol (ICMP) and the Resource Reservation Protocol (RSVP).

Minimizing the Impact of Control Plane Traffic on the CPU

In many cases, you can disable the reception and transmission of certain types of packets on an interface to minimize the amount of CPU load that is required to process unneeded packets. These types of packets fall into a category known as process-switched traffic. This traffic must be handled by the CPU and hence results in a performance impact on the CPU of the network device.

Process-switched traffic falls into two primary categories:

- **Receive adjacency traffic:** This traffic contains an entry in the Cisco Express Forwarding (CEF) table whereby the next router hop is the device itself, which is indicated by the term "receive" in the **show ip cef** command-line interface (CLI) output. This indication is the case for any IP address that requires direct handling by the Cisco IOS device CPU, which includes interface IP addresses, multicast address space, and broadcast address space.

Example 6-40 provides sample output generated when issuing the **show ip cef** command. Any of the IP addresses/subnets for which "receive" is listed as the next hop indicates that packets destined for this address space will end up hitting the control plane and CPU.

Example 6-40 *show ip cef Output*

```
LAB-Router-1# show ip cef
Prefix              Next Hop          Interface
0.0.0.0/0           no route
0.0.0.0/8           drop
0.0.0.0/32          receive
10.2.2.0/24         192.168.10.2      GigabitEthernet0/1
127.0.0.0/8         drop
192.168.10.0/24     attached          GigabitEthernet0/1
192.168.10.0/32     receive           GigabitEthernet0/1
192.168.10.1/32     receive           GigabitEthernet0/1
192.168.10.2/32     attached          GigabitEthernet0/1
192.168.10.255/32   receive           GigabitEthernet0/1
```

```
192.168.15.0/24      attached        Loopback1
192.168.15.0/32      receive         Loopback1
192.168.15.1/32      receive         Loopback1
192.168.15.255/32    receive         Loopback1
192.168.30.0/24      192.168.10.2    GigabitEthernet0/1
192.168.100.0/24     attached        Loopback0
192.168.100.0/32     receive         Loopback0
192.168.100.1/32     receive         Loopback0
192.168.100.255/32   receive         Loopback0
192.168.200.0/24     192.168.10.2    GigabitEthernet0/1
224.0.0.0/4          drop

LAB-Router-1#
```

- **Data plane traffic requiring special processing by the CPU:** The following types of data plane traffic require special processing by the CPU, resulting in a performance impact on the CPU:

  - **Access control list (ACL) logging:** ACL logging traffic consists of any packets that are generated due to a match (permit or deny) of an access control entry (ACE) on which the log keyword is used.

  - **Unicast Reverse Path Forwarding (Unicast RPF):** Unicast RPF, used in conjunction with an ACL, can result in the process switching of certain packets.

  - **IP options:** Any IP packets with options included must be processed by the CPU.

  - **Fragmentation:** Any IP packet that requires fragmentation must be passed to the CPU for processing.

  - **Time-To-Live (TTL) expiry:** Packets that have a TTL value less than or equal to 1 require "Internet Control Message Protocol Time Exceeded (ICMP Type 11, Code 0)" messages to be sent, which results in CPU processing.

  - **ICMP unreachables:** Packets that result in ICMP unreachable messages due to routing, maximum transmission unit (MTU), or filtering are processed by the CPU.

  - **Traffic requiring an ARP request:** Destinations for which an ARP entry does not exist require processing by the CPU.

  - **Non-IP traffic:** All non-IP traffic is processed by the CPU.

Details about CoPP

Control Plane Policing (CoPP) can be used to identify the type and rate of traffic that reaches the control plane of the Cisco IOS device. Control Plane Policing can be performed through the use of granular classification ACLs, logging, and the use of the **show policy-map control-plane** command.

CoPP is a Cisco IOS-wide feature designed to allow users to manage the flow of traffic handled by the route processor of their network devices. CoPP is designed to prevent unnecessary traffic from overwhelming the route processor that, if left unabated, could affect system performance. Route processor resource exhaustion, in this case, refers to all resources associated with the punt path and route processors, such as Cisco IOS process memory and buffers and ingress packet queues.

As just discussed, more than just control plane packets can be punted to be processed by the CPU and affect the route processor and system resources. Management plane traffic, as well as certain data plane exception IP packets and some services plane packets, may also require the use of route processor resources. Even so, it is common practice to identify the resources associated with the punt path and route processors as the control plane.

In Example 6-41, only Telnet and DNS traffic from trusted hosts (that is, devices in the 192.168.1.0/24 subnet) is permitted to reach the Cisco IOS device CPU. In addition, certain types of ICMP traffic destined to the network infrastructure (that is, devices with IP addresses in the 10.1.1.0/24 subnet) will be rate-limited to 5000 packets per second (pps).

> **NOTE**   When you're constructing ACLs to be used for CoPP, traffic that is "permitted" translates to traffic that will be inspected by CoPP, and traffic that is "denied" translates to traffic that CoPP bypasses. Please refer to this white paper on CoPP: http://www.cisco.com/web/about/security/intelligence/coppwp_gs.html. Specifically, see the following excerpt from the section, "Access List Construction":
>
> "There are several caveats and key points to keep in mind when constructing your access lists. The log or log-input keywords must never be used in access-lists that are used within MQC policies for CoPP. The use of these keywords may cause unexpected results in the functionality of CoPP. The use of the deny rule in access lists used in MQC is somewhat different to regular interface ACLs. Packets that match a deny rule are excluded from that class and cascade to the next class (if one exists) for classification. This is in contrast to packets matching a permit rule, which are then included in that class and no further comparisons are performed."

Example 6-41   *Control Plane Policing Configuration*

```
access-list 101 permit icmp any 10.1.1.0 0.0.0.255 echo
access-list 101 permit icmp any 10.1.1.0 0.0.0.255 echo-reply
access-list 101 permit icmp any 10.1.1.0 0.0.0.255 time-exceeded
access-list 101 permit icmp any 10.1.1.0 0.0.0.255 ttl-exceeded
access-list 123 deny   tcp 192.168.1.0 0.0.0.255 any eq telnet
access-list 123 deny   udp 192.168.1.0 0.0.0.255 any eq domain
access-list 123 permit tcp any any eq telnet
access-list 123 permit udp any any eq domain
access-list 123 deny ip any any
!
!
class-map match-all ICMP
 match access-group 101
class-map match-all UNDESIRABLE-TRAFFIC
 match access-group 123
!
```

```
policy-map COPP-INPUT-POLICY
 class UNDESIRABLE-TRAFFIC
  drop
 class ICMP
  police 50000 5000 5000 conform-action transmit  exceed-action drop
!
control-plane
 service-policy input COPP-INPUT-POLICY
 !
```

To display the CoPP currently configured on the device, issue the **show policy-map control-plane** command, as demonstrated in Example 6-42.

Example 6-42 *Verifying the Control Plane Policing Configuration*

```
LAB-Router-1# show policy-map control-plane

 Control Plane

  Service-policy input: COPP-INPUT-POLICY

    Class-map: UNDESIRABLE-TRAFFIC (match-all)
      0 packets, 0 bytes
      5 minute offered rate 0000 bps, drop rate 0000 bps
      Match: access-group 123
      drop

    Class-map: ICMP (match-all)
      0 packets, 0 bytes
      5 minute offered rate 0000 bps, drop rate 0000 bps
      Match: access-group 101
      police:
          cir 50000 bps, bc 5000 bytes, be 5000 bytes
        conformed 0 packets, 0 bytes; actions:
          transmit
        exceeded 0 packets, 0 bytes; actions:
          drop
        violated 0 packets, 0 bytes; actions:
          drop
        conformed 0000 bps, exceeded 0000 bps, violated 0000 bps

    Class-map: class-default (match-any)
      3 packets, 551 bytes
      5 minute offered rate 0000 bps, drop rate 0000 bps
      Match: any
LAB-Router-1#
```

Details about CPPr

CPPr is the other feature, similar to Control Plane Policing, that can help to mitigate the effects on the CPU of traffic that requires processing by the CPU.

CPPr can restrict traffic with finer granularity by dividing the aggregate control plane into three separate control plane categories known as sub-interfaces. The three sub-interfaces are as follows:

- Host sub-interface

- Transit sub-interface

- CEF-Exception sub-interface

In addition to providing three more granular buckets in which to place packets destined to the device's control plane, the CPPr feature also provides the following:

- **Port-filtering feature:** Enables the policing and dropping of packets that are sent to closed or nonlistening TCP or UDP ports

- **Queue-thresholding feature:** Limits the number of packets for a specified protocol that are allowed in the control plane IP input queue

Securing Routing Protocols

By default, network devices send routing information to and from their routing peers in the clear, making this information visible to all interested parties. Failure to secure the exchange of routing information allows an attacker to introduce false routing information into the network. By using password authentication with routing protocols between routers, you can enhance the overall security of the network. However, because this authentication is sent as clear text, it can be simple for an attacker to subvert this security control.

If you add Message Digest 5 (MD5) algorithm hash capabilities to the authentication process, routing updates no longer contain cleartext passwords, and the entire contents of the routing update are more resistant to tampering. However, MD5 authentication is still susceptible to brute-force and dictionary attacks if weak passwords are chosen. You are advised to use passwords with sufficient randomization. Because MD5 authentication is much more secure when compared to password authentication, these examples are specific to MD5 authentication.

Implementing Routing Update Authentication on OSPF

MD5 authentication for OSPF requires configuration at either the interface level (that is, for each interface in which OSPF will be used) or within the router OSPF process itself.

The authentication type must be the same for all routers and access servers in an area. The authentication password for all OSPF routers on a network must be the same if they are to communicate with each other via OSPF. Use the **ip ospf authentication-key** interface command to specify this password.

If you enable MD5 authentication with the **message-digest** keyword, you must configure a password with the **ip ospf message-digest-key** interface command.

To remove the authentication specification for an area, use the **no** form of this command with the authentication keyword.

Example 6-43 shows the portion of a configuration required to implement OSPF router authentication using MD5.

Example 6-43   *OSPF MD5 Authentication Configuration*

```
interface GigabitEthernet0/1
 ip address 192.168.10.1 255.255.255.0
 ip ospf authentication message-digest
 ip ospf message-digest-key 1 md5 LAB
!
router ospf 65000
 router-id 192.168.10.1
 area 0 authentication message-digest
 network 10.1.1.0 0.0.0.255 area 10
 network 192.168.10.0 0.0.0.255 area 0
!
```

NOTE   The same configuration (with the exception of the interface IP address and router ID) must be identical on the other OSPF peer.

Implementing Routing Update Authentication on EIGRP

The addition of authentication to your routers' EIGRP messages ensures that your routers accept routing messages only from other routers that know the same pre-shared key. Without this authentication configured, if someone introduces another router with different or conflicting route information onto the network, the routing tables on your routers could become corrupt and a denial-of-service attack could ensue. Thus, when you add authentication to the EIGRP messages sent between your routers, it prevents someone from purposely or accidentally adding another router to the network and causing a problem.

As with OSPF, MD5 authentication for EIGRP requires configuration at the interface level (that is, for each interface in which EIGRP will be used); however, there is no specific configuration required within the router EIGRP process itself. In addition, unlike OSPF, EIGRP authentication also makes use of a key chain that is configured in global configuration mode. The key chain consists of a key number and a key string.

Example 6-44 shows the portion of a configuration required to implement EIGRP router authentication using MD5.

Example 6-44   *EIGRP MD5 Authentication Configuration*

```
key chain LAB
 key 1
  key-string LAB-SECURITY
!
!
interface Loopback0
 ip address 192.168.100.1 255.255.255.0
!
```

```
interface GigabitEthernet0/1
  ip address 192.168.10.1 255.255.255.0
  ip authentication mode eigrp 65000 md5
  ip authentication key-chain eigrp 65000 LAB
 !
router eigrp 65000
 network 192.168.10.0
 network 192.168.100.0
 !
```

NOTE   The same configuration (with the exception of the interface IP address) needs to be identical on the other EIGRP peer.

Implementing Routing Update Authentication on RIP

The Cisco implementation of RIPv2 supports two modes of authentication: plaintext authentication and MD5 authentication. Plaintext authentication mode is the default setting in every RIPv2 packet, when authentication is enabled. Plaintext authentication should not be used when security is an issue because the unencrypted authentication password is sent in every RIPv2 packet.

NOTE   RIP Version 1 (RIPv1) does not support authentication. If you are sending and receiving RIPv2 packets, you can enable RIP authentication on an interface.

As with both OSPF and EIGRP, MD5 authentication for RIPv2 requires configuration at the interface level (that is, for each interface in which RIP will be used). However, as with EIGRP, no specific configuration is required within the router RIP process. Also, like EIGRP, RIPv2 authentication makes use of a key chain, which is configured in global configuration mode. The key chain consists of a key number and a key string.

Example 6-45 shows the portion of a configuration required to implement RIPv2 router authentication using MD5.

Example 6-45   *RIPv2 MD5 Authentication Configuration*

```
key chain LAB
 key 1
  key-string LAB-SECURITY
 !

 !
interface Loopback0
 ip address 192.168.100.1 255.255.255.0
 !
 !
```

```
interface GigabitEthernet0/1
 ip address 192.168.10.1 255.255.255.0
 ip rip authentication mode md5
 ip rip authentication key-chain LAB
!
router rip
 version 2
 network 192.168.10.0
 network 192.168.100.0
!
```

NOTE The same configuration (with the exception of the interface IP address and network statements) needs to be identical on the other RIP peer.

Implementing Routing Update Authentication on BGP

Routing protocols each support a different set of cryptographic algorithms. BGP supports only HMAC-MD5 and HMAC-SHA1-12.

Peer authentication with HMAC-MD5 creates an MD5 digest of each packet sent as part of a BGP session. Specifically, portions of the IP and TCP headers, TCP payload, and a secret key are used to generate the digest. It is recommended to use HMAC-SHA1-12.

The created digest is then stored in TCP option Kind 19, which was created specifically for this purpose by RFC 2385. The receiving BGP speaker uses the same algorithm and secret key to regenerate the message digest. If the received and computed digests are not identical, the packet is discarded.

NOTE The TCP option Kind 19 (MD5 Signature Option) has been obsoleted by TCP option Kind 29.

Peer authentication with MD5 is configured with the password option to the neighbor BGP router configuration command.

As shown in Example 6-46, MD5 authentication for BGP is much simpler than the other routing protocols (OSPF, EIGRP, and RIPv2) discussed earlier. All that is required is one additional neighbor statement within the router BGP process.

Example 6-46 shows the portion of a configuration required to implement BGP router authentication using MD5.

Example 6-46   *BGP MD5 Authentication Configuration*

```
interface Loopback1
 ip address 192.168.15.1 255.255.255.0
!
interface GigabitEthernet0/1
 ip address 192.168.10.1 255.255.255.0
!
```

```
router bgp 65000
 bgp log-neighbor-changes
 network 192.168.15.0
 neighbor 192.168.10.2 remote-as 65100
 neighbor 192.168.10.2 password LAB-SECURITY
 !
```

NOTE   A similar configuration must be in place on the other BGP peer.

To verify that MD5 authentication is used between the BGP peers, issue the **show ip bgp neighbors | include Option Flags** command and look for **md5** in the output, as demonstrated in Example 6-47.

Example 6-47   *Verifying MD5 Authentication Between BGP Peers*

```
LAB-Router-1#show ip bgp neighbors | include Option Flags

Option Flags: nagle, path mtu capable, md5, Retrans timeout
LAB-Router-1#
```

TIP   BGP keychains enable keychain authentication between two BGP peers. BGP is able to use the keychain feature to implement hitless key rollover for authentication. Key rollover specification is time based, and in the event of clock skew between the peers, the rollover process is impacted. The configurable tolerance specification allows for the accept window to be extended (before and after) by that margin. This accept window facilitates a hitless key rollover for applications (for example, routing and management protocols). The key rollover does not impact the BGP session, unless there is a keychain configuration mismatch at the endpoints resulting in no common keys for the session traffic (send or accept).

Exam Preparation Tasks

As mentioned in the section "How to Use This Book" in the Introduction, you have a couple of choices for exam preparation: the exercises here, Chapter 12, "Final Preparation," and the exam simulation questions in the Pearson Test Prep Software Online.

Review All Key Topics

Review the most important topics in this chapter, noted with the Key Topic icon in the outer margin of the page. Table 6-5 lists these key topics and the page numbers on which each is found.

Table 6-5 Key Topics for Chapter 6

Key Topic Element	Description	Page Number
List	Understanding Layer 2 best practices	322
List	Layer 2 Security Toolkit	324
Section	CDP and LLDP	327
Section	DHCP Snooping	328
Section	Dynamic ARP Inspection	330
Section	The Network Foundation Protection Framework	333
Section	Best Practices for Securing the Management Plane	334
Section	Best Practices for Securing the Control Plane	336
Section	Best Practices for Protecting the Data Plane	337
Section	Additional Data Plane Protection Mechanisms	338
Section	Management Plane Best Practices	339
Section	Password Recommendations	341
Section	Using AAA to Verify Users	342
Section	Router Access Authentication	342
Section	The AAA Method List	343
Section	Custom Privilege Levels	344
Paragraph	Understanding syslog and the different syslog severity levels	345
Section	Understanding NTP	346
Section	Protecting Cisco IOS, Cisco IOS-XE, Cisco IOS-XR, and Cisco NX-OS Files	346
Section	User Authentication with AAA	349
Section	RBAC Privilege Level/Parser View	356
Section	SSH and HTTPS	360
Section	Configuring Syslog Support	363
Section	Configuring NTP	363
Section	Securing the Network Infrastructure Device Image and Configuration Files	364
Table 6-4	Identifying the notable differences and some similarities between IPv4 and IPv6	366
Section	Best Practices Common to Both IPv4 and IPv6	372
Section	Threats Common to Both IPv4 and IPv6	373
Section	IPv6 Best Practices	376
Section	Minimizing the Impact of Control Plane Traffic on the CPU	379

Key Topic Element	Description	Page Number
Section	Details about CoPP	380
Section	Details about CPPr	383
Section	Securing Routing Protocols	383

Define Key Terms

Define the following key terms from this chapter and check your answers in the glossary:

VLANs, 802.1Q, root port, designated port, nondesignated port, port security, BPDU Guard, Root Guard, Dynamic ARP Inspection, IP Source Guard, DHCP snooping, CDP, LLDP, management plane, control plane, data plane

Review Questions

1. You have been asked to restrict users without having to create custom privilege levels. Which of the following features or functionality would you deploy to accomplish this task?

   a. Parser views (or "views")

   b. AAA profiles

   c. DAI

   d. All of these answers are correct.

2. The concept of _____ is to create a set of permissions or limited access and assign that set of permissions to users or groups. Those permissions are used by individuals for their given roles, such as a role of administrator, a role of a help desk person, and so on.

   a. ABAC

   b. RBAC

   c. Dynamic groups

   d. Downloadable ACLs

3. Which feature can protect against Address Resolution Protocol (ARP) spoofing, ARP poisoning (which is advertising incorrect IP-to-MAC-address mapping information), and resulting Layer 2 man-in-the-middle attacks?

   a. DHCP spoofing

   b. Dynamic ARP Inspection (DAI)

   c. IP Source Guard

   d. All of these answers are correct.

4. Which of the following statements is not true?

    a. CoPP is applied to a logical control plane interface (not directly to any Layer 3 interface) so that the policy can be applied globally to the router.

    b. The benefit of CPPr is that you can rate-limit and filter this type of traffic with a more fine-toothed comb than CoPP.

    c. The host sub-interface that handles traffic to one of the physical or logical interfaces of the router is one of the sub-interfaces of CPPr.

    d. CPPr is not applied to a physical interface, so regardless of the logical or physical interface on which the packets arrive, the router processor can still be protected.

5. DHCP snooping is a security feature that acts like a firewall between untrusted hosts and trusted DHCP servers. Which of the following are activities performed by DHCP snooping?

    a. Validates DHCP messages received from untrusted sources and filters out invalid messages.

    b. Rate-limits DHCP traffic from trusted and untrusted sources.

    c. Builds and maintains the DHCP snooping binding database, which contains information about untrusted hosts with leased IP addresses.

    d. Utilizes the DHCP snooping binding database to validate subsequent requests from untrusted hosts.

    e. All of these answers are correct.

6. Your switch might be connected to other switches that you do not manage. If you want to prevent your local switch from learning about a new root switch through one of its local ports, you can configure which of the following features?

    a. Dynamic Root Inspection

    b. Root Guard

    c. DHCP Guard

    d. Port Security

    e. A and C

7. Which of the following prevents spoofing of Layer 2 information by hosts?

    a. Dynamic ARP Inspection

    b. BPDU Guard

    c. Root Guard

    d. All of these answers are correct.

8. Which of the following prevents spoofing of Layer 3 information by hosts?

    a. Dynamic ARP Inspection

    b. BPDU Guard

    c. IP Source Guard

    d. All of these answers are correct.

9. Which of the following limits the amount of broadcast or multicast traffic flowing through the switch?

    a. Root Guard

    b. BPDU Guard

    c. Storm Control

    d. DHCP snooping

10. CDP operates at _____ and may provide attackers information we would rather not disclose.

    a. Layer 2

    b. Layer 3

    c. Layer 4

    d. Layer 7

6

CHAPTER 7

Cisco Next-Generation Firewalls and Cisco Next-Generation Intrusion Prevention Systems

This chapter covers the following topics:

Introduction to Cisco Next-Generation Firewalls (NGFW) and Next-Generation Intrusion Prevention System (NGIPS)

Comparing Network Security Solutions That Provide Firewall Capabilities

Deployment Modes of Network Security Solutions and Architectures That Provide Firewall Capabilities

High Availability and Clustering

Implementing Access Control

Cisco Firepower Intrusion Policies

Cisco Advanced Malware Protection (AMP)

The following SCOR 350-701 exam objectives are covered in this chapter:

- Domain 2.0 Network Security

  - 2.1 Compare network security solutions that provide intrusion prevention and firewall capabilities

  - 2.2 Describe deployment models of network security solutions and architectures that provide intrusion prevention and firewall capabilities

  - 2.5 Implement segmentation, access control policies, AVC, URL filtering, and malware protection

  - 2.6 Implement management options for network security solutions such as intrusion prevention and perimeter security

"Do I Know This Already?" Quiz

The "Do I Know This Already?" quiz allows you to assess whether you should read this entire chapter thoroughly or jump to the "Exam Preparation Tasks" section. If you are in doubt about your answers to these questions or your own assessment of your knowledge of the topics, read the entire chapter. Table 7-1 lists the major headings in this chapter and their corresponding "Do I Know This Already?" quiz questions. You can find the answers in Appendix A, "Answers to the 'Do I Know This Already?' Quizzes and Q&A Sections."

Table 7-1 "Do I Know This Already?" Section-to-Question Mapping

Foundation Topics Section	Questions
Introduction to Cisco Next-Generation Firewalls (NGFW) and Next-Generation Intrusion Prevention System (NGIPS)	1
Comparing Network Security Solutions That Provide Firewall Capabilities	2
Deployment Modes of Network Security Solutions and Architectures That Provide Firewall Capabilities	3–5
High Availability and Clustering	6
Implementing Access Control	7–8
Cisco Firepower Intrusion Policies	9
Cisco Advanced Malware Protection (AMP)	10

CAUTION The goal of self-assessment is to gauge your mastery of the topics in this chapter. If you do not know the answer to a question or are only partially sure of the answer, you should mark that question as wrong for purposes of the self-assessment. Giving yourself credit for an answer you incorrectly guess skews your self-assessment results and might provide you with a false sense of security.

1. Which of the following Cisco firewalls is designed for very large enterprises and service providers?

   a. Cisco Zone-Based Firewall (ZBFW)

   b. Cisco Firepower 9300 appliances

   c. Cisco FTD running Cisco Unified Computing System (UCS) E-Series blades installed on Cisco ISR routers

   d. Cisco Firepower 2140

2. Cisco IOS Zone-Based Firewall (ZBFW) can be deployed to provide firewall services in small and medium-sized organizations. Which of the following is not true about zone-based firewalls?

   a. With ZBFWs, an interface can be assigned to only one security zone.

   b. Zone-based firewalls cannot be implemented in an SD-WAN solution.

   c. ZBFWs support zone pairs, which are a container that associates a source zone with a destination zone and that applies a firewall policy to the traffic that flows between the two zones.

   d. ZBFWs support a security policy, similar to a localized security policy, that defines the conditions that the data traffic flow from the source zone must match to allow the flow to continue to the destination zone.

3. You were hired to deploy a Cisco ASA to provide separation of management and policies on a shared appliance. Which operational mode is best for this scenario?

   a. Routed mode

   b. Transparent mode

   c. Multiple context mode

   d. ERSPAN with tap mode

4. Which of the following statements is not true about firewalls deployed in Layer 3 (routed) mode?

   a. Routed firewalls do not provide a way to filter packets that traverse from one host to another in the same LAN segment.

   b. The Layer 3 firewalls require a new network segment to be created when they are inserted into a network, which requires quite a bit of planning, network downtime, and reconfiguration of network devices.

   c. Layer 3 firewalls support EtherType ACLs.

   d. In Cisco FTD deployments, ERSPAN interfaces are only supported in routed mode.

5. Which of the following statements are true about the inline interface sets and passive interfaces in Cisco FTD deployments?

   a. Inline sets and passive interfaces are only supported on physical interfaces and EtherChannels.

   b. Inline sets cannot use redundant interfaces or VLANs.

   c. Inline sets and passive interfaces are supported in intra-chassis and inter-chassis clustering.

   d. All of these answers are correct.

6. Which of the following are requirements for failover configurations?

   a. The two participant devices must be configured in the same firewall mode (for example, routed or transparent).

   b. The two participant devices must be running the same software version.

   c. You can configure different Cisco FTD devices in groups (or domains) in the Cisco FMC. Devices configured for failover must be in the same domain or group on the Cisco FMC.

   d. All of these answers are correct.

7. In Cisco ASA deployments, an access control list (ACL) is a collection of security rules or policies that allows or denies packets after looking at the packet headers and other attributes. Each permit or deny statement in the ACL is referred to as an access control entry (ACE). These ACEs classify packets by inspecting Layer 2 through Layer 7 headers for a number of parameters, including which of the following?

   a. Layer 2 protocol information such as EtherTypes

   b. Layer 3 header information such as source and destination IP addresses

   c. Layer 4 header information such as source and destination TCP or UDP ports

   d. All of these answers are correct.

8. You were tasked to configure a Cisco ASA to permit SMTP traffic from hosts in 192.168.88.0/25 to an email server (10.2.2.2). Which of the following access control entries (ACEs) in an ACL will accomplish this task?

   a. access-list my-list extended permit eq 25 tcp 192.168.88.0 255.255.255.128 host 10.2.2.2

   b. access-list my-list extended permit tcp 192.168.88.0 255.255.255.192 host 10.2.2.2 eq 25

   c. access-list my-list extended permit tcp 192.168.88.0 255.255.255.128 host 10.2.2.2 eq 25

   d. access-list my-list extended permit tcp host 10.2.2.2 192.168.88.0 255.255.255.128 eq 25

9. Which of the following is true about Cisco Firepower Intrusion Policies?

   a. Both network analysis and intrusion policies are invoked by a parent access control policy, but at different times.

   b. As the system analyzes traffic, the network analysis (decoding and preprocessing) phase occurs before and separately from the intrusion prevention (additional preprocessing and intrusion rules) phase.

   c. The Cisco FTD has several similarly named network analysis and intrusion policies (for example, Balanced Security and Connectivity) that complement and work with each other.

   d. All of these answers are correct.

10. Which of the following sandboxing technologies provides a dynamic analysis that includes an external kernel monitor, dynamic disk analysis that illuminates any modifications to the physical disk (such as the master boot record), monitoring user interaction, video capture and playback, process information, artifacts, and network traffic?

    a. Threat Grid

    b. Talos

    c. Cisco Threat Response (CTR)

    d. None of the above

Foundation Topics

Introduction to Cisco Next-Generation Firewalls (NGFW) and Next-Generation Intrusion Prevention Systems (NGIPS)

Cisco next-generation security products provide protection throughout the attack continuum. Devices such as the Cisco ASA with FirePOWER Services, available on the Cisco ASA 5500-X Series and ASA 5585-X Adaptive Security Appliances; Firepower Threat Defense (FTD); and Cisco Advanced Malware Protection (AMP) provide a security solution that helps discover threats and enforce and harden policies before an attack takes place. In addition, you can detect, block, and defend against attacks that have already taken place with next-generation intrusion prevention systems (NGIPSs), Email Security, and Web Security Appliance with AMP. These solutions provide the capabilities to contain and remediate an attack to minimize data loss and additional network degradation. Email Security and Web Security Appliances are covered in Chapter 10, "Content Security."

You may have seen the terms *FirePOWER* and *Firepower* being used by Cisco in different instances. What is the difference between FirePOWER and Firepower? Cisco uses the term FirePOWER (uppercase POWER) when referring to the Cisco ASA FirePOWER Services module and uses Firepower (lowercase power) when referring to the FTD unified image and newer software.

Cisco Firewall History and Legacy

Cisco started in the firewall business with a legacy firewall called Centri Firewall. Through acquisitions it then released a very popular firewall called the PIX (Private Internet Exchange) Firewall. Cisco also purchased a company called WheelGroup, which introduced the Cisco legacy IDS and IPS systems. In the early 2000s, Cisco released the Cisco Adaptive Security Appliance (ASA), one of the most popular firewalls of all time.

The members of the Cisco ASA family come in many shapes and sizes, but they all provide a similar set of features. Typically, smaller model numbers represent smaller capacity for throughput. The Cisco ASA also comes in a virtual form—the Cisco Adaptive Security Virtual Appliance (ASAv). For an up-to-date list of all the models available by Cisco, visit https://www.cisco.com/c/en/us/products/security/firewalls/index.html.

Introducing the Cisco ASA

The Cisco ASA family provides a very comprehensive set of features and next-generation security capabilities. For example, it provides capabilities such as simple packet filtering (normally configured with access control lists [ACLs]) and stateful inspection. The Cisco ASA family also provides support for application inspection/awareness. A Cisco ASA device can listen in on conversations between devices on one side of the firewall and devices on the other side. The benefit of listening in is that the firewall can pay attention to application layer information.

The Cisco ASA family also supports network address translation (NAT), the capability to act as a Dynamic Host Configuration Protocol (DHCP) server or client, or both. The Cisco ASA family supports most of the interior gateway routing protocols, including Routing Information Protocol (RIP), Enhanced Interior Gateway Routing Protocol (EIGRP), and Open Shortest Path First (OSPF). It also supports static routing. A Cisco ASA device also can be implemented as a traditional Layer 3 firewall, which has IP addresses assigned to each of its routable interfaces. The other option is to implement a firewall as a transparent (Layer 2) firewall, in which case the actual physical interfaces are not configured with individual IP addresses, but a pair of interfaces that operate like a bridge. Traffic that is going across this two-port bridge is still subject to the rules and inspection that can be implemented by the ASA. In addition, a Cisco ASA device is often used as a headend or remote-end device for VPN tunnels for both remote-access VPN users and site-to-site VPN tunnels. The Cisco ASA family supports IPsec and SSL-based remote-access VPNs. The SSL VPN capabilities include support for clientless SSL VPN and full AnyConnect SSL VPN tunnels.

The Cisco ASA family also provides a basic botnet traffic-filtering feature. A botnet is a collection of computers that have been compromised and are willing to follow the instructions of someone who is attempting to centrally control them (for example, 200,000 machines all commanded to send a flood of ping requests to the IP address by the person controlling these devices). Often, users of these computers have no idea that their computers are participating in a coordinated attack. An ASA device works with an external system at Cisco that provides information about the Botnet Traffic Filter Database and so can protect against such attacks.

The Cisco ASA FirePOWER Module

Cisco introduced the Cisco ASA FirePOWER module as part of the integration of the Sourcefire technology.

The Cisco ASA FirePOWER module can be managed by the Firepower Management Center (FMC), formerly known as the FireSIGHT Management Center. The Firepower Management Center and the Cisco ASA FirePOWER module require additional licenses. In all Cisco ASA models except the 5506-X, 5508-X, and 5516-X, the licenses are installed in the FirePOWER module. There are no additional licenses required in a Cisco ASA device. FirePOWER Services running on the Cisco ASA 5506-X, 5508-X, and 5516-X can be managed using Adaptive Security Device Manager (ASDM), and the licenses can be installed using ASDM. In all Cisco ASAs with FirePOWER Services managed by a Firepower Management Center, the license is installed on the Firepower Management Center and used by the module.

 ## Cisco Firepower Threat Defense (FTD)

Cisco FTD is unified software that includes Cisco ASA features, legacy FirePOWER Services, and new features. FTD can be deployed on Cisco Firepower 1000 Series, Firepower 2100 Series, Firepower 4100 Series, and Firepower 9000 Series appliances to provide next-generation firewall (NGFW) services.

In addition to being able to run on the Cisco Firepower appliances, FTD can also run natively on the ASA 5506-X, ASA 5506H-X, ASA 5506W-X, ASA 5508-X, ASA 5512-X, ASA 5515-X, ASA 5516-X, ASA 5525-X, ASA 5545-X, and ASA 5555-X. It is not supported in the ASA 5505 or the 5585-X.

Cisco Firepower 1000 Series

The Cisco Firepower 1000 Series appliances are next-generation firewalls that run the Cisco FTD software and features designed for small business and home offices. There are three models:

- **Cisco Firepower 1010:** A desktop firewall with eight 1 Gigabit Ethernet ports, and scales up to 650 Mbps of NGFW throughput.

- **Cisco Firepower 1120:** A rack-mount firewall with eight 1 Gigabit Ethernet ports and four SFP ports. The Firepower 1120 scales to up to 1.5 Gbps of NGFW throughput.

- **Cisco Firepower 1140:** A rack-mount firewall with eight 1 Gigabit Ethernet ports and four SFP ports. The Firepower 1140 scales to up to 2.2 Gbps of NGFW throughput.

Cisco Firepower 2100 Series

The Cisco Firepower 2100 Series appliances are next-generation firewalls that run the Cisco FTD software and features designed for many use cases, including Internet edge and data center. The Cisco Firepower 2100 Series appliances scale up to 8.5 Gbps. There are four models at the time of writing:

- **Cisco Firepower 2110:** A rack-mount 1 RU firewall with twelve 1 Gigabit Ethernet ports and four SFP ports. The Firepower 2110 scales up to 2 Gbps of NGFW throughput.

- **Cisco Firepower 2120:** A rack-mount 1 RU firewall with twelve 1 Gigabit Ethernet ports and four SFP ports. The Firepower 2120 scales up to 3 Gbps of NGFW throughput.

- **Cisco Firepower 2130:** A rack-mount 1 RU firewall with up-to twenty-four 1 Gigabit Ethernet ports or twelve 1 Gigabit Ethernet and twelve 10 Gigabit Ethernet ports. The Firepower 2130 scales up to 5 Gbps of NGFW throughput.

- **Cisco Firepower 2140:** A rack-mount 1 RU firewall with up-to twenty-four 1 Gigabit Ethernet ports or twelve 1 Gigabit Ethernet and twelve 10 Gigabit Ethernet ports. The Firepower 2140 scales up to 8.5 Gbps of NGFW throughput.

Cisco Firepower 4100 Series

The Cisco Firepower 4100 Series appliances are next-generation firewalls that run the Cisco FTD software and features. There are seven models at the time of writing:

- **Cisco Firepower 4110:** A rack-mount 1 RU firewall with 1, 10, or 40 Gbps interfaces scaling up to 35 Gbps of firewall throughput and 11 Gbps of threat inspection throughput.

- **Cisco Firepower 4120:** A rack-mount 1 RU firewall with 1, 10, or 40 Gbps interfaces scaling up to 60 Gbps of firewall throughput and 19 Gbps of threat inspection throughput.

- **Cisco Firepower 4140:** A rack-mount 1 RU firewall with 1, 10, or 40 Gbps interfaces scaling up to 70 Gbps of firewall throughput and 27 Gbps of threat inspection throughput.

- **Cisco Firepower 4150:** A rack-mount 1 RU firewall with 1, 10, or 40 Gbps interfaces scaling up to 75 Gbps of firewall throughput and 39 Gbps of threat inspection throughput.

- **Cisco Firepower 4115:** A rack-mount 1 RU firewall with 1, 10, or 40 Gbps interfaces scaling up to 80 Gbps of firewall throughput and 26 Gbps of threat inspection throughput.

- **Cisco Firepower 4125:** A rack-mount 1 RU firewall with 1, 10, or 40 Gbps interfaces scaling up to 80 Gbps of firewall throughput and 35 Gbps of threat inspection throughput.

- **Cisco Firepower 4145:** A rack-mount 1 RU firewall with 1, 10, or 40 Gbps interfaces scaling up to 80 Gbps of firewall throughput and 45 Gbps of threat inspection throughput.

NOTE Cisco continues to enhance and introduce new Firepower appliances and platforms. For the most up-to-date list of appliances refer to: https://cisco.com/go/security.

Cisco Firepower 9300 Series

The Cisco Firepower 9300 appliances are designed for very large enterprises or service providers. They can scale beyond 1.2 Tbps and are designed in a modular way, supporting Cisco ASA software, Cisco FTD software, and Radware DefensePro DDoS mitigation software. The Radware DefensePro DDoS mitigation software is available and supported directly from Cisco on Cisco Firepower 4150 and Cisco Firepower 9300 appliances. Radware's DefensePro DDoS mitigation software provides real-time analysis to protect the enterprise or service provider infrastructure against network and application downtime due to distributed denial-of-service (DDoS) attacks.

Cisco FTD for Cisco Integrated Services Routers (ISRs)

Cisco FTD can run on Cisco Unified Computing System (UCS) E-Series blades installed on Cisco ISR routers. Both the FMC and FTD are deployed as virtual machines. There are two internal interfaces that connect a router to an UCS E-Series blade. On ISR G2, Slot0 is a Peripheral Component Interconnect Express (PCIe) internal interface, and UCS E-Series Slot1 is a switched interface connected to the backplane Multi Gigabit Fabric (MGF). In Cisco ISR 4000 Series routers, both internal interfaces are connected to the MGF.

A hypervisor is installed on the UCS E-Series blade, and the Cisco FTD software runs as a virtual machine on it. FTD for ISRs is supported on the following platforms:

- Cisco ISR G2 Series: 2911, 2921, 2951, 3925, 3945, 3925E, and 3945E

- Cisco ISR 4000 Series: 4331, 4351, 4451, 4321, and 4431

Legacy IPS were traditionally used in network infrastructures to protect against known security threats. Often, two concepts were used: IPS and intrusion detection systems (IDS). IDS mostly detect and generate alerts for various attacks or intrusion attempts, whereas IPS can also prevent and mitigate attacks. The remainder of this section focuses on IPS, but you should note that there are no significant differences between the methodologies used in IPS and IDS for attack detection.

Introduction to Cisco's NGIPS

Legacy IPSs depend mostly on matching signature-based patterns to identify and potentially prevent malicious activity. These are some of the basic characteristics of legacy IPSs:

- They are sometimes deployed behind a firewall when providing IPS functionality (inline). Often, an IPS is also placed in the network without a firewall in front of it.

- They often look for attempts to exploit a vulnerability and not for the existence of a vulnerability.

- Legacy IPSs often generate large amounts of event data that are difficult to correlate.

- They focus on individual indicators/events without focusing on contextual information to take action.

- Legacy IPSs require manual tuning for better efficacy.

Thus, legacy IPSs suffer from certain shortcomings, including the following:

- They often need to be operated in conjunction with other products or tools (firewalls, analytics, and correlation tools).

- They are sometimes not very effective and may be ignored.

- Their operation costs and the operating resources they need are high.

- They can leave infrastructures imperfectly covered against attackers.

NGIPSs supplement legacy IPS functionality with more capabilities, such as the following:

- **Application awareness and control:** NGIPSs provide visibility into Layer 7 applications and can protect against Layer 7 threats.

- **Content awareness of the information traversing the infrastructure:** For example, knowledge about files transferred between two hosts can be used to identify viruses transferred and the trajectory of a virus infection in a system.

- **Contextual awareness:** Helps better understand alerts and automatically deduce comprehensive information about the events taking place, which makes the NGIPS less complex and means it requires less tuning.

- **Host and user awareness:** The infrastructure offers more conclusive information about the events taking place.

- **Automated tuning and recommendations:** This allows an administrator to follow recommendations and tune signatures specifically to his environment.

- **Impact and vulnerability assessment of the events taking place:** The impact of a security event identified by the system can be evaluated based on the information available for the environment. For example, a Windows system that is identified to secure a vulnerability cannot be severely impacted by an attempt to exploit the vulnerability against it.

Thus, it is clear that in the threat landscape of both today and in the future, NGIPS functionality has an important role in protecting and providing coverage against known attacks and new types of exploits.

Modern networks constantly evolve, as do miscreants and their attack methods. People and machines that could misbehave reside inside and outside a network infrastructure. Devices are communicating in many different forms. The interconnected infrastructure with attackers that could be located anywhere is called the "any-to-any challenge." Almost all modern environments face this challenge. Cisco is a leader in NGIPS, offering Cisco Firepower NGIPS products that can provide protection against constantly evolving attack surfaces in these environments.

Modern security tools need to integrate directly into the network fabric in order to maximize performance and efficiency. Responses need to be comprehensive and simple. Protection must be continuous. Network controls should not be implemented disparately and individually. To abide by these modern security requirements, Cisco follows a new security model that looks at the actions needed before, during, and after attacks that apply to mobile devices, virtual machines, endpoints, or more. The Cisco Firepower NGIPS functionality operates mostly in the "during phase" of the attack continuum, but all phases are covered by the integrated capabilities of the Cisco Firepower product portfolio.

The Cisco Firepower NGIPS engine is based on well-defined open source Snort. Snort, originally created by SourceFire, is an open source IPS tool that is widely used in the industry. The Cisco Snort IPS rules are developed by the Cisco Talos team and are open for inspection. They are built based on collective security intelligence by Cisco Talos and a variety of other sources. The rule set offers broad product and threat coverage. In addition, third-party rules can be integrated and customized in the Cisco NGIPS.

The following are some of the most important capabilities of Cisco NGIPS:

- **Threat containment and remediation:** Cisco Firepower NGIPS provides protection against known and new threats. Its features include file analysis, packet- and flow-based inspection, and vulnerability assessment.

- **Application visibility:** Cisco Firepower NGIPS offers deep inspection and control of application-specific information for better efficacy.

- **Identity management:** NGIPS policies can be enforced by using contextual user information.

- **Security automation:** Cisco Firepower NGIPS includes automated event impact assessment and policy tuning.

- **Logging and traceability management:** This can be used in retrospective analysis.

- **High availability and stacking:** Cisco Firepower NGIPS provides redundancy and performance by leveraging multiple devices.

- **Network behavioral analysis:** Key behavioral indicators and threat scores help analysts prioritize and recover from attacks.

- **Access control and segmentation:** Access policies can be applied to separate traffic profiles in the network.

- **Real-time contextual awareness:** NGIPS discovers and provides information about applications, users, devices, operating systems, vulnerabilities, services, processes, files, and threat data related to IT environments.

Surveying the Cisco Firepower Management Center (FMC)

Cisco FTD devices, Cisco Firepower NGIPS devices, and the Cisco ASA FirePOWER modules can be managed by the Firepower Management Center (FMC), formerly known as the FireSIGHT Management Center. The Cisco FMC, Cisco FTD, and the Cisco ASA FirePOWER module require additional licenses.

NOTE In all Cisco ASA models except the 5506-X, 5508-X, and 5516-X, the licenses are installed in the FirePOWER module. There are no additional licenses required in a Cisco ASA device. FirePOWER Services running on the Cisco ASA 5506-X, 5508-X, and 5516-X can be managed using Adaptive Security Device Manager (ASDM), and the licenses can be installed using ASDM. In all Cisco ASAs with FirePOWER Services managed by a Firepower Management Center, the license is installed on the Firepower Management Center and used by the module.

When you add a managed device to the Cisco FMC, you must provide an IP addresses of the managed device along with a registration key for authentication. The Cisco FMC and the managed device use the registration key and a NAT ID (instead of IP addresses in the case that the device is behind NAT) to authenticate and authorize for initial registration. For instance, when you add a device to the Cisco FMC, and you do not know the device IP address (or the device is behind a NAT/PAT device), you specify only the NAT ID and the registration key on the FMC and leave the IP address blank.

You typically use the NAT ID for NAT environments; however, you can also use the NAT ID to simplify adding many devices to the Cisco FMC. On the other hand, the NAT ID must be unique per device.

The Cisco FMC provides very detailed analytics and statistics of what's happening in your network. You can select from many prebuilt dashboards or create your own. Figure 7-1 shows the Cisco FMC Summary Dashboard. In the Summary Dashboard, statistics and data about the top attackers, top targets, intrusion events, events by application protocols, and other elements are displayed. You can customize all dashboards in the Cisco FMC.

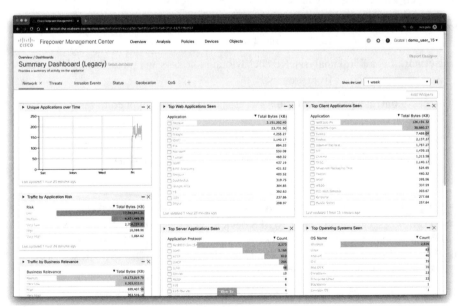

Figure 7-1 *The Cisco FMC Summary Dashboard*

Figure 7-2 shows the Cisco FMC Connection Summary Dashboard. The Cisco FMC Connection Summary Dashboard displays information about the allowed and denied connections by application and over time.

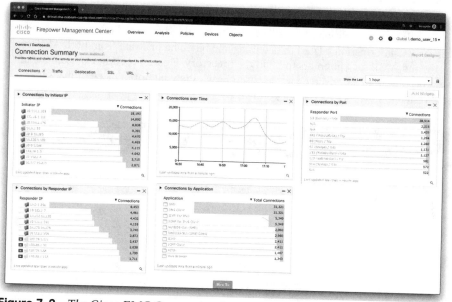

Figure 7-2 *The Cisco FMC Connection Summary Dashboard*

Figure 7-3 shows the Cisco FMC Content Explorer providing statistics of traffic and intrusions over time, indicators of compromise, and other network information.

Figure 7-3 *The Cisco FMC Content Explorer*

The Network File Trajectory feature maps how hosts transferred files, including malware files, across your network. A trajectory charts file transfer data, the disposition of the file, and whether a file transfer was blocked or the file was quarantined. You can determine which hosts and users may have transferred malware, which hosts are at risk, and observe file transfer trends.

You can track the transmission of any file with a Cisco cloud-assigned disposition. The system can use information related to detecting and blocking malware from both AMP for Networks and AMP for Endpoints to build the trajectory. Figure 7-4 shows an example of the Network File Trajectory screen in the Cisco FMC, showing how a piece of malware infected different hosts across the network.

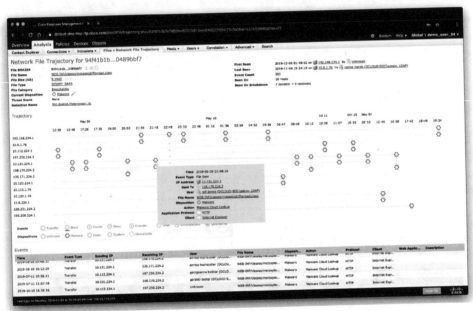

Figure 7-4   *The Cisco FMC Network File Trajectory*

Exploring the Cisco Firepower Device Manager (FDM)

The Cisco Firepower Device Manager (FDM) is used to configure small Cisco FTD deployments. To access the Cisco FDM, you just need to point your browser at the firewall in order to configure and manage the device.

Figure 7-5 shows the Cisco FDM Device Setup Wizard.

Figure 7-6 shows the Cisco FDM main dashboard. The Cisco FDM main dashboard includes information about the configured interfaces, configured routes, and updates (Geolocation, Rules, Vulnerability Database [VDB], Security Intelligence Feeds, and so on). It also provides information about the smart licenses that are enabled on the system, backup and restore configuration, and troubleshooting information.

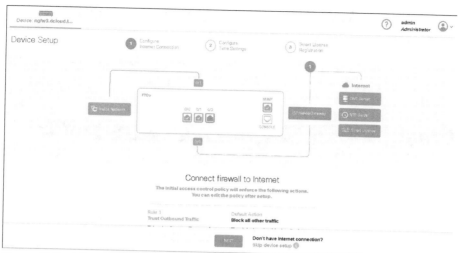

Figure 7-5 *The Cisco FDM Device Setup Wizard*

Figure 7-6 *The Cisco FDM Main Dashboard*

The Cisco FDM allows you to configure network address translation (NAT), access control, and intrusion policies. Figure 7-7 shows the manual NAT configuration in the Security Policies screen of the Cisco FDM. NAT will be discussed in detail later in this chapter.

Figure 7-8 shows an example of a NAT rule in the Cisco FDM.

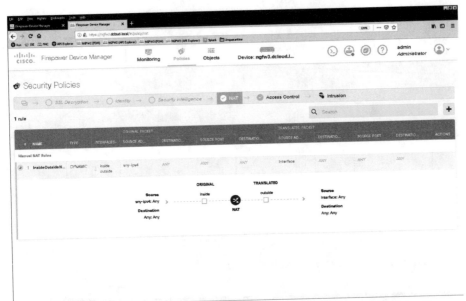

Figure 7-7   *Manual NAT Configuration in the Cisco FDM Security Policies Screen*

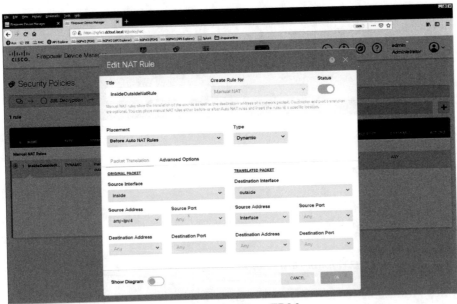

Figure 7-8   *Example of a NAT Rule in the Cisco FDM*

Figure 7-9 shows an example of how an access control rule is applied to the inside and outside security zones.

A *security zone* is a collection of one or more inline, passive, switched, or routed interfaces (or ASA interfaces) that you can use to manage and classify traffic in different policies. Interfaces in a single zone could span multiple devices. Furthermore, you can also enable multiple zones on a single device to segment your network and apply different policies.

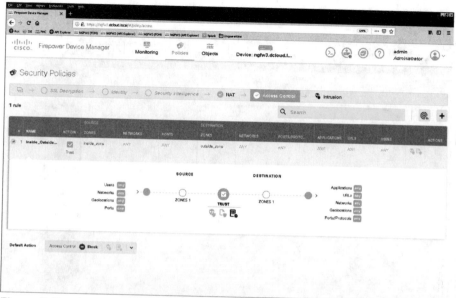

Figure 7-9 *Access Control Rule Configured to Protect Traffic in Different Security Zones*

NOTE   You must assign at least one interface to a security zone in order to match traffic against such a security zone. Interfaces can only belong to one security zone.

You can also use zones in different places in the system's web interface. This includes access control policies, network discovery rules, and event searches—for instance, to configure an access control rule that applies only to a specific source or destination zone or to restrict network discovery to traffic to or from a specific zone.

Cisco FTD creates security zones upon device registration, depending on the detection mode you selected for the device during its initial configuration. For instance, Cisco FTD creates a *passive* zone in passive deployments, while in inline deployments, Cisco FTD creates External and Internal zones. You can also create security zones when you are configuring interfaces on a Cisco Firepower device.

Figure 7-10 shows the configuration of a source- and destination-based access control rule.

Figure 7-11 shows the configuration of an application-based access control rule.

TIP   The Cisco FTD can be installed in a Firepower appliance that runs its own operating system called FXOS (Firepower eXtensible Operating System) to control basic operations of the device. The FTD logical device is installed on a module on such a device (chassis). You can use the FXOS Firepower Chassis Manager (FCM) or the FXOS CLI to configure the Cisco Firepower device functions. On the other hand, the FCM GUI is not available when the Cisco FTD software is installed on the Firepower 2100 series (or similar), just the FXOS CLI.

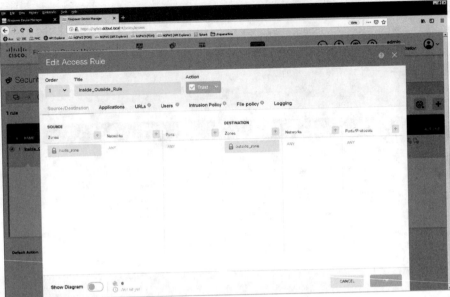

Figure 7-10 *A Source- and Destination-Based Access Control Rule*

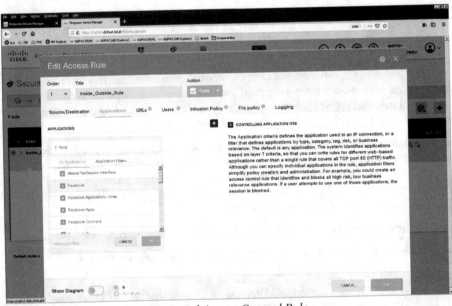

Figure 7-11 *An Application-Based Access Control Rule*

Cisco Defense Orchestrator

The Cisco Defense Orchestrator (CDO) is a solution that allows you to manage your firewalls from the cloud. You can write a policy once and enforce it consistently across multiple Cisco ASA and Cisco FTD devices. In addition, you can compare, filter, edit, and create new policies, all from a central point (the cloud). The Cisco Defense Orchestrator allows you to analyze access control policies and objects to identify errors and inconsistencies. You can also create standard policy templates that give you consistent, effective protection across your enterprise environment.

Figure 7-12 shows the Cisco Defense Orchestrator main dashboard.

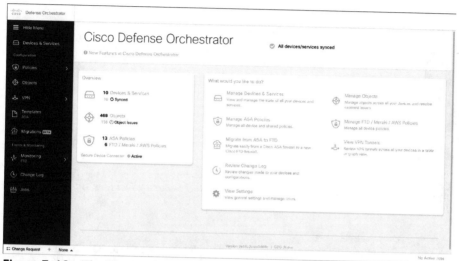

Figure 7-12   *The Cisco Defense Orchestrator Main Dashboard*

Figure 7-13 shows examples of Cisco ASA policies configured in Cisco Defense Orchestrator.

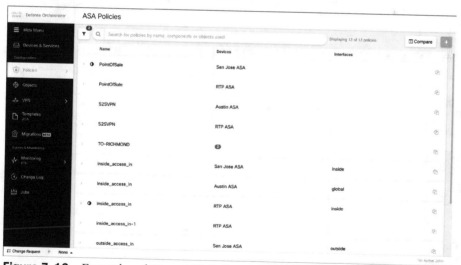

Figure 7-13   *Examples of Cisco ASA Policies Configured in Cisco Defense Orchestrator*

The Cisco Defense Orchestrator can also manage and analyze Cisco FTD, Meraki security devices, and virtual firewalls in AWS. Figure 7-14 shows examples of policies of Cisco FTD, Meraki security devices, and virtual firewalls in AWS.

Figure 7-14   *Examples of Policies of Cisco FTD, Meraki Security Devices, and Virtual Firewalls in AWS*

TIP   There are several prerequisites for a successful deployment of Cisco FTD in AWS. For instance, you must configure and enable two management interfaces during launch. Then, once you are ready to pass traffic, you must have two traffic interfaces and two management interfaces (a total of four interfaces). For the most up-to-date requirements, refer to the following link: https://www.cisco.com/c/en/us/td/docs/security/firepower/quick_start/aws/ftdv-aws-qsg.html.

Figure 7-15 shows a detailed example of Cisco FTD policies in Cisco Defense Orchestrator.

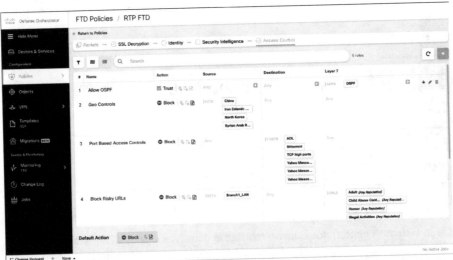

Figure 7-15   *A Detailed Example of Cisco FTD Policies in Cisco Defense Orchestrator*

Comparing Network Security Solutions That Provide Firewall Capabilities

There are different Cisco solutions that provide intrusion prevention and firewall capabilities. You already learned a few details about the Cisco ASA and the Cisco Firepower next-generation firewalls (NGFW). Next-generation firewalls also provide intrusion prevention capabilities. On the other hand, there is one more firewall solution: Cisco IOS Zone-Based Firewall (ZBFW). The Cisco IOS Zone-Based Firewall is a stateful firewall used in Cisco IOS devices. ZBFW is the successor of the legacy IOS firewall or the Context-Based Access Control (CBAC) feature.

Cisco ASA and FTD devices are considered dedicated firewall devices; however, Cisco integrated the firewall functionality into the router, which in fact will make the firewall a cost-effective device. The ZBFW includes features that are not available in CBAC (including the assignment of the router interfaces to different security zones to control IP traffic). Zone-based firewalls can also be implemented in an SD-WAN solution.

In an SD-WAN configuration, zone deployments and configurations include the following components:

- **Source zone:** A group of VPNs where the data traffic flows originate. A VPN can be part of only one Source zone.

- **Destination zone:** A grouping of VPNs where the data traffic flows terminate. A VPN can be part of only one Destination zone.

- **Firewall policy:** A security policy, similar to a localized security policy, that defines the conditions that the data traffic flow from the source zone must match to allow the flow to continue to the destination zone. Firewall policies can match IP prefixes, IP ports, the protocols TCP, UDP, and ICMP, and applications. Matching flows for prefixes, ports, and protocols can be accepted or dropped, and the packet headers can be logged. Nonmatching flows are dropped by default. Matching applications can only be denied.

- **Zone pair:** A container that associates a source zone with a destination zone and applies a firewall policy to the traffic that flows between the two zones.

Matching flows that are accepted can be processed in two different ways:

- **Inspect:** The packet's header can be inspected to determine its source address and port.

- **Pass:** The packet can pass to the destination zone without the packet's header being inspected at all.

In ZBFW deployments a zone must be configured before interfaces can be assigned to the zone. In addition, an interface can be assigned to only one security zone. All traffic to and from a given interface in a device with ZBFW configured is implicitly blocked when the interface is assigned to a zone, except traffic to and from other interfaces in the same zone as well as traffic to any interface on the router (that is, the self-zone). IP traffic is implicitly allowed to flow by default among interfaces that are members of the same zone. When you configure ZBFW, in order to permit traffic to and from a zone member interface, a policy allowing or inspecting traffic must be configured between that zone and any other zone.

The self-zone is the only exception to the default "deny all" policy. All traffic to any router interface is allowed until traffic is explicitly denied. Traffic cannot flow between a zone member interface and any interface that is not a zone member. Pass, inspect, and drop actions can only be applied between two zones.

> **TIP**   Additional information and a set of resources about ZBFW can be obtained from the following GitHub repository: https://github.com/The-Art-of-Hacking/h4cker/blob/master/SCOR/zbfw.md.

Deployment Modes of Network Security Solutions and Architectures That Provide Firewall Capabilities

Cisco ASA and Cisco FTD devices can be configured in different deployment modes. Let's start with the Cisco ASA. The Cisco ASA protects an internal (or protected) network from external threats sourced through (or from) an untrusted interface. Each interface is assigned a name to designate its role on the network. The most secure network is typically labeled as the inside network, whereas the least secure network is designated as the outside network. For semi-trusted networks, you can define them as demilitarized zones (DMZ) or any logical interface name. You must use the interface name to set up the configuration features that are linked to an interface.

The Cisco ASA also uses the concept of assigning security levels to the interfaces. The higher the security level, the more protected the interface. Consequently, the security level is used to reflect the level of trust of this interface with respect to the level of trust of another interface on the Cisco ASA. The security level can be between 0 and 100. Therefore, the safest network is placed behind the interface with a security level of 100, whereas the least protected network is placed behind an interface with a security level of 0. A DMZ interface should be assigned a security level between 0 and 100.

When an interface is configured with a **nameif** command, the Cisco ASA automatically assigns a preconfigured security level. If an interface is configured with the name "inside," the Cisco ASA assigns a security level of 100. For all the other interface names, the Cisco ASA assigns a security level of 0.

Cisco ASA enables you to assign the same security level to more than one interface. If communication is required between the hosts on interfaces at the same security level, use the **same-security-traffic permit inter-interface** global configuration command. Additionally, if an interface is not assigned a security level, it does not respond at the network layer.

The most important parameter under the interface configuration is the assignment of an IP address. This is required if an interface is to be used to pass traffic in a Layer 3 firewall, also known as routed mode. An address can be either statically or dynamically assigned. For a static IP address, configure an IP address and its respective subnet mask.

The Cisco ASA also supports interface address assignment through a Dynamic Host Configuration Protocol (DHCP) server. Assigning an address via DHCP is a preferred method if an ISP dynamically allocates an IP address to the outside interface. You can also inform the Cisco ASA to use the DHCP server's specified default gateway as the default route if the Obtain Default Route Using DHCP option is enabled.

Routed vs. Transparent Firewalls

Traditionally, network firewalls have been deployed to filter traffic passing through them. These firewalls usually examine the upper-layer headers (Layer 3 or above) and the data payload in the packets. The packets are then either allowed or dropped based on the configured access control lists (ACLs). These firewalls, commonly referred as routed firewalls, segregate protected networks from unprotected ones by acting as an extra hop in the network design. They route packets from one IP subnet to another subnet by using the Layer 3 routing table. In most cases, these firewalls translate addresses to protect the original IP addressing scheme used in the network.

If a Cisco ASA is deployed in transparent mode, the IP address is assigned in global configuration mode or on a bridge virtual interface (BVI), depending on the version of code. Assigning an interface address through DHCP is not supported if used with failover.

Routed firewalls do not provide a way to filter packets that traverse from one host to another in the same LAN segment. The Layer 3 firewalls require a new network segment to be created when they are inserted into a network, which requires quite a bit of planning, network downtime, and reconfiguration of network devices. To avoid these issues, stealth or transparent firewalls have been developed to provide LAN-based protection. You can place a transparent firewall between the LAN and the next-hop Layer 3 device (usually a router) without having to readdress the network devices.

By using transparent firewalls (also known as Layer 2 firewalls or stealth firewalls), you can optionally inspect Layer 2 traffic and filter unwanted traffic. Figure 7-16 shows SecretCorp's network running a transparent firewall. SecretCorp.org (aka SecretCorp) wants to inspect all traffic before it hits the default gateway. When the host 192.168.10.2 sends traffic destined to www.cisco.com, the firewall makes sure that the packets are allowed before passing them to the default gateway, 192.168.10.1. In this case, the default gateway router is responsible for translating the 192.168.10.0/27 subnet to 209.165.200.224/27 to reach the Internet.

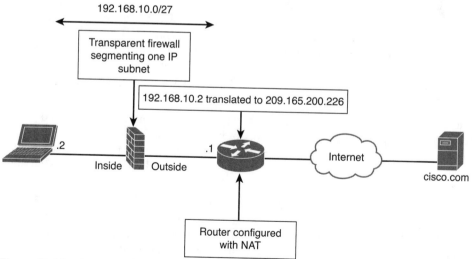

Figure 7-16 *A Firewall Configured in Transparent (Layer 2) Mode*

Security Contexts

Security contexts enable a physical firewall to be partitioned into multiple standalone firewalls. Each standalone firewall acts and behaves as an independent entity with its own configuration, interfaces, security policies, routing table, and administrators. In Cisco ASA, these "virtual" firewalls are known as security contexts.

The following are some sample scenarios in which security contexts are useful in network deployments:

■ You act as a service provider and you want to provide firewall services to customers; however, you do not want to purchase additional physical firewalls for each client.

■ You manage an educational institution and you want to segregate student networks from faculty networks for improved security while using one physical security appliance.

■ You administer a large enterprise with different departmental groups, and each department wants to implement its own security policies.

■ You have overlapping networks in your organization and you want to provide firewall services to all those networks without changing the addressing scheme.

■ You currently manage many physical firewalls and you want to integrate security policies from all firewalls into one physical firewall.

■ You manage a data center environment and you want to provide end-to-end virtualization to reduce operational costs and increase efficiency.

A transparent firewall can coexist with these modes to provide a great deal of flexibility for network deployments. You can set the Cisco ASA firewall mode (either routed or transparent) per context in multimode.

Single-Mode Transparent Firewalls

In a single-mode transparent firewall (SMTF), the Cisco ASA acts as a secured bridge that switches traffic from one interface to another. You do not configure IP addresses on either the inside interface or the outside interface. Rather, you must specify a global IP address that is primarily used for management purposes—Telnet and Secure Shell (SSH). The transparent firewall also uses the management IP address when it needs to source packets such as ARP requests and syslog messages.

This is the simplest form of configuration because it does not require configuration of security contexts, dynamic routing protocols, or interface-specific addresses. The configuration only requires you to define the ACLs, inspection rules, and optionally NAT policies to determine which traffic is allowed. The next section talks about how a packet flows through an SMTF.

The admin context provides connectivity to the network resources such as the AAA or syslog server. It is recommended that you assign the management interface(s) of the Cisco ASA to the admin context. You must assign IP addresses to the allocated interfaces as you would with any other context. The Cisco ASA uses the configured IP addresses to retrieve configurations for other contexts if those configurations are stored on a network share, or to provide remote management of the device through SSH or Telnet. A system administrator with access to the admin context can switch into the other contexts to manage them. The

Cisco ASA also uses the admin context to send the syslog messages that relate to the physical system. Essentially, the admin context is also used by the system context to perform functions that may involve Layer 2 or Layer 3 functionality. This includes file copying and management functionality such as generating syslogs or SNMP traps.

Contexts include the admin and user contexts.

The admin context must be created before you define other contexts. Additionally, it must reside on the local disk. You can designate a new admin context at any time by using the **admin-context** command.

When a Cisco ASA is converted from single mode to multiple mode, the network-related configuration of the single-mode Cisco ASA is saved into the admin context. The security appliance, by default, names this context admin.

The admin context configuration is similar to a user context. Aside from its relationship to the system execution space, it can be used as a regular context. However, using it as a regular context is not recommended because of its system significance.

Each user or customer context acts as a virtual firewall with its own configuration that contains almost all the options that are available in a standalone firewall. A virtual firewall supports a number of features that are available in a standalone firewall, such as the following:

- IPS functionality

- Dynamic routing

- Packet filtering

- Network address translation (NAT)

- Site-to-site VPN

- IPv6 and device management

When packets traverse the Cisco ASA in multiple-context mode, they are classified and forwarded to the correct context. One of the benefits of using virtualization is the sharing of resources, such as the physical interfaces between the security contexts.

This brings up the question of how the Cisco ASA should determine which security context packets should be processed when packets are received on an interface allocated to multiple contexts. Cisco ASA makes this determination through the use of a packet classifier, identifying packets at the ingress interface. The Cisco ASA designates an assortment of packet-classifying criteria to identify the correct security context before forwarding packets. After packets are sent to a security context, they are processed based on the security policies configured in that context.

Cisco ASA uses a number of criteria to classify packets before forwarding traffic to the correct security context. Depending on how you want to implement a security appliance, the packet classifying criteria could be different. You can deploy the security contexts in either a shared interface environment or in a non-shared interface environment.

If all contexts in the Cisco ASA use unique physical or logical sub-interfaces, the packet classification becomes easier because the Cisco ASA labels these packets based on the source interface.

The Cisco ASA enables you to share one or more interfaces between the security contexts. In this deployment model, the Cisco ASA can use either a destination IP address or a unique MAC address on the Cisco ASA to classify the packet to the correct context.

If you share an interface between multiple security contexts, then, depending on the version, the interface may use the same MAC address across all virtual firewalls. With this approach, the ingress packets on the physical interface, regardless of the security context, have the same MAC address. When packets are received by the Cisco ASA for a particular security context, the classifier does not know to which security context to send the packets.

To address this challenge, the classifier uses the packet's destination IP address to identify which security context should receive packets in a shared environment. However, the firewall cannot classify traffic based on the routing table of the contexts because multiple mode allows for overlapping networks, and the routing table might be the same for two contexts. The classifier, in this case, relies strictly on the NAT table of each security context to learn about the subnets located behind each security context.

In a multimode transparent firewall (MMTF), Cisco ASA acts in a similar fashion to how it performs in single mode, with two major exceptions:

- Packets are handled in different contexts. Because each context acts and behaves as an independent entity, you must configure an IP address to the bridge virtual interface (BVI) in each context for administration and management purposes.

- An interface cannot be shared between multiple contexts in this mode.

Surveying the Cisco FTD Deployment Modes

Cisco FTD devices can be configured in routed and transparent mode, just like the Cisco ASA devices.

Figure 7-17 shows a Cisco FTD in routed mode.

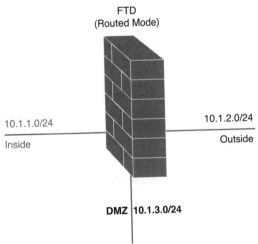

Figure 7-17 *Cisco FTD in Routed Mode*

Figure 7-18 shows a Cisco FTD in transparent (Layer 2) mode.

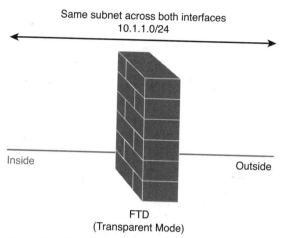

Figure 7-18 *Cisco FTD in Transparent (Layer 2) Mode*

Cisco FTD Interface Modes

Cisco FTD devices can also operate as next-generation firewalls and next-generation intrusion prevention devices in different interfaces. NGFW inherits operational modes from ASA and adds Firepower features. NGIPS operates as standalone Firepower with limited ASA data plane functionality.

You can configure IPS-only passive interfaces, passive ERSPAN interfaces, and inline sets in Cisco FTD devices. When interfaces are configured in IPS-only mode, they bypass many firewall checks and only support IPS (intrusion) security policies. Typically, you deploy IPS-only interfaces if you have a separate firewall protecting these interfaces and do not want the overhead of firewall functions.

NOTE   ERSPAN interfaces are only supported in routed mode.

Inline sets and passive interfaces are only supported on physical interfaces and EtherChannels. Inline sets cannot use redundant interfaces or VLANs. Inline sets and passive interfaces are supported in intra-chassis and inter-chassis clustering.

Bidirectional Forwarding Detection (BFD) echo packets are not allowed through the FTD when using inline sets. If there are two neighbors on either side of the FTD running BFD, then the FTD will drop BFD echo packets because they have the same source and destination IP address.

Figure 7-19 shows the Cisco FTD deployment modes, and Figure 7-20 shows the Cisco FTD interface modes.

Cisco FTD interface modes can be mixed on a single device. Figure 7-21 demonstrates intra-BVI communication. In this mode, no additional routing is needed, as long as the destination IP address is in the same subnet as the BVI interface.

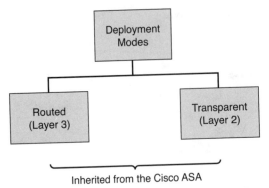

Figure 7-19 *Cisco FTD Deployment Modes*

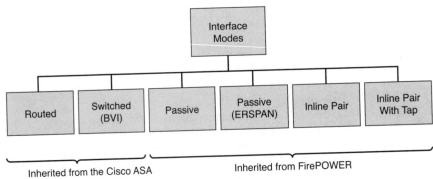

Figure 7-20 *Cisco FTD Interface Modes*

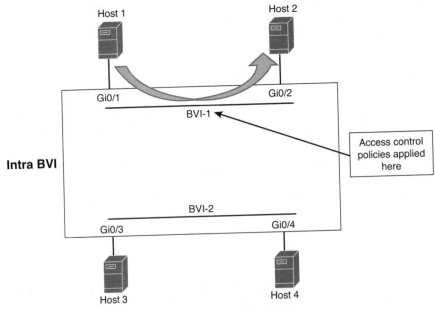

Figure 7-21 *Cisco FTD Intra-BVI Communication*

Figure 7-22 demonstrates inter-BVI communication. In this mode, no additional routing is needed, as long as the destination IP address is in the directly connected subnet as the BVI interfaces.

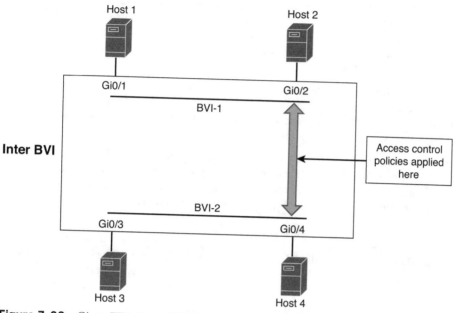

Figure 7-22   *Cisco FTD Inter-BVI Communication*

Figure 7-23 demonstrates traffic between a regular routed (Layer 3) interface and a BVI.

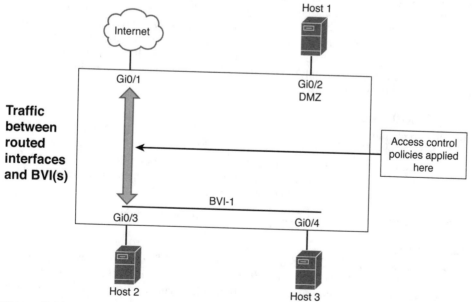

Figure 7-23   *Traffic Between a Regular Routed (Layer 3) Interface and a BVI*

Inline Pair

As in legacy IPS, Cisco NGFW and Cisco NGIPS can operate in two main modes: inline and passive (monitoring) mode. Inline mode is used for prevention. Inline devices can be placed between two assets that communicate or at a point that aggregates traffic from various sources—between a switch and a firewall or between a router and a switch—to block and mitigate threats. Two of the interfaces of the device are used in an inline pair for traffic to enter and exit the device after being inspected. Based on the configured policies, traffic can be dropped, allowed, or reset.

The caveat with *inline interfaces* is that in the event of a software failure or a loss of power, all traffic is dropped. The fail-open capability can be used to allow traffic to bypass the device rules and policies, and consequently, avoid traffic loss. Fail-open should not be used when the security policy doesn't allow traffic to go unaccounted for or uninspected.

Inline mode offers two more modes of operation: routed and switched. In routed mode, the device operates at Layer 3, as a router would. As you learned earlier in this chapter, in switched mode, the device is in transparent mode and doesn't show in the path as a Layer 3 hop. It uses two interfaces in a VLAN pair and bridges them together.

> **TIP**   Most of the legacy ASA features (NAT, routing, ACLs, and so on) are not available for flows going through an inline pair. A few traditional ASA checks can be applied along with full Snort engine checks. You will learn more about Snort later in this chapter.

Figure 7-24 shows a Cisco FTD configured with 10 interfaces. There are two inline sets. One of the inline sets includes three pairs of interfaces, and the other includes two pairs of interfaces. You can configure the Cisco FTD in an inline NGIPS deployment or transparently on a network segment by binding two ports together. This allows the Cisco FTD to be deployed without the need to perform additional configuration in adjacent network devices. Cisco FTD inline interfaces receive all traffic unconditionally. Furthermore, all the traffic received on inline interfaces is retransmitted out of an inline set unless explicitly dropped. When you add multiple inline interface pairs to the same inline interface set, the Cisco FTD categorizes the inbound and outbound traffic as part of the same traffic flow. If you configure passive interfaces (promiscuous mode only), this traffic categorization is achieved by including the interface pairs in the same security zone.

Inline Pair with Tap

You can also configure an inline pair with a "tap," where you have two physical interfaces internally bridged. A few firewall engine checks are applied along with full Snort engine checks to a copy of the actual traffic. Most of the traditional firewall features, such as NAT, routing, and ACLs, are not available for flows going through an inline pair. Inline pair with tap is supported in routed or transparent deployment modes.

Figure 7-25 shows an example of an inline pair with tap configuration.

Passive Mode

Monitoring (passive) mode is the mode where the Cisco NGFW or NGIPS device does not usually prevent attacks. The device uses one interface to silently inspect traffic and identify malicious activity without interrupting traffic flow. It is usually connected to a switch's span

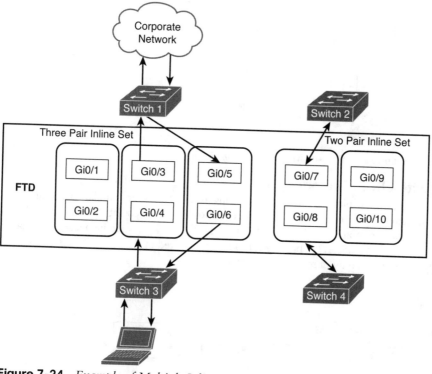

Figure 7-24   *Example of Multiple Inline Interface Sets*

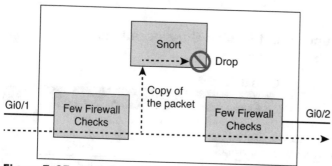

Figure 7-25   *Example of Inline Pair with Tap*

port, a mirrored port, or a network tap interface. Even though in monitoring mode the device doesn't block traffic, there is an option to reset malicious connections, but that should not be considered as a mitigation mechanism as it can't guarantee attack prevention. Passive mode is supported in routed or transparent deployment modes. A few firewall engine checks and full Snort engine checks are applied to a copy of the IP traffic.

Figure 7-26 shows an example of a passive mode deployment.

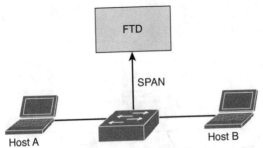

Figure 7-26 *Example of Passive Mode Deployment*

Passive with ERSPAN Mode

You can configure one physical interface operating as a sniffer—very similar to a traditional remote intrusion detection system (IDS). A Generic Routing Encapsulation (GRE) tunnel between the capture point and the Cisco FTD carries the packets to be inspected. An example of a passive with ERSPAN deployment mode is illustrated in Figure 7-27. Passive with ERSPAN is supported only in routed deployment mode.

Figure 7-27 *Example of Passive ERSPAN Deployment Mode*

Additional Cisco FTD Deployment Design Considerations

Table 7-2 provides additional design considerations.

Table 7-2 Deployment Mode Design Considerations

Design Consideration	Recommendation/Comment
Management	Local (FDM) or Central (FMC). FMC recommended for multiple appliances, enhanced visual analysis, central configuration, alerting, and reporting.
Standalone or resilient	Resilient recommended. Has impact on the number of interfaces required.
Link speed(s) / types	Up/downstream speeds, internal/DMZ connections.
Routed or Transparent mode	Routed mode recommended for Edge. Transparent/NGIPS only for customers who already have a third-party firewall.
Number of interfaces	Internal, External, HA, or DMZs.
Traffic profile	Clear or Encrypted, Streaming, Hosting Services.
Application control	Typical requirement at the edge. Opportunity to discuss OpenAppID.
URL filtering	Good for remote/branch locations. Central/HQ may already have provision.

Design Consideration	Recommendation/Comment
Deep inspection	Cisco strength. Industry-leading protection. Can have an impact on performance, so size accordingly.
File and malware protection	Cisco differentiator. Advanced Malware Protection (AMP) can enhance existing AMP solutions or lead to upsell or further expansion.

High Availability and Clustering

You can deploy Cisco next-generation firewalls in high-availability (failover) mode or in a cluster. Let's start with failover. High availability (failover) is supported in Cisco ASA and Cisco FTD devices. When you configure failover, you must have two identical Cisco FTD or Cisco ASA devices connected to each other through a dedicated failover link. You can also configure a state link (to pass stateful firewall information between both devices).

Cisco ASA supports Active-Standby failover, where one unit is the active unit and passes traffic. Cisco ASA also supports Active-Active failover, where both units are passing traffic at the same time. Cisco FTD supports only Active-Standby failover. Figure 7-28 illustrates the concept of Active-Standby failover.

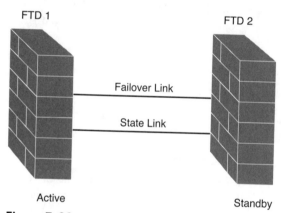

Figure 7-28 *Active-Standby Failover*

In Figure 7-28, two Cisco FTD devices are deployed. The standby FTD (FTD 2) does not actively pass traffic; however, it synchronizes configuration and other state information from the active unit via the state link. When a failover occurs, the active FTD device (FTD 1) fails over to the standby device (FTD 2), which then becomes active.

The failover and the stateful failover links are dedicated connections between the two units. Cisco recommends using the same interface on both devices in a failover link or a stateful link. Configuring two separate interfaces on each of the devices—one for failover and another for stateful link—is also recommended. For instance, if Gigabit Ethernet 1 is used for the failover link in the first Cisco FTD device, then you should use the same interface (Gigabit Ethernet 1) in the second participant device. The same goes for the stateful link. If you have Gigabit Ethernet 2 for the stateful link, you should match the same interface in the second Cisco FTD device (that is, Gigabit Ethernet 2 in FTD 2).

Both of the devices in the Cisco FTD failover pair constantly communicate over the failover link to determine the operating status of each device. In a failover configuration, the health of the active unit (hardware failures, interfaces, software, and so on) is monitored to determine if specific failover conditions are met, and then, subsequently, failover occurs.

The following is the type of information being exchanged over the failover link:

- The firewall state (active or standby)

- Hello messages (keepalives)

- Network link status

- MAC address exchange

- Configuration replication and synchronization

TIP   You can use a physical, redundant, or EtherChannel interface that is not being used as the failover link and another unused interface to be the state link. On the other hand, you cannot specify an interface that is currently configured with a name (outside, inside, dmz-1, and so on). Sub-interfaces are not supported for failover links. Interfaces configured for failover links or state links can only be used for that purpose.

The following are requirements for failover configurations:

- The two participant devices must be configured in the same firewall mode (for example, routed or transparent).

- The two participant devices must be running the same software version.

- You can configure different Cisco FTD devices in groups (or domains) in the Cisco FMC. Devices configured for failover must be in the same domain or group on the Cisco FMC.

- The two participant devices must have the same Network Time Protocol (NTP) configuration (and time must be synchronized).

- Failover participant devices must be fully deployed with no uncommitted changes in the Cisco FMC.

- DHCP or PPPoE must not be configured on any of their interfaces.

Additionally, Cisco FTD devices configured for failover must have the same licenses. When you configure failover, the Cisco FMC releases any unnecessary licenses assigned to the standby device and replaces them with identical licenses assigned to the active and standby device.

TIP You can connect the failover link in two ways. The first way is using a switch with no other device on the same network segment (broadcast domain or VLAN) as the failover interfaces of the Cisco FTD. The second way is using an Ethernet cable to connect both Cisco FTD devices directly (without an external switch). If you do not use a switch between the Cisco FTD devices, when the interface fails, the link goes down on both peers. This may introduce difficulties when troubleshooting because you cannot easily determine which unit has the failed interface and caused the link to come down.

The stateful failover link (also known as the state link) is used to pass connection state information. As a best practice, make sure that the bandwidth of the stateful failover link is at least the same bandwidth as the data interfaces.

TIP For optimum performance, the latency for the state link should be less than 10 milliseconds and no more than 250 milliseconds in order to avoid performance degradation due to retransmission of failover messages.

Clustering

Clustering lets you group multiple Cisco FTD units together as a single logical device. Clustering is supported on the Cisco Firepower 9300 and the Cisco Firepower 4100 series. Figure 7-29 illustrates a cluster of four FTD devices connected to two switches, providing all the convenience of "a single logical entity" (management and integration into a network) while achieving the increased throughput and redundancy of multiple devices.

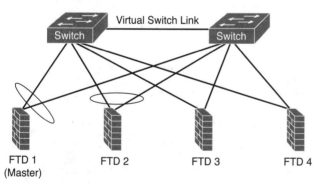

Figure 7-29 *FTD Cluster Example*

The cluster in Figure 7-29 consists of multiple devices acting as a single logical unit. When you deploy a cluster, the FTD devices create a cluster-control link (by default, port-channel 48) for unit-to-unit communication. You can also configure intra-chassis clustering in the Cisco Firepower 9300. The intra-chassis cluster link utilizes the Firepower 9300 backplane for cluster communications. For inter-chassis clustering, you need to manually assign one or more physical interfaces in an EtherChannel configuration for communications between chassis. When you deploy the cluster, the Firepower chassis supervisor pushes a minimal bootstrap configuration to each unit that includes the cluster name and cluster control link interface settings.

NOTE   The Cisco FTD cluster assigns data interfaces to the cluster as spanned interfaces.

In the example illustrated in Figure 7-29, the FTD cluster members work together sharing the security policy and traffic flows. FTD 1 is the master unit. The master unit in a cluster configuration is determined automatically, and all other cluster members are considered "slave units." The master is either the first unit joining the cluster or based on a configured priority. You must perform all configuration on the master unit only; the configuration is then replicated to the other units. A new master is elected only upon a departure of the existing one. The master unit handles all management and centralized functions.

The Cisco FTD cluster automatically creates the cluster control link using the port-channel 48 interface. Both data and control traffic are sent over the cluster control link. The cluster control link should be sized to match the expected throughput of each cluster participant (each Cisco FTD device). This is done to make sure that the cluster-control link can handle the worst-case scenarios of data throughput. The cluster control link traffic includes state update and forwarded packets. The amount of traffic at any given time on the cluster control link varies. The amount of forwarded traffic depends on the load-balancing efficacy or whether there is a lot of traffic for centralized features.

TIP   All cluster participant units should be connected to a single management network (separate from the cluster control link). When enabling clustering, you must assign a management-type interface to the cluster. This interface will allow you to connect directly to each Cisco FTD device by connecting to a logical interface separate from the other interfaces on the device.

Figure 7-30 demonstrates the Cisco FTD cluster unit state transition.

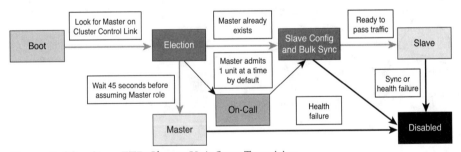

Figure 7-30   *Cisco FTD Cluster Unit State Transition*

The following are the steps illustrated in Figure 7-30:

1.   After the unit boots, it looks for a master over the cluster control link.

2.   If there are no master units, it waits 45 seconds before assuming the master role.
     If a master already exists, it becomes a slave and the configuration is synchronized.

3.   If there is a synchronization or health issue, the unit is considered disabled in the cluster.

> **TIP** You can obtain additional detailed information and tips about clustering at the following GitHub repository: https://h4cker.org/scor/clustering.html.

Implementing Access Control

Access control is the crossroads for all traffic traversing the firewall. In the Cisco ASA, access control is done by configuring access control lists (ACLs). The Cisco FTD supports different access control policies that extend beyond traditional ACLs. The following sections cover how to implement access control lists in Cisco ASA and how access control is done in Cisco FTD deployments.

Implementing Access Control Lists in Cisco ASA

An ACL is a collection of security rules or policies that allows or denies packets after looking at the packet headers and other attributes. Each permit or deny statement in the ACL is referred to as an access control entry (ACE). These ACEs classify packets by inspecting Layer 2 through Layer 7 headers for a number of parameters, including the following:

- Layer 2 protocol information such as EtherTypes

- Layer 3 protocol information such as ICMP, TCP, or UDP

- Layer 3 header information such as source and destination IP addresses

- Layer 4 header information such as source and destination TCP or UDP ports

- Layer 7 information such as application and system service calls

After an ACL has been properly configured, apply it to an interface to filter traffic. The Cisco ASA filters packets in both the inbound and outbound direction on an interface. When an inbound ACL is applied to an interface, the Cisco ASA analyzes packets against the ACEs after receiving them. If a packet is permitted by the ACL, the firewall continues to process the packet and eventually passes the packet out the egress interface.

ACLs can also be configured outbound. When an outbound ACL is configured, the Cisco ASA inspects the packets that are exiting out of the specific interface, but not on incoming packets.

ACLs include a five-tuple:

- Source IP address (or subnet)

- Source port

- Destination IP address (or subnet)

- Destination port

- Protocol

ACLs are also identified by either a number or a name. Figure 7-31 shows an example of an ACL allowing HTTPS (TCP port 443) traffic from 10.2.2.2 to 10.1.1.2.

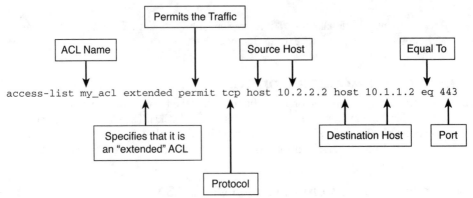

Figure 7-31 *An Access Control List Example in the Cisco ASA*

If a packet is denied by the ACL, the Cisco ASA discards the packet and generates a syslog message indicating that such an event has occurred. In Figure 7-32, the Cisco ASA administrator has configured an ACL that permits only HTTP traffic destined for 10.1.1.2 and has applied the ACL inbound on the outside interface. All other traffic is dropped at the outside interface by the Cisco ASA.

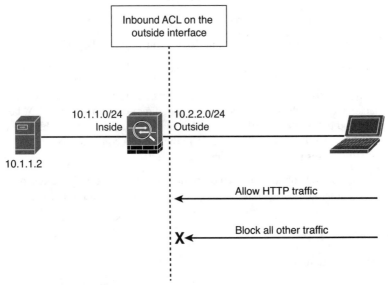

Figure 7-32 *Inbound ACL Example*

If an outbound ACL is applied on an interface, the Cisco ASA processes the packets by sending them through the different processes (NAT, QoS, and VPN) and then applies the configured ACEs before transmitting the packets out on the wire. The Cisco ASA transmits the packets only if they are allowed to go out by the outbound ACL on that interface. If the packets are denied by any one of the ACEs, the Cisco ASA discards the packets and generates a syslog message indicating that such an event has occurred. In Figure 7-33, the Cisco ASA administrator has configured an ACL that permits only HTTP traffic destined for 10.1.1.2 and has applied the ACL outbound on the inside interface. All other traffic gets dropped at the interface by the Cisco ASA.

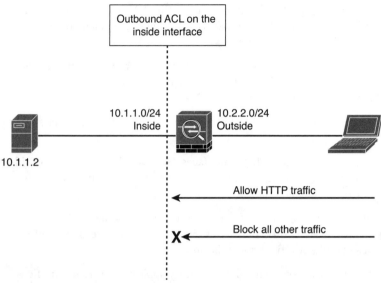

Figure 7-33   *Outbound ACL Example*

The Cisco ASA also allows you to define a global ACL. All inbound traffic, regardless of the interface, is subject to inspection by a global ACL. If both an inbound interface ACL and a global ACL are configured, the Cisco ASA applies the interface ACL first, then the global ACL, and eventually the implicit deny in case there is no specific match.

So, should you use an interface ACL or a global ACL? An interface-specific ACL gives you more control to filter traffic if you know which interface it enters or leaves. You can apply an interface ACL in both inbound and outbound directions. However, if you have multiple interfaces connected to a network and you are not sure where packets may enter into the security appliance, consider applying a global ACL.

The following are some important characteristics of an ACL:

- When a new ACE is added to an existing ACL, it is appended to the end of the ACL, unless a specific line number is specified.

- When a packet enters the security appliance, the ACEs are evaluated in sequential order. Hence, the order of an ACE is critical. For example, if you have an ACE that allows all IP traffic to pass through, and then you create another ACE to block all IP traffic, the packets are never evaluated against the second ACE because all packets match the first ACE.

- There is an implicit deny at the end of all ACLs. If a packet is not matched against a configured ACE, it is dropped and a syslog with message ID of 106023 is generated.

- By default, you do not need to define an ACE to permit traffic from a high-security-level interface to a low-security-level interface. However, if you want to restrict traffic flows from a high-security-level interface destined to a low-security-level interface, you can define an ACL. If you configure an ACL for traffic originating from a high-security-level interface to a low-security-level interface, it disables the implicit permit from that interface. All traffic is now subject to the entries defined in that ACL.

■ An ACL must explicitly permit traffic traversing the Cisco ASA from a lower- to a higher-security-level interface of the firewall. The ACL must be applied to the lower-security-level interface or globally.

■ The ACLs (extended or IPv6) must be applied to an interface to filter traffic that is passing through the security appliance. Beginning with Cisco ASA Software version 9.0(1), you can use a single ACL to filter both IPv4 and IPv6 traffic.

■ You can bind one extended ACL and one EtherType ACL in each direction of an interface at the same time.

■ You can apply the same ACL to multiple interfaces. However, doing so is not considered to be a good security practice because correlating ACL hit counts to a specific interface's traffic would be impossible.

■ You can use ACLs to control both traffic through the Cisco ASA and traffic to the security appliance. The ACLs that control traffic to the appliance are applied differently than ACLs that filter traffic through the appliance.

■ When TCP or UDP traffic flows through the security appliance, the return traffic is automatically allowed to pass through because the connections are bidirectional.

■ Other protocols such as ICMP are considered unidirectional connections, so you need to allow ACL entries in both directions. However, when you enable the ICMP inspection engine, the inspection engine keeps track of the ICMP messages and then allows replies (such as ping packets).

The Cisco ASA supports four different types of ACLs to provide a flexible and scalable solution to filter unauthorized packets into the network:

■ Standard ACLs

■ Extended ACLs (Figure 7-31 illustrated an example of an extended ACL)

■ EtherType ACLs

■ Webtype ACLs

Standard ACLs are used to identify packets based on their destination IP addresses. These ACLs can be used in scenarios such as split tunneling for the remote-access VPN tunnels and route redistribution within route maps. These ACLs, however, cannot be applied to an interface for filtering traffic. A standard ACL can be used only if the Cisco ASA is running in routed mode, by acting as an extra Layer 3 hop in the network.

Extended ACLs, the most commonly deployed ACLs, classify packets based on the following attributes:

■ Source and destination IP addresses

■ Layer 3 protocols

■ Source and/or destination TCP and UDP ports

- Destination ICMP type for ICMP packets

- User identity attributes such as Active Directory (AD) username or group membership

An extended ACL can be used for interface packet filtering, QoS packet classification, packet identification for NAT and VPN encryption, and a number of other features. These ACLs can be set up on the Cisco ASA in both routed mode and transparent firewall mode.

EtherType ACLs are used to filter IP- and non-IP-based traffic by checking the Ethernet type code field in the Layer 2 header. IP-based traffic uses an Ethernet type code value of 0x800, whereas Novell IPX uses 0x8137 or 0x8138, depending on the Netware version. An EtherType ACL can be configured only if the Cisco ASA is running in transparent mode.

Like all ACLs, the EtherType ACL has an implicit deny at the end of it. However, this implicit deny does not affect the IP traffic passing through the security appliance. As a result, you can apply both EtherType and extended ACLs to each direction of an interface. If you configure an explicit deny at the end of an EtherType ACL, it blocks IP traffic even if an extended ACL allows those packets to pass through.

A Webtype ACL allows Cisco ASA administrators to restrict traffic coming through the SSL VPN tunnels. In cases where a Webtype ACL is defined but there is no match for a packet, the default behavior is to drop the packet because of the implicit deny. On the other hand, if no ACL is defined, the Cisco ASA allows traffic to pass through it.

Access control lists on a Cisco ASA can be used to filter out not only packets passing through the appliance but also packets destined to the appliance.

Through-the-box traffic filtering refers to traffic that is passing through the security appliances from one interface to another interface. As mentioned earlier, an ACL is a collection of access control entries. When new connections are being established through the security appliance, they are subjected to the ACL configured on the interfaces that would pass traffic. The packets are either allowed or dropped based on the configured action on each ACE. An ACE can be as simple as permitting all IP traffic from one network to another, or as complicated as permitting or denying traffic originating from a unique source IP address on a particular port destined to a specific port on the destination address in a specific time period.

The Implementing and Operating Cisco Security Core Technologies (SCOR 350-701) exam does not cover detailed configuration and troubleshooting. The CCNP Security concentration exams and the CCIE Security exam focus on configuration and troubleshooting. However, you may find some questions with high-level configuration scenarios in the SCOR exam. The following are a few examples of ACL configurations to enhance the learning.

The topology illustrated in Figure 7-34 is used in the next examples.

Cisco ASA protects the protected network from external threats. Each interface is assigned a name to designate its role on the network. The most secure network is typically labeled as the inside network, whereas the least secure network is designated as the outside network. For semi-trusted networks, you can define them as demilitarized zones (DMZ) or any logical interface name. You must use the interface name to set up the configuration features that are linked to an interface.

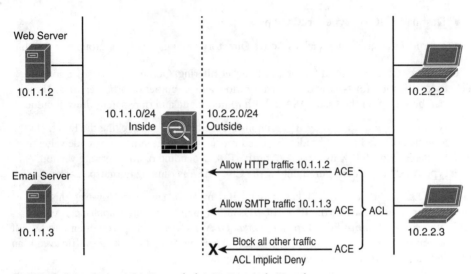

Figure 7-34 *Cisco ASA Extended ACL Example Topology*

As discussed earlier, the Cisco ASA also uses the concept of assigning security levels to the interfaces. The higher the security level, the more protected the interface. Consequently, the security level is used to reflect the level of trust of this interface with respect to the level of trust of another interface on the Cisco ASA. The security level can be between 0 and 100. Therefore, the safest network is placed behind the interface with a security level of 100, whereas the least protected network is placed behind an interface with a security level of 0. A DMZ interface should be assigned a security level between 0 and 100.

> **TIP** Cisco ASA enables you to assign the same security level to more than one interface. If communication is required between the hosts on interfaces at the same security level, use the **same-security-traffic permit inter-interface** global configuration command. Additionally, if an interface is not assigned a security level, it does not respond at the network layer.

When an interface is configured with a **nameif** command, the Cisco ASA automatically assigns a preconfigured security level. If an interface is configured with the name "inside," the Cisco ASA assigns a security level of 100. For all the other interface names, the Cisco ASA assigns a security level of 0. This is particularly important when configuring ACLs in the Cisco ASA.

In Example 7-1, an extended ACL called **outside_access_in** is set up with five ACEs. The first two ACEs allow HTTP traffic destined for 10.1.1.2 from the two client machines, whereas the next two ACEs allow SMTP access to 209.165.202.132 from both machines. The last ACE explicitly drops and logs all other IP packets. The ACL is then applied to the **outside** interface in the inbound direction.

You can apply only one extended ACL in each direction of an interface. That means you can apply an inbound and an outbound extended ACL simultaneously on an interface. Similarly, you can apply an extended ACL and an EtherType ACL in the same direction, if running in transparent firewall mode.

Example 7-1 *ACL Example Allowing HTTP and SMTP Traffic*

```
omar-asa# configure terminal
omar-asa(config)# access-list outside_access_in extended permit tcp host
10.2.2.2 host 10.1.1.2 eq http
omar-asa(config)# access-list outside_access_in extended permit tcp host
10.2.2.3 host 10.1.1.2 eq http
omar-asa(config)# access-list outside_access_in extended permit tcp host
10.2.2.2 host 209.165.202.132 eq smtp
omar-asa(config)# access-list outside_access_in extended permit tcp host
10.2.2.3 host 209.165.202.132 eq smtp
omar-asa(config)# access-list outside_access_in extended deny ip any any log
omar-asa(config)# access-group outside_access_in in interface outside
```

The Cisco ASA does not block the return TCP or UDP traffic on the lower-security-level interface if the traffic is originated from a host on the higher-security-level interface, and vice versa. For other connectionless protocols, such as GRE or ESP, you must permit the return traffic in the ACL applied on that interface. For ICMP, you can either allow the return traffic in the ACL or enable ICMP inspection.

Cisco ASA Application Inspection

The Cisco ASA mechanisms that are used for stateful application inspection enforce the secure use of applications and services in your network. The stateful inspection engine keeps information about each connection traversing the security appliance's interfaces and makes sure they are valid. Stateful application inspection examines not only the packet header but also the contents of the packet up through the application layer.

Several applications require special handling of data packets when they pass through the Layer 3 devices. These include applications and protocols that embed IP addressing information in the data payload of the packet as well as open secondary channels on dynamically assigned ports. The Cisco ASA application inspection mechanisms recognize the embedded addressing information, which allows network address translation (NAT) to work and update any other fields or checksums.

Using application inspection, the Cisco ASA identifies the dynamic port assignments and allows data exchange on these ports during a specific connection.

Cisco ASA provides the Modular Policy Framework (MPF) to provide application security or perform Quality of Service (QoS) functions. The MPF offers a consistent and flexible way to configure the Cisco ASA application inspection and other features in a manner similar to that used for the Cisco IOS Software Modular QoS CLI.

As a general rule, the provisioning of inspection policies requires the following steps:

Step 1.   Configure traffic classes to identify interesting traffic.

Step 2.   Associate actions to each traffic class to create service policies.

Step 3.   Activate the service policies on an interface or globally.

7

You can complete these policy provisioning steps by using these three main commands of the MPF:

- **class-map:** Classifies the traffic to be inspected. Various types of match criteria in a class map can be used to classify traffic. The primary criterion is the use of an access control list (ACL). Example 7-2 demonstrates this.

- **policy-map:** Configures security or QoS policies. A policy consists of a **class** command and its associated actions. Additionally, a policy map can contain multiple policies.

- **service-policy:** Activates a policy map globally (on all interfaces) or on a targeted interface.

Example 7-2   *Cisco ASA Application Inspection Example*

```
omar-asa(config)# access-list tftptraffic permit udp any any eq 69
omar-asa(config)# class-map TFTPclass
omar-asa(config-cmap)# match access-list tftptraffic
omar-asa(config-cmap)# exit
omar-asa(config)# policy-map tftppolicy
omar-asa(config-pmap)# class TFTPclass
omar-asa(config-pmap-c)# inspect tftp
omar-asa(config-pmap-c)# exit
omar-asa(config-pmap)# exit
omar-asa(config)# service-policy tftppolicy global
```

In Example 7-2, an ACL named **tftptraffic** is configured to identify all TFTP traffic. This ACL is then used as a match criterion in a class map named **TFTPclass**.

A policy map named **tftppolicy** is configured that has the class map **TFTPclass** mapped to it. The policy map is set up to inspect all TFTP traffic from the UDP packets that are being classified in the class map. Finally, the service policy is applied globally. The Cisco ASA contains a default class map named **inspection_default** and a policy map named **global_policy**.

To-the-Box Traffic Filtering in the Cisco ASA

To-the-box traffic filtering, also known as management access rules, applies to traffic that terminates on the Cisco ASA. Some management-specific protocols such as SSH and Telnet have their own control list, where you can specify which hosts and networks are allowed to connect to the security appliance. However, they do not provide full protection from other types of traffic, such as IPsec. Before you implement management access rules, consult these guidelines:

- Traffic filtering requires you to configure an ACL and then apply the ACL to the appropriate interface, using the **control-plane** keyword at the end.

- The ACL cannot be applied to an interface designated as a **management-only** interface.

- Management-specific protocols provide their own control-plane protection and have higher precedence than a to-the-box traffic-filtering ACL. For example, if you allow a host to establish an SSH session (by defining its IP address in the **ssh** command) and then block its IP address in the management access rule, the host can establish an SSH session to the security appliance.

If you want to use the CLI to define a policy, use the **control-plane** keyword at the end of the **access-group** command. This declares that it is a management access rule to block traffic destined to the security appliance. In Example 7-3, a control-plane ACL called **outside_access_in_1** is configured to block all IP traffic destined to the security appliance. This ACL is then applied to the outside interface in the inbound direction using the **control-plane** keyword.

Example 7-3 *Cisco ASA To-the-Box Traffic Filtering*

```
omar-asa# configure terminal
omar-asa(config)# access-list outside_access_in_1 remark Blocking all
Management Traffic on the Outside Interface
omar-asa(config)# access-list outside_access_in_1 extended deny ip any any
omar-asa(config)# access-group outside_access_in_1 in interface outside
 control-plane
```

Object Grouping and Other ACL Features

The Cisco ASA provides many advanced packet-filtering features to suit any network environments. These features include the following:

- Object grouping

- Standard ACLs

- Time-based ACLs

- Downloadable ACLs

Object grouping is a way to group similar items together to reduce the number of ACEs. Without object grouping, the configuration on the Cisco ASA may contain thousands of lines of ACEs, which becomes hard to manage. The Cisco ASA follows the multiplication factor rule when ACEs are defined. For example, if three outside hosts need to access two internal servers running HTTP and SMTP services, the Cisco ASA will have 12 host-based ACEs, calculated as follows:

Number of ACEs = (2 internal servers) * (3 outside hosts) * (2 services) = 12

If you use object grouping, you can reduce the number of ACEs to just a single entry. Object grouping clusters network objects such as internal servers into one group and clusters outside hosts into another. The Cisco ASA can also combine both TCP services into a service-based object group. All these groups can be linked to each other in one ACE.

Although the number of viewable ACEs is reduced when object groups are used, the actual number of ACEs is not. Use the **show access-list** command to display the expanded ACEs in the ACL. Always be mindful when using large object groups in ACLs that seemingly minor changes may multiply out to add thousands of ACEs.

The Cisco ASA supports nesting an object group into another one. This hierarchical grouping can further reduce the number of configured ACEs in Cisco ASA.

The Cisco ASA supports six different types of objects that can group similar items or services:

- Protocol

- Network

- Service

- Local user group

- Security group

- ICMP type

Standard ACLs

As mentioned earlier in this chapter, standard ACLs are used when the source network in the traffic is not important. These ACLs are used by processes, such as OSPF and VPN tunnels, to identify traffic based on the destination IP addresses.

You can define standard ACLs by using the **access-list** command and the **standard** keyword after the ACL name. In Example 7-4, the Cisco ASA identifies traffic destined for host 192.168.88.10 and network 192.168.20.0/24 and discards all other traffic explicitly. The ACL name is **Dest-Net**.

Example 7-4   *Cisco ASA Standard ACL Example*

```
omar-asa(config)# access-list Dest-Net standard permit host 192.168.88.10
omar-asa(config)# access-list Dest-Net standard permit 192.168.20.0
255.255.255.0
omar-asa(config)# access-list Dest_Net standard deny any
```

After a standard ACL is defined, it must be applied to a process for implementation. In Example 7-5, a route map called OSPFMAP is set up to use the standard ACL configured in the previous example.

Example 7-5   *Route Map Using a Standard ACL*

```
omar-asa(config)# route-map OSPFMAP permit 10
omar-asa(config-route-map)# match ip address Dest_Net
```

Time-Based ACLs

The Cisco ASA can also enforce ACLs that are time based. These rules, commonly referred to as time-based ACLs, prevent users from accessing the network services when the traffic arrives outside the preconfigured time intervals. The Cisco ASA relies on the system's clock when evaluating time-based ACLs. Consequently, it is important to ensure that the system clock is accurate, and thus the use of Network Time Protocol (NTP) is highly recommended. You can use time-based ACLs with the extended, IPv6, and Webtype ACLs.

Time-based ACLs apply only to new connections; therefore, existing connections are not affected when time-based ACLs become active. The Cisco ASA enables you to specify two different types of time restrictions:

- **Absolute:** Using the **absolute** function, you can specify the values based on a start and/or an end time. This function is useful in cases where a company hires consultants for a period of time and wants to restrict access when they leave. In this case, you can set an absolute time and specify the start time and the end time. After the time period expires, the consultants cannot pass traffic through the security appliance. The start and end times are optional. If no start time is provided, the Cisco ASA assumes that the ACL needs to be applied right away. If no end time is configured, the Cisco ASA applies the ACL indefinitely. Additionally, only one instance of the **absolute** parameter is allowed to be set up in a given time range.

■ **Periodic:** Using the periodic function, you can specify the values based on the recurring events. The Cisco ASA provides many easy-to-configure parameters to suit an environment. Time-based ACLs using this option are useful when an enterprise wants to allow user access during the normal business hours on the weekdays and wants to deny access over the weekends. Cisco ASA enables you to configure multiple instances of the **periodic** parameter.

If both **absolute** and **periodic** parameters are configured in a time range, the **absolute** time parameters are evaluated first, before the **periodic** time value.

In **periodic** time ranges, you can configure a day of the week such as **Monday**, specify the keyword **weekdays** for a workweek from Monday to Friday, or specify the keyword **weekend** for Saturday and Sunday. The Cisco ASA can further the restrictions on the users by setting the optional 24-hour format hh:mm time specifications.

ICMP Filtering in the Cisco ASA

If you deploy interface ACLs to block all ICMP traffic, the security appliance, by default, does not restrict the ICMP traffic that is destined to its own interface. Depending on an organization's security policy, an ICMP policy can be defined on the Cisco ASA to block or restrict the ICMP traffic that terminates at a security appliance's interface. The Cisco ASA enables you to filter ICMP traffic to their interfaces by either deploying the control plane ACLs or defining the ICMP policy.

You can define an ICMP policy by using the **icmp** command, followed by an action (permit or deny), source network, ICMP type, and the interface where you want to apply this policy. As shown in Example 7-6, an ICMP policy is applied to the outside interface to block the ICMP echo packets sourced from any IP address. The second **icmp** statement permits all other ICMP types that are destined for the Cisco ASA's IP address.

Example 7-6 *Cisco ASA ICMP Filtering Example*

```
omar-asa(config)# icmp deny any echo outside
omar-asa (config)# icmp permit any outside
```

The ICMP commands are processed in sequential order, with an implicit deny at the end of the list. If an ICMP packet is not matched against a specific entry in the ICMP list, the packet is dropped. If there is no ICMP list defined, all ICMP packets are allowed to be terminated on the security appliance. In other words, by default, you are able to ping the ASA's interfaces. You can also use the control-plane ACLs to manage ICMP traffic that is destined to the security appliance. However, ICMP traffic filtering, discussed in this section, takes precedence over the control-plane traffic filtering mechanism.

Network Address Translation in Cisco ASA

Another core security feature of a firewall is its capability to mask the network address on the trusted side from the untrusted networks. This technique, commonly referred to as address translation, allows an organization to hide the internal addressing scheme from the outside by displaying a different IP address space. Address translation is useful in the following network deployments:

■ You use a private addressing scheme internally and want to assign global routable addresses to those hosts.

■ You change to a service provider that requires you to modify your addressing scheme. Rather than redesigning the entire IP infrastructure, you implement translation on the border appliance.

■ For security reasons, you do not want to advertise the internal addressing scheme to the outside hosts.

■ You have multiple internal networks that require Internet connectivity through the security appliance, but only one global address (or a few) is available for translation.

■ You have overlapping networks in your organization and you want to provide connectivity between the two without modifying the existing addressing scheme.

The Cisco ASA supports two types of address translation: namely, network address translation (NAT) and port address translation (PAT).

NAT defines a one-to-one address mapping when a packet passes through the Cisco ASA and matches criteria for translation. The Cisco ASA either assigns a static IP address (static NAT) or allocates an address from a pool of addresses (dynamic NAT).

Cisco ASA can translate an internal address to a global address when packets are destined for the public network. With this method, also known as inside NAT, the Cisco ASA converts the global address of the return traffic to the original internal address. Inside NAT is used when traffic originates from a higher-security-level interface, such as the inside interface, and is destined for a lower-security-level interface, such as the outside interface. In Figure 7-35, a host on the internal network, 192.168.10.10, sends traffic to a host on the outside network, 209.165.201.1. The Cisco ASA converts the source IP address to 209.165.200.226 while keeping the destination IP address intact. When the web server responds to the global IP address, 209.165.200.226, the Cisco ASA reverts the global IP address to the original internal IP address of 192.168.10.10.

Optionally, the hosts on the lower-security-level interface can be translated when traffic is destined for a host on the higher-security-level interface. This method, known as outside NAT, is useful when you want a host on the outside network to appear as one of the internal IP addresses. In Figure 7-36, a host on the outside network, 209.165.201.1, sends traffic to a host on the inside network, 192.168.10.10, by using its global IP address as the destination address. Cisco ASA converts the source IP address to 192.168.10.100 while changing the destination IP address to 192.168.10.10. Because both the source and destination IP addresses are changing, this is also known as bidirectional NAT.

TIP   If the packets are denied by the interface ACLs, the Cisco ASA does not build the corresponding address translation table entry.

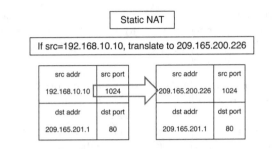

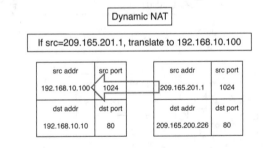

Figure 7-35 *Static NAT Example*

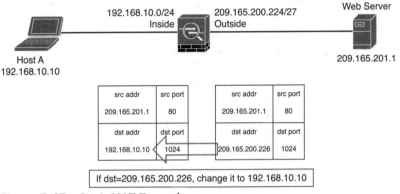

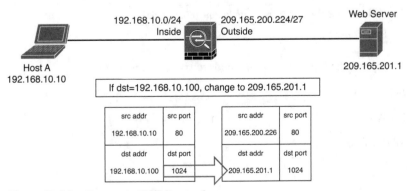

Figure 7-36 *Dynamic NAT Example*

7

Port address translation (PAT) defines a many-to-one address mapping when a packet passes through the Cisco ASA and matches criteria for translation. The Cisco ASA creates the translation table by looking at the Layer 4 information in the header to distinguish between the inside hosts using the same global IP address.

Figure 7-37 illustrates an appliance set up for PAT for the inside network of 192.168.10.0/24. However, only one global address is available for translation. If two inside hosts, 192.168.10.10 and 192.168.10.20, require connectivity to an outside host, 209.165.201.1, the Cisco ASA builds the translation table by evaluating the Layer 4 header information. In this case, because both inside hosts have the same source port number, the Cisco ASA assigns a random source port number to keep both entries unique from each other. This way, when the response from the web server returns to the security appliance, the Cisco ASA knows which inside host to forward the packets.

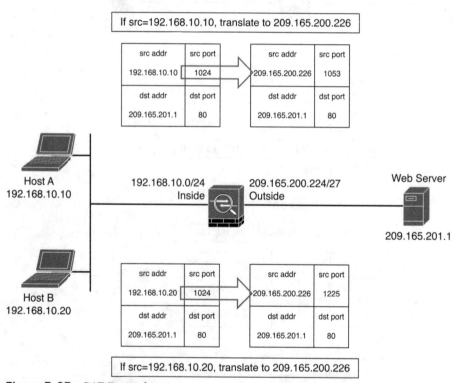

Figure 7-37 *PAT Example*

The Cisco ASA supports the following four methods to translate an address:

- Static NAT/PAT
- Dynamic NAT/PAT
- Policy NAT/PAT
- Identity NAT

Static NAT defines a fixed translation of an inside host or subnet address to a global routable address or subnet. The Cisco ASA uses the one-to-one methodology by assigning one global IP address to one inside IP address. Thus, if 100 hosts residing on the inside network require address translation, the Cisco ASA should be configured for 100 global IP addresses. Additionally, the inside hosts are assigned the same IP address whenever the Cisco ASA translates the packets going through it. This is a recommended solution in scenarios in which an organization provides services such as email, web, DNS, and FTP for outside users. Using static NAT, the servers use the same global IP address for all the inbound and outbound connections.

Static PAT, also known as port redirection, is useful when the Cisco ASA needs to statically map multiple inside servers to one global IP address. Port redirection is applied to traffic when it passes through the Cisco ASA from a lower-security-level interface to a higher-security-level interface. The outside hosts connect to the global IP address on a specific TCP or UDP port, which the Cisco ASA redirects to the appropriate internal server, as shown in Figure 7-38.

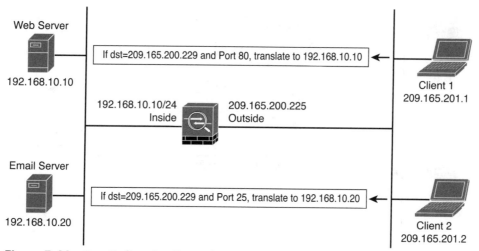

Figure 7-38 *Port Redirection Example*

The Cisco ASA redirects traffic destined for 209.165.200.229 on TCP port 80 to 192.168.10.10. Similarly, any traffic destined for 209.165.200.229 on TCP port 25 is redirected to 192.168.10.20. The Cisco ASA allows the use of either a dedicated IP address or the global interface's IP address for port redirection. When port redirection is set up to use the public interface's IP address, the Cisco ASA uses the same address for the following:

■ Address translation for the traffic traversing through the security appliance

■ Traffic destined for the security appliance

Dynamic NAT assigns a random IP address from a preconfigured pool of global IP addresses. The Cisco ASA uses a one-to-one methodology by allocating one global IP address to an inside IP address. Hence, if 100 hosts reside on the inside network, then you have at least 100 addresses in the pool of addresses. This is a recommended solution in scenarios in which an organization uses protocols that don't contain Layer 4 information, such as Generic Routing Encapsulation (GRE), Reliable Datagram Protocol (RDP), and Data Delivery Protocol (DDP). After the Cisco ASA has built a dynamic NAT entry for an inside host, any outside machine can connect to the assigned translated address, assuming that the Cisco ASA allows the inbound connection.

With dynamic PAT, the Cisco ASA builds the address translation table by looking at the Layer 3 and Layer 4 header information. It is the most commonly deployed scenario because multiple inside machines can get outside connectivity through one global IP address. In dynamic PAT, the Cisco ASA uses the source IP addresses, the source ports, and the IP protocol information (TCP or UDP) to translate an inside host. As with static PAT, you have the option of using either a dedicated public address or the IP address of an interface for translations. As shown in Figure 7-39, two inside machines are accessing an external web server, using the IP address of the outside interface.

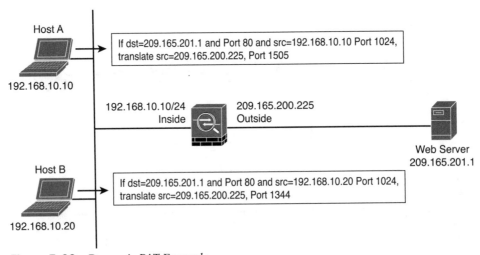

Figure 7-39 *Dynamic PAT Example*

The Cisco ASA supports up to 65,535 PAT translations using a single address. You can also use extended PAT, which allows you to have 65,535 PAT translations per service as opposed to per address.

Policy NAT/PAT translates the IP address of the packets passing through the Cisco ASA only if those packets match a defined criterion or policy. You define the policy by identifying interesting traffic through the use of ACLs or by using manual NAT by identifying specific traffic. If traffic matches the defined entries in the ACL, then the original source or destination address can be translated to a different address.

In many scenarios, you want to bypass address translation so that the security appliance does not change the source or the destination address. You may want to bypass address translation, also known as identity NAT, if you already have address translation defined for the inside network so that hosts can get Internet connectivity. However, you do not want to change their addresses if they send traffic to a specific host or network.

TIP Address translation not only masquerades the original IP address; it also provides protection against TCP connection hijacking for hosts with weak SYN implementation. When a packet enters the higher-security-level interface and is destined for a lower-security-level interface during the TCP three-way handshake, the security appliance randomizes the original sequence numbers used by the hosts. When the host 192.168.10.10 sends a TCP SYN HTTP packet to host 209.165.201.1 with an initial sequence number (ISN) of 12345678, the Cisco ASA changes the source IP address to 209.165.200.226 and also modifies the ISN to a randomly generated value of 95632547. In some deployment scenarios, such as Border Gateway Protocol (BGP) peering with MD5 authentication, it is recommended to turn off the randomization of TCP packets. When two routers establish BGP peering with each other, the TCP header and data payload are 128-bit hashed, using the BGP password. When the sequence number is changed, the peering router fails to authenticate the packets because of the mismatched hash. For more information about BGP MD5 authentication, consult RFC 2385.

TCP Intercept is a security feature that protects the TCP-based servers from TCP SYN attacks by filtering the bogus denial-of-service (DoS) traffic. The TCP Intercept feature is configured using the MPF to set connection limits. The security appliance also protects network resources from an unexpected increase in the number of connections by setting maximum limits. This is applicable for both TCP- and UDP-based connections.

Cisco ASA Auto NAT

The Cisco ASA allows you to define your NAT policies within a network object; thus it is also known as Network Object NAT. A network object can be a host, a range of IP addresses, a subnet, or a network. This mode is typically used when you need to translate the source address of a new object or a predefined object in the configuration. NAT configuration is added with the object definition. The key point is that it is mainly used when you need to translate the source address of an object. This is a popular choice of address translation because most policies require translation of the source address.

You can also enable the Cisco ASA to perform NAT automatically with the "Auto NAT" feature. Auto NAT is helpful when you want to translate the source address of an object regardless of the destination address. In this mode, you define an object and add an address translation policy within the object definition.

NOTE Manual NAT is required when you need to define policies involving destination addresses, such as policy-based or identity NAT-based address translations. You can define very specific translation policies that hosts need to communicate with each other. This mode can also be used when you need to translate both the source and destination addresses of traffic flowing from one entity to another entity.

Implementing Access Control Policies in the Cisco Firepower Threat Defense

As you learned in previous sections in this chapter, access control policy is the crossroads for all traffic in a firewall. The Cisco FTD provides different ways to configure access control policies. Traditionally, policies tend to proliferate and settings diverge. In addition, the defaults for new policies change over time and inconsistencies lead to problems with detection. At the end of the day, complexity is the enemy of security. It is recommended that you take a consistent approach to make things simple and not over-complicate your access control and intrusion policies.

The Cisco FTD combines the features provided by the traditional Cisco ASA firewall and next-generation Firepower services, including different advanced security technologies, such as network discovery, application control, file control, security intelligence, and a Snort-based intrusion prevention system. When you configure the Cisco FTD to block from an ingress to egress interface, the access control is performed by either the Firewall engine or the Firepower engine. Subsequently, if two hosts experience any connectivity issues while sending traffic through a Cisco FTD device, it is essential to analyze packets from both engines to determine the root cause of the problem. For example, to investigate any registration or communications issues between FTD and the FMC, capturing traffic from the Firepower management interfaces is one of the key troubleshooting steps.

Figure 7-40 provides a high-level overview of how traffic flows through a Cisco FTD. When the Firewall engine receives a packet on an ingress interface, it redirects it to the Firepower engine for inspection and policy checks. The Firepower engine will block the traffic if it is not compliant with the configured security policies or it will forward the packet via the egress interface if it complies with the configured policies.

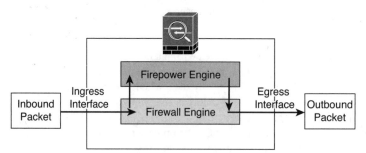

Figure 7-40 *Cisco FTD Flow of Traffic High-Level Overview*

The Cisco FTD supports a wide range of block actions including the following:

- simple blocking
- blocking with reset
- interactive blocking
- interactive blocking with reset

NOTE A block action cannot drop any suspicious packet if the interfaces are not set up properly.

You can choose any interface mode to apply a policy, regardless of the underlying deployment mode (routed or transparent). On the other hand, the capability of an interface mode defines whether the Cisco FTD is able to block any suspicious traffic it detects.

Let's take a look at how to configure an access control policy to block any Telnet (TCP port 23) traffic traversing the firewall.

To add a new access control policy, navigate to the **Policies > Access Control > Access Control** page in the Cisco FMC. Click the **New Policy** button to add the new policy. The screen shown in Figure 7-41 is displayed.

Figure 7-41   *Creating an Access Control Policy in Cisco FMC*

In Figure 7-41, the policy is named **Block Telnet** and the default action is to block all traffic. There are two available devices. The policy is applied to a virtual FTD device (**vFTD**). When you click **Save**, the new policy is added to the configuration.

The next step is to add an access control rule. To add an access control rule, click **Add Rule** and the screen in Figure 7-42 will appear.

The access control rule shown in Figure 7-42 is named **Block Telnet Rule**, and the action is set to **Block with reset**. The rule is configured to block TCP port 23 (Telnet) traversing all interfaces.

TIP Cisco FTD matches traffic to access control rules in the order you specify. In most cases, the system handles network traffic according to the first access control rule where all the rule's conditions match the traffic. Each rule also has an action, which determines whether you monitor, trust, block, or allow matching traffic. When you allow traffic, you can specify that the system first inspect it with intrusion or file policies to block any exploits, malware, or prohibited files before they reach your assets or exit your network. Rules in an access control policy are numbered, starting at 1. If you are using policy inheritance, rule 1 is the first rule in the outermost policy. The system matches traffic to rules in top-down order by ascending rule number. With the exception of Monitor rules, the first rule that traffic matches is the rule that handles that traffic. Rules can also belong to a section and a category, which are organizational only and do not affect rule position. Rule position goes across sections and categories.

Figure 7-42 *Creating an Access Control Rule for the New Policy in Cisco FMC*

 Cisco Firepower Intrusion Policies

Cisco delivers several intrusion policies with the Firepower System. These policies are designed by the Cisco Talos Security Intelligence and Research Group, which sets the intrusion and preprocessor rule states and advanced settings. You cannot modify these policies. However, you can change the action to take for a given rule, as described in "Changing Intrusion Rule Actions."

For access control rules that allow traffic, you can select one of the following intrusion policies to inspect traffic for intrusions and exploits. An intrusion policy examines decoded packets for attacks based on patterns and can block or alter malicious traffic.

Network analysis and intrusion policies work together as part of the Cisco FTD intrusion detection and prevention feature. The term *intrusion detection* generally refers to the process of passively analyzing network traffic for potential intrusions and storing attack data for security analysis. The term *intrusion prevention* includes the concept of intrusion detection but adds the ability to block or alter malicious traffic as it travels across your network. Figure 7-43 shows the details about network analysis and intrusion policies.

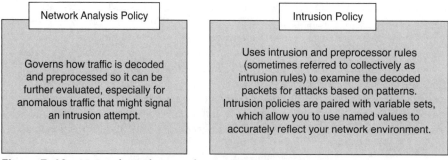

Figure 7-43 *Network Analysis and Intrusion Policies*

Both network analysis and intrusion policies are invoked by a parent access control policy, but at different times. As the system analyzes traffic, the network analysis (decoding and preprocessing) phase occurs before and separately from the intrusion prevention (additional preprocessing and intrusion rules) phase. Together, network analysis and intrusion policies provide broad and deep packet inspection. They can help you detect, alert on, and protect against network traffic that could threaten the availability, integrity, and confidentiality of hosts and their data. The Cisco FTD has several similarly named network analysis and intrusion policies (for example, Balanced Security and Connectivity) that complement and work with each other. By using system-provided policies, you can take advantage of the experience of the Cisco Talos Security Intelligence and Research Group. For these policies, Talos sets intrusion and preprocessor rule states, as well as provides the initial configurations for preprocessors and other advanced settings.

Figure 7-44 shows an example of an intrusion policy.

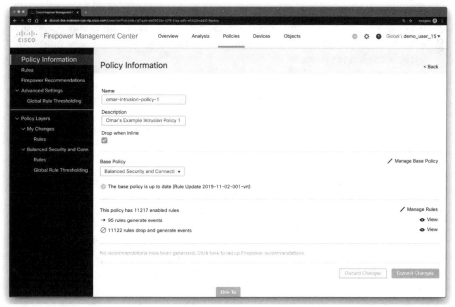

Figure 7-44 *Intrusion Policy Example*

The policy in Figure 7-44 is called **omar-intrusion-policy-1**, and it is configured to drop noncompliant traffic inline. The policy is configured with a base policy called **Balanced Security and Connectivity**.

The following are a few facts about intrusion policies:

■ In an inline deployment (that is, where relevant configurations are deployed to devices using routed, switched, or transparent interfaces, or inline interface pairs), the system can block traffic without further inspection at almost any step in the illustrated process. Security Intelligence, the SSL policy, network analysis policies, file policies, and intrusion policies can all either drop or modify traffic. Only the network discovery policy, which passively inspects packets, cannot affect the flow of traffic.

■ Common issues in intrusion policies include when you deploy a flat policy structure, inconsistent rule sets, and when you configure difficult rules for a new administrator to understand.

■ Every user-created policy depends on another policy as its "Base."

■ All policies trace their roots back to a "Talos-provided-base" policy. These base policies are as follows:

   ■ **Connectivity Over Security:** This policy is built for organizations where connectivity (being able to get to all resources) takes precedence over network infrastructure security. The intrusion policy enables far fewer rules than those enabled in the Security Over Connectivity policy. Only the most critical rules that block traffic are enabled. Select this policy if you want to apply some intrusion protection but you are fairly confident in the security of your network.

   ■ **Balanced Security and Connectivity:** This policy is designed to balance overall network performance with network infrastructure security. This policy is appropriate for most networks. Select this policy for most situations where you want to apply intrusion prevention.

   ■ **Security over Connectivity:** This policy is built for organizations where network infrastructure security takes precedence over user convenience. The intrusion policy enables numerous network anomaly intrusion rules that could alert on or drop legitimate traffic. Select this policy when security is paramount or for traffic that is high risk.

   ■ **No Rules Active:** No rules are active at this point.

   ■ **Maximum Detection:** This policy is built for organizations where network infrastructure security is given even more emphasis than is given by the Security Over Connectivity policy, with the potential for even greater operational impact. For example, the intrusion policy enables rules in a large number of threat categories, including malware, exploit kit, old and common vulnerabilities, and known in-the-wild exploits. If you select this policy, carefully evaluate whether too much legitimate traffic is being dropped.

■ When the system identifies a possible intrusion, it generates an intrusion or preprocessor event (sometimes collectively called intrusion events). Managed devices transmit their events to the Firepower Management Center, where you can view the aggregated data and gain a greater understanding of the attacks against your network assets. In an inline deployment, managed devices can also drop or replace packets that you know to be harmful. Each intrusion event in the database includes an event header and contains information about the event name and classification; the source and destination IP addresses; ports; the process that generated the event; and the date and time of the event, as well as contextual information about the source of the attack and its target. For packet-based events, the system also logs a copy of the decoded packet header and payload for the packet or packets that triggered the event.

The preceding sections explain how to configure various aspects of an IPS policy. After all the changes have been made, the policy doesn't take effect unless it is committed. When you click the Commit Changes button (refer to Figure 7-44), you might be prompted to insert a comment that describes the change. The comment is a configurable option under the intrusion policy preferences. Comments are stored in the audit logs and thus are searchable. Alternatively, you can choose to discard the changes, in which case you remove the changes since the last commit. Finally, if you choose to leave the Intrusion Policy screen with uncommitted changes, the changes are not lost. They remain uncommitted and can be revisited later to be committed or discarded.

After a policy is committed, Firepower performs a validation check to make sure there are no incorrect or invalid settings. For example, enabling a rule that uses a disabled preprocessor would be flagged after the commit.

Variables

Variables are used in multiple locations in Cisco Firepower NGIPS. IPS rules use preconfigured variables representing networks and ports. For example, the inside (protected) network is represented by the variable $HOME_NET, and the outside (unprotected) network is represented by the variable $EXTERNAL_NET. Other variables are used in more specialized rules. For example, $HTTP_SERVERS represents the web servers and $HTTP_PORTS the ports used by these servers.

There are two types of variables:

- **System default variables that are the preconfigured variables in the system:** These include $AIM_SERVERS, $DNS_SERVERS, $EXTERNAL_NET, $FILE_DATA_PORTS, $GTP_PORTS, $HOME_NET, $HTTP_PORTS, $HTTP_SERVERS, $ORACLE_PORTS, $SHELLCODE_PORTS, and more.

- **Policy variables that override default variable:** They are used in specific policies.

> **NOTE**   You can manage variables in Cisco FMC's Variable Set section of the Objects tab. Here, you can edit or create the default and other variable sets to be used in an IPS policy.

When creating a network variable, you can use more than one address; just separate addresses with commas—for example, [192.168.1.2, 10.10.10.2, 172.18.1.2]. You can also use CIDR representation—for example, [192.168.1.2, 10.10.10.2, 172.18.1.0/24]. In addition, you can exclude addresses from a list by using an exclamation point—for example, [192.168.1.2, !10.10.10.2, 172.18.1.0/24].

When creating port variables, you can use ranges, such as [25-121]. The following are some other possibilities:

- For ports less than a number, you can use the format [-1024].

- For ports more than a number, you can use the format [1024-].

- To exclude ports, you can use the format [!25].

- To list ports, you can use the format [25, !21, 80-].

When you accurately define variables, the processing of the rules using them is optimized and the right systems are monitored for illegitimate traffic. By leveraging variables, you can more efficiently change the variable value without changing all the rules for which that variable is used. For example, if multiple rules are defined to protect the web servers, you would need only to define or update the variable $HTTP_SERVERS instead of updating all the rules.

Platform Settings Policy

Cisco Firepower deployments can take advantage of platform settings policies. These policies are a shared set of parameters that define the aspects of a Cisco Firepower device that are likely to be similar to other managed devices. Examples of these platform settings policies are time and date settings, external authentication, and other common administrative features.

Shared platform policies allow you to configure multiple managed devices at once. This, in turn, allows you to be more consistent and to streamline your management efforts.

When you perform any changes to a platform settings policy, it takes effect in all the managed devices where the policy is applied.

Cisco NGIPS Preprocessors

IPS pattern matching is almost impossible because of the different protocols and their intricacies. For example, matching a file pattern in SMTP email traffic is not performed the same way as matching it when flowing through HTTP with compression enabled. For that reason, the Cisco NGIPS offers a variety of preprocessors that normalize traffic so that it can be matched against the defined Snort rules.

The preprocessors attempt to make the streams of packets as much as possible like the reassembled packets that will be seen by the endpoints receiving them. For example, the preprocessors perform checksum calculation, stream and fragment reassembly, stateful inspection, and more. Each preprocessor has a variety of settings to be configured in order to minimize false positives and false negatives. In addition, customized ones perform dedicated resources to detect specific suspicious attack activity to avoid burdening the system with these tasks.

The following preprocessors are available in the Cisco Firepower NGIPS:

- **DCE/RPC:** The DCE/RPC preprocessor monitors DCE/RPC and SMB protocol streams and messages for anomalous behavior or evasions.

- **DNS:** The DNS preprocessor inspects DNS responses for overflow attempts and for obsolete and experimental DNS record types.

- **FTP and Telnet:** The FTP and Telnet preprocessor normalizes FTP and Telnet streams before they are passed to the IPS engine. It requires TCP stream preprocessing.

- **HTTP:** The HTTP preprocessor normalizes HTTP requests and responses for IPS processing, separates HTTP messages to improve IPS rule performance, and detects URI-encoding attacks.

- **Sun RPC:** The Sun RPC preprocessor normalizes and reassembles fragmented SunRPC records so the rules engine can process the complete record.

- **SIP:** The SIP preprocessor normalizes and inspects SIP messages to extract the SIP header and body for further rule processing and generates events when identifying out-of-order calls and SIP message anomalies.

- **GTP:** The GTP preprocessor normalizes General Packet Radio Service (GPRS) Tunneling Protocol (GTP) command channel signaling messages and sends them to the rules engine for inspection.

- **IMAP and POP:** The IMAP and POP preprocessors monitor server-to-client email traffic and alert on anomalies. They also extract and decode email attachments and allow further processing by the NGIPS rules engine and other Cisco Firepower features.

- **SMTP:** The SMTP preprocessor inspects SMTP traffic for anomalous behavior and extracts and decodes email attachments for further processing.

- **SSH:** The SSH preprocessor detects SSH buffer overflow attempts and monitors for illegal SSH versions.

- **SSL:** The SSL preprocessor monitors the SSL handshake transactions. After the SSL session is encrypted, the SSL preprocessor stops inspecting. It requires TCP stream preprocessing. The SSL preprocessor can reduce the amount of false positives and save detection resources from the IPS system.

- **SCADA:** There are two supervisory control and data acquisition (SCADA) protocols for which the Cisco Firepower NGIPS offers preprocessors: DNP3 and Modbus. These protocols monitor and control industrial facilities. The SCADA preprocessors monitor the DNP and Modbus protocols for anomalies and decode their messages for further rule inspection.

- **Network:** Multiple network and transport layer preprocessors detect attacks exploiting the following:

  - Checksum verification

  - Ignoring VLAN headers

  - Inline normalization

  - IP defragmentation

  - Packet decoding

  - TCP stream

  - UDP stream

  The packet decoder normalizes packet headers and payloads for further processing. The inline normalization preprocessor normalizes traffic to prevent evasion techniques in inline deployments. The rest of the network preprocessors detect anomalous network or transport layer behavior.

- **Threat detection:** The threat detection preprocessors detect specific threats (such as port scan detection, rate-based attack prevention, and sensitive data detection).

7

NOTE   The FMC offers a wealth of settings for each preprocessor that are available in the IPS policy Advanced Settings view.

Cisco Advanced Malware Protection (AMP)

The Cisco Advanced Malware Protection (AMP) solution enables you to detect and block malware, continuously analyze for malware, and get retrospective alerts. Traditionally, Cisco has the concept of AMP for Networks and AMP for Endpoints.

NOTE   AMP for Endpoints will be covered in Chapter 11, "Endpoint Protection and Detection."

Cisco AMP for Networks provides next-generation security services that go beyond point-in-time detection. It provides continuous analysis and tracking of files and also retrospective security alerts so that a security administrator can take action during and after an attack. The file trajectory feature of Cisco AMP for Networks tracks file transmissions across the network, and the file capture feature enables a security administrator to store and retrieve files for further analysis.

Cisco acquired a security company called ThreatGRID that provides cloud-based and on-premises malware analysis solutions. Cisco integrated Cisco AMP and Threat Grid to provide a solution for advanced malware analysis with deep threat analytics. The Cisco AMP Threat Grid integrated solution analyzes millions of files and correlates them with hundreds of millions of malware samples. This provides a look into attack campaigns and how malware is distributed. This solution provides a security administrator with detailed reports of indicators of compromise and threat scores that help prioritize mitigations and recover from attacks.

Cisco AMP has the following features:

- **File reputation:** AMP allows you to analyze files inline and block or apply policies.

- **File sandboxing:** AMP allows you to analyze unknown files to understand true file behavior.

- **File retrospection:** AMP allows you to continue to analyze files for changing threat levels.

There are major architectural benefits to the AMP solution, which leverages a cloud infrastructure for the heavy lifting. The architecture of AMP can be broken down into three main components: the AMP cloud, AMP client connectors, and intelligence sources. AMP client connectors include AMP for Networks, AMP for Endpoints, and AMP for Content Security.

The AMP cloud contains many different analysis tools and technologies to detect malware in files, including the Threat Grid analysis solution. Cisco's research teams, including the Cisco Talos Security Intelligence and Research Group, feed information about malware into the AMP cloud. Threat intelligence from Cisco products, services, and third-party relationships is also sent to the AMP cloud. The following are some examples of threat intelligence sources:

- **Snort, ClamAV, and Immunet AV open source communities:** Users of these open source projects contribute threat information daily.

- **Talos:** The Cisco Talos Security Intelligence and Research Group is a team of leading threat researchers that contributes to the threat information ecosystem of Cisco security products. Talos team members get threat information from a variety of sources and their own internal research efforts. Talos maintains the official rule sets of Snort. org, ClamAV, SenderBase.org, and SpamCop. Talos is also the primary team that contributes to the Cisco Collective Security Intelligence (CSI) ecosystem. A variety of sources contribute to the vast amount of data provided to Cisco through submitted malware samples, data from the web, and email traffic monitored by Cisco products and other third-party sources.

TIP You can subscribe to the official Talos blog, obtain access to many tools they have created, and access the Talos Reputation Center at https://talosintelligence.com.

- **Threat Grid:** This deep threat analysis solution leverages many identification techniques, including sandboxing. Threat Grid is built as a cloud architecture and is used to do deep analysis of file samples submitted to the AMP Threat Grid cloud. The analysis results are fed into the AMP cloud and can be used to update file disposition (the result).

The most critical item of the Cisco AMP architecture is the AMP cloud itself. The AMP cloud has two deployment methods—public and private—and regardless of the deployment chosen, the role of the cloud is the same.

The AMP cloud houses all the detection signatures. A major benefit of storing these signatures in the cloud is that it reduces the client connector size and reduces the processing requirements on the client, since the bulk of the work is handled in the cloud.

An interesting and fairly unique feature is that AMP administrators can create custom signatures in the cloud, and then those custom signatures are pushed to the connectors. In addition, the cross-referencing of files and signatures is done in the AMP cloud, so the cloud can be self-updating, without having to constantly communicate updates to the connectors.

The AMP cloud is also responsible for large-scale data processing, or big data. The data comes to the AMP cloud from multiple sources, including honeypots, threat feeds, open source communities, AV solutions such as Immunet AV and ClamAV, and more. File samples are provided to the AMP cloud, where they are processed. If the disposition of a sample file is deemed to be malicious, it is stored in the cloud and reported to the client connectors that see the same file.

Advanced analytic engines, including Threat Grid, are part of the AMP cloud and are constantly correlating the incoming data. The analytical results are used to update the AMP signatures. In addition to the advanced analytics, machine-learning engines are employed to further refine signatures and reevaluate detections that have already been performed. The cloud is not just a repository of signatures; the decision making is performed in real time, evolving constantly based on the data received.

Prevention involves keeping malware at bay. With prevention, speed is critical. It requires real-time, dynamic decisions to be made from real-world data. The data must have high accuracy, with low false positives and false negatives. Prevention could also be viewed as the "security control mode."

As illustrated in Figure 7-45, the AMP cloud's prevention framework is made up of seven core components: 1-to-1 signatures, Ethos, Spero, indicators of compromise (IOCs), device flow correlation, advanced analytics, and dynamic analysis.

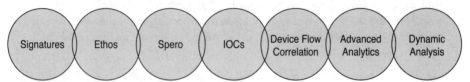

Figure 7-45   *The AMP Cloud's Protection Framework*

1-to-1 signatures are a traditional technology that is used all over the security industry in various forms. With these signatures, a hash is created of a file, and that hash is compared to a database. If a match is found, the specific file is known, and a verdict—clean or malicious—is returned. If the hash has not been seen before, the cloud returns a verdict of unknown. The benefit of this method is that it can quickly identify and block malicious files. The downside is that a simple change to a file also changes the hash, thereby evading the signature.

AMP differentiates itself from other 1-to-1 signature solutions by storing the signature database in the cloud instead of on the client. The database is quite large, and many solutions cut corners by including only a subset of the signatures in the full database. Storing the database in the cloud allows AMP to leverage the entire database. Comparing the files to the database can be quite resource intensive. AMP does the comparison in the cloud, freeing those resources from the client connector. AMP is also able to collect, process, and detect in near real time.

The next component of the protection framework is the Ethos engine. Ethos is a "fuzzy fingerprinting" engine that uses static or passive heuristics. The engine creates generic file signatures that can match polymorphic variants of a threat. This is useful because when a threat morphs or a file is changed, the structural properties of that file often remain the same, even though the content has changed.

Unlike most other signature tools, Ethos uses distributed data mining to identify suitable files. It uses in-field data for sources, which provide a highly relevant collection from which to generate the signatures. Ethos is completely automated and provides rapid generation of the generic signatures that are based on in-field data instead of relying on individual "rock star" engineers to generate a limited number of generic signatures.

Spero is a machine-learning-based technology that proactively identifies threats that were previously unknown. It uses active heuristics to gather execution attributes, and because the underlying algorithms come up with generic models, it can identify malicious software based on its general appearance rather than based on specific patterns or signatures.

An indicator of compromise (IOC) is any observed artifact on a system or a network that could indicate an intrusion. There may be artifacts left on a system after an intrusion or a breach, and they can be expressed in a language that describes the threat information, known as an IOC. The sets of information describe how and where to detect the signs of the intrusion or breach. IOCs can be host-based and/or network-based artifacts, but the scan actions are carried out on the host only.

IOCs are very high-confidence indicators, and they may describe numerous specific items, including FileItem, RegistryItem, EventLogItem, ProcessItem, and ServiceItem.

TIP   In Chapter 1, "Cybersecurity Fundamentals," you learned about different standards for sharing threat intelligence and indicators of compromise. As a refresher, the Structured Threat Information eXpression (STIX) and the Trusted Automated eXchange of Indicator Information (TAXII) are some of the most popular standards for threat intelligence exchange. STIX details can contain data such as the IP addresses or domain names of command-and-control servers (often referred to C2 or CnC), malware hashes, and so on. STIX was originally developed by MITRE and is now maintained by OASIS. You can obtain more information at https://oasis-open.github.io/cti-documentation. You can review numerous examples of STIX content, objects, and properties at https://oasis-open.github.io/cti-documentation/stix/examples.

TAXII is an open transport mechanism that standardizes the automated exchange of cyber-threat information. TAXII was originally developed by MITRE and is now maintained by OASIS. You can also obtain detailed information about TAXII at https://oasis-open.github.io/cti-documentation.

Device flow correlation provides a kernel-level view into network I/O. It allows for blocking or alerting on network activity, traced back to the initiating process itself. It enables internal and external networks to be monitored, leverages IP reputation data, and offers URL/domain logging. The flow points are extra telemetry data and are not file disposition specific.

Cisco provides intelligence on many malicious destinations, including generic command-and-control (CnC or C2) servers, phishing hosts, zero-access C2 servers, and more.

Advanced analytics consists of a set of multifaceted engines that provide big data context beyond a single host and beyond a single file. Advanced analytics highlights files executed in an environment, from least common to most. This can aid in identifying previously undetected threats that may have only been seen by a small number of users.

Dynamic analysis performed by Cisco AMP Threat Grid integration is not a single tool. It is a full solution for dynamic malware analysis and threat intelligence. It performs high-speed, automated analysis with adjustable runtimes while not exposing any tags or other indicators that malware could use to detect that it is being observed.

Threat Grid provides video playbacks, a glovebox for malware interaction and operational troubleshooting, a process graph for visual representation of process lineage, and a threat score with behavior indicators.

7

It searches and correlates all data elements of a single sample against billions of sample artifacts collected and analyzed over years, leveraging global and historic context. This enables an analyst to better understand the relevancy of a questionable sample as it pertains to the analyst's own environment.

Threat Grid was architected from the ground up as a cloud solution with an API designed to integrate with existing IT security solutions and to create custom threat intelligence feeds. It can automatically receive submissions from other solutions and pull the results into your environment.

Many think that Threat Grid is a sandboxing solution. It is much more than just that, however; sandboxing is a piece of the solution, and Threat Grid's sandboxing functions are performed in a way that evades detection by malware. Threat Grid uses an outside-in approach, with no presence in the virtual machine. The sandboxing's dynamic analysis includes an external kernel monitor, dynamic disk analysis that illuminates any modifications to the physical disk (such as the master boot record), monitoring user interaction, video capture and playback, process information, artifacts, and network traffic.

Threat Grid supports the following samples and object types:

- Executable files (.EXE) and libraries (.DLL)

- Java archives (.JAR)

- Portable document format (.PDF)

- Office documents (.RTF, .DOC, .DOCX, .XLS, .XLSX, .PPT, .PPTX)

- ZIP containers (.ZIP)

- Quarantine containers

- URLs

- HTML documents

- Flash

Retrospection means taking a look at what has already transpired; it involves tracking system behavior regardless of disposition, focusing on uncovering malicious activity. Retrospection could be viewed as the "incident response mode," using continuous analysis to reactively act on a file that was assigned a clean disposition once but was later found to have a bad disposition.

The retrospective framework is designed to show the trajectory of a malicious file, with a goal of 30 days of telemetry data. Even files that are originally given a clean verdict are tracked, and if a clean file is later found to be malicious, all connectors that have seen the file are notified to quarantine the file retrospectively.

Figure 7-46 shows the network file trajectory for a confirmed piece of malware in the Cisco FMC.

Figure 7-46   *The Network File Trajectory for a Confirmed Piece of Malware in the Cisco FMC*

Security Intelligence, Security Updates, and Keeping Firepower Software Up to Date

You learned that Security Intelligence uses reputation intelligence to quickly block connections to or from IP addresses, URLs, and domain names. This is called Security Intelligence blacklisting. Security Intelligence is an early phase of access control, before the system performs more resource-intensive evaluation. Blacklisting improves performance by quickly excluding traffic that does not require inspection. Although you can configure custom blacklists, Cisco provides access to regularly updated intelligence feeds. Sites representing security threats such as malware, spam, botnets, and phishing appear and disappear faster than you can update and deploy custom configurations.

> **TIP**   You can refine Security Intelligence blacklisting with whitelists and monitor-only blacklists. These mechanisms exempt traffic from being blacklisted, but do not automatically trust or fast-path matching traffic. Traffic whitelisted or monitored at the Security Intelligence stage is intentionally subject to further analysis with the rest of access control.

Security Intelligence Updates

The following are some best practices around security intelligence updates:

- The default update frequency is 2 hours. In some scenarios, you may want to reduce this for more agile blacklisting.

- The minimum intervals should be as follows:

  - **Talos network feed:** 5 minutes

  - **Talos URL/DNS feed:** 30 minutes

  - **Custom feed:** 30 minutes

- Use custom feeds. Everybody has a custom blacklist, and feeds are the best way to implement these in Cisco FTD and Firepower devices. Security Intelligence lists and feeds are not the same. Security intelligence lists are okay, but remember you need to deploy policy to update them.

Keeping Software Up to Date

Keeping software up to date is not exciting but it is important. The following are the different software update types in Cisco FTD and Firepower devices:

- **Snort rules updates (SRUs):** Contain the latest Snort rules. SRUs are released on Tuesday and Thursday; however, they can also be released out-of-band for critical updates. Each SRU contains a complete rule set.

- **Vulnerability database (VDB) updates:** Updates for known vulnerabilities, identified with a Common Vulnerability and Exposures (CVE) identifier. These vulnerability reports are used to set impact on intrusion events.

- **Geolocation updates:** These are updated weekly.

- **Firepower software patches and updates.**

Exam Preparation Tasks

As mentioned in the section "How to Use This Book" in the Introduction, you have a couple of choices for exam preparation: the exercises here, Chapter 12, "Final Preparation," and the exam simulation questions in the Pearson Test Prep Software Online.

Review All Key Topics

Review the most important topics in this chapter, noted with the Key Topic icon in the outer margin of the page. Table 7-3 lists these key topics and the page numbers on which each is found.

Table 7-3 Key Topics for Chapter 7

Key Topic Element	Description	Page Number
Section	Cisco Firepower Threat Defense (FTD)	397
List	An introduction to Cisco's Next-Generation Intrusion Prevention System (NGIPS)	399
List	List of the supplemental capabilities introduced by Cisco NGIPS	400

Key Topic Element	Description	Page Number
List	List of the most important capabilities of Cisco NGIPS	401
Section	Surveying the Cisco Firepower Management Center (FMC)	401
Section	Exploring the Cisco Firepower Device Manager (FDM)	404
Section	Cisco Defense Orchestrator	408
Section	Comparing Network Security Solutions That Provide Firewall Capabilities	411
Section	Deployment Modes of Network Security Solutions and Architectures That Provide Firewall Capabilities	412
Section	Routed vs. Transparent Firewalls	413
Section	Security Contexts	414
Section	Single-Mode Transparent Firewalls	414
Section	Surveying the Cisco FTD Deployment Modes	416
Section	Cisco FTD Interface Modes	417
Section	Inline Pair	420
Section	Inline Pair with Tap	420
Section	Passive Mode	420
Section	Passive with ERSPAN Mode	422
Section	Additional Cisco FTD Deployment Design Considerations	422
Section	High Availability and Clustering	423
Section	Implementing Access Control Lists in Cisco ASA	427
List	Defining the five-tuple	427
List	Types of ACL supported in the Cisco ASA	430
Paragraph	Exploring the different security levels of Cisco ASA interfaces	432
Section	Cisco ASA Application Inspection	433
Section	To-the-Box Traffic Filtering in the Cisco ASA	434
Section	Standard ACLs	436
Section	ICMP Filtering in the Cisco ASA	437
Section	Network Address Translation in Cisco ASA	437
List	Understanding the different methods to translate an address in the Cisco ASA	440
Section	Cisco ASA Auto NAT	443
Section	Implementing Access Control Policies in the Cisco Firepower Threat Defense	443
Section	Cisco Firepower Intrusion Policies	446
Section	Variables	449
Section	Platform Settings Policy	450

Key Topic Element	Description	Page Number
Section	Cisco NGIPS Preprocessors	450
Section	Cisco Advanced Malware Protection (AMP)	452
Tip	Exploring standards for threat intelligence exchange, including STIX and TAXII	455
Paragraph	Understanding retrospection and retrospective analysis	456

Define Key Terms

Define the following key terms from this chapter and check your answers in the glossary:

Cisco FMC, Cisco FDM, security zone, zone-based firewall (ZBFW), Ethos, Spero, indicator of compromise (IOC), Structured Threat Information eXpression (STIX), Trusted Automated eXchange of Indicator Information (TAXII)

Review Questions

1. When you add a Cisco FTD or Firepower device to FMC, what information is required?

   a. Serial number

   b. The initial access intrusion policies configured in the FTD or Firepower device

   c. The initial access control policies configured in the FTD or Firepower device

   d. A registration key

2. Which of the following is a best practice for security intelligence updates?

   a. The default security intelligence update frequency is 2 hours. In some scenarios, you may want to increase this update frequency to avoid network bandwidth consumption.

   b. The default security intelligence update frequency is 2 hours. In some scenarios, you may want to reduce this for more agile blacklisting.

   c. Configure Snort Rule Updates to be done twice a week.

   d. Configure vulnerability database updates every second Tuesday of the month (Microsoft Patch Tuesday).

3. Variables are used in multiple locations in Cisco Firepower NGIPS. IPS rules use preconfigured variables representing networks and ports. Which of the following are system default variables that are preconfigured in Cisco Firepower devices?

   a. $FILE_DATA_PORTS

   b. $HOME_NET

   c. $HTTP_SERVERS

   d. All of these answers are correct.

4. IPS pattern matching is almost impossible because of the different protocols and their intricacies. For example, matching a file pattern in SMTP email traffic is not performed the same way as matching it when flowing through HTTP with compression enabled. For that reason, the Cisco NGIPS offers a variety of _____ that normalize traffic so that it can be matched against the defined Snort rules.

    a. preprocessors

    b. intrusion policies

    c. variables

    d. access control rules

5. You are tasked to deploy an intrusion policy designed to balance overall network performance with network infrastructure security. Which of the following base policies would you deploy in situations where you want to apply intrusion prevention?

    a. Connectivity Over Security

    b. Balanced Security and Connectivity

    c. Balanced Security and Connectivity over Security

    d. Maximum Detection

6. You are tasked to configure NAT and translate the source address of an object regardless of the destination address. Which of the following NAT configuration features would you deploy to accomplish this task?

    a. Auto NAT

    b. Destination NAT

    c. Identity NAT

    d. Port redirection

7. Which of the following statements is not true about standard ACLs?

    a. Standard ACLs are used when the source network in the traffic is not important.

    b. Standard ACLs are used by processes, such as OSPF and VPN tunnels, to identify traffic based on the destination IP addresses.

    c. Standard ACLs can be used with route maps.

    d. Standard ACLs provide a way to group similar items together to reduce the number of ACEs.

8. Cisco ASA provides the Modular Policy Framework (MPF) to provide application security or perform Quality of Service (QoS) functions. The MPF offers a consistent and flexible way to configure the Cisco ASA application inspection and other features in a manner similar to that used for the Cisco IOS Software Modular QoS CLI. Which of the following are commands associated with the MPF?

    a. class-map

    b. policy-map

    c. service-policy

    d. All of these answers are correct.

9. Which of the following is not a step in the Cisco FTD cluster unit transition?

   a. After the unit boots, it looks for a master over the cluster control link.

   b. If there are no master units, it waits 45 seconds before assuming the master role. If a master already exists, it becomes a slave and the configuration is synchronized.

   c. If there is a synchronization or health issue, the unit is considered disabled in the cluster.

   d. After the master boots, it creates a logical interface and assigns an IP address to that interface to communicate to the other units in the cluster.

10. Which of the following can be deployed to generate, transport, and consume threat intelligence information?

   a. A TAXII server

   b. A STIX server

   c. A C2 or CnC

   d. A Cybox server

CHAPTER 8

Virtual Private Networks (VPNs)

This chapter covers the following topics:

Virtual Private Network (VPN) Fundamentals

Deploying and Configuring Site-to-Site VPNs in Cisco Routers

Configuring Site-to-Site VPNs in Cisco ASA Firewalls

Configuring Remote-Access VPNs in the Cisco ASA

Configuring Clientless Remote-Access SSL VPNs in the Cisco ASA

Configuring Client-Based Remote-Access SSL VPNs in the Cisco ASA

Configuring Remote-Access VPNs in FTD

Configuring Site-to-Site VPNs in FTD

The following SCOR 350-701 exam objectives are covered in this chapter:

- **Domain 1.0: Security Concepts**

    - 1.3 Describe functions of the cryptography components, such as hashing, encryption, PKI, SSL, IPsec, NAT-T IPv4 for IPsec, pre-shared key, and certificate-based authorization

    - 1.4 Compare site-to-site VPN and remote-access VPN deployment types such as sVTI, IPsec, Cryptomap, DMVPN, FLEXVPN, including high availability considerations, and AnyConnect

- **Domain 2.0: Network Security**

    - 2.9 Configure and verify site-to-site VPN and remote-access VPN

        - 2.9.a Site-to-site VPN utilizing Cisco routers and IOS

        - 2.9.b Remote-access VPN using Cisco AnyConnect Secure Mobility client

        - 2.9.c Debug commands to view IPsec tunnel establishment and troubleshooting

"Do I Know This Already?" Quiz

The "Do I Know This Already?" quiz allows you to assess whether you should read this entire chapter thoroughly or jump to the "Exam Preparation Tasks" section. If you are in doubt about your answers to these questions or your own assessment of your knowledge of the topics, read the entire chapter. Table 8-1 lists the major headings in this chapter and their corresponding "Do I Know This Already?" quiz questions. You can find the answers in Appendix A, "Answers to the 'Do I Know This Already?' Quizzes and Q&A Sections."

Table 8-1 "Do I Know This Already?" Section-to-Question Mapping

Foundation Topics Section	Questions
Virtual Private Network (VPN) Fundamentals	1–2
Deploying and Configuring Site-to-Site VPNs in Cisco Routers	3–4
Configuring Site-to-Site VPNs in Cisco ASA Firewalls	5
Configuring Remote-Access VPNs in the Cisco ASA	6
Configuring Clientless Remote-Access SSL VPNs in the Cisco ASA	7
Configuring Client-Based Remote-Access SSL VPNs in the Cisco ASA	8
Configuring Remote-Access VPNs in FTD	9
Configuring Site-to-Site VPNs in FTD	10

CAUTION   The goal of self-assessment is to gauge your mastery of the topics in this chapter. If you do not know the answer to a question or are only partially sure of the answer, you should mark that question as wrong for purposes of the self-assessment. Giving yourself credit for an answer you incorrectly guess skews your self-assessment results and might provide you with a false sense of security.

1. Which of the following VPN protocols do not provide encryption?

   a. Point-to-Point Tunneling Protocol (PPTP)

   b. Layer 2 Forwarding (L2F) Protocol

   c. Layer 2 Tunneling Protocol (L2TP)

   d. Generic Routing Encapsulation (GRE)

   e. All of these answers are correct.

2. You are hired to configure a site-to-site VPN between a Cisco FTD device and a Cisco IOS-XE router. Which of the following encryption and hashing protocols will you select for optimal security?

   a. AES-192, SHA, Diffie-Hellman Group 21

   b. IDEA, SHA, Diffie-Hellman Group 2

   c. AES-192, SHA, Diffie-Hellman Group 5

   d. AES-256, SHA, Diffie-Hellman Group 21

3. Which of the following technologies groups many spokes into a single mGRE interface?

   a. DMVPN

   b. GETVPN

   c. FlexVPN

   d. GRE over IPsec

4. You are hired to deploy site-to-site VPN tunnels in a Cisco router where the VPN peers are third-party devices from different vendors. These devices have IKEv2 enabled. Which of the following technologies will you choose?

a. DMVPN

b. GETVPN

c. FlexVPN

d. GRE over IPsec

5. An IPsec transform (proposal) set specifies what type of encryption and hashing to use for the data packets after a secure connection has been established. This provides data authentication, confidentially, and integrity. The IPsec transform set is negotiated during quick mode. Which of the following commands is used to create an IPsec proposal (transform set) in a Cisco ASA?

a. crypto ipsec ikev2 ipsec-proposal mypolicy

b. crypto ipsec ikev2 transform_set mypolicy

c. crypto ikev2 mypolicy 1

d. crypto isakmp policy mypolicy

6. Refer to the following configuration snippet:

```
tunnel-group SecretCorp_TG general-attributes
 address-pool pool_1
 default-group-policy SecretCorp_GP
 authentication-server-group LOCAL
```

Which VPN implementation type does this configuration snippet apply to?

a. Remote Access VPN in Cisco routers

b. Remote Access VPN in Cisco ASA

c. Site-to-site VPN in Cisco ASA

d. Site-to-site VPN in Cisco routers

7. Which of the following are key points you need to take into consideration before you choose your SSL VPN deployment mode?

a. Before designing and implementing the SSL VPN solution for your corporate network, you need to determine whether your users connect to your corporate network from public shared computers, such as workstations made available to guests in a hotel or computers in an Internet kiosk. In this case, using a clientless SSL VPN is the preferred solution to access the protected resources.

b. The SSL VPN functionality on the ASAs requires that you have appropriate licenses. Make sure that you have the appropriate license for your SSL VPN deployment.

c. Network security administrators need to determine the size of the SSL VPN deployment, especially the number of concurrent users that will connect to gain network access. If one Cisco ASA is not enough to support the required number of users, clustering or load balancing must be considered to accommodate all the potential remote users.

d. All of these answers are correct.

8. Which of the following are AnyConnect deployment modes? (Select all that apply.)

   a. Web-enabled mode, where the AnyConnect client is downloaded to a user computer through a browser. The user opens a browser and references the IP address or the FQDN of a Cisco ASA or Cisco FTD device to establish an SSL VPN tunnel, and the client is downloaded to the user's system.

   b. FlexVPN mode, where the AnyConnect client is downloaded to a user computer through a browser. The user opens a browser and references the IP address or the FQDN of a Cisco ASA or Cisco FTD device to establish an SSL VPN tunnel, and the client is downloaded to the user's system.

   c. Standalone mode. With this method, the client is downloaded as a standalone application from a file server or directly from Cisco.com.

   d. All of these answers are correct.

9. Which of the following statements are true about Cisco FTD VPN deployments?

   a. Rapid Threat Containment is supported by Cisco FTD using RADIUS Change of Authorization (CoA) or RADIUS dynamic authorization.

   b. Double authentication is supported using an additional AAA server for secondary authentication.

   c. Remote access VPN can be configured on both FMC and FDM.

   d. All of these answers are correct.

10. Which of the following statements are true about site-to-site VPN deployments in Cisco FTD?

   a. A site-to-site VPN connection in Cisco FTD devices can only be made across domains by using an extranet peer for the endpoint not in the current domain.

   b. A VPN topology cannot be moved between domains.

   c. Network objects with a "range" option are not supported in VPN.

   d. All of these answers are correct.

Foundation Topics

8

Virtual Private Network (VPN) Fundamentals

Organizations deploy VPNs to provide data integrity, authentication, and data encryption to ensure confidentiality of the packets sent over an unprotected network or the Internet. VPNs were originally designed to avoid the cost of unnecessary leased lines. However, they now play a critical role of security and in some cases privacy. Individuals use VPNs to connect to their corporate network, but also use them for privacy.

Many different protocols have been used throughout the years for VPN implementations, including the following:

- Point-to-Point Tunneling Protocol (PPTP)

- Layer 2 Forwarding (L2F) Protocol

- Layer 2 Tunneling Protocol (L2TP)

- Generic Routing Encapsulation (GRE) Protocol

- Multiprotocol Label Switching (MPLS) VPN

- Internet Protocol Security (IPsec)

- Secure Sockets Layer (SSL)

L2F, L2TP, GRE, and MPLS VPNs do not provide data integrity, authentication, and data encryption. On the other hand, you can combine L2TP, GRE, and MPLS with IPsec to provide these benefits. Many organizations use IPsec as their preferred protocol because it supports all three of these features.

VPN implementations are categorized into two distinct groups:

- **Site-to-site VPNs:** Enable organizations to establish VPN tunnels between two or more network infrastructure devices in different sites so that they can communicate over a shared medium such as the Internet. Many organizations use IPsec, GRE, and MPLS VPN as site-to-site VPN protocols.

- **Remote-access VPNs:** Enable users to work from remote locations such as their homes, hotels, and other premises as if they were directly connected to their corporate network.

> **NOTE** Typically, site-to-site VPN tunnels are terminated between two or more network infrastructure devices, whereas remote-access VPN tunnels are formed between a VPN head-end device and an end-user workstation or hardware VPN client.

Figure 8-1 illustrates a site-to-site IPsec tunnel between two sites (corporate headquarters and a branch office).

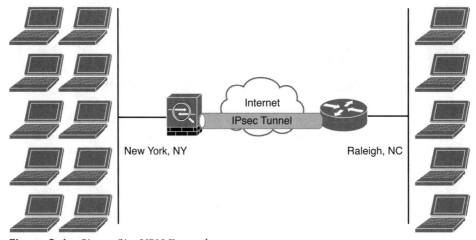

Figure 8-1 *Site-to-Site VPN Example*

Cisco IPsec VPN solutions have evolved over the years to very robust and comprehensive technologies. Figure 8-2 lists the different Cisco IPsec site-to-site VPN technologies.

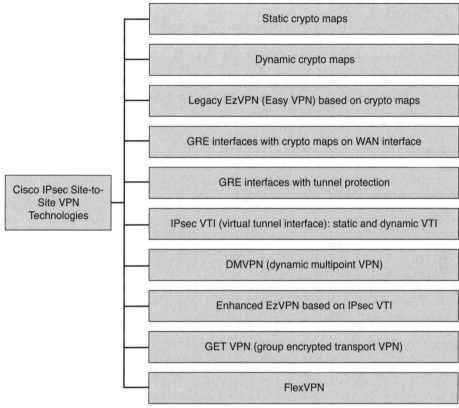

Figure 8-2 *Site-to-Site VPN Technologies*

Figure 8-3 shows an example of a remote-access VPN. In this case, a telecommuter employs an IPsec VPN while a remote user from a hotel employs an SSL VPN to connect to the corporate headquarters.

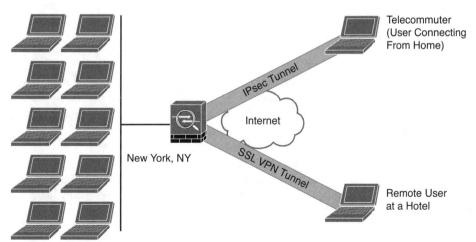

Figure 8-3 *Remote-Access VPN Example*

An Overview of IPsec

IPsec uses the Internet Key Exchange (IKE) protocol to negotiate and establish secured site-to-site or remote-access VPN tunnels. IKE is a framework provided by the Internet Security Association and Key Management Protocol (ISAKMP).

> **TIP** IKE is defined in RFC 2409, "The Internet Key Exchange (IKE)." IKE version 2 (IKEv2) is defined in RFC 5996, "Internet Key Exchange Protocol Version 2 (IKEv2)."

IKEv1 Phase 1

Within Phase 1 negotiation, several attributes are exchanged:

- Encryption algorithms
- Hashing algorithms
- Diffie-Hellman groups
- Authentication method
- Vendor-specific attributes

The following are the traditional encryption algorithms used in IKE:

- Data Encryption Standard (DES): 64 bits long (should be avoided in favor of AES)
- Triple DES (3DES): 168 bits long (should be avoided in favor of AES)
- Advanced Encryption Standard (AES): 128 bits long
- AES 192: 192 bits long
- AES 256: 256 bits long

Hashing algorithms include the following:

- Secure Hash Algorithm (SHA)
- Message digest algorithm 5 (MD5)

> **TIP** Cisco has an excellent resource that provides an overview of all cryptographic algorithms. The same document outlines the algorithms that should be avoided and the ones that are recommended. The document can be accessed at https://tools.cisco.com/security/center/resources/next_generation_cryptography.

The common authentication methods in VPNs are pre-shared keys (where peers use a shared secret to authenticate each other) and digital certificates with the use of Public Key Infrastructure (PKI). Typically, small and medium-sized organizations use pre-shared keys as their authentication mechanism. Several large organizations employ digital certificates for scalability, centralized management, and additional security mechanisms.

You can establish a Phase 1 security association (SA) in main mode or aggressive mode. In main mode, the IPsec peers complete a six-packet exchange in three round trips to negotiate the ISAKMP SA, whereas aggressive mode completes the SA negotiation in three packet exchanges. Main mode provides identity protection if pre-shared keys are used. Aggressive mode offers identity protection only if digital certificates are employed.

NOTE Cisco products that support IKEv1 typically use main mode for site-to-site tunnels and use aggressive mode for remote-access VPN tunnels. This is the default behavior when pre-shared keys are employed as the authentication method.

Figure 8-4 illustrates the six-packet exchange in main mode negotiation.

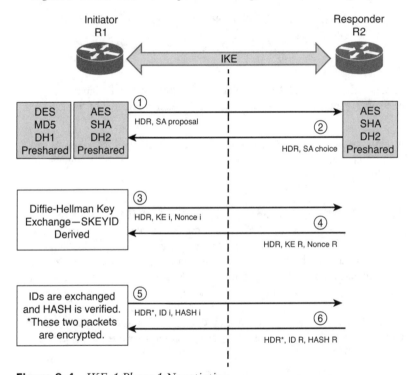

Figure 8-4 *IKEv1 Phase 1 Negotiation*

In Figure 8-4, two routers are configured to terminate a site-to-site VPN tunnel between them. The router labeled as R1 is the initiator, and R2 is the responder. The following steps are illustrated in Figure 8-3:

1. R1 (the initiator) has two ISAKMP proposals configured. In the first packet, R1 sends its configured proposals to R2.

2. R2 evaluates the received proposal. Because it has a proposal that matches the offer of the initiator, R2 sends the accepted proposal back to R1 in the second packet.

3. Diffie-Hellman exchange and calculation is started. Diffie-Hellman is a key agreement protocol that enables two users or devices to authenticate each other's pre-shared

keys without actually sending the keys over the unsecured medium. R1 sends the Key Exchange (KE) payload and a randomly generated value called a nonce.

4. R2 receives the information and reverses the equation, using the proposed Diffie-Hellman group/exchange to generate the SKEYID. The SKEYID is a string derived from secret material that is known only to the active participants in the exchange.

5. R1 sends its identity information. The fifth packet is encrypted with the keying material derived from the SKEYID. The asterisk in Figure 1-8 is used to illustrate that this packet is encrypted.

6. R2 validates the identity of R1, and R2 sends its own identity information to R1. This packet is also encrypted.

TIP IKE uses UDP port 500 for communication. UDP port 500 is employed to send all the packets described in the previous steps.

IKEv1 Phase 2

Phase 2 is used to negotiate the IPsec SAs. This phase is also known as quick mode. The ISAKMP SA protects the IPsec SAs because all payloads are encrypted except the ISAKMP header.

A single IPsec SA negotiation always creates two security associations—one inbound and one outbound. Each SA is assigned two unique security parameter index (SPI) values—one by the initiator and the other by the responder.

TIP The security protocols (AH and ESP) are Layer 3 protocols and do not have Layer 4 port information. If an IPsec peer is behind a PAT device, the ESP or AH packets are typically dropped. To work around this, many vendors, including Cisco, use a feature called IPsec pass-through. The PAT device that is capable of IPsec pass-through builds the Layer 4 translation table by looking at the SPI values on the packets. Many industry vendors, including Cisco, implement another feature called NAT Traversal (NAT-T). With NAT-T, the VPN peers dynamically discover whether an address translation device exists between them. If they detect a NAT/PAT device, they use UDP port 4500 to encapsulate the data packets, subsequently allowing the NAT device to successfully translate and forward the packets.

Another interesting point is that if the VPN router needs to connect multiple networks over the tunnel, it must negotiate twice as many IPsec SAs. Remember, each IPsec SA is unidirectional, so if three local subnets need to go over the VPN tunnel to talk to the remote network, then six IPsec SAs are negotiated. IPsec can use quick mode to negotiate these multiple Phase 2 SAs, using the single pre-established ISAKMP (IKEv1 Phase 1) SA. The number of IPsec SAs can be reduced, however, if source and/or destination networks are summarized.

In addition to generating the keying material, quick mode also negotiates identity information. The Phase 2 identity information specifies which network, protocol, and/or port number to encrypt. Hence, the identities can vary anywhere from an entire network to a single host address, allowing a specific protocol and port.

Figure 8-5 illustrates the Phase 2 negotiation between the two routers that just completed Phase 1.

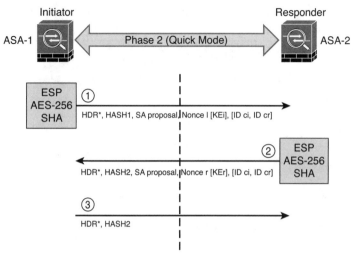

Figure 8-5 *IPsec Phase 2 Negotiation*

The following are the steps illustrated in Figure 8-5:

1. ASA-1 sends the identity information, IPsec SA proposal, nonce payload, and (optional) Key Exchange (KE) payload if Perfect Forward Secrecy (PFS) is used. PFS is employed to provide additional Diffie-Hellman calculations.

2. ASA-2 evaluates the received proposal against its configured proposal and sends the accepted proposal back to ASA-1, along with its identity information, nonce payload, and the optional KE payload.

3. ASA-1 evaluates the ASA-2 proposal and sends a confirmation that the IPsec SAs have been successfully negotiated. This starts the data encryption process.

IPsec uses two different protocols to encapsulate the data over a VPN tunnel:

- **Encapsulation Security Payload (ESP):** IP Protocol 50

- **Authentication Header (AH):** IP Protocol 51

ESP is defined in RFC 4303, "IP Encapsulating Security Payload (ESP)," and AH is defined in RFC 4302, "IP Authentication Header."

IPsec can use two modes with either AH or ESP:

- **Transport mode:** Protects upper-layer protocols, such as User Datagram Protocol (UDP) and TCP

- **Tunnel mode:** Protects the entire IP packet

Transport mode is used to encrypt and authenticate the data packets between the peers. A typical example is the use of GRE over an IPsec tunnel. Tunnel mode is employed to encrypt and authenticate the IP packets when they are originated by the hosts connected behind the VPN device.

8

Figure 8-6 illustrates the differences between IPsec transport mode versus tunnel mode.

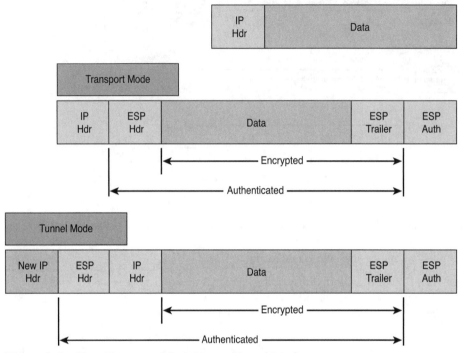

Figure 8-6 *IPsec Transport Mode Versus Tunnel Mode*

> **TIP** Tunnel mode is the default mode in Cisco IPsec devices.

NAT Traversal (NAT-T)

The security protocols (AH and ESP) are Layer 3 protocols and do not have Layer 4 port information. If an IPsec peer is behind a PAT device, the ESP or AH packets are typically dropped. To work around this, many vendors, including Cisco Systems, use a feature called IPsec pass-through. The PAT device that is capable of IPsec pass-through builds the Layer 4 translation table by looking at the SPI values on the packets.

Many industry vendors, including Cisco Systems, implement another feature called NAT Traversal (NAT-T). With NAT-T, the VPN peers dynamically discover whether an address translation device exists between them. If they detect a NAT/PAT device, they use UDP port 4500 to encapsulate the data packets, subsequently allowing the NAT device to successfully translate and forward the packets.

> **TIP** IKEv2 enhances the IPsec interoperability between vendors by offering built-in technologies such as Dead Peer Detection (DPD), NAT Traversal (NAT-T), and Initial Contact.

IKEv2

IKE version 2 (IKEv2) is defined in RFC 5996 and enhances the function of performing dynamic key exchange and peer authentication. IKEv2 simplifies the key exchange flows and introduces measures to fix vulnerabilities present in IKEv1. Both IKEv1 and IKEv2 protocols operate in two phases. IKEv2 provides a simpler and more efficient exchange.

Phase 1 in IKEv2 is IKE_SA, consisting of the message pair IKE_SA_INIT. IKE_SA is comparable to IKEv1 Phase 1. The attributes of the IKE_SA phase are defined in the Key Exchange Policy. Phase 2 in IKEv2 is CHILD_SA. The first CHILD_SA is the IKE_AUTH message pair. This phase is comparable to IKEv1 Phase 2. Additional CHILD_SA message pairs can be sent for rekey and informational messages. The CHILD_SA attributes are defined in the data policy.

The following differences exist between IKEv1 and IKEv2:

- IKEv1 Phase 1 has two possible exchanges: main mode and aggressive mode. There is a single exchange of a message pair for IKEv2 IKE_SA.

- IKEv2 has a simple exchange of two message pairs for the CHILD_SA. IKEv1 uses an exchange of at least three message pairs for Phase 2. In short, IKEv2 has been designed to be more efficient than IKEv1, since fewer packets are exchanged and less bandwidth is needed compared to IKEv1.

- IKEv2 supports the use of next-generation encryption protocols and anti-DoS capabilities.

- Despite IKEv1 supporting some of the authentication methods used in IKEv2, IKEv1 does not allow the use of Extensible Authentication Protocol (EAP). EAP allows IKEv2 to provide a solution for remote-access VPN, as well.

TIP   IKEv1 and IKEv2 are incompatible protocols; subsequently, you cannot configure an IKEv1 device to establish a VPN tunnel with an IKEv2 device.

Figure 8-7 illustrates the attributes negotiated in IKEv1 exchanges in comparison to IKEv2.

Because the message exchanges within IKEv2 are fewer than IKEv1 and cryptographically expensive key material is exchanged in the first two messages (SA_INIT), an attacker could cause a denial-of-service (DoS) condition on an IKEv1-enabled device by sending many SA_INIT requests with a spoofed source address. Subsequently, the gateway would generate key material and dedicate resources to connections that would not be established. The good news is that IKEv2 has the ability to use a stateless cookie. The use of this stateless cookie results in a zero state being assigned to an IKEv2 session if the VPN gateway is under attack.

In IKEv1 implementations, the lifetime of the IKE Phase 1 security association (SA) is negotiated in the first pair of messages (MM1 and MM2). The lifetime configured on the responder must be equal to or less than the lifetime proposed by the initiator in order for Phase 1 to be established. This introduced incompatibilities among different vendors because IKEv1 sessions may not be able to be established if the IKEv1 lifetimes don't match. In IKEv2, the lifetime is a locally configured value that is not negotiated between peers. This means that a device that is a participant of a VPN tunnel will delete or rekey a session

8

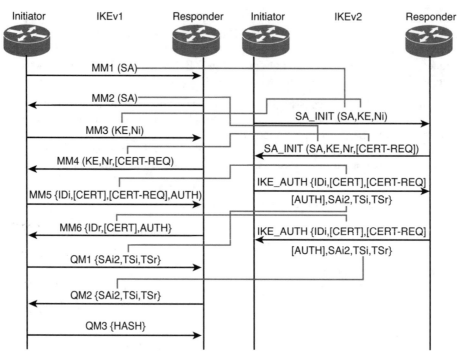

Figure 8-7   *Attributes Negotiated in IKEv1 Exchanges in Comparison to IKEv2*

when its local lifetime expires. Each IKEv2 peer can dictate which entity will initiate a rekey. Consequently, the incompatibility issues faced many times in IKEv1 implementations are not present in IKEv2.

IKE also supports three authentication methods: pre-shared keys (PSK), digital signatures, and EAP. IKEv2 clients can authenticate via an already established standardized mechanism, in contrast to IKEv1 (since IKEv1 required non-standard extensions). EAP is a standard that allows the use of a number of different authentication methods to validate the identities of IPsec VPN peers. This is why it is used natively in IKEv2 implementations.

SSL VPNs

SSL-based VPNs leverage the SSL protocol. SSL is a legacy protocol and has been replaced by Transport Layer Security (TLS). However, most of the TLS-based VPNs are still being referred to as "SSL VPNs." The Internet Engineering Task Force (IETF) created TLS to consolidate the different SSL vendor versions into a common and open standard.

One of the most popular features of SSL VPN is the capability to launch a browser such as Google Chrome, Microsoft Internet Explorer, or Firefox and simply connect to the address of the VPN device, as opposed to running a separate VPN client program to establish an IPsec VPN connection. In most implementations, a clientless solution is possible. Users can access corporate intranet sites, portals, and email from almost anywhere (even from an airport kiosk). Because most people allow SSL (TCP port 443) over their firewalls, it is unnecessary to open additional ports.

The most successful application running on top of SSL is HTTP because of the huge popularity of the World Wide Web. All the most popular web browsers in use today support HTTPS (HTTP over SSL/TLS). This ubiquity, if used in remote-access VPNs, provides some appealing properties:

- **Secure communication using cryptographic algorithms:** HTTPS/TLS offers confidentiality, integrity, and authentication.

- **Ubiquity:** The ubiquity of SSL/TLS makes it possible for VPN users to remotely access corporate resources from anywhere, using any PC, without having to preinstall a remote-access VPN client.

- **Low management cost:** The clientless access makes this type of remote-access VPN free of deployment costs and free of maintenance problems at the end-user side. This is a huge benefit for the IT management personnel, who would otherwise spend considerable resources to deploy and maintain their remote-access VPN solutions.

- **Effective operation with a firewall and NAT:** SSL VPN operates on the same port as HTTPS (TCP/443). Most Internet firewalls, proxy servers, and NAT devices have been configured to correctly handle TCP/443 traffic. Consequently, there is no need for any special consideration to transport SSL VPN traffic over the networks. This has been viewed as a significant advantage over native IPsec VPN, which operates over IP protocol 50 (ESP) or 51 (AH), which in many cases needs special configuration on the firewall or NAT devices to let traffic pass through.

As SSL VPN evolves to fulfill another important requirement of remote-access VPNs—namely, the requirement of supporting any application—some of these properties are no longer applicable, depending on which SSL VPN technology the VPN users choose. But overall, these properties are the main drivers for the popularity of SSL VPN in recent years and are heavily marketed by SSL VPN vendors as the main reasons for IPsec replacement.

Today's SSL VPN technology uses TLS as secure transport and employs a heterogeneous collection of remote-access technologies such as reverse proxy, tunneling, and terminal services to provide users with different types of access methods that fit different environments.

HTTPS provides secure web communication between a browser and a web server that supports the HTTPS protocol. SSL VPN extends this model to allow VPN users to access corporate internal web applications and other corporate application servers that might or might not support HTTPS, or even HTTP. SSL VPN does this by using several techniques that are collectively called reverse proxy technology.

A reverse proxy is a proxy server that resides in front of the application servers, normally web servers, and functions as an entry point for Internet users who want to access the corporate internal web application resources. To the external clients, a reverse proxy server appears to be the true web server. Upon receiving the user's web request, a reverse proxy relays the user request to the internal web server to fetch the content on behalf of the user and relays the web content to the user with or without additional modifications to the data being presented to the user.

Many web server implementations support reverse proxy. One example is the mod_proxy module in Apache. With so many implementations, you might wonder why you need an SSL

VPN solution to have this functionality. The answer is that SSL VPN offers much more functionality than traditional reverse proxy technologies:

- SSL VPN can transform complicated web and some non-web applications that simple reverse proxy servers cannot handle. The content transformation process is sometimes called "webification." For example, SSL VPN solutions enable users to access Windows or UNIX file systems. The SSL VPN gateway must be able to communicate with internal Windows or UNIX servers and webify the file access in a web browser–presentable format for the VPN users.

- SSL VPN supports a wide range of business applications. For applications that cannot be webified, SSL VPN can use other resource access methods to support them. For users who demand ultimate access, SSL VPN provides network-layer access to directly connect a remote system to the corporate network, in the same manner as an IPsec VPN.

- SSL VPN provides a true remote-access VPN package, including user authentication, resource access privilege management, logging and accounting, endpoint security, and user experience.

The reverse proxy mode in SSL VPN is also known as clientless web access or clientless access because it does not require any client-side applications to be installed on the client machine. Client-based SSL VPN provides a solution where you can connect to the corporate network by just pointing your web browser to the Cisco ASA without the need of additional software being installed in your system.

Cisco AnyConnect Secure Mobility

There is a new wave of technological adoption and security threats—mobile devices that allow employees to work from anywhere at any time. Mobility is not completely new for enterprises. As discussed earlier in this chapter, remote-access and telecommuting solutions have existed for quite some time. However, the rapid proliferation of mobile devices increases on a daily basis. Every organization must embrace mobility to remain competitive and evolve to a new model of efficient workloads, especially because many organizations have spent millions of dollars enabling remote-access VPNs to their networks. In some environments, smartphones, tablets, and other mobile devices have surpassed traditional PC-based devices.

The Cisco AnyConnect Secure Mobility solution is designed to secure connections from these mobile devices. The combination of the Cisco AnyConnect Secure Mobility Client, the Cisco ASA, Cisco ISE, Cisco AMP, Cisco Umbrella, and Cisco content security appliances provides a complete, secure mobility solution.

The Cisco AnyConnect Secure Mobility Client is built on SSL VPN technology that enables security to the network behind the Cisco ASA and also provides corporate policy enablement when users are not connected to the corporate Cisco ASA.

Figure 8-8 shows the Cisco AnyConnect Secure Mobility Client.

Figure 8-8   *The Cisco AnyConnect Secure Mobility Client*

Deploying and Configuring Site-to-Site VPNs in Cisco Routers

As you recall from Figure 8-2, many technologies have been used for site-to-site VPN and have evolved through the years—from static traditional crypto maps (traditional site-to-site VPNs in Cisco IOS and Cisco IOS-XE devices) to DMVPN, GETVPN, and FlexVPN. The sections that follow discuss all of these technologies in more detail.

Traditional Site-to-Site VPNs in Cisco IOS and Cisco IOS-XE Devices

Some people refer to the traditional (original) configuration of site-to-site VPNs in Cisco IOS and Cisco IOS-XE devices as "crypto maps." However, a *crypto map* is a Cisco IOS and/or Cisco IOS-XE software configuration command that performs a number of functions related to setting up an IPsec SA. When you configure a crypto map, the networks you want to be protected by the IPsec tunnel are referenced with access control lists. The IPsec Phase 2 security policy, protocol, mode, and algorithms are defined by a *transform set* (these settings include to whom the session will be established and is defined by the peer statement). Crypto maps are applied to an interface. A crypto map can be applied on a physical or tunnel interface (with certain restrictions).

So, let's take a look at how this all is configured in Cisco routers. Let's assume that you were hired by SecretCorp to establish a site-to-site VPN tunnel between two routers (R1 in London and R2 in Raleigh, North Carolina). The topology shown in Figure 8-9 is used to illustrate this scenario. The goal is for the devices in 10.1.1.0/24 be able to communicate to the devices in 192.168.1.0/24.

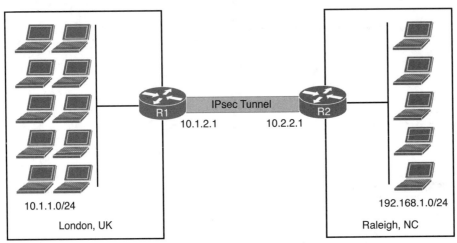

Figure 8-9 *Site-to-Site IPsec VPN Configuration Topology*

Example 8-1 shows R1's configuration. The highlighted comments explain each section of the configuration. In this example, IKEv1 and crypto maps are used.

Example 8-1 *R1's Configuration*

```
!--- Create an ISAKMP policy for IKEv1 Phase 1
crypto isakmp policy 10
 hash sha
 authentication pre-share
!--- Define the pre-shared key and the remote peer address
crypto isakmp key superSecretKey! address 10.2.2.1
!
!--- Create the Phase 2 policy in a transform-set.
!--- The transform set is configured with AES-256 and SHA-HMAC.
crypto ipsec transform-set myset esp-aes 256 esp-sha-hmac
!
!--- Create the crypto map and define the peer IP address, transform
!--- set, and an access control list (ACL) for the tunnel.
crypto map mymap 10 ipsec-isakmp
 set peer 10.2.2.1
 set transform-set myset
 match address 100
```

```
!
interface GigabitEthernet0/0
 ip address 10.1.1.0 255.255.255.0
!
!--- Apply the crypto map on the "outside" interface.
interface GigabitEthernet0/1
 ip address 10.1.2.1 255.255.255.0
 crypto map mymap
!
!--- Add a route for the default gateway (10.1.2.2)
ip route 0.0.0.0 0.0.0.0 10.1.2.2
!
!--- Create an ACL for the traffic to be encrypted over the tunnel.
access-list 100 permit ip 10.1.1.0 0.0.0.255 192.168.1.0 0.0.0.255
!
```

Example 8-2 shows R2's configuration.

Example 8-2   *R2's configuration*

```
!--- Create an ISAKMP policy for IKEv1 Phase 1
crypto isakmp policy 10
 hash sha
 authentication pre-share
!--- Define the pre-shared key and the remote peer address
crypto isakmp key superSecretKey! address 10.1.2.1
!
!--- Create the Phase 2 policy in a transform-set.
!--- The transform set is configured with AES-256 and SHA-HMAC.
crypto ipsec transform-set myset esp-aes 256 esp-sha-hmac
!
!--- Create the crypto map and define the peer IP address, transform
!--- set, and an access control list (ACL) for the tunnel.
crypto map mymap 10 ipsec-isakmp
 set peer 10.1.2.1
 set transform-set myset
 match address 100
!
interface GigabitEthernet0/0
 ip address 192.168.1.0 255.255.255.0
!
!--- Apply the crypto map on the "outside" interface.
interface GigabitEthernet0/1
 ip address 10.2.2.1 255.255.255.0
 crypto map mymap
```

8

```
!
!--- Add a route for the default gateway (10.2.2.2)
ip route 0.0.0.0 0.0.0.0 10.2.2.2
!
access-list 100 permit ip 192.168.1.0 0.0.0.255 10.1.1.0 0.0.0.255
!
```

Tunnel Interfaces

Cisco IOS also has an alternative to crypto maps: tunnel interfaces with tunnel protection. This is accomplished by creating a logical interface that represents the source and destination endpoints of the tunnel. The tunnel interface connects the transport network and the overlay network (the traffic that will transverse within the tunnel). Cisco IOS and Cisco IOS-XE tunnel interfaces support different types of encapsulation (or modes):

- Generic Routing Encapsulation (GRE) protocol

- IP-in-IP

- Distance Vector Multicast Routing Protocol (DVMRP)

- IPv6-in-IPv4

The most common type of encapsulation used with these IPsec implementations is GRE over IPsec.

Generic Routing Encapsulation (GRE) over IPsec is covered next, as well as the reasons why this is needed when using non-IP protocols or multicast.

GRE over IPsec

GRE (protocol 47) is defined by RFC 2784 and extended by RFC 2890. GRE provides a simple mechanism to encapsulate packets of any protocol (the payload packets) over any other protocol (the delivery protocol) between two endpoints. In a GRE tunnel implementation, the GRE protocol adds its own header (4 bytes plus options) between the payload (data) and the delivery header.

> **TIP** GRE supports IPv4 and IPv6. The outer IP header is either IPv4 or IPv6, depending on whether the endpoint addresses are defined as IPv4 or IPv6.

The following is the overhead of a GRE packet compared to the original packet:

- 4 bytes (+ GRE options) for the GRE header,

- 20 bytes (+ IP options) for the outer IPv4 header (GRE over IPv4), or

- 40 bytes (+ extension headers) for the outer IPv6 header (GRE over IPv6).

In GRE over IPsec, the original packets are first encapsulated within GRE, which results in a new IP packet being created inside the network infrastructure device. This GRE packet is then selected for encryption and encapsulated into IPsec, as shown in Figure 8-10.

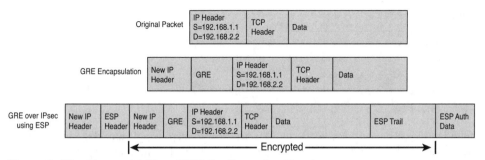

Figure 8-10   *Site-to-site IPsec VPN Configuration Topology*

The actual encapsulation depends on whether tunnel or transport mode is used. Figure 8-10 shows a representation of GRE over IPsec in tunnel mode. When you deploy GRE over IPsec in tunnel mode, the plaintext IPv4 or IPv6 packet is encapsulated into GRE. Then that packet is encapsulated into another IPv4 or IPv6 packet containing the tunnel source and destination IP addresses. This is protected by IPsec for confidentiality and/or integrity assurance, with finally an additional outer IP header being used as the tunnel source and tunnel destination to route the traffic to the destination.

In contrast, with GRE over IPsec transport mode, a plaintext IPv4 or IPv6 packet is GRE-encapsulated and then protected by IPsec for confidentiality and/or integrity protection; the outer IP header with the GRE tunnel source and destination addresses helps route the packet correctly.

The following is the IPsec tunnel mode overhead compared to the original packet:

- 40 bytes (more if IP options are present) for the outer and inner IPv4 headers, or

- 80 bytes (more if extension headers are present) for the outer and inner IPv6 headers, plus the GRE (4 byte) and encryption overhead.

> **NOTE**   In multicast deployments, it is possible that if a device has multiple peers and as a result has security associations from multiple sources, these can conflict with each other. For multicast traffic, there is the issue of multiple destination systems associated with a single SA. A method or system is required to coordinate among all multicast devices the ability to select an SPI or SPIs on behalf of each multicast group and then communicate the group's IPsec information to all of the legitimate members of that multicast group via some mechanism.

Traditional GRE over IPsec configuration is pretty straightforward. Let's use the same topology shown in Figure 8-9 and configure a GRE over IPsec tunnel between R1 and R2. Traditional GRE is configured with Tunnel interfaces. Example 8-3 shows the Tunnel interface configuration of R1.

Example 8-3   *R1's GRE Tunnel Interface Configuration*

```
interface Tunnel0
 ip address 192.168.16.2 255.255.255.0
 tunnel source GigabitEthernet0/1
 tunnel destination 10.2.2.1
 crypto map mymap
```

In traditional GRE over IPsec configurations, the crypto map is applied to the Tunnel interface, as demonstrated in Example 8-3. R2's configuration will mimic and reciprocate R1's configuration.

One of the main use cases for GRE over IPsec is to be able to carry routing protocol information (that is, multicast packets in the case of OSPF) over the VPN tunnel. Additional technologies and standards have been developed for this purpose, as well. The following are examples of such standards and technologies:

- Multicast Group Security Architecture (RFC 3740)

- Group Domain of Interpretation (RFC 6407)

- Cisco's GETVPN (covered later in this chapter)

More About Tunnel Interfaces

As you know by now, the original implementation of IPsec VPNs used on Cisco IOS was known as crypto maps. The concept of configuring a crypto map was closely aligned to the IPsec protocol, with traffic that was required to be encrypted being defined in an access control list. This list was then referenced within an entry in the crypto map along with the IPsec cryptographic algorithms within the transform set. This configuration could become overly complex, and administrators introduced many errors when long access control lists were used.

Cisco introduced the concept of logical tunnel interfaces. These logical interfaces are basically doing the same as traditional crypto maps but they are user configurable. The attributes used by this logical tunnel interface are referenced from the user-configured IPsec profile used to protect the tunnel. All traffic traversing this logical interface is protected by IPsec. This technique allows for traffic routing to be used to send traffic with the logical tunnel being the next hop and results in simplified configurations with greater flexibility for deployments. When the requirement for a user to configure what traffic was to be protected is removed, it reduces the chances of misconfigurations, which frequently occurred with manual configurations of access control lists with crypto maps.

On Cisco routers, every access control entry (ACE) in an access control list (ACL) consumes a ternary content-addressable memory (TCAM) entry. TCAM has a limited number of entries. Consequently, crypto map implementations that contain a large number of access control entries and the device TCAM can become exhausted. Tunnel protection uses only a single TCAM entry and allows for a larger number of IPsec security associations to be established compared to using crypto maps.

TIP   IPsec tunnels can be set up statically or dynamically using virtual interfaces of type VTI (Virtual-Tunnel Interface) or GRE over IPsec. These types of interfaces already existed in legacy IKEv1, and their use has been extended in FlexVPN. FlexVPN is covered later in this chapter.

The **tunnel mode** command was introduced to simplify IPsec and GRE configurations. The difference between the two tunnel interface types is the sequence of encapsulation. In short, VTI (**tunnel mode ipsec {ipv4 | ipv6}**) carries IPv4 or IPv6 traffic directly within IPsec tunnel mode, while GRE over IPsec (**tunnel mode gre {ip | ipv6}**) first encapsulates traffic within GRE and a new IP header before encapsulating the resulting GRE over IP packet within IPsec transport mode.

You can use static virtual interfaces (IPsec or GRE over IPsec) on routers configured as initiators and responders. Example 8-4 includes a static Tunnel interface configuration.

Example 8-4   *A Static Tunnel Interface Configuration (IPsec Mode)*

```
interface Tunnel0
 ip address 192.168.16.2 255.255.255.0
 tunnel source GigabitEthernet0/0
 tunnel mode ipsec ipv4
 tunnel destination 209.165.200.225
 tunnel protection ipsec profile default
```

NOTE   When you configure FlexVPN client profile feature (**crypto ikev2 client flexvpn**), the peer IP address can optionally be set to **"dynamic"**. FlexVPN will be discussed later in this chapter. In this instance, the interface still remains in the configuration, and the peer address is populated at runtime, based on the client profile configuration and tracking states.

You can also use dynamic tunnel interfaces. When you configure dynamic interfaces (IPsec tunnel or GRE over IPsec), the tunnel interface is in a responder-only mode. In the case of IKEv2 implementations, the tunnel negotiation triggers the cloning of a virtual template into a virtual-access interface (**interface Virtual-Access** *number*) that represents the point-to-point link to the remote peer that is protected by IPsec.

8

NOTE   This virtual-access interface remains up as long as the IPsec tunnel is up and gets deleted along with the corresponding IKE and IPsec security associations when the IPsec session is torn down.

Example 8-5 shows a sample virtual template (dynamic VTI) configuration.

Example 8-5 *A Dynamic Virtual-Template Configuration*

```
interface Loopback1
 ip address 10.3.3.3 255.255.255.255


interface Virtual-Template1 type tunnel
 ip unnumbered Loopback1
 tunnel mode ipsec ipv4
 tunnel protection ipsec profile default
```

TIP When you configure native IPsec tunnel encapsulation (IPsec without GRE), a statically configured tunnel interface is often referred to as an sVTI (static VTI) and a virtual template as a dVTI (dynamic VTI). A single site-to-site tunnel is typically sVTI-to-sVTI, while a hub-and-spoke setup is typically sVTI-to-dVTI, with the dVTI on the hub. Furthermore, a feature called multi-SA dVTI provides support for interoperating with non-VTI-compatible peers (for example, legacy crypto maps), where the dVTI will present a mirror image of the traffic selectors requested by the initiator and can terminate multiple IPsec security associations from the same peer on the same virtual-access interface.

Multipoint GRE (mGRE) Tunnels

A particular type of GRE encapsulation is multipoint GRE (mGRE). mGRE interfaces are configured with the **tunnel mode gre multipoint** command. A single static GRE tunnel interface is used as the endpoint for multiple site-to-site tunnels. DMVPN is based on this mode and uses a single interface on each hub as well as on each spoke to terminate all static and dynamic tunnels. FlexVPN does not rely on multipoint interfaces.

An mGRE interface is basically a GRE tunnel interface that acts as if it is directly connected to a group of remote mGRE interfaces. DMVPN uses mGRE and the Next Hop Resolution Protocol (NHRP), acting as a resolution mechanism between a peer's tunnel address (the IP address configured on the peer's mGRE interface) and the mGRE endpoint address on that peer, called the Non-Broadcast Multiple Access (NBMA) address.

DMVPN

DMVPN is a technology created by Cisco that aims to reduce the hub router configuration. In legacy hub-and-spoke IPsec configuration, each spoke router has a separate block of configuration lines on the hub router that define the crypto map characteristics, the crypto ACLs, and the GRE tunnel interface. When deploying DMVPN, you configure a single mGRE tunnel interface, a single IPsec profile, and no crypto access lists on the hub router. The main benefit is that the size of the configuration on the hub router remains the same even if spoke routers are added at a later point.

DMVPN groups many spokes into a single mGRE interface. This allows you to not have to configure a distinct physical or logical interface for each spoke.

DMVPN also uses Next Hop Resolution Protocol (NHRP), which is a client and server protocol (the hub is the server and the spokes are the clients). The hub (or server) maintains an NHRP database of the public interface addresses of the each spoke. Each spoke registers its real address when it boots and queries the NHRP database for real addresses of the destination spokes to build direct tunnels.

DMVPN also eliminates the need for spoke-to-spoke configuration for direct tunnels. When a spoke router tries to transmit a packet to another spoke router, it uses NHRP to dynamically determine the required destination address of the target spoke router. Then the two spoke routers dynamically create an IPsec tunnel between them so data can be directly transferred. This is illustrated in Figure 8-11.

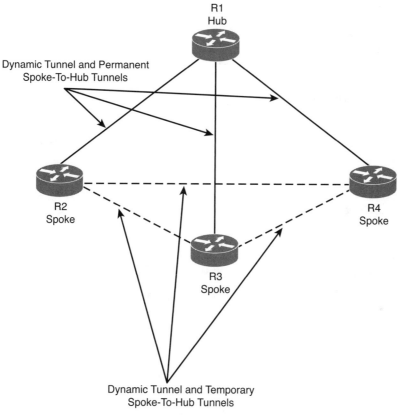

Figure 8-11   *DMVPN Example*

In Figure 8-11, each spoke (R2, R3, and R4) has a permanent IPsec tunnel to the hub (R1), not to the other spokes within the network. Each spoke registers as clients of R1 (the NHRP server). If a spoke wants to send traffic to another spoke, it queries the NHRP server (R1), and after the originating spoke "learns" the peer address of the target spoke, it initiates a dynamic IPsec tunnel to the target spoke.

TIP   DMVPN routers can be configured behind a NAT device. DMVPN supports NAT-T (often also referred to as NAT-Transparency-aware DMVPN).

Example 8-6 shows an example of a hub configuration for DMVPN.

Example 8-6   *A Hub Configuration for DMVPN*

```
!The ISAKMP policy
crypto isakmp policy 1
 encryption aes
 authentication pre-share
 group 14
! A dynamic ISAKMP key and IPsec profile
crypto isakmp key supersecretkey address 0.0.0.0
crypto ipsec transform-set trans2 esp-aes esp-sha-hmac
mode transport
!
crypto ipsec profile my_hub_vpn_profile
 set transform-set trans2
!
! The tunnel interface with NHRP
Interface Tunnel0
 ip address 10.0.0.1 255.255.255.0
 ip nhrp authentication anothersupersecretkey
 ip nhrp map multicast dynamic
 ip nhrp network-id 99
 ip nhrp holdtime 300
 tunnel source GigabitEthernet0/0
 tunnel mode gre multipoint
! This line must match on all nodes that want to use this mGRE tunnel.
 tunnel key 100000
 tunnel protection ipsec profile my_hub_vpn_profile
!
interface GigabitEthernet0/0
 ip address 172.16.0.1 255.255.255.0
!
interface GigabitEthernet0/1
 ip address 192.168.0.1 255.255.255.0
!
router eigrp 1
 network 10.0.0.0 0.0.0.255
 network 192.168.0.0 0.0.0.255
```

Example 8-7 shows an example of a spoke configuration in a DMVPN deployment.

Example 8-7   *A Spoke Configuration Example in a DMVPN Deployment*

```
crypto isakmp policy 1
 encr aes
 authentication pre-share
 group 14
crypto isakmp key supersecretkey address 0.0.0.0
!
```

```
crypto ipsec transform-set trans2 esp-aes esp-sha-hmac
 mode transport
!
crypto ipsec profile my_spoke_vpn_profile
 set transform-set trans2
!
interface Tunnel0
 ip address 10.0.0.2 255.255.255.0
 ip nhrp authentication anothersupersecretkey
 ip nhrp map 10.0.0.1 172.17.0.1
 ip nhrp map multicast 172.17.0.1
 ip nhrp network-id 99
 ip nhrp holdtime 300
! Configures the hub router as the NHRP next-hop server.
 ip nhrp nhs 10.0.0.1
  tunnel source GigabitEthernet0/0
 tunnel mode gre multipoint
 tunnel key 100000
 tunnel protection ipsec profile my_spoke_vpn_profile
!
interface GigabitEthernet0/0
 ip address dhcp hostname Spoke1
!
interface GigabitEthernet0/1
 ip address 192.168.1.1 255.255.255.0
!
router eigrp 1
 network 10.0.0.0 0.0.0.255
 network 192.168.1.0 0.0.0.255
```

GETVPN

Cisco's Group Encrypted Transport VPN (GETVPN) provides a collection of features and capabilities to protect IP multicast group traffic or unicast traffic over a private WAN. GETVPN combines the keying protocol Group Domain of Interpretation (GDOI) and IPsec. GETVPN enables the router to apply encryption to "native" (non-tunneled) IP multicast and unicast packets and removes the requirement to configure tunnels to protect multicast and unicast traffic. GETVPN allows Multiprotocol Label Switching (MPLS) networks to maintain full-mesh connectivity, natural routing path, and Quality of Service (QoS).

GETVPN incorporates Multicast Rekeying. Multicast Rekeying and GETVPN are based on GDOI, as defined in RFC 3547. GDOI is defined as the ISAKMP Domain of Interpretation (DOI) for group key management. The GDOI protocol operates between a group member and a group controller or key server (GCKS), which establishes SAs among authorized group members.

In addition to the existing IKE, IPsec, and multicast technologies, a GETVPN solution relies on the following core building blocks to provide the required functionality:

■ GDOI (RFC 6407)

■ Key servers (KSs)

■ Cooperative (COOP) KSs

■ Group members (GMs)

■ IP tunnel header preservation

■ Group security association

■ Rekey mechanism

■ Time-based anti-replay (TBAR)

■ G-IKEv2

■ IP-D3P

Figure 8-12 shows a sample topology of a GETVPN implementation.

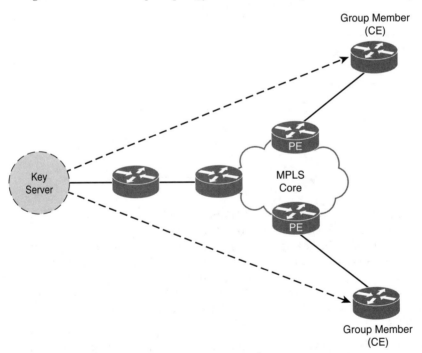

Figure 8-12 *GETVPN Example*

In the example illustrated in Figure 8-12, the group members are customer edge (CE) routers connected to provider edge (PE) routers. The group members register with the key server to get the SAs that are necessary to communicate with the group. The group member sends the group ID to the key server to get the respective policy and keys for this group. These keys are refreshed periodically, and before the current IPsec SAs expire, so that there is no loss of traffic.

The key server creates and maintains the keys for the group. During the registration process of a new group member participant, the key server downloads this policy and the keys to the group member. The key server is also in charge of rekeying before existing keys expire. In short, the key server has two major roles: servicing registration requests and sending rekeys.

GETVPN group members can register at any time. When they register the new group member, it receives the most current policy and keys. The key server verifies the group ID that the group member is attempting to join. If this ID is a valid group ID, the key server sends the SA policy to the group member. After the group member acknowledges that it can handle the downloaded policy, the key server downloads the respective keys.

Figure 8-13 includes details about the two types of keys that group members can retrieve from the key server.

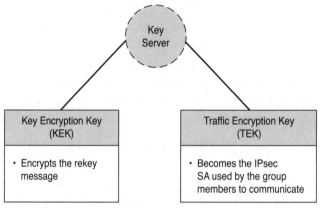

Figure 8-13 *GETVPN Keys*

The key server will send rekey messages if an impending IPsec SA expiration occurs or if the policy has changed on the key server. The rekey messages could also be retransmitted intermittently to account for possible packet loss (if the rekey messages fail for some reason).

The following are the minimum requirements of a basic GETVPN key server configuration:

- IKE policy

- RSA key for re-keying (used to secure the re-key messages)

- IPsec Phase 2 polices

- Traffic classification ACL (to specify which traffic should be encrypted)

The following are the minimum requirements of a basic GETVPN group member configuration:

- IKE policy
- GDOI crypto map
- Crypto map applied to an interface

> **TIP** The following GitHub repository lists several additional resources about GETVPN, including sample configurations: https://github.com/The-Art-of-Hacking/h4cker/blob/master/SCOR/GETVPN.md.

FlexVPN

As you learned earlier in this chapter, FlexVPN is an IKEv2-based solution that provides several benefits beyond traditional site-to-site VPN implementations. The following are some of the benefits of FlexVPN deployments:

- Standards-based solution that can interoperate with non-Cisco IKEv2 implementations.
- Supports different VPN topologies, including point-to-point, remote-access, hub-and-spoke, and dynamic mesh (including per-user or per-peer policies).
- Combines all these different VPN technologies using one command-line interface (CLI) set of configurations. FlexVPN supports unified configuration and show commands, the underlying interface infrastructure, and features across different VPN topologies.
- Support for dynamic overlay routing.
- Integration with Cisco IOS Authentication, Authorization, and Accounting (AAA) infrastructure.
- Supports GRE and native IPsec encapsulations by automatically detecting the encapsulation protocol.
- Supports IPv4 and IPv6 overlay and underlay by automatically detecting the IP transport type.

> **TIP** Since FlexVPN is a based on IKEv2, it provides all the IKEv2 protocol features (including configuration exchange, IKEv2 redirect for server load balancing, cookie challenge for DoS mitigation, NAT Traversal, and IKEv2 fragmentation). It also supports Cisco-specific IKE features such as IKEv2 call admission control, session deletion on certificate expiry, and revocation to all the supported VPN topologies.

Figure 8-14 illustrates the FlexVPN IKE authentication and configuration generation process between a FlexVPN client (IKEv2 initiator) and the FlexVPN server.

FlexVPN implementations support EAP only as a remote authentication method (leveraging an external EAP server). The FlexVPN server does not support EAP as a local authentication method and acts a pass-through authenticator relaying EAP messages between an EAP server and the IKEv2 clients (otherwise known as EAP supplicants or peers). The FlexVPN server authenticates the client or supplicant using a digital certificate-based authentication method. This authentication process is illustrated in Figure 8-15.

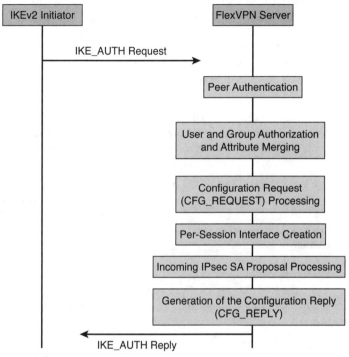

Figure 8-14 *FlexVPN IKE Authentication and Configuration Generation Process*

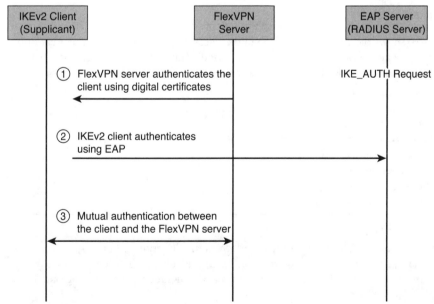

Figure 8-15 *FlexVPN High-Level IKEv2 and EAP Authentication Process*

The EAP messages between the IKEv2 client and the FlexVPN server are embedded within the IKEv2 EAP payload and are transported within the IKE_AUTH request and response messages. The EAP messages between the FlexVPN server and the RADIUS-based EAP server are embedded within the RADIUS EAP-Message attribute. Several other things happen during the second step illustrated in Figure 8-15. The EAP server (RADIUS server) uses an EAP-Request message to request information from the IKEv2 client (EAP supplicant), and the IKEv2 client uses an EAP-Response message to provide the requested information. Subsequently, the RADIUS/EAP server uses EAP-Success or EAP-Failure messages to communicate the authentication result.

EAP is configured in the IKEv2 profile. In order to configure EAP as a remote authentication method, the local authentication method must be certificate-based. You can define the AAA authentication method list to specify the RADIUS server hosting the EAP server, as demonstrated in Example 8-8.

Example 8-8   *Defining the AAA Parameters on the FlexVPN Server*

```
aaa new-model
aaa authentication login my_radius_list group my_radius_group
aaa group server radius my_radius_group
 server name radius_server
```

Example 8-9 shows how to define the RADIUS server. The RADIUS server has the IP address 10.1.2.3. UDP port 1645 is used for RADIUS authentication, and UDP port 1646 is used for RADIUS accounting.

Example 8-9   *Defining the RADIUS Server*

```
radius server radius_server
 address ipv4 10.1.2.3 auth-port 1645 acct-port 1646
 key radius_key
```

Example 8-10 shows the EAP configuration in the FlexVPN server (remote authentication method with the **authentication remote eap** command). This configuration assumes that you have AAA enabled using a RADIUS-based external EAP server.

Example 8-10   *FlexVPN Server EAP Configuration*

```
crypto ikev2 profile ikev2_profile
 aaa authentication eap my_radius_list
 authentication remote eap query-identity
 authentication remote eap timeout 120
 authentication remote rsa-sig
 aaa authentication eap my_radius_list
```

NOTE   You can configure a **timeout** option with the **authentication remote eap** command (in seconds) to define the duration for which the FlexVPN server will wait for an EAP-Response from the IKEv2 client after sending the EAP-Request, before timing out. The default timeout is 90 seconds.

Let's take a look at the FlexVPN deployment example illustrated in Figure 8-16. R1 is configured as a FlexVPN server, and R2 is configured as a FlexVPN client. The server interacts with the RADIUS server (10.1.2.3) for authentication. This is a scalable management of per-peer pre-shared keys using a RADIUS server (AAA-based pre-shared keys).

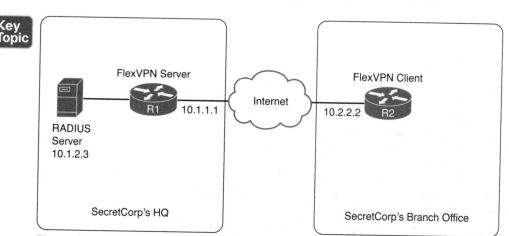

Figure 8-16 *FlexVPN Configuration Example Topology*

Example 8-11 shows the configuration of the FlexVPN server (R1). The highlighted lines explain each configuration section.

Example 8-11 *FlexVPN Server's (R1) Configuration*

```
! AAA configuration. R1 is configured with AAA authorization to use
! the RADIUS server (10.1.2.3) to retrieve the IKEv2 pre-shared keys.

aaa new-model
aaa group server radius radius_group1
 server name radius_server1
!
aaa authorization network aaa_psk_list group radius_group1
!
radius server radius_server1
 address ipv4 10.1.2.3 auth-port 1645 acct-port 1646
 key radius_server1_key
! The IKEv2 name mangler is configured to derive the AAA username from
! the hostname portion of the peer IKEv2 identity of type FQDN.
! When each branch router is configured with a unique local FQDN identity,
! the name mangler will yield a unique AAA username for the pre-shared key
! lookup on the RADIUS server.
! The IKEv2 profile is configured to match all the branch routers, based on
! the domain portion (secretcorp.org) of the peer FQDN identity.
! The profile is configured to use an AAA-based keyring that would retrieve
```

```
! the pre-shared keys, using AAA authorization from the RADIUS
! server specified in the referenced AAA method list.
! The referenced IKEv2 name mangler will yield a unique AAA username for
! pre-shared key lookup on the RADIUS server that is derived from the
! username portion the peer FQDN identity.
crypto ikev2 name-mangler aaa_psk_name_mangler
 fqdn hostname
!
crypto ikev2 profile default
 match identity remote fqdn domain example.com
 identity local fqdn hq.example.com
 authentication local pre-share
 authentication remote pre-share
 keyring aaa aaa_psk_list name-mangler aaa_psk_name_mangler
```

Example 8-12 shows the FlexVPN client's (R2) configuration. R2 is configured with a unique local IKEv2 identity and a local IKEv2 keyring with a symmetric pre-shared key for authentication with the hub router at headquarters.

Example 8-12 *FlexVPN Client's (R2) Configuration*

```
crypto ikev2 keyring local_keyring
peer hub-router
  address 10.1.1.1
  pre-shared-key branch1-hub-key

crypto ikev2 profile default
match identity remote fqdn hq.secretcorp.org
identity local fqdn rtp-branch.secretcorp.org
authentication local pre-share
authentication remote pre-share
keyring local local_keyring
```

Debug and Show Commands to Verify and Troubleshoot IPsec Tunnels

When you configure IPsec VPNs in supported Cisco devices, you can use a plethora of **debug** and **show** commands for IP connectivity, IKEv1, IKEv2, IPsec, GRE encapsulation, RADIUS authentication, and many other related technologies. Furthermore, these technologies are fairly complex and often are made up of a number of components themselves.

Figure 8-17 shows different methods to obtain valuable information from a Cisco infrastructure device for troubleshooting.

> **TIP** When you are troubleshooting any VPN implementations (or any other deployments for that matter), having a good background and knowledge of how all related protocols work will always help.

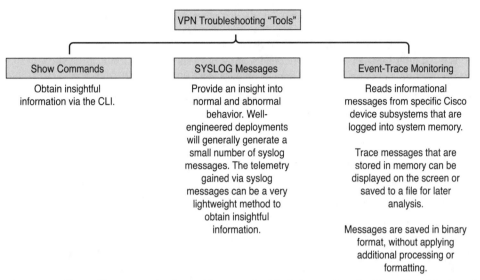

Figure 8-17 *Different Methods to Obtain Valuable Information for IPsec Troubleshooting*

Figure 8-18 shows the different IPsec VPN's functions, protocols, and related troubleshooting commands.

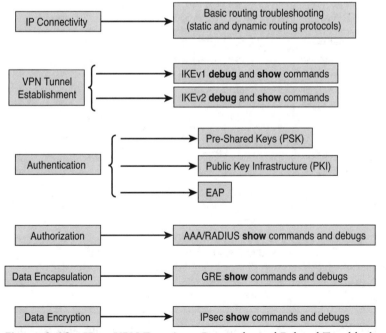

Figure 8-18 *IPsec VPN Functions, Protocols, and Related Troubleshooting Commands*

Let's start by going over some of the most popular **show** commands for troubleshooting IPsec VPNs in Cisco routers. In legacy VPN implementations, you can use the **show crypto isakmp sa** command to troubleshoot and display information about the security associations (SAs) built between the IPsec peers. Example 8-13 shows the output of the **show crypto isakmp sa** command for a connection between 10.1.1.1 and 10.2.2.2.

Example 8-13   *The Output of the show crypto isakmp sa Command*

```
R2# show crypto isakmp sa
dst       src       state     conn-id    slot
10.1.1.1  10.2.2.2   QM_IDLE    1          0
```

In newer IKEv2 implementations, you can display the IKE SAs with the **show crypto ikev2 sa** command, as shown in Example 8-14.

Example 8-14   *Displaying the IKE SAs with the show crypto ikev2 sa Command*

```
R1# show crypto ikev2 sa
Tunnel-id   Local          Remote          fvrf/ivrf         Status
2           10.1.1.1/500   10.2.2.2/500    (none)/(none)     READY
      Encr: 3DES, Hash: SHA96, DH Grp:2, Auth: PSK
      Life/Active Time: 86400/361 sec
```

You can also display more detailed information about the IKEv2 SAs with the **show crypto ikev2 sa detailed** command, as demonstrated in Example 8-15.

Example 8-15   *The Output of the show crypto ikev2 sa detailed Command*

```
R1# show crypto ikev2 sa detailed
Tunnel-id   Local          Remote          fvrf/ivrf         Status
2           10.1.1.1/500   10.2.2.2/500    (none)/(none)     READY
      Encr: 3DES, Hash: SHA96, DH Grp:2, Auth: PSK
      Life/Active Time: 86400/479 sec
      CE id: 0, Session-id: 2, MIB-id: 2
      Status Description: Negotiation done
      Local spi: ACF1453548BE731C
      Remote spi: 45CB158F05817B3A
      Local id:      10.1.1.1          Remote id:      10.2.2.2
      Local req mess id:   3           Remote req mess id: 0
      Local next mess id:  3           Remote next mess id: 1
      Local req queued:   3            Remote req queued: 0
      Local window:    5               Remote window: 5
      DPD configured for 0 seconds
      NAT-T is not detected
      Redirected From: 10.1.1.88
```

To display information about the active IKEv2 sessions, use the **show crypto ikev2 session** command, as shown in Example 8-16.

Example 8-16   *Output of the show crypto ikev2 session Command*

```
R1# show crypto ikev2 session
Session-id:1, Status:UP-ACTIVE, IKE count:1, CHILD count:1
Tunnel-id   Local           Remote          fvrf/ivrf        Status
1           10.1.1.1/500    10.2.2.2/500    (none)/(none)    READY
      Encr: 3DES, Hash: SHA96, DH Grp:2, Auth: PSK
      Life/Active Time: 86400/65 sec
Child sa: local selector  10.0.0.1/0 - 10.0.0.1/65535
          remote selector 10.0.0.2/0 - 10.0.0.2/65535
          ESP spi in/out: 0x9430B91/0x6C060041
          CPI in/out: 0x9FA4/0xC669
```

> **NOTE**   You can also obtain more detailed information about the IKEv2 sessions with the **show crypto ikev2 session detailed** command.

The following are some additional **show** commands to display IKEv2 statistics:

- show crypto ikev2 stats
- show crypto ikev2 stats exchange
- show crypto ikev2 stats ext-service
- show crypto ikev2 stats priority-queue
- show crypto ikev2 stats timeout

> **TIP**   The following GitHub repository includes a collection of Cisco IPsec VPN sample configurations and troubleshooting guides that can be used to enhance your learning: https://github.com/The-Art-of-Hacking/h4cker/blob/master/SCOR/IPSEC-VPNs.md.

You can also use numerous **debug** commands to troubleshoot IPsec implementations, such as the following:

- debug crypto isakmp (for legacy IKE implementations)
- debug crypto ikev2 (for IKEv2 implementations)
- debug crypto ikev2 internal
- debug radius authentication (useful to troubleshoot authentication communication between the router and the RADIUS server)
- debug crypto ipsec

Example 8-17 includes the output of the **debug crypto ikev2, debug crypto ikev2 internal**, and **debug radius authentication** commands during a FlexVPN tunnel establishment between R1 and R2.

Example 8-17 *Troubleshooting a FlexVPN Deployment Using debug crypto ikev2, debug crypto ikev2 internal, and debug radius authentication*

```
IKEv2:(SESSION ID = 69,SA ID = 1):Received Packet [From 10.1.1.1:59040/To
10.2.2.2:500/VRF i0:f0]
Initiator SPI : 3A1C6331D3A4726E - Responder SPI : 123A08C78505C625 Message id: 1
IKEv2 IKE_AUTH Exchange REQUEST
Payload contents:
IKEv2-INTERNAL:Parse Vendor Specific Payload: (CUSTOM) VID IDi CERTREQ CFG SA
IKEv2:(SESSION ID = 69,SA ID = 1):Generating EAP request
IKEv2:(SESSION ID = 69,SA ID = 1):Building packet for encryption.
Payload contents:
 VID IDr CERT CERT AUTH EAP
IKEv2:(SESSION ID = 69,SA ID = 1):Sending Packet [To 10.1.1.1:59040/From
10.2.2.2:4500/VRF i0:f0]
Initiator SPI : 3A1C6331D3A4726E - Responder SPI : 123A08C78505C625 Message id: 1
IKEv2 IKE_AUTH Exchange RESPONSE
IKEv2:(SESSION ID = 69,SA ID = 1):Received Packet [From 10.1.1.1:59040/To
10.2.2.2:4500/VRF i0:f0]
Initiator SPI : 3A1C6331D3A4726E - Responder SPI : 123A08C78505C625 Message id: 2
IKEv2 IKE_AUTH Exchange REQUEST


Payload contents:
 EAP


IKEv2:(SESSION ID = 69,SA ID = 1):Processing EAP response
IKEv2:Use authen method list TEST
IKEv2:pre-AAA: client sent User1 as EAP-Id response
IKEv2:sending User1 [EAP-Id] as username to AAA
IKEv2:(SA ID = 1):[IKEv2 -> AAA] Authentication request sent
RADIUS(00000049): Send Access-Request to 10.1.2.3:1812 id 1645/36, len 196
RADIUS:  authenticator 5A 04 CC CF DA 86 4C D3 - 4E B8 8A F3 75 47 52 81
RADIUS:  Service-Type        [6]   6   Login                   [1]
RADIUS:  Vendor, Cisco       [26]  26
RADIUS:   Cisco AVpair       [1]   20  "service-type=Login"
RADIUS:  Vendor, Cisco       [26]  28
RADIUS:   Cisco AVpair       [1]   22  "isakmp-phase1-id=router1"
RADIUS:  Calling-Station-Id  [31]  16  "10.2.2.2"
RADIUS:  Vendor, Cisco       [26]  63
RADIUS:   Cisco AVpair       [1]   57  "audit-session-
id=L2L0127842111ZO2L40A6869A2ZH1194B127221"
RADIUS:  User-Name           [1]   4   "omar"
RADIUS:  EAP-Message         [79]  9
RADIUS:   02 3B 00 07 01 70 31                [ ;omar]
RADIUS:  Message-Authenticato[80]  18
RADIUS:   41 41 1E 2B 3B B9 4A 55 F0 1C 04 0A D4 60 62 86        [ AA+;JU`b]
```

```
RADIUS:  NAS-IP-Address      [4]   6   10.1.2.17
RADIUS: Received from id 1645/36 172.16.1.3:1812, Access-Challenge, len 1159
RADIUS:  authenticator 52 73 15 7B 6F A2 46 38 - 7D 6B EB 3D 81 88 71 D0
RADIUS:  Service-Type        [6]   6   NAS Prompt                 [7]
RADIUS:  EAP-Message         [79]  24


IKEv2:(SESSION ID = 69,SA ID = 1):Generating EAP request
IKEv2:(SESSION ID = 69,SA ID = 1):Building packet for encryption.
Payload contents:
 EAP
IKEv2:(SESSION ID = 69,SA ID = 1):Sending Packet [To 10.1.1.1:59040/From
10.2.2.2:4500/VRF i0:f0]
Initiator SPI : 3A1C6331D3A4726E - Responder SPI : 123A08C78505C625 Message id: 2
IKEv2 IKE_AUTH Exchange RESPONSE
```

In Example 8-17, all the IKEv2-related debug messages start with the "IKEv2" label. All the RADIUS debug messages start with the "RADIUS" label. The router starts by negotiating the IKEv2 exchange from 10.1.1.1 to 10.2.2.2. It then generates the EAP request, authenticates the peers, and sends an authentication request to the RADIUS server (for username **omar**). After the user is authenticated, the IKEv2 SAs are established.

Enabling detailed debugging (with **debug** commands) could potentially increase the load on busy systems. The event-trace monitoring feature allows logging information to be stored in binary files so that you can later retrieve them without adding any more stress on the infrastructure device.

You can use the **show monitor event-trace crypto ikev2** command to obtain and view errors, events, or exceptions in IKEv2 negotiations, as demonstrated in Example 8-18.

Example 8-18   *The show monitor event-trace crypto ikev2 Command*

```
R1# show monitor event-trace crypto ikev2 ?
  error      Show IKEv2 Errors
  event      Show IKEv2 Events
  exception  Show IKEv2 Exceptions
```

Each of the **show monitor event-trace crypto ikev2** command options have several other options, as shown in Example 8-19.

Example 8-19   *The show monitor event-trace crypto ikev2 Command Options*

```
R1# show monitor event-trace crypto ikev2 error ?
  all        Show all the traces in current buffer
  back       Show trace from this far back in the past
  clock      Show trace from a specific clock time/date
  from-boot  Show trace from this many seconds after booting
  latest     Show latest trace events since last display
  parameters Parameters of the trace
```

Example 8-20 demonstrates the gathering of information pertaining to IKEv2 using the event-trace monitoring feature by invoking the **show monitor event-trace crypto ikev2 error all** command.

Example 8-20 *Output of the show monitor event-trace crypto ikev2 error all Command*

```
R1# show monitor event-trace crypto ikev2 error all
SA ID:1 SESSION ID:1 Remote: 10.2.2.2/500 Local:
                     10.1.1.1/500
SA ID:1 SESSION ID:1 Remote: 10.2.2.2/500 Local:
                     10.1.1.1/500  : Auth exchange failed
SA ID:1 SESSION ID:1 Remote: 10.2.2.2/500 Local:
                     10.1.1.1/500  Negotiation aborted due to ERROR: Auth
                     exchange failed
ID:0 SESSION ID:0    Optional profile description not
                     updated in PSH
```

In Example 8-20, the IKEv2 negotiation failed because of an authentication failure.

You can also use the following commands to take advantage of the event-trace feature to troubleshoot other IPsec VPN functions:

- **show monitor event-trace crypto ipsec** (for IPsec Phase 2 information)

- **show monitor event-trace crypto pki error all**, **show monitor event-trace crypto pki event all**, and **show monitor event-trace crypto pki event internal all** (for PKI troubleshooting)

- **show monitor event-trace dmvpn** (for PKI troubleshooting)

- **show monitor event-trace gdoi** (for GDOI and GETVPN troubleshooting)

> **TIP** The following document includes additional details about the Cisco IOS and Cisco IOS-XE Event Tracer: https://www.cisco.com/c/en/us/td/docs/ios-xml/ios/bsm/configuration/15-2mt/bsm-event-tracer.html.

Configuring Site-to-Site VPNs in Cisco ASA Firewalls

There are several ways to configure static site-to-site VPNs in the Cisco ASA (via ASDM and via the CLI). The configuration is very similar to the traditional Cisco IOS site-to-site IPsec VPN configuration. The configuration process includes the following steps:

Step 1. Enable ISAKMP.

Step 2. Create ISAKMP policy.

Step 3. Set the tunnel type.

Step 4. Define the IPsec policy.

Step 5. Configure the crypto map.

Step 6. Configure traffic filtering (optional).

Step 7.   Bypass NAT (optional).

Step 8.   Enable Perfect Forward Secrecy (optional).

TIP   This chapter focuses on IKEv2 because of its security and scalability enhancements. You can still use IKEv1 if necessary. In most cases, replace the IKEv2 keyword with IKEv1 when configuring IKEv1. The Cisco ASA allows you to migrate your existing IKEv1 configuration to IKEv2 by using the **migrate** command:

```
migrate {l2l | remote-access {ikev2 | ssl} | overwrite}
```

The options are as follows:

■ **l2l:** This option converts current IKEv1 site-to-site tunnels to IKEv2.

■ **remote-access:** This option converts the existing IKEv1 or SSL remote-access configuration to IKEv2.

■ **overwrite:** This option allows you to convert the existing IKEv1 configuration and removes the underlying IKEv2 configuration.

Step 1: Enable ISAKMP in the Cisco ASA

IKE Phase 1 configuration starts by enabling ISAKMP (version 1 or version 2) on the interface that terminates the VPN tunnels. Typically, it is enabled on the outside, Internet-facing interface. If ISAKMP is not enabled on an interface, the Cisco ASA does not listen for the ISAKMP traffic (UDP port 500) on that interface. So even if you have a fully configured IPsec tunnel, if ISAKMP is not enabled on the outside interface, the Cisco ASA will not respond to a tunnel initialization request.

Example 8-21 shows the CLI output if you want to enable IKEv2 on the outside interface. If you want to use IKEv1, use the **crypto ikev1 enable outside** command.

Example 8-21   *Enabling IKEv2 in the Cisco ASA*

```
omar-asa(config)# crypto ikev2 enable outside
```

Step 2: Create the ISAKMP Policy

After you enable ISAKMP on the interface, create a Phase 1 policy that matches the other end of the VPN connection. The Phase 1 policy negotiates the encryption and other parameters that are useful in authenticating the remote peer and establishing a secure channel for both VPN peers to use for communication.

In the ISAKMP policy, you can specify the following attributes:

■ **Priority:** Enter a number between 1 and 65535. If multiple ISAKMP policies are configured, the Cisco ASA checks the ISAKMP policy with the lowest priority number first. If there is no match, it checks the policy with the next highest priority number, and so on, until all policies have been evaluated. A priority value of 1 is evaluated first, and a priority value of 65535 is evaluated last.

- **Encryption:** Select the respective encryption type. Choosing AES 256-bit encryption is recommended because the performance impact is pretty much the same as using a weaker encryption algorithm, such as DES.

- **D-H Group:** Choose the appropriate Diffie-Hellman (D-H) group from the drop-down list. The D-H group is used to derive a shared secret to be used by the two VPN devices.

- **Integrity Hash:** A hashing algorithm provides data integrity by verifying that the packet has not been changed in transit. You have the option to use SHA-1 or MD5. Using SHA-1 is recommended because it provides better security than MD5 and results in fewer hash collisions.

- **Pseudo Random Function PRF Hash (IKEv2 only):** Choose the PRF that you want to use to construct the keying material for all the cryptographic algorithms used in the SAs.

- **Authentication (IKEv1 only):** Choose the appropriate authentication type. The authentication mechanism establishes the identity of the remote IPsec peer. You can use pre-shared keys for authenticating a small number of IPsec peers, whereas RSA signatures are useful if you are authenticating a large number of peers.

- **Lifetime:** Specify the lifetime within which a new set of ISAKMP keys can be renegotiated. You can specify a finite lifetime between 120 and 2,147,483,647 seconds. You can also check Unlimited in case the remote peer does not propose a lifetime. Cisco recommends that you use the default lifetime of 86,400 seconds for IKE rekeys.

Example 8-22 shows how to configure an ISAKMP policy for AES-256 encryption, SHA integrity and PRF hashing, D-H group 5, with 86,400 seconds as the ISAKMP key lifetime.

Example 8-22   *Configuring an IKEv2 Policy in the Cisco ASA*

```
omar-asa(config)# crypto ikev2 policy 1
omar-asa(config-isakmp-policy)# encryption aes-256
omar-asa(config-isakmp-policy)# integrity sha
omar-asa(config-isakmp-policy)# group 5
omar-asa(config-isakmp-policy)# prf sha
omar-asa(config-isakmp-policy)# lifetime seconds 86400
```

If one of the ISAKMP attributes is not configured, the Cisco ASA adds that attribute with its default value. To remove an ISAKMP policy, use the **clear config crypto ikev2 policy** command, followed by the policy number to be removed.

Step 3: Set Up the Tunnel Groups

A tunnel group, also known as a connection profile, defines a site-to-site or a remote-access tunnel and is used to map the attributes that are assigned to a specific IPsec peer. The remote-access connection profile is used to terminate all types of remote-access VPN tunnels, such as IPsec, L2TP over IPsec, and SSL VPN.

For the site-to-site IPsec tunnels, the IP address of the remote VPN device is used as the tunnel group name. For an IPsec device whose IP address is not defined as the tunnel group, the Cisco ASA tries to map the remote device to the default site-to-site group called DefaultL2LGroup, given that the pre-shared key between the two devices matches.

> **TIP** If you use ASDM, **DefaultL2LGroup** is shown in the ASDM configuration, but if you look at the Cisco ASA configuration via the CLI, it does not appear unless a default attribute within that tunnel group is modified.

If you are using pre-shared keys as an authentication mechanism, configure a long alphanumeric key with special characters. It is difficult to decode a complex pre-shared key, even if someone tries to break it by using brute force.

Example 8-23 shows how to configure a site-to-site tunnel group on Cisco ASA if the peer's public IP address is 209.165.201.1. You define the pre-shared key under the **ipsec-attributes** of the tunnel group by using the **pre-shared-key** command.

Example 8-23 *Setting Up the Tunnel Group*

```
omar-asa(config)# tunnel-group 209.165.201.1 type ipsec-l2l
omar-asa(config)# tunnel-group 209.165.201.1 ipsec-attributes
omar-asa(config-tunnel-ipsec)# ikev2 remote-authentication pre-shared-key mY!PsK3y
omar-asa(config-tunnel-ipsec)# ikev2 local-authentication pre-shared-key mY!PsK3y
```

Step 4: Define the IPsec Policy

An IPsec transform set specifies what type of encryption and hashing to use for the data packets after a secure connection has been established. This provides data authentication, confidentially, and integrity. The IPsec transform set is negotiated during quick mode. You can configure the following attributes within the IPsec policy (transform set):

- **Set Name:** Specify the name for this transform set. This name has local significance and is not transmitted during IPsec tunnel negotiations.

- **Encryption:** Choose the appropriate encryption type. As mentioned earlier in the IKE configuration, it is recommended that you select AES 256-bit encryption because the performance impact is pretty much the same as using a weaker encryption algorithm such as DES.

- **Integrity Hash (IKEv2 only):** Choose the appropriate hash type. A hashing algorithm provides data integrity by verifying that the packet has not been changed in transit. You have the option to use MD5, SHA-1, SHA-256, SHA-384, SHA-512, or None. It is recommended to use SHA-512 because it provides stronger security than all the other options.

- **ESP Authentication (IKEv1 only):** Choose the appropriate ESP authentication method. Refer to the Cisco Next-Generation Cryptography whitepaper at the following link for the most up-to-date recommendations about encryption and hashing algorithms: https://tools.cisco.com/security/center/resources/next_generation_cryptography.

- **Mode (IKEv1 only):** Choose the appropriate encapsulation mode. You have the option to use transport mode or tunnel mode. Transport mode is used to encrypt and authenticate the data packets that are originated by the VPN peers. Tunnel mode is used to encrypt and authenticate the IP packets when they are originated by the hosts connected behind the VPN device. In a typical site-to-site IPsec connection, tunnel mode is always used.

Example 8-24 shows how an IPsec (IKEv2) policy can be configured through the CLI.

Example 8-24   *Configuring the IPsec Policy in the Cisco ASA*

```
omar-asa(config)# crypto ipsec ikev2 ipsec-proposal mypolicy
omar-asa(config-ipsec-proposal)# protocol esp encryption aes-256
omar-asa(config-ipsec-proposal)# protocol esp integrity sha-512
```

Step 5: Create the Crypto Map in the Cisco ASA

After you have configured both ISAKMP and IPsec policies, create a crypto map so that these policies can be applied to a static site-to-site IPsec connection. A crypto map instance is considered complete when it has the following three parameters:

- At least one IPsec policy (transform set)

- At least one VPN peer

- An encryption ACL

A crypto map uses a priority number (or sequence number) to define an IPsec instance. Each IPsec instance defines a VPN connection to a specific peer. You can have multiple IPsec tunnels destined to different peers. If the Cisco ASA terminates an IPsec tunnel from another VPN peer, a second VPN tunnel can be defined, using the existing crypto map name with a different priority number. Each priority number uniquely identifies a site-to-site tunnel. However, the Cisco ASA evaluates the site-to-site tunnel with the lowest priority number first.

Using ASDM, you can configure the following attributes in a Cisco ASA crypto map:

- **Interface:** You must choose an interface that terminates the IPsec site-to-site tunnel. As mentioned earlier, it is usually the outside, Internet-facing interface. You can apply only one crypto map per interface. If there is a need to configure multiple site-to-site tunnels, you must use the same crypto map with different priority numbers.

- **Policy Type:** If the remote IPsec peer has a static IP address, choose static. For a remote static peer, the local Cisco ASA can both initiate and respond to an IPsec tunnel request. Choosing dynamic is useful if the remote peer receives a dynamic IP address on its outside interface. In this scenario, where the peer is marked as dynamic, it is the peer's responsibility to initiate the VPN connection; it cannot be initiated by the hub.

- **Priority:** You must specify the priority number for this site-to-site connection. If multiple site-to-site connections are defined for a particular crypto map, Cisco ASA checks the connection with the lowest priority number first. A priority value of 1 is evaluated first; a priority value of 65535 is evaluated last.

- **IPsec Proposals (Transform Sets):** Choose a predefined IPsec policy (transform set). You can choose multiple transform sets, in which case the Cisco ASA sends all the configured transform sets to its peer if it is initiating the connection. If the Cisco ASA is responding to a VPN connection from the peer, it matches the received transform sets with the locally configured transform sets and chooses one to use for the VPN connection. You can add up to 11 transform sets with varying algorithms.

- **Connection Type:** If you want either VPN peer to initiate an IPsec tunnel, choose Bidirectional from the drop-down menu.

- **IP Address of Peer to Be Added:** Specify the IP address of the remote VPN peer. Typically, it is the public IP address of the remote VPN device.

- **Enable Perfect Forward Secrecy:** If you choose to enable Perfect Forward Secrecy (PFS), enable this option and specify the Diffie-Hellman group you want to use. PFS is discussed later in this chapter.

A crypto map is not complete until you define an ACL for the interesting traffic that needs to be encrypted. When a packet enters the Cisco ASA, it gets routed based on the destination IP address. When it leaves the interface, which is set up for a site-to-site tunnel, the encryption engine intercepts that packet and matches it against the encryption access control entries (ACE) to determine whether it needs to be encrypted. If a match is found, the packet is encrypted and then sent out to the remote VPN peer.

An ACL can be as simple as permitting all IP traffic from one network to another or as complicated as permitting traffic originating from a unique source IP address on a particular port destined to a specific port on the destination address. Deploying complicated crypto ACLs using TCP or UDP ports is not recommended. Many IPsec vendors do not support port-level encryption ACLs.

Encryption ACLs also perform a security check for the inbound encrypted traffic. If a cleartext packet matches one of the encryption ACEs, the Cisco ASA drops that packet and generates a syslog message indicating this incident.

TIP   Each ACE creates two unidirectional IPsec SAs. If you have 100 entries in your ACL, then the ASA creates 200 IPsec SAs. Using host-based encryption ACEs is not recommended because that results in numerous ACEs, with double the number of SAs. The Cisco ASA uses system resources to maintain the SAs, which may affect overall performance.

Example 8-25 shows that the Cisco ASA is configured to protect all IP traffic sourced from 192.168.10.0 with a mask of 255.255.255.0 and destined to 10.10.10.0 with a mask of 255.255.255.0. The ACL name is **outside_cryptomap**.

Example 8-25   *Configuring a Crypto Map in the Cisco ASA*

```
omar-asa# configure terminal
omar-asa(config)# access-list outside_cryptomap line 1 remark ACL to encrypt traffic
from omar-asa to NY-asa
omar-asa(config)# access-list outside_cryptomap line 2 extended permit ip
192.168.10.0 255.255.255.0 10.10.10.0 255.255.255.0
omar-asa(config)# crypto map outside_map 1 match address outside_cryptomap
omar-asa(config)# crypto map outside_map 1 set  peer  209.165.201.1
omar-asa(config)# crypto map outside_map 1 set  ikev2 ipsec-proposal mypolicy
omar-asa(config)# crypto map outside_map interface outside
```

8

The Cisco ASA does not allow IP traffic from the remote private network to connect directly to the ASA's inside interface by default. Many enterprises prefer to manage the Cisco ASA using its inside interface from the management network using the VPN connection. You can configure this feature by using the **management-access** command.

Step 6: Configure Traffic Filtering (Optional)

Like a traditional firewall, the Cisco ASA can protect the trusted (inside) network by blocking new inbound connections from the outside, unless the ACL explicitly permits these connections. However, by default, the Cisco ASA allows all inbound connections from the remote VPN network to the inside network without an ACL explicitly allowing them. What that means is that even if the inbound ACL on the outside interface denies the decrypted traffic to pass through, the Cisco ASA still allows it.

This default behavior can be changed if you want the outside interface ACL to inspect the IPsec protected traffic. If you want to configure the Cisco ASA so that specific traffic is allowed or blocked from two hosts or networks, you must follow these steps:

Step 1.   Define an inbound ACE on the Cisco ASA outside interface ACL.

Step 2.   Disable the **vpn sysopt** feature that allows new inbound connections initiated from over the VPN to bypass all access list checks.

Example 8-26 shows how the outside ACL (**outside_acl**) allows traffic from remote host **10.10.10.10** to go to local host **192.168.10.10** on TCP port **23**. The **no sysopt connection permit-vpn** command enables the Cisco ASA to subject all new inbound connections through the firewall to the configured interface ACLs. You can see this from the CLI with the command **show run all sysopt**.

Example 8-26   *Configuring Traffic Filtering over the VPN Tunnel*

```
omar-asa(config)# access-list outside_acl extended permit tcp host 10.10.10.10 host
192.168.10.10 eq 23
omar-asa(config)# access-group outside_acl in interface outside
omar-asa(config)# no sysopt connection permit-vpn
```

TIP   The **sysopt connection permit-vpn** command is a global command. It is enabled by default and allows the Cisco ASA to bypass the ACL check for all the VPN tunnels, including remote-access IPsec tunnels and SSL VPN tunnels. You can still control traffic by defining authorization ACLs on group policies and user policies.

Step 7: Bypass NAT (Optional)

In most cases, you do not want to change the IP addresses for the traffic going over the tunnel. If NAT is configured on the Cisco ASA to change the source or destination IP addresses for non-VPN traffic, you can set up the policies to bypass address translation for traffic destined over the VPN tunnels. You learned about the different NAT implementations in Chapter 7, "Cisco Next-Generation Firewalls and Cisco Next-Generation Intrusion Prevention Systems."

To bypass address translation, you must identify traffic that needs to go over the VPN tunnel and then apply the NAT exemption rule. Example 8-27 shows a NAT exempt policy configuration for IPsec encryption. A network object group is created for the internal network of 192.168.10.0/24 network (called **192.168-Net**). A separate network object group for the 10.10.10.0/24 subnet (called **10.10-Net**) is also configured. Both of the network object groups are associated with the **nat** command to "bypass NAT" or create an exception for traffic originating from the 192.168.10.0/24 network not to be "translated" when communicating to devices over VPN.

Example 8-27   *Configuring a NAT Exempt Policy for IPsec Encryption*

```
omar-asa(config)# object network 192.168-Net
omar-asa(config-network-object)# subnet 192.168.10.0 255.255.255.0
omar-asa(config-network-object)# object network 10.10-Net
omar-asa(config-network-object)# subnet 10.10.10.0 255.255.255.0
omar-asa(config-network-object)# exit
omar-asa(config)# nat (inside,outside) source static 192.168-Net 10.10-Net
destination static 192.168-Net 10.10-Net
```

If you do not define a NAT exemption policy and NAT is performed for VPN traffic, then the crypto ACL should match the post-NAT (or global) IP addresses.

Step 8: Enable Perfect Forward Secrecy (Optional)

Perfect Forward Secrecy (PFS) is a cryptographic technique where the newly generated keys are unrelated to any previously generated key. With PFS enabled, the Cisco ASA creates a new set of keys that is used during the IPsec Phase 2 negotiations. Without PFS, the Cisco ASA uses Phase 1 keys in the Phase 2 negotiations. The Cisco ASA uses Diffie-Hellman groups 1, 2, and 5 for PFS to generate the keys.

Example 8-28 shows how to enable PFS D-H group 5 for a peer that uses sequence number 10, using the CLI.

Example 8-28   *Enabling PFS*

```
omar-asa(config)# crypto map outside_map 10 set pfs group5
```

Additional Attributes in Cisco Site-to-Site VPN Configurations

Cisco ASA provides many advanced features to suit your site-to-site VPN implementations. These features include the following:

- **OSPF updates over IPsec:** Open Shortest Path First (OSPF) uses the multicast methodology to communicate with its neighbors. IPsec, on the other hand, does not allow encapsulation of the multicast traffic. Cisco ASA solves this problem by enabling you to statically define the neighbors so that unicast OSPF packets can be sent to the remote VPN peer.

- **Reverse route injection:** Reverse route injection (RRI) is a way to distribute remote network information into the local network with the help of a routing protocol. With RRI, the Cisco ASA automatically adds static routes about the remote private networks across the tunnel to its routing table and then announces these routes to its neighbors on the local private network, using OSPF.

8

■ **NAT Traversal (NAT-T):** As you learned earlier in this chapter, traditionally, the IPsec tunnels fail to pass traffic if there is a PAT device between the peers. By default, IPsec devices use the Encapsulated Security Payload (ESP) protocol, which does not have any Layer 4 information, and therefore the PAT device ends up dropping the IPsec packet. You can use NAT Traversal (NAT-T) to encapsulate the ESP packets into a UDP port connection on port 4500 so that any intermediate PAT device would have no trouble translating the encrypted packets. NAT-T is dynamically negotiated if both VPN peers are NAT-T capable or if there is a NAT or PAT device between the peers. If both conditions are met, VPN peers start their communication using ISAKMP (UDP port 500), and as soon as a NAT or PAT device is detected, they switch to UDP port 4500 to complete the rest of their negotiations. NAT-T is globally enabled on the Cisco ASA by default. In many cases, the NAT/PAT devices time out the NAT-T encrypted connection on UDP port 4500 entries if there is no active traffic passing through them. NAT-T keepalives are used so that the Cisco ASA can send periodic keepalive messages to prevent the entries from timing out on the intermediary devices. You can specify a NAT-T keepalive range between 10 and 3600 seconds, with a default keepalive timeout of 20 seconds.

■ **Tunnel default gateway:** A Layer 3 device typically has a default gateway that it uses to route packets when the destination address is not found in its routing table. The tunnel default gateway is used to route the packets if they reach the Cisco ASA over an IPsec tunnel and their destination IP address is not found in the routing table. The tunneled traffic can be either remote-access or site-to-site VPN traffic. The tunnel default gateway next-hop address is generally the IP address of the inside router. The tunnel default gateway feature is important if you do not want to define routes to your internal networks on the Cisco ASA and you prefer the tunneled traffic to be sent to the internal router for routing.

■ **Management access:** As briefly mentioned earlier in the chapter, Cisco ASA does not allow the remote private network to manage the Cisco ASA if the traffic traverses a VPN tunnel. The traffic accesses the inside (or any interface other than the interface through which the VPN traffic entered the firewall) of the Cisco ASA. This is true even if the inside interface's IP address is included in the encryption ACL. Many enterprises want to monitor the status of the inside interface over the tunnel to check the appliance's health. To solve this, enable the Management Access feature on the inside interface of the appliance. With this feature turned on, remote devices can use management applications such as SNMP polling, ASDM, Telnet, SSH, ping, HTTPS requests access, syslog messages, and NTP requests.

■ **Fragmentation policies:** The outbound maximum transmission unit (MTU) of an Ethernet interface is typically 1500 bytes. IPsec appends headers when it encrypts the data packets. Therefore, when an original packet is the same size or larger than the outbound interface's MTU, the packet must be fragmented so that IPsec can successfully add its headers. Most of the VPN devices perform fragmentation after encryption. Therefore, the other end of the VPN tunnel is responsible for defragmenting and then decrypting the packet. The problem with this approach is that packet reassembly is typically done at the processor level, which utilizes extra CPU cycles for this additional task. If the packets are fragmented before they are encrypted, then the other side of the tunnel is responsible only for decrypting the packets, and the destination host is responsible for defragmenting them. Thus, you save the CPU overhead of defragmentation on the Cisco ASA by delegating this task to the end hosts. The Cisco ASA, by default, allows fragmentation to occur before packets are encrypted. However, if the do-not-fragment (DF) bit is set on the packets, the Cisco ASA retains the DF bit and the original packet does not get fragmented. Therefore, if large packets with the DF bit set try to pass through the Cisco ASA, they are dropped. This may be something you do not want to occur in your network.

Configuring Remote Access VPNs in the Cisco ASA

Remote-access VPN services provide a way to connect home and mobile users to the corporate network. Until a decade ago, the only way to provide this service was through dialup connections using analog modems. Corporations had to maintain a huge pool of modems and access servers to accommodate remote users. Additionally, they were billed for providing toll-free and long-distance phone services. With the rapid growth of Internet technologies, most remote users have migrated from dialup connections to broadband DSL or cable-modem connections. In response, corporations have moved these users to remote-access VPNs for faster communication.

There are many remote-access VPN protocols available to provide secure network access. The commonly used remote-access VPN protocols include the following:

- Point-to-Point Tunneling Protocol (PPTP)

- Layer 2 Tunneling Protocol (L2TP)

- Layer 2 Forwarding (L2F) Protocol

- IPsec

- L2TP over IPsec

- Secure Sockets Layer (SSL) VPN

The Cisco ASA supports the native IPsec and L2TP over IPsec to provide VPN services in a secure manner. The Cisco IPsec VPN solution uses Cisco VPN Client, whereas L2TP over IPsec uses the built-in VPN client on Microsoft Windows and Android operating systems. AnyConnect supports the Internet Key Exchange version 2 (IKEv2) protocol; it does not support IKE version 1.

Cisco ASA allows mobile and remote users to establish an IPsec VPN tunnel by using the following:

- Cisco AnyConnect Secure Mobility Client (SSL VPN or IKEv2)

- Built-in clients in operating systems such as macOS and Apple iOS products (iPhones, iPads, and iPods)

The Cisco ASA supports VPN connections from Android mobile devices when using the L2TP over IPsec protocol and the native Android VPN client.

8

The Cisco ASA IKEv2 support uses Cisco's IKEv2 implementation toolkit, which is common in AnyConnect, Cisco ASA, and Cisco IOS devices. It includes a few extensions for fragmentation and client redirection support and uses a proprietary EAP method (AnyConnect EAP). However, it uses the same authentication methods supported previously with IPsec and SSL VPN.

Configuring IPsec Remote Access VPN in the Cisco ASA

Let's suppose you are hired by SecretCorp to configure the Cisco ASA illustrated in Figure 8-19 to allow remote-access IPsec VPN clients to connect to the corporate network.

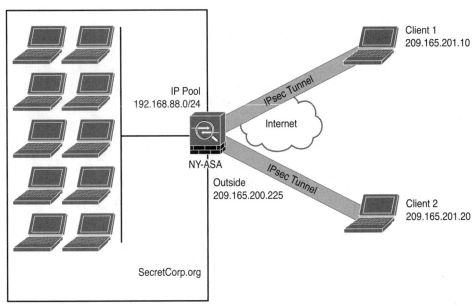

Figure 8-19 *SecretCorp's Remote Access IPsec VPN Deployment*

At the end of the day, the configuration steps are very similar to the ones you have learned throughout this chapter:

Step 1. Enable ISAKMP (IKEv1).

Step 2. Create the IKEv1 (ISAKMP) policy.

Step 3. Set up tunnel and group policies.

Step 4. Define the IPsec policy.

Step 5. Configure user authentication.

Step 6. Assign an IP address.

Step 7. Create a crypto map.

Step 8. Configure traffic filtering (optional).

Step 9. Bypass NAT (optional).

Step 10. Set up split tunneling (optional).

Step 11. Define DNS and WINS addresses (optional).

Example 8-29 shows the Cisco ASA IKEv2 policy, the IPsec policy, and a dynamic crypto map to terminate the remote VPN clients.

Example 8-29   *Cisco ASA Remote Access IPsec VPN IKEv2 Policy, IPsec Policy, and Dynamic Crypto Map*

```
crypto ikev2 policy 1
 encryption aes-256
 integrity sha
 group 5
 prf sha
 lifetime seconds 86400
crypto ikev2 enable outside
!
crypto ikev2 remote-access trustpoint HeadEnd
crypto ipsec ikev2 ipsec-proposal AES256
 protocol esp encryption aes-256
 protocol esp integrity sha-1 md5
!
crypto dynamic-map Anyconnect 65535 set ikev2 ipsec-proposal AES256
crypto map outside_map 65535 ipsec-isakmp dynamic Anyconnect
crypto map outside_map interface outside
```

Example 8-30 shows how the group policy (**SecretCorp_GP**) is configured for the IKEv2 tunnels.

Example 8-30   *Configuring the Group Policy in the Cisco ASA*

```
group-policy SecretCorp_GP internal
group-policy SecretCorp_GP attributes
 vpn-tunnel-protocol ikev2
```

After defining the group policy, you must configure an IP address pool to assign IP addresses to the remote access VPN clients, as demonstrated in Example 8-31.

Example 8-31   *Creating the IP Pool for the VPN Clients*

```
ip local pool VPN_Pool_1 192.168.88.1-192.168.88.254 mask 255.255.255.0
```

Create a tunnel group, assign the IP pool, and associate the previously configured group policy, as demonstrated in Example 8-32.

Example 8-32   *Creating the Tunnel Group for the Remote Access VPN Clients*

```
tunnel-group SecretCorp_TG type remote-access
tunnel-group SecretCorp_TG general-attributes
 address-pool VPN_Pool_1
 default-group-policy SecretCorp_GP
 authentication-server-group LOCAL
```

8

You also need to specify the authentication server group under the tunnel group. In Example 8-32, local authentication is used. In large deployments, an external AAA server (such as a RADIUS server) should be used.

The next step is to configure the tunnel group IPsec attributes and enable IKEv2 authentication, as shown in Example 8-33.

Example 8-33   *Configuring the Tunnel Group IPsec Attributes*

```
tunnel-group SecretCorp_TG ipsec-attributes
 ikev2 remote-authentication certificate
 ikev2 local-authentication certificate SecretCorpHeadEnd
 tunnel-group-map enable rules
 tunnel-group-map CERT_MAP 10 SecretCorp_TG
crypto ca certificate map CERT_MAP 10
 issuer-name co SecretCorp-ROOT-CA
```

In Example 8-33, authentication is done using certificates. Detailed information about certificate configuration and enrollment in the Cisco ASA can be obtained from https://www.cisco.com/c/en/us/support/docs/security-vpn/public-key-infrastructure-pki/200339-Configure-ASA-SSL-Digital-Certificate-I.html.

NOTE   The CCNP Security concentration exam "Implementing Secure Solutions with Virtual Private Networks (SVPN)" and the CCIE Security lab focus on configuration and troubleshooting. Sample configurations throughout this chapter are included to provide you an overview of the different deployment options.

The final step is to configure a NAT exemption rule for the IP addresses that will be assigned to the VPN clients, as demonstrated in Example 8-34.

Example 8-34   *Configuring a NAT Exemption Rule for the VPN Clients*

```
object network NETWORK_OBJ_192.168.88.0_24
 subnet 192.168.88.0 255.255.255.0
nat (inside,outside) source static any any destination static NETWORK_
OBJ_192.168.88.0_24 NETWORK_OBJ_192.168.88.0_24 no-proxy-arp route-lookup
```

Configuring Clientless Remote Access SSL VPNs in the Cisco ASA

The Cisco ASA supports the following SSL VPN modes:

- **Clientless:** The remote client needs only an SSL-enabled browser to access resources on the private network of the Cisco ASAs. SSL clients can access internal resources such as HTTP, HTTPS, and even Windows file shares over the SSL tunnel.

- **Thin client:** The remote client needs to install a small Java-based applet to establish a secure connection to the TCP-based internal resources. SSL clients can access TCP-based internal resources such as HTTP, HTTPS, SSH, and Telnet servers.

- **Full tunnel:** The remote client needs to install an SSL VPN client first to give full access to the internal private network over an SSL tunnel. Using the full tunnel client mode, remote machines to send all IP unicast traffic such as TCP-, UDP-, or even ICMP-based traffic. SSL clients can access internal resources such as HTTP, HTTPS, DNS, SSH, and Telnet servers. Most customers prefer using the full tunnel mode option because a VPN client can be automatically pushed to a user after a successful authentication.

Typically, clientless and thin client solutions are grouped under one umbrella and classified as clientless SSL VPN.

Cisco ASA Remote-Access VPN Design Considerations

The following are several key points that you need to take into consideration before you choose your SSL VPN deployment mode:

- Before you implement the SSL VPN services in Cisco ASA, you must analyze your current environment and determine which features and modes might be useful in your implementation.

- Before designing and implementing the SSL VPN solution for your corporate network, you need to determine whether your users connect to your corporate network from public shared computers, such as workstations made available to guests in a hotel or computers in an Internet kiosk. In this case, using a clientless SSL VPN is the preferred solution to access the protected resources.

- Network security administrators need to determine the size of the SSL VPN deployment, especially the number of concurrent users that will connect to gain network access. If one Cisco ASA is not enough to support the required number of users, clustering or load balancing must be considered to accommodate all the potential remote users.

- The Cisco ASA supports load-balancing the clientless SSL VPN sessions, because the SSL VPN load-balancing configuration is identical to the remote-access IPsec load-balancing configuration.

- The SSL VPN functionality on the ASAs requires that you have appropriate licenses. Make sure that you have the appropriate license for your SSL VPN deployment.

The infrastructure requirements for SSL VPNs include, but are not limited to, the following options:

- **ASA placement:** If you are installing a new Cisco ASA, determine the location that best fits your requirements. If you plan to place it behind an existing corporate firewall, make sure you allow appropriate SSL VPN ports to pass through the firewall.

- **User account:** Before SSL VPN tunnels are established, users must authenticate themselves to either the local database or to an external authentication server. The supported external servers include RADIUS, RADIUS one-time password (OTP), RSA SecurID, Active Directory/Kerberos, generic Lightweight Directory Access Protocol (LDAP), and Duo integration with SAML authentication. Make sure that SSL VPN users have accounts and appropriate access.

> **TIP**   A detailed example of integrating Duo with ASA or Cisco Firepower VPN implementations can be accessed at https://duo.com/docs/cisco.

8

■ **Administrative privileges:** Administrative privileges on the local workstation are required for all connections with port forwarding if you want to use host mapping.

Pre-SSL VPN Configuration Steps

After analyzing the deployment considerations and selecting the SSL VPN as the remote-access VPN solution, you must follow the configuration steps described in this section to properly set up the SSL VPN so that it can be enabled on a Cisco ASA. These tasks include the following:

■ Enroll digital certificates (recommended)

■ Set up tunnel and group policies

■ Set up user authentication

Enrollment is the process of obtaining a certificate from a certificate authority (CA). Even though the Cisco ASA can generate self-signed certificates, use of an external CA is highly recommended. The enrollment process can be broken into three steps, as described in the following sections.

Obtain a CA/root certificate before requesting an identity certificate from the CA server. Make sure you have received a CA certificate from the server in Base64 format. Define a **trustpoint** and then use the **crypto ca authenticate** command to import the CA certificate, as demonstrated in Example 8-35.

Example 8-35 *Importing the CA Certificate Manually*

```
omar-asa(config)# crypto ca trustpoint SecretCorpSSLCert
omar-asa(config-ca-trustpoint)# enrollment terminal
omar-asa(config)# crypto ca authenticate SecretCorpSSLCert
Enter the base 64 encoded CA certificate.
End with the word "quit" on a line by itself

---BEGIN CERTIFICATE---
MIIC0jCCAnygAwIBAgIQI1s45kcfzKZJQnk0zyiQcTANBgkqhkiG9w0BAQUFADCB
hjEeMBwGCSqGSIb3DQEJARYPamF6aWJAY21zY28uY29tMQswCQYDVGEwJVUzEL
MAkGA1UECBMCTkMxDDAKBgNVBAcTA1JUUDEWMBQGA1UEChMNQ21zY28gU31zdGVt
czEMMAoGA1UECxMDVEFDMRYwFAYDVDEw1KYXppYkNBU2VydmVyMB4XDTA0MDYy
---END CERTIFICATE---

quit

INFO: Certificate has the following attributes:
Fingerprint:    2224ced6 55a0095e 243a5ad1 a62ed6a9
Do you accept this certificate? [yes/no]: yes
Trustpoint CA certificate accepted.
% Certificate successfully imported
```

Before requesting an identity certificate, you must generate the RSA key pair through ASDM or through the CLI. If you already have the RSA keys generated that you want to use for SSL encryption, you can skip creating new ones. After generating the RSA keys, request an identity certificate to be used for SSL VPN. In Example 8-36, the CLI configuration for manual enrollment is shown. The **enrollment terminal** subcommand in SecretCorpSSLCert configuration is used to declare manual enrollment of the CA server. This **trustpoint** uses the **SecreCorpSSLCert** RSA key.

Example 8-36   *Manual Certificate Enrollment*

```
omar-asa# configure terminal
omar-asa(config)# domain-name secretcorp.org
omar-asa(config)# crypto key generate rsa label SecretCorpSSLRSA
The name for the keys will be: omar-asa.secretcorp.org


% The key modulus size is 1024 bits
% Generating 1024 bit RSA keys, keys will be non-exportable...[OK]


omar-asa(config)# crypto ca trustpoint SecretCorpSSLCert
omar-asa(ca-trustpoint)# keypair SecretCorpSSLRSA
omar-asa(ca-trustpoint)# id-usage ssl-ipsec
omar-asa(ca-trustpoint)# no fqdn
omar-asa(ca-trustpoint)# subject-name CN=omar-asa
omar-asa(ca-trustpoint)# enrollment terminal
omar-asa(ca-trustpoint)# crypto ca enroll SecretCorpSSLCert
```

NOTE   After you submit a certificate request, the certificate should be in a pending state until the CA administrator approves it.

After the identity certificate is approved by the CA server administrator, use the **crypto ca import** command to import the Base64-encoded ID certificate, as demonstrated in Example 8-37.

Example 8-37   *Importing the ID Certificate*

```
omar-asa(config)# crypto ca import SecretCorpSSLCert certificate

% The fully-qualified domain name in the certificate will be: omar-asa.secretcorp.
org
Enter the base 64 encoded certificate.
End with the word "quit" on a line by itself


----BEGIN CERTIFICATE----
MIIECDCCA7KgAwIBAgIKHJGvRQAAAAAADTANBgkqhkiG9w0BAQUFADCBhjEeMBwG
CSqGSIb3DQEJARYPamF6aWJAY2lzY28uY29tMQswCQYDVGEwJVUzELMAkGA1UE
CBMCTkMxDDAKBgNVBAcTA1JUUDEWMBQGA1UEChMNQ2lzY28gU3lzdGVtczEMMAoG
A1UECxMDVEFDMRYwFAYDVDEw1KYXppYkNBU2VydmVyMB4XDTA0MDkwMjAyNTgw
NVoXDTA1MDkwMjAzMDgwNVowLzEQMA4GA1UEBRMHNDZmZjUxODEbMBkGCSqGSIb3
SGzFQHtnqURciJBtay9RNnMpZmZYpfOHzmeFmQ==
----END CERTIFICATE----
```

You must activate the identity certificate to the interface terminating the VPN tunnels, as demonstrated in Example 8-38. In Example 8-38, the interface terminating the VPN tunnels is the **outside** interface.

Example 8-38   *Activating the Identity Certificate*

```
omar-asa(config)# ssl trust-point SecretCorpSSLCert outside
```

Understanding the Remote Access VPN Attributes and Policy Inheritance Model

As you learned earlier in this chapter, the Cisco ASA uses an inheritance model when it pushes network and security policies to the end-user sessions. Using this model, you can configure policies at the following three policy locations:

- Under the default group policy

- Under the user's assigned group policy

- Under the specific user's policy

In the inheritance model, a user receives the attributes and policies from the user policy, which inherits its attributes and policies from the user group policy, which in turn receives its attributes and policies from the default group policy, as illustrated in Figure 8-20. A user with ID **derek** receives a traffic ACL and an assigned IP address from the user policy, the domain name from the user group policy, and Windows Internet Naming Server (WINS) information along with the number of simultaneous logins from the default group policy.

> **NOTE**   The policy **DfltGrpPolicy** is a special group name, used solely for the default group policy.

After these policies are defined, they must be bound to a tunnel group where users terminate their sessions. This way, a user who establishes his or her VPN session to a tunnel group inherits all the policies mapped to that tunnel. The tunnel group defines a VPN connection profile, of which each user is a member.

Configuring Clientless SSL VPN Group Policies

You need to configure the user group and default group policies. This group policy allows only clientless SSL VPN tunnels to be established and strictly rejects all the other tunneling protocols. If you would rather assign attributes to a default group policy, modify DfltGrp-Policy (System Default). Any attribute that is modified under DfltGrpPolicy is propagated to any user group policy that inherits that attribute. A group policy name other than DfltGrp-Policy is treated as a user group policy.

Example 8-39 demonstrates how to define a user group policy called **ClientlessGroupPolicy**. This policy allows only the clientless tunnels to be terminated on the group, by using the **webvpn** keyword.

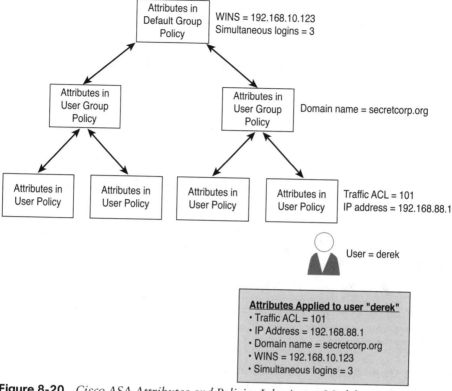

Figure 8-20 *Cisco ASA Attributes and Policies Inheritance Model*

Example 8-39 *Clientless SSL VPN Group Policy Definition*

```
omar-asa(config)# group-policy ClientlessGroupPolicy internal
omar-asa(config)# group-policy ClientlessGroupPolicy attributes
omar-asa(config-group-policy)# vpn-tunnel-protocol webvpn
```

NOTE The default and user group policies are set up to allow both Cisco IPsec VPN and SSL VPN tunnels. However, you can restrict a policy to solely accept clientless SSL VPN, or vice versa.

The user, group, and default group polices can be applied to clientless, AnyConnect, and IPsec-based remote-access VPN tunnels. The clientless SSL VPN–specific attributes are discussed in detail in the next few sections of this chapter.

Configuring the Tunnel Group for Clientless SSL VPN

You can configure a tunnel group, also known as a connection profile, as demonstrated in Example 8-40.

Example 8-40 *Configuring the Tunnel Group for Clientless SSL VPN*

```
omar-asa(config)# tunnel-group SecretCorpClientlessTunnel type remote-access
omar-asa(config)# tunnel-group SecretCorpClientlessTunnel general-attributes
omar-asa(config-tunnel-general)# default-group-policy ClientlessGroupPolicy
omar-asa(config-tunnel-general)# exit
omar-asa(config)# dns server-group DefaultDNS
omar-asa(config-dns-server-group)# domain-name ssecretcorp.org
omar-asa(config-dns-server-group)# name-server 192.168.3.4
```

After configuring a connection profile, you can define a URL that users can employ to connect to this tunnel group. This is beneficial if you want to create a specific URL for each connection profile you create and distribute the URL accordingly so that users do not have to decide to which connection profile they should connect.

You define a specific URL by modifying the connection profile. Example 8-41 demonstrates how to specify a **group-url** of **https://sslvpn.secretcorp.org/SecretCorpClientless** for SecretCorpClientlessTunnel.

Example 8-41 *Specifying the Tunnel Group URL*

```
omar-asa(config)# tunnel-group SecretCorpClientlessTunnel webvpn-attributes
omar-asa(config-tunnel-webvpn)# group-url https://sslvpn.secretcorp.org/
SecretCorpClientless enable
```

Configuring User Authentication for Clientless SSL VPN

Cisco ASA supports a number of authentication mechanisms and databases:

- RADIUS
- NT domain
- Kerberos
- SDI
- LDAP
- Digital certificates
- Smart cards
- SAML
- Local databases

For small organizations, a local database can be set up for user authentication. For medium to large SSL VPN deployments, it is highly recommended that you use an external authentication server, such as RADIUS or Kerberos, as the user authentication database. If you are deploying the SSL VPN feature for a few users, use the local database.

The Cisco ASA supports SAML 2.0 for Clientless VPN. With SAML 2.0, VPN users are able to input their credentials only one time when they switch between Clientless VPN and other SAAS applications outside of the private network. For instance, an organization may have

Duo enabled for single sign-on (SSO) multifactor authentication and has accounts on Service-Now, GitHub, Microsoft ADFS, or Dropbox that have been SAML 2.0 SSO enabled. When you configure the Cisco ASA to support SAML 2.0 SSO as a Service Provider (SP), end users are able to sign in once and have access to all these services, including Clientless VPN.

NOTE Client-based SSL VPN implementations with AnyConnect also support SAML. AnyConnect will be covered later in this chapter.

In Example 8-42, two accounts, **hannah** and **adminuser**, are configured for user authentication. The **hannah** account, with a password of **C1$c0123**, is used for SSL VPN user authentication, whereas **adminuser**, with a password of **@dm1n123**, is used to manage the Cisco ASA.

Example 8-42 *Local User Accounts*

```
omar-asa(config)# username hannah password C1$c0123
omar-asa(config)# username adminuser password @dm1n123
```

Many organizations use either a RADIUS server or Kerberos to leverage their existing Active Directory infrastructure for user authentication. Before configuring an authentication server on Cisco ASA, you must specify authentication, authorization, and accounting (AAA) server groups. User passwords are sent as encrypted values from the Cisco ASA to the RADIUS server. This protects this critical information from an intruder. The Cisco ASA hashes the password, using the shared secret that is defined on the Cisco ASA and the RADIUS server.

NOTE You can optionally modify the authentication and accounting port numbers if your RADIUS server does not use the default ports. The Cisco ASA uses UDP ports 1645 and 1646 as defaults for authentication and accounting, respectively. Most of the RADIUS servers use the official IANA assigned ports 1812 and 1813 as authentication and accounting ports.

After defining the authentication server group, you need to bind it to the SSL VPN process under a tunnel group. Example 8-43 shows how a RADIUS server can be defined. The RADIUS group name is **Radius**, and it is located toward the inside interface at 192.168.10.123. The shared secret is **R4diuzkeyz!**.

Example 8-43 *Configuring the Cisco ASA to Authenticate Users Using a RADIUS Server*

```
omar-asa(config)# aaa-server Radius protocol radius
omar-asa(config)# aaa-server Radius (inside) host 192.168.10.123
omar-asa(config-aaa-server-host)# key R4diuzkeyz!
omar-asa(config-aaa-server-host)# exit
```

TIP   For large VPN deployments (both IPsec and SSL VPNs), you can control user access and policy mapping from an external authentication server. Pass the user group policy name as a RADIUS or LDAP attribute to the Cisco ASA. By doing so, you guarantee that a user always gets the same policy, regardless of the tunnel group name to which the user connects. If you are using RADIUS as the authentication and authorization server, specify the user group policy name as attribute 25 (class attribute). Append the keyword **OU=** as the value of the **class** attribute. For example, if you define a user group policy called engineering group, you can enable attribute 25 and specify **OU=engineering** as its value. The Cisco ASA also supports the double authentication feature, which requires a user to provide two separate sets of logon credentials at the logon page. For example, you can choose to authenticate users on a primary authentication server such as Active Directory and on a secondary authentication server such as RADIUS. When both authentications succeed, the user is allowed to establish the SSL VPN tunnel.

Enabling Clientless SSL VPN

The clientless configuration of SSL VPN describes the mandatory steps for enabling SSL VPNs and setting up the user interface for clientless SSL VPN users. The following sections focus on the clientless SSL VPN users who want to access internal corporate resources but do not have an SSL VPN client loaded on their workstations. These users typically access protected resources from shared workstations or even from hotels or Internet cafes. The clientless configuration on Cisco ASA can be broken down into the following subsections:

- Enable clientless SSL VPN on an interface

- Configure SSL VPN portal customization

- Configure bookmarks

- Configure WebType ACLs

- Configure application access

- Configure client-server plug-ins

The first step in setting up a clientless SSL VPN on the Cisco ASA is to enable an SSL VPN on the interface that will terminate the user session. If SSL VPN is not enabled on the interface, the Cisco ASA does not accept any connections, even if SSL VPN is globally enabled.

Example 8-44 shows that SSL VPN functionality is enabled on the outside interface. Historically, SSL VPN has been referred to as WebVPN. Thus, why you see the **webvpn** configuration command.

Example 8-44   *Enabling SSL VPN on the Outside Interface*

```
omar-asa(config)# webvpn
omar-asa(config-webvpn)# enable outside
```

After SSL VPN is enabled on an interface, the Cisco ASA is ready to accept the connections. However, you still need to go through other configuration steps to successfully accept user connections and to allow traffic to pass through.

> **TIP**   You can customize the initial SSL VPN logon page based on your organization's security policies. Cisco ASA also enables you to modify the user web portal by offering a number of options. The Cisco ASA enables you to upload images and unique XML data to fully customize the logon page. Portal customization can only be achieved through ASDM. You cannot modify the configuration through the CLI because the attributes are defined and stored in XML.
>
> Using portal customization, you can design and present the SSL VPN page in any way you like. ASA allows you to create the default logon page and the logon page for a group of users. For example, if you want contractors to access a few applications, you can customize a web portal to include those applications and then map the portal to the group policy that contractors use. This way, when a user who belongs to the contractor group policy attempts to log in, they see only applications that are listed in their portal.
>
> If you want to use customization through XML, Cisco ASA contains a customization template. Export the template to a workstation and modify its content. You can import the customized content into the Cisco ASA as a new customized object. XML customization is outside the scope of the exam and this book.

Configuring WebType ACLs

The Cisco ASA enables network administrators to further their clientless SSL VPN security by configuring WebType access control lists to manage access to the following:

- Web

- Telnet

- SSH

- Citrix

- FTP

- File and email servers

- All types of traffic

These ACLs affect only the clientless SSL VPN traffic and are processed in sequential order until a match is found. If an ACL is defined but no match exists, the default behavior on the Cisco ASA is to drop packets. On the other hand, if no WebType ACL is defined, Cisco ASA allows all traffic to pass through it.

Moreover, this feature allows these ACLs to be downloaded from a RADIUS server such as the Cisco Identity Services Engine (ISE) through the use of vendor-specific attributes (VSA). This permits central control and management of user access into the corporate network because ACL definitions are offloaded to an external server.

The Cisco ASA allows you to filter based on the following protocols for all types of URLs:

- CIFS

- Citrix

- FTP

- HTTP/HTTPS

- IMAP4/POP3/SMTP

- NFS

- Smart tunnel

- SSH

- Telnet

- VNC

- RDP

If you want to include all URLs that are not explicitly matched in the ACL, include an asterisk (*) as a wildcard.

Example 8-45 shows a WebType ACL called Restrict being configured to allow http://internal.secretcorp.org. This ACL is then applied to ClientlessGroupPolicy.

Example 8-45   *Defining a WebType ACL*

```
omar-asa(config)# access-list Restrict webtype permit url
http://internal.secretcorp.org
omar-asa(config)# group-policy ClientlessGroupPolicy attributes
omar-asa(config-group-policy)# webvpn
omar-asa(config-group-webvpn)# filter value Restrict
```

TIP   Web ACLs do not block a user from accessing the resources outside the SSL VPN tunnel. For instance, if a user opens another tab with a web browser and accesses a different site, that traffic is not sent to the ASA, and therefore the security policies configured on the ASA will have no effect.

Configuring Application Access in Clientless SSL VPNs

Cisco ASA allows clientless SSL VPN users to access applications that reside on the protected network. Application access supports only applications that use TCP ports such as SSH, Outlook, and Remote Desktop. The Cisco ASA allows the following two methods to configure application access:

- Port forwarding

- Smart tunnels

Using port forwarding, the clientless SSL VPN users can access corporate resources over the known and fixed TCP ports such as Telnet, SSH, Terminal Services, SMTP, and so on. The port-forwarding feature requires you to install Oracle's Java Runtime Environment (JRE) and configure applications on the end user's PC. If users are establishing the SSL VPN tunnel from public computers, such as Internet kiosks or web cafes, they might not be able to use this feature because JRE installation requires administrative rights on the client computer. Port forwarding is supported only on the 32-bit-based operating systems.

As discussed earlier, port forwarding provides access to applications that use static TCP ports. It modifies the HOSTS files on a host so that traffic can be redirected to a forwarder that encapsulates traffic over the SSL VPN tunnel. Additionally, with port forwarding, the Cisco ASA administrator needs to know to which addresses and ports the SSL VPN users will connect and requires the SSL VPN users to have admin rights to modify the HOSTS file. To overcome some of the challenges related to port forwarding, Cisco ASA presents a new method to tunnel application-specific traffic called smart tunnels. Smart tunnels define which applications can be forwarded over the SSL VPN tunnel, whereas port forwarding defines which TCP ports can be forwarded over the tunnel.

Smart tunnels do not require administrators to preconfigure the addresses of the servers running the applications or the ports for those applications. In fact, smart tunnels work at the application layer by establishing a Winsock 2 connection between the client and the server. A stub is loaded into each process for the application that needs to be tunneled and then intercepts socket calls through the Cisco ASA. Thus, the principal benefit of smart tunnels over port forwarding is that users do not need to have administrative rights to use this feature.

Smart tunnels provide better performance than port forwarding, and the user experience is simpler because users don't need to configure their applications for a loopback address and for a specific local port.

Smart tunnels require browsers with ActiveX, Java, or JavaScript support. Both 32- and 64-bit-based operating systems are supported.

Configuring Client-Based Remote-Access SSL VPNs in the Cisco ASA

As you learned earlier in this chapter, clientless VPN application does not provide full network access to your remote users. If you want your users to have full network connectivity from their remote workstations, similar to what they would have with remote-access IPsec but by using SSL VPN, you can implement the full tunnel mode functionality on the Cisco ASA. Using the full tunnel client mode, remote machines can send all IP unicast traffic, including TCP-, UDP-, or even ICMP-based packets. SSL clients can access internal resources via HTTP, HTTPS, SSH, or Telnet, to name a few.

Many enterprises are in the process of migrating from an existing IPsec-based deployment. Their main motivation is that the Cisco AnyConnect Secure Mobility Client is easy to deploy and maintain, has a smaller package size, requires no machine reboots during client installation, and is easy to configure.

In the full tunnel mode, Cisco AnyConnect Secure Mobility Client can be pushed to or installed on the remote workstations after a successful authentication. After it is installed, you can choose to keep the client installed permanently and thus reduce the connection time for the remote user.

Setting Up Tunnel and Group Policies

As discussed earlier in this chapter, the Cisco ASA uses an inheritance model when it pushes network and security policies to the end-user sessions. Using this model, you can configure policies at the following three policy locations:

- Under the default group policy

- Under the user's assigned group policy
- Under the specific user's policy

In the inheritance model, a user receives the attributes and policies from the user policy, which inherits its attributes and policies from the user group policy, which in turn receives its attributes and policies from the default group policy.

After defining these policies, you must bind them to a tunnel group where users terminate their sessions. This way, a user who establishes her VPN session to a tunnel group inherits all the policies mapped to that tunnel. The tunnel group defines a VPN connection profile, of which each user is a member. Configure the user group and default group policies.

The user, group, and default group polices can be applied to clientless, Cisco AnyConnect Secure Mobility Client, and IPsec-based remote-access VPN tunnels. The Cisco AnyConnect Secure Mobility Client SSL VPN–specific attributes are discussed in detail in the next few sections of this chapter.

Example 8-46 illustrates how to define a user group policy called Cisco **AnyConnectGroup-Policy**. This policy allows only the AnyConnect tunnels to be terminated on the group, by using the **svc** keyword.

Example 8-46 *Group Policy Definition*

```
omar-asa(config)# group-policy AnyConnectGroupPolicy internal
omar-asa(config)# group-policy AnyConnectGroupPolicy attributes
omar-asa(config-group-policy)# vpn-tunnel-protocol svc
```

Configure a tunnel group, also known as connection profile, as demonstrated in Example 8-47.

Example 8-47 *Tunnel Group Definition*

```
omar-asa(config)# tunnel-group SecretCorpAnyConnect type remote-access
omar-asa(config)# tunnel-group SecretCorpAnyConnect general-attributes
omar-asa(config-tunnel-general)# default-group-policy AnyConnectGroupPolicy
```

After configuring a connection profile, you can define a specific URL for users to connect to this tunnel group. This is beneficial if you want to create a specific URL for each connection profile you design. Distribute the URL accordingly so that users do not have to decide to which connection profile to connect.

Example 8-48 shows how a RADIUS server is defined. The RADIUS group name is **secret-corp-radius-group**, and it is located toward the inside interface at 192.168.1.123. The shared secret is **R4diuzkeyz!**. The RADIUS server is added to **SecretCorpAnyConnect**.

Example 8-48 *Defining the RADIUS Server for Client-Based SSL VPN*

```
omar-asa(config)# aaa-server Radius protocol radius
omar-asa(config)# aaa-server secretcorp-radius-group (inside) host 192.168.1.123
omar-asa(config-aaa-server-host)# key R4diuzkeyz!
omar-asa(config-aaa-server-host)# exit
omar-asa(config)# tunnel-group SecretCorpAnyConnect general-attributes
omar-asa(config-tunnel-general)# authentication-server-group secretcorp-radius-group
```

Deploying the AnyConnect Client

Cisco AnyConnect Secure Mobility Client VPN client can be installed on a user's computer using one of these methods:

- **Web-enabled mode:** With this method, the client is downloaded to a user's computer through a browser. The user opens a browser and references the IP address or the FQDN of Cisco ASA to establish an SSL VPN tunnel. The user is presented with the standard SSL VPN logon page and is prompted for credentials.

- **Standalone mode:** With this method, the client is downloaded as a standalone application from a file server or directly from Cisco.com. The Windows Installer is executed to install the client to the workstation. If the client is not preconfigured, the user needs to specify the IP address or FQDN of the Cisco ASA, the tunnel group to connect to, the username, and the associated password.

In the case of mobile devices (such as iPhones, iPads, and Android devices), Cisco AnyConnect Secure Mobility Client can be downloaded directly from the Apple App Store or from Google Play, respectively.

The configuration of Cisco AnyConnect Secure Mobility Client VPN client is a two-step process:

Step 1. Load the Cisco AnyConnect Secure Mobility Client package.

Step 2. Define Cisco AnyConnect Secure Mobility Client VPN client attributes.

After loading the Cisco AnyConnect Secure Mobility Client package in the Cisco ASA's configuration, you can define client parameters such as the IP address that the client should receive via ASDM. Before a Cisco AnyConnect Secure Mobility Client SSL VPN tunnel is functional, you have to configure the following two required actions:

- Enabling Cisco AnyConnect Secure Mobility Client connections

- Address pool definition

Optionally, you can define other attributes to enhance the functionality of the Cisco AnyConnect Secure Mobility Client configuration, including the following:

- Split tunneling

- DNS and WINS assignment

- Keeping SSL VPN client installed

- DTLS

- Configuring traffic filters

- Configuring a tunnel group

During the SSL VPN tunnel negotiations, an IP address is assigned to the VPN adapter of the Cisco AnyConnect Secure Mobility Client. The client uses this IP address to access

8

resources on the protected side of the tunnel. Cisco ASA supports three different methods to assign an IP address back to the client:

- Local address pool

- DHCP server

- RADIUS server

Many organizations prefer assigning an IP address from the local pool of addresses for flexibility. You assign the IP address by configuring an address pool and then linking the pool to a group policy. You can either create a new pool of addresses or select a preconfigured address pool.

Example 8-49 demonstrates how SSL VPN (WebVPN) is enabled on the outside interface (similarly to the clientless SSL VPN configuration you learned earlier in this chapter). In Example 8-49, a local IP pool is configured to assign IP addresses to the VPN clients. The pool is then assigned to the **AnyConnectGroupPolicy** group policy.

Example 8-49   *Enabling SSL VPN and Configuring the IP Pool*

```
omar-asa(config)# webvpn
omar-asa(config-webvpn)# enable outside
omar-asa(config)# ip local pool SSLVPNPool 192.168.88.1-192.168.88.100 mask
255.255.255.0
omar-asa(config)# group-policy AnyConnectGroupPolicy attributes
omar-asa(config-group-policy)# address-pools value SSLVPNPool
```

Understanding Split Tunneling

After the tunnel is established, the default behavior of the Cisco AnyConnect Secure Mobility Client is to encrypt traffic to all the destination IP addresses. This means that if an SSL VPN user wants to browse to http://www.cisco.com over the Internet, as illustrated in Figure 8-21, the packets are encrypted and sent to Cisco ASA. After decrypting them, the Cisco ASA searches its routing table and forwards the packet to the appropriate next-hop IP address in clear text. These steps are reversed when traffic returns from the web server and is destined to the SSL VPN client.

The behavior illustrated in Figure 8-21 might not always be desirable for the following two reasons:

- Traffic destined to the nonsecure networks traverses over the Internet twice: once encrypted and once in clear text.

- The Cisco ASA handles extra VPN traffic destined to the nonsecure subnet. The Cisco ASA analyzes all traffic leaving and coming from the Internet.

With split tunneling, the Cisco ASA notifies the Cisco AnyConnect Secure Mobility Client about the secured subnets. The VPN client, using the secured routes, encrypts only those packets that are destined for the networks behind the security appliance. With split tunneling, the remote computer is susceptible to threat actors, who can potentially take control over the computer and direct traffic over the tunnel. To mitigate this behavior, a personal firewall is highly recommended on the Cisco AnyConnect Secure Mobility Client workstations.

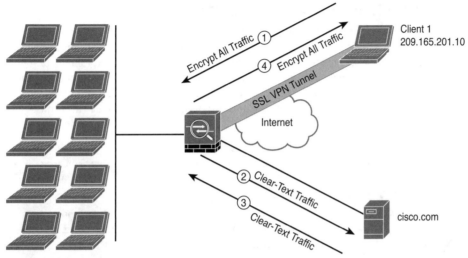

Figure 8-21   *VPN Traffic with No Split Tunneling*

Split tunneling can be configured under a user policy, user group policy, or default group policy. In split tunneling configurations, you define the traffic that will be encrypted by creating an ACL and assigning it to the group policy attributes, as demonstrated in Example 8-50.

Example 8-50   *Configuring Split Tunneling*

```
omar-asa(config)# access-list SplitTunnelList standard permit 192.168.0.0
255.255.255.0
omar-asa(config)# group-policy AnyConnectGroupPolicy attributes
omar-asa(config-group-policy)# split-tunnel-policy tunnelspecified
omar-asa(config-group-policy)# split-tunnel-network-list value SplitTunnelList
```

Cisco AnyConnect Secure Mobility Client supports split DNS functionality for Windows and macOS operating systems. It tunnels any DNS queries that match specified domain names to the private corporate DNS server. True split DNS allows tunnel access only to DNS requests that match the domains pushed to the client by the ASA. This DNS traffic is encrypted; however, if the DNS requests do not match the domains sent by the ASA, Cisco AnyConnect Secure Mobility Client permits the client operating system's DNS resolver to submit the hostname in the clear for DNS resolution.

Understanding DTLS

The Datagram Transport Layer Security (DTLS), defined in RFC 6347, provides security and privacy for UDP packets. This allows UDP-based applications to send and receive traffic in a secure fashion without concern about packet tampering and message forgery. Thus, applications can avoid the delays associated with TCP but still communicate securely by using DTLS.

Cisco AnyConnect Client supports both SSL and DTLS transport protocols. DTLS is enabled, by default, on the security appliance. If it is enabled and UDP is blocked or filtered, communication between the client and the security appliance reverts to the TCP-based SSL protocol.

Configuring Remote Access VPNs in FTD

Cisco FTD supports remote access SSL and IPsec-IKEv2 VPNs. Many organizations use the Cisco AnyConnect Secure Mobility Client (or AnyConnect for short) in combination with Cisco FTD devices to provide secure SSL and IPsec-IKEv2 and additional protections for remote users. AnyConnect is the only client supported on endpoint devices for remote VPN connectivity to Cisco FTD devices. The client gives remote users the benefits of an SSL or IPsec-IKEv2 VPN client without the need for network administrators to install and configure clients on remote computers.

> **TIP**   The AnyConnect client for Windows, Mac, and Linux is deployed from the Cisco FTD secure gateway upon connectivity. The AnyConnect apps for Apple iOS and Android devices are installed from the respective platform app stores.
>
> You can obtain additional information about all the different licenses and AnyConnect modules at https://www.cisco.com/c/en/us/products/collateral/security/anyconnect-secure-mobility-client/datasheet-c78-733184.html?cachemode=refresh. You can also engage in discussions with and learn from others in the VPN and AnyConnect Cisco Community at https://community.cisco.com/t5/vpn-and-anyconnect/bd-p/6001-discussions-vpn.

The following are the features related to remote access VPN supported in Cisco FTD devices:

- SSL and IPsec-IKEv2 remote access using the Cisco AnyConnect Secure Mobility Client.

- Cisco FMC support for all combinations such as IPv6 over an IPv4 tunnel.

- Configuration support on both FMC and FDM.

- Support for multiple interfaces and multiple AAA servers.

- Support for both FMC and FTD high-availability environments.

- Rapid Threat Containment support using RADIUS Change of Authorization (CoA) or RADIUS dynamic authorization.

- AAA server authentication using self-signed or CA-signed identity certificates.

- AAA username and password-based remote authentication using RADIUS server or LDAP or AD.

- RADIUS group and user authorization attributes as well as RADIUS accounting.

- Double authentication support using an additional AAA server for secondary authentication.

- NGFW Access Control policy integration using VPN Identity.

- VPN Tunneling address assignment.

- Split tunneling and Split DNS support.

- Client Firewall ACLs.

- Session timeouts for maximum connect and idle time.

- VPN dashboard widget in FMC showing VPN users by various characteristics, such as duration and client application.

- Remote-access VPN events, including authentication information such as username and OS platform.

- Tunnel statistics available using the FTD Unified CLI.

Using the Remote Access VPN Policy Wizard

You can use the Remote Access VPN Policy Wizard in the Firepower Management Center (FMC) to easily configure SSL and IPsec-IKEv2 remote-access VPNs with basic capabilities. You can then enhance the policy configuration (if desired) and deploy it to your Cisco FTD devices.

The following are the steps to configure SSL and IPsec-IKEv2 remote-access VPNs with the Remote Access VPN Policy Wizard.

Step 1. To start the Remote Access VPN Policy Wizard in the Cisco FMC, navigate to **Devices > VPN > Remote Access**. Click **Add** to create a new Remote Access VPN Policy. The wizard walks you through a basic policy configuration. You must complete all the steps in order for the wizard to create a new policy. Keep in mind that the policy is not saved if you cancel before completing the wizard.

Step 2. After you start the wizard, the screen shown in Figure 8-22 is displayed. Enter a name and an optional description for the new policy. Select **VPN Protocols** and **Target Devices**, as demonstrated in Figure 8-22. You can select SSL or IPsec-IKEv2, or both the VPN protocols. The Cisco FTD devices selected in the screen shown in Figure 8-22 will function as your remote-access VPN gateways for the VPN client users. You can select the devices from the list or add a new device.

Figure 8-22 *The FMC Remote Access VPN Policy Wizard Policy Assignment*

Step 3. Configure the Connection Profile and Group Policy settings, as demonstrated in Figure 8-23. A connection profile specifies a set of parameters that define how the remote users connect to the VPN device. The parameters include settings and attributes for authentication, address assignments to VPN clients, and group policies. The Cisco FTD provides a default connection profile named **DefaultWEBVPNGroup** when you configure a remote access VPN policy. The Connection Profile Name used in this example is **secretcorp-RA-VPN-policy-1**.

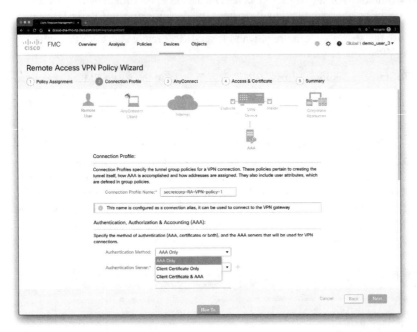

Figure 8-23 *Configuring the Remote Access VPN Connection Profile in Cisco FTD Devices*

Step 4. You can also set the AAA method (AAA, certificates, or both) and the AAA servers that will be used for the VPN connections, as shown in Figure 8-23. If you do not have an authentication server defined, you can click the **plus sign (+)** to define a new AAA server. The screen shown in Figure 8-24 is displayed after you click the **plus sign (+)** to define a new AAA server.

Step 5. Figure 8-25 shows the authentication server that you configured (RADIUS-1 in this example). You can select the same RADIUS server or a different one for authorization (Authorization Server) or accounting (Accounting Server). In the example shown in Figure 8-25, the same RADIUS server (RADIUS-1) is selected as the authentication, authorization, and accounting server.

Step 6. A group policy is a set of attribute and value pairs, stored in a group policy object, that define the remote-access VPN experience for VPN users. You can configure different attributes in the group policy such as user authorization profile, IP addresses, AnyConnect settings, VLAN mapping, and user session settings. The RADIUS authorization server assigns the group policy, or it is obtained from the current connection profile. Figure 8-26 shows the client address assignment section for the group policy.

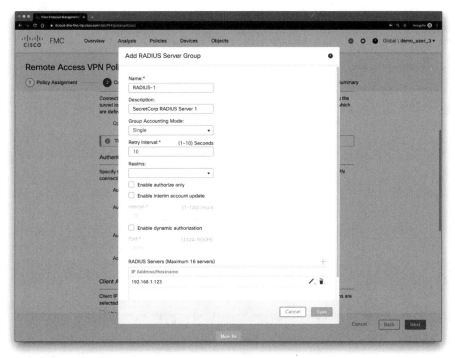

Figure 8-24 *Adding a New AAA Server (RADIUS Server Group)*

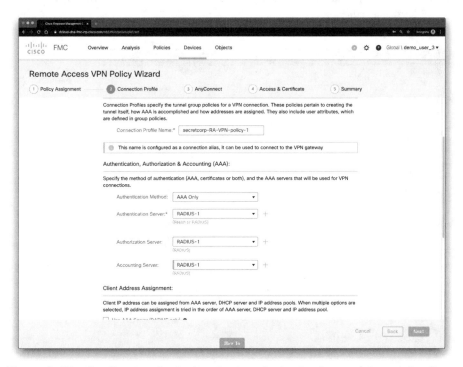

Figure 8-25 *Configuring the Authentication, Authorization, and Accounting Servers in the Connection Policy*

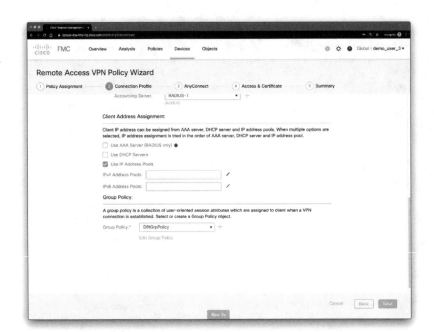

Figure 8-26   *The Client Address Assignment Section for the Group Policy*

Step 7.   Click the **pencil icon** to add an **Address Pool**. The screen shown in Figure 8-27 is displayed.

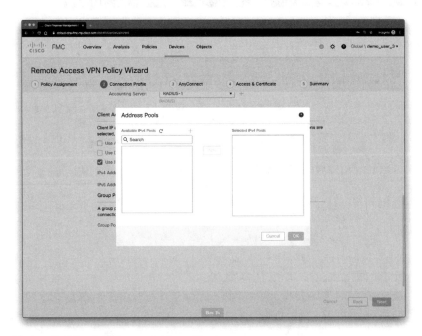

Figure 8-27   *Adding an Address Pool*

Step 8.   To add an IPv4 pool click the **plus sign** next to **Available IPv4 Pools.** The screen shown in Figure 8-28 is displayed. In Figure 8-28, the IPv4 address pool is called Pool-1 and the pool will assign IP addresses from 10.10.10.1 through 10.10.10.100. Clients addresses are configured with a 24-bit (/24 or 255.255.255.0) subnet mask.

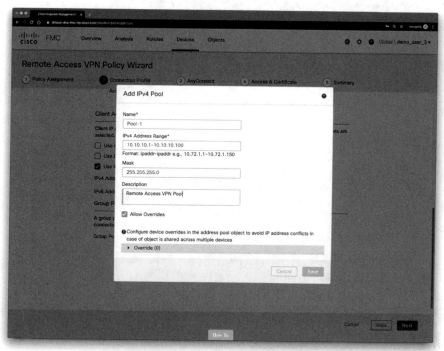

Figure 8-28   *Adding an IPv4 Pool*

Once you click **Save**, the screen shown in Figure 8-29 is displayed. Add the newly created pool (**Pool-1** in this example) to the Selected IPv4 Pools section on the right of the screen, as shown in Figure 8-29.

Step 9.   Select the AnyConnect Client Image that the VPN users will use to connect to the remote access VPN. After you configure the remote access VPN policy, VPN users can enter the IP address of the configured device interface in their browser to download and install the AnyConnect client. The AnyConnect Client Image page shown in Figure 8-30 is displayed.

Click **Add new AnyConnect Image** to add the AnyConnect software image for VPN clients. The screen shown in Figure 8-31 is displayed. Enter a name for the AnyConnect file and browse your local computer to locate the AnyConnect image to be uploaded to the Cisco FTD device. In Figure 8-31, the user-friendly name for the file is **MAC-AnyConnect**. A Mac OS X AnyConnect image is used in this example. You can also enter an optional description, as shown in Figure 8-31.

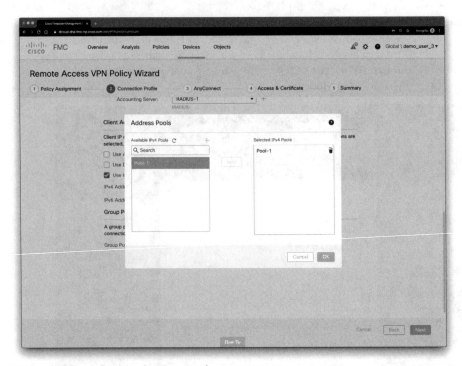

Figure 8-29   *Selecting the IPv4 Pool*

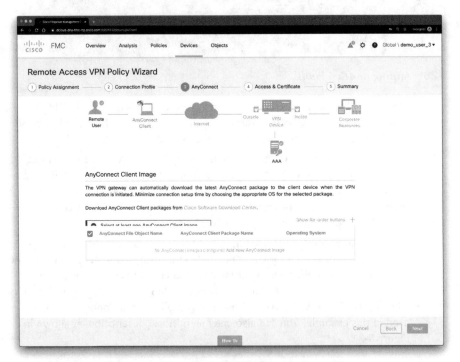

Figure 8-30   *Adding AnyConnect Images to the Cisco FTD Device(s)*

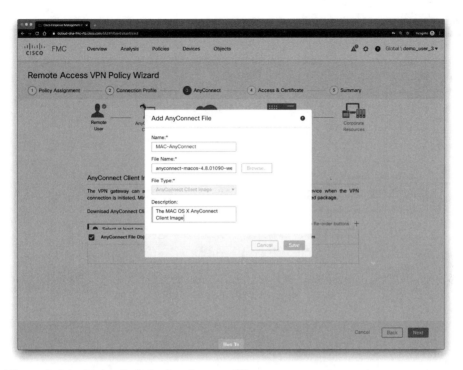

Figure 8-31 *Uploading the AnyConnect File*

You can click additional AnyConnect images by clicking the **plus sign (+)** shown in Figure 8-32.

Step 10.   Select the Network Interface and Identity Certificate. Interface objects segment your network to help you manage and classify traffic flow. A security zone object combines (groups) interfaces. These groups may span multiple Cisco FTD devices. In addition, you can configure multiple zones interface objects on a single device. There are two types of interface objects: security zones and interface groups. An interface can belong to only one security zone. In contrast, an interface can belong to multiple interface groups (and to one security zone). You can add the network interface for incoming VPN connections in the wizard screen shown in Figure 8-33.

You can perform digital certificate enrollment for the Cisco FTD device in the wizard. To enroll and obtain a new certificate, click the **plus sign** next to **Certificate Enrollment** in the screen shown in Figure 8-33. The screen shown in Figure 8-34 is displayed.

Step 11.   View the Summary of the Remote Access VPN policy configuration. The Summary page displays all the remote access VPN settings you have configured so far and provides links to the additional configurations that need to be performed before deploying the remote-access VPN policy on the selected devices. After you complete the wizard, it returns to the policy listing page. You can then set up the DNS configuration for the VPN clients, configure access control for VPN users, and enable NAT exemption (if necessary) to

Figure 8-32   *Adding Additional AnyConnect Client Images*

Figure 8-33   *Adding the Network Interface for Incoming VPN Connections*

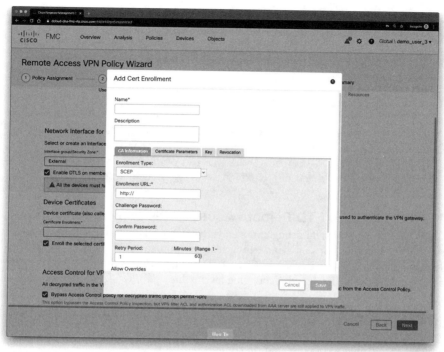

Figure 8-34 *Adding Certificate Enrollment*

complete a basic RA VPN Policy configuration. After completing these additional configuration steps, deploy the configuration and establish VPN connections.

TIP   Your remote-access VPN policy can include the AnyConnect Client Image and an AnyConnect Client Profile for distribution to connecting endpoints. Or, the client software can be distributed using other methods. See the "Deploy AnyConnect" chapter in the appropriate version of the Cisco AnyConnect Secure Mobility Client Administrator Guide:

https://www.cisco.com/c/en/us/support/security/anyconnect-secure-mobility-client/products-installation-and-configuration-guides-list.html

Without a previously installed client, remote users enter the IP address in their browser of an interface configured to accept SSL or IPsec-IKEv2 VPN connections. Unless the Cisco FTD or Cisco ASA is configured to redirect http:// requests to https://, remote users must enter the URL in the form https://address. After the user enters the URL, the browser connects to that interface and displays the login screen. After a user logs in, if the secure gateway identifies the user as requiring the VPN client, it downloads the client that matches the operating system of the remote computer. After downloading, the client installs and configures itself, establishes a secure connection, and either remains or uninstalls itself (depending on the security appliance configuration) when the connection stops. In the case of a previously installed client, after login, the Cisco FTD security gateway examines the client version and upgrades it as necessary.

When the AnyConnect negotiates a connection with the security appliance, the client connects using TLS and, optionally, using the Datagram Transport Layer Security (DTLS) protocol to prevent latency and bandwidth problems associated with some SSL connections. DTLS also improves the performance of real-time applications that are sensitive to packet delays.

> **TIP**   An AnyConnect client profile is a group of configuration parameters, stored in an XML file that the VPN client uses to configure its operation and appearance. These parameters (XML tags) include the names and addresses of host computers and settings to enable more client features. You can configure a profile using the AnyConnect Profile Editor. This editor is a convenient GUI-based configuration tool that is available as part of the AnyConnect software package. The profile editor is an independent program that you run outside of the Cisco FMC.

Troubleshooting Cisco FTD Remote Access VPN Implementations

The Cisco FTD provides several debug commands that can be useful to troubleshoot remote access VPN connections. The command **debug feature [subfeature] [level]** can be used to enable debugging for specific features in Cisco FTD devices. The **level** option might not be available for all features. The default debugging level is 1.

When you have multiple clients connecting via VPN, troubleshooting can be difficult given the size of the logs. To ease the troubleshooting task, you can use the **debug webvpn condition** command to set up filters to target your debug process more precisely. The following is the syntax of the **debug webvpn condition** command:

```
debug webvpn condition {group name | p-ipaddress ip_address
[{subnet subnet_mask | prefix length}] | reset | user name}
```

When you use the **group name** keyword, the debug filters on a group policy (not a tunnel group or connection profile). The **p-ipaddress ip_address [{subnet subnet_mask | prefix length}]** filters on the public IP address of the client. The subnet mask (for IPv4) or prefix (for IPv6) is optional. You can use the **reset** option to reset all filters, and you can use the **no debug webvpn condition** command to turn off a specific conditional debug.

Example 8-51 shows an example of enabling a conditional debug to filter messages for the connections associated with the user **hannah**.

Example 8-51   *Enabling Conditional Debugs to Filter Connections by VPN Users*

```
firepower# debug webvpn condition user hannah
firepower# show webvpn debug-condition
INFO: Webvpn conditional debug is turned ON
INFO: User name filters:
INFO: hannah
firepower# debug webvpn
INFO: debug webvpn  enabled at level 1.
firepower# show debug
debug webvpn  enabled at level 1
INFO: Webvpn conditional debug is turned ON
INFO: User name filters:
INFO: hannah
```

Configuring Site-to-Site VPNs in FTD

Cisco FTD devices support site-to-site IPsec tunnels using both IKEv1 and IKEv2. Similarly to the Cisco ASA, the Cisco FTD also supports certificates and automatic or pre-shared keys for authentication. IPv4 and IPv6 are both supported in Cisco FTD site-to-site VPN deployments.

The following are the steps to configure site-to-site IPsec in Cisco FTD devices:

Step 1. To configure a site-to-site VPN in a Cisco FTD, you have to create a new site-to-site VPN topology by navigating to **Devices > VPN > Site-to-Site VPN** in the Cisco FMC. The screen shown in Figure 8-35 is displayed, allowing you to add a new VPN topology by clicking the **Firepower Device** or **Firepower Threat Defense Device** link.

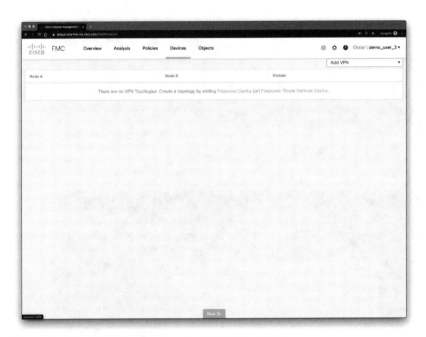

Figure 8-35   *VPN Topologies for Site-To-Site VPN*

Step 2. Let's create a new VPN topology by clicking the **Firepower Device** link. The screen shown in Figure 8-36 is displayed.

Step 3. Enter a unique name for the new topology and specify a topology type, as demonstrated in Figure 8-36. Point-to-point (PTP) deployments establish a VPN tunnel between two endpoints. Hub and Spoke (Star) topologies establish a group of VPN tunnels connecting a hub endpoint to a group of spoke nodes. Full Mesh (Mesh) deployments establish a group of VPN tunnels among a set of endpoints.

Step 4. Specify the node pairs by clicking the plus sign (+) shown in Figure 8-36. After you click the plus sign, the screen shown in Figure 8-37 is displayed.

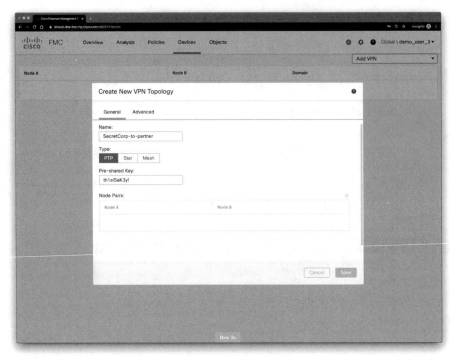

Figure 8-36   *Creating a New VPN Topology for Site-to-Site VPN*

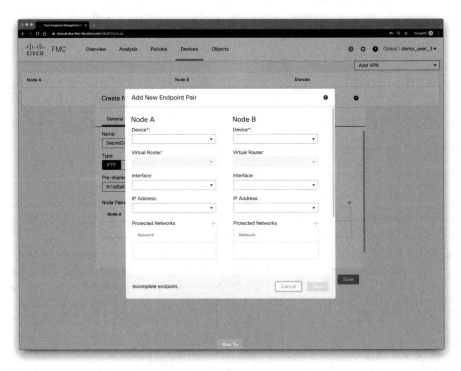

Figure 8-37   *Adding the New Endpoint Pair*

Step 5.   Select the devices you want to configure to establish the site-to-site VPN tunnel, their associated interfaces, and the IP addresses. Also specify the protected networks at each VPN endpoint device.

> **NOTE**   A site-to-site VPN connection in Cisco FTD devices can only be made across domains by using an extranet peer for the endpoint not in the current domain. A VPN topology cannot be moved between domains. Network objects with a "range" option are not supported in VPN.

Exam Preparation Tasks

As mentioned in the section "How to Use This Book" in the Introduction, you have a couple of choices for exam preparation: the exercises here, Chapter 12, "Final Preparation," and the exam simulation questions in the Pearson Test Prep Software Online.

Review All Key Topics

Review the most important topics in this chapter, noted with the Key Topic icon in the outer margin of the page. Table 8-2 lists these key topics and the page numbers on which each is found.

Table 8-2   Key Topics for Chapter 8

Key Topic Element	Description	Page Number
List	Identifying the different protocols that have been used throughout the years for VPN implementations	467
List	Identifying the different VPN implementation modes	468
Section	An Overview of IPs	470
List	Understanding the protocols used to encapsulate the data over an IPsec VPN tunnel	473
Section	NAT Traversal (NAT-T)	474
List	Comparing the differences between IKEv1 and IKEv2	475
Section	SSL VPNs	476
Section	Cisco AnyConnect Secure Mobility	478
Section	Traditional Site-to-Site VPNs in Cisco IOS and Cisco IOS-XE Devices	479
Section	Tunnel Interfaces	482
Section	GRE over IPsec	482
Section	More About Tunnel Interfaces	484
Tip	Understanding what are static and dynamic VTIs	486
Section	Multipoint GRE (mGRE) Tunnels	486
Section	DMVPN	486
Section	GETVPN	489
Section	FlexVPN	492

8

Key Topic Element	Description	Page Number
Figure 8-16	An example of a FlexVPN deployment	495
Section	Debug and Show Commands to Verify and Troubleshoot IPsec Tunnels	496
Section	Configuring Site-to-Site VPNs in Cisco ASA Firewalls	502
Section	Configuring IPsec Remote Access VPN in the Cisco ASA	512
List	Identifying SSL VPN modes	514
Section	Cisco ASA Remote-Access VPN Design Considerations	515
Section	Pre-SSL VPN Configuration Steps	516
Section	Understanding the Remote Access VPN Attributes and Policy Inheritance Model	518
Section	Deploying the AnyConnect Client	527
Section	Understanding Split Tunneling	528
Section	Configuring Remote Access VPN in FTD	530
Section	Using the Remote Access VPN Policy Wizard	531
Section	Troubleshooting Cisco FTD Remote Access VPN Implementations	540
Section	Configuring Site-to-Site VPNs in FTD	541

Define Key Terms

Define the following key terms from this chapter and check your answers in the glossary:

PPTP, L2F, GRE, Diffie-Hellman, SKEYID, DMVPN, GDOI, GETVPN, FlexVPN, PFS, NAT Traversal (NAT-T), split tunneling, DTLS

Review Questions

1. With NAT-T, the VPN peers dynamically discover whether an address translation device exists between them. If they detect a NAT/PAT device, they use _____ to encapsulate the data packets, subsequently allowing the NAT device to successfully translate and forward the packets.

    a. UDP port 4500

    b. UDP port 500

    c. TCP port 4500

    d. TCP port 443

2. GDOI is defined as the ISAKMP Domain of Interpretation (DOI) for group key management. The GDOI protocol operates between a group member and a group controller or key server (GCKS), which establishes SAs among authorized group members. Which of the following technologies use GDOI to establish SAs between authorized peers (group members)?

    a. FlexVPN

    b. GETVPN

    c. DMVPN

    d. None of these answers is correct.

3. The EAP messages between the IKEv2 client and the FlexVPN server are embedded within the IKEv2 EAP payload and are transported within the IKE_AUTH request and response messages. The EAP messages between the FlexVPN server and the RADIUS-based EAP server are embedded within which of the following?

   a. The RADIUS EAP-Message attribute

   b. The RADIUS AV-Pair attribute

   c. The RADIUS Accounting packet

   d. The EAP-Failure message

4. Refer to the following configuration snippet:

```
crypto ikev2 keyring local_keyring
peer hub-router
 address 10.1.1.1
 pre-shared-key branch1-hub-key
 !
crypto ikev2 profile default
match identity remote fqdn hq.secretcorp.org
identity local fqdn rtp-branch.secretcorp.org
authentication local pre-share
authentication remote pre-share
keyring local local_keyring
```

Which VPN technology is used in the configuration snippet?

   a. DMVPN

   b. GRE over IPsec

   c. GETVPN

   d. FlexVPN

5. You configured a site-to-site VPN tunnel between two Cisco routers. IKE Phase 1 is established; however, the tunnel Phase 2 negotiation is failing. Which of the following commands will you use to troubleshoot IPsec Phase 2 negotiations? (Select all that apply.)

   a. show crypto isakmp sa

   b. show crypto ipsec sa

   c. show crypto ikev2 sa detailed

   d. show crypto ikev2 session

6. Enabling debugs could potentially increase the load on busy network infrastructure devices. Which of the following is a feature that allows logging information to be stored in binary files so that you can later retrieve them without adding any more stress on the infrastructure device?

   a. Event-trace monitoring

   b. FMC health debugs

   c. Diagnostics and Reporting Tool (DART)

   d. Debug Binary Decomposition (DBD)

8

7. Which of the following are examples of the differences that exist between IKEv1 and IKEv2?

   a. IKEv1 Phase 1 has two possible exchanges: main mode and aggressive mode. There is a single exchange of a message pair for IKEv2 IKE_SA.

   b. IKEv2 has a simple exchange of two message pairs for the CHILD_SA. IKEv1 uses an exchange of at least three message pairs for Phase 2. In short, IKEv2 has been designed to be more efficient than IKEv1, since fewer packets are exchanged and less bandwidth is needed compared to IKEv1.

   c. Despite that IKEv1 supports some of the authentication methods that are used in IKEv2, IKEv1 does not allow the use of Extensible Authentication Protocol (EAP).

   d. All of these answers are correct.

8. Which of the following statements is true?

   a. IKEv1 and IKEv2 are incompatible protocols; consequently, you cannot configure an IKEv1 device to establish a VPN tunnel with an IKEv2 device.

   b. IKEv1 and IKEv2 are compatible protocols; consequently, you can configure an IKEv1 device to establish a VPN tunnel with an IKEv2 device.

   c. IKEv1 and IKEv2 are incompatible protocols; however, they both support EAP for authentication.

   d. In IKEv1 implementations, fewer packets are exchanged and less bandwidth is needed compared to IKEv2.

9. You were hired to deploy a VPN solution that will provide connectivity to kiosks at a retail store that are used by all customers to find information about the services and products offered. Which VPN solution will you implement?

   a. FlexVPN with AnyConnect

   b. GETVPN

   c. Clientless SSL VPN

   d. None of these answers is correct.

10. IPsec tunnels can be set up statically or dynamically using virtual interfaces of type VTI (Virtual-Tunnel Interface) or GRE over IPsec. These types of interfaces already existed in legacy IKEv1, and their use has been extended to which of the following solutions?

   a. Clientless SSL VPN

   b. MPLS VPNs

   c. AnyConnect

   d. FlexVPN

CHAPTER 9

Securing the Cloud

This chapter covers the following topics:

What Is Cloud and What Are the Cloud Service Models?

DevOps, Continuous Integration (CI), Continuous Delivery (CD), and DevSecOps

Describing the Customer vs. Provider Security Responsibility for the Different Cloud Service Models

Cisco Umbrella

Cisco Email Security in the Cloud

Cisco Cloudlock

Stealthwatch Cloud

AppDynamics Cloud Monitoring

Cisco Tetration

The following SCOR 350-701 exam objectives are covered in this chapter:

- Domain 3.0 Securing the Cloud
  - 3.1 Identify security solutions for cloud environments
    - 3.1.a Public, private, hybrid, and community clouds
    - 3.1.b Cloud service models: SaaS, PaaS, IaaS (NIST 800-145)
  - 3.2 Compare the customer vs. provider security responsibility for the different cloud service models
    - 3.2.a Patch management in the cloud
    - 3.2.b Security assessment in the cloud
    - 3.2.c Cloud-delivered security solutions such as firewall, management, proxy, security intelligence, and CASB
  - 3.3 Describe the concept of DevSecOps (CI/CD pipeline, container orchestration, and security)
  - 3.4 Implement application and data security in cloud environments

- 3.5 Identify security capabilities, deployment models, and policy management to secure the cloud

- 3.6 Configure cloud logging and monitoring methodologies

- 3.7 Describe application and workload security concepts

"Do I Know This Already?" Quiz

The "Do I Know This Already?" quiz allows you to assess whether you should read this entire chapter thoroughly or jump to the "Exam Preparation Tasks" section. If you are in doubt about your answers to these questions or your own assessment of your knowledge of the topics, read the entire chapter. Table 9-1 lists the major headings in this chapter and their corresponding "Do I Know This Already?" quiz questions. You can find the answers in Appendix A, "Answers to the 'Do I Know This Already?' Quizzes and Q&A Sections."

Table 9-1 "Do I Know This Already?" Section-to-Question Mapping

Foundation Topics Section	Questions
What Is Cloud and What Are the Cloud Service Models?	1
DevOps, Continuous Integration (CI), Continuous Delivery (CD), and DevSecOps	2–3
Describing the Customer vs. Provider Security Responsibility for the Different Cloud Service Models	4
Cisco Umbrella	5
Cisco Email Security in the Cloud	6
Cisco Cloudlock	7
Stealthwatch Cloud	8
AppDynamics Cloud Monitoring	9
Cisco Tetration	10

CAUTION The goal of self-assessment is to gauge your mastery of the topics in this chapter. If you do not know the answer to a question or are only partially sure of the answer, you should mark that question as wrong for purposes of the self-assessment. Giving yourself credit for an answer you incorrectly guess skews your self-assessment results and might provide you with a false sense of security.

1. Which of the following is a cloud computing model that provides everything except applications? Services provided by this model include all phases of the system development life cycle (SDLC) and can use application programming interfaces (APIs), website portals, or gateway software. These solutions tend to be proprietary, which can cause problems if the customer moves away from the provider's platform.

   a. IaaS

   b. PaaS

   c. SaaS

   d. Hybrid clouds

2. Which of the following is a software and hardware development and project management methodology that has at least five to seven phases that follow in strict linear order, where each phase cannot start until the previous phase has been completed?

   a. Agile

   b. Waterfall

   c. DevOps

   d. CI/CD

3. Which of the following is a software development practice where programmers merge code changes in a central repository multiple times a day?

   a. Continuous Integration (CI)

   b. Agile Scrum

   c. Containers

   d. None of these answers is correct.

4. In which of the following cloud models is the end customer responsible for maintaining and patching applications and making sure that data is protected, but not the virtual network or operating system?

   a. PaaS

   b. SaaS

   c. IaaS

   d. IaaS and PaaS

5. Which technology is used by Cisco Umbrella to scale and to provide reliability of recursive DNS services?

   a. Umbrella Investigate

   b. Multicast

   c. BGP Route Reflectors

   d. Anycast

6. Which Cisco Email Security feature is used to detect spear phishing attacks by examining one or more parts of the SMTP message for manipulation, including the "Envelope-From," "Reply To," and "From" headers?

   a. Forged Email Detection (FED)

   b. Forged Email Protection (FEP)

   c. Sender Policy Framework (SPF)

   d. Domain-based Message Authentication, Reporting, and Conformance (DMARC)

7. Which of the following is a cloud access security broker (CASB) solution provided by Cisco?

   a. Tetration

   b. Stealthwatch Cloud

   c. Cloudlock

   d. Umbrella

8. The Cisco Stealthwatch Cloud Sensor appliance can be deployed in which two different modes?

   a. Processing network metadata from a SPAN or a network TAP

   b. Processing metadata out of NetFlow or IPFIX flow records

   c. Processing data from Tetration

   d. PROCESSING data from Cloudlock

   e. Processing data from Umbrella

9. AppDynamics provides cloud monitoring and supports which of the following platforms?

   a. Kubernetes

   b. Azure

   c. AWS Lambda

   d. All of these answers are correct.

10. Which statement is not true about Cisco Tetration?

   a. Tetration uses software agents or can obtain telemetry information from Cisco's network infrastructure devices.

   b. You can use the Application Dependency Mapping (ADM) functionality to provide insight into the kind of complex applications that run in a data center, but not in the cloud.

   c. ADM enables network admins to build tight network security policies based on various signals such as network flows, processes, and other side information like load balancer configs and route tags.

   d. Tetration's Vulnerability Dashboard supports CVSS versions 2 and 3.

Foundation Topics

What Is Cloud and What Are the Cloud Service Models?

In Chapter 1, "Cybersecurity Fundamentals," you learned that the National Institute of Standards and Technology (NIST) created Special Publication (SP) 800-145, "The NIST Definition of Cloud Computing," to provide a standard set of definitions for the different aspects of cloud computing. The SP 800-145 document also compares the different cloud services and deployment strategies. In short, the advantages of using a cloud-based service include the use of distributed storage, scalability, resource pooling, access to applications and resources from any location, and automated management.

According to NIST, the essential characteristics of cloud computing include the following:

- On-demand self-service

- Broad network access

- Resource pooling

- Rapid elasticity

- Measured service

9

Cloud deployment models include the following:

- **Public cloud:** Open for public use

- **Private cloud:** Used just by the client organization on-premises (on-prem) or at a dedicated area in a cloud provider

- **Community cloud:** Shared between several organizations

- **Hybrid cloud:** Composed of two or more clouds (including on-prem services)

Cloud computing can be broken into the following three basic models:

- **Infrastructure as a Service (IaaS):** IaaS describes a cloud solution where you are renting infrastructure. You purchase virtual power to execute your software as needed. This is much like running a virtual server on your own equipment, except you are now running a virtual server on a virtual disk. This model is similar to a utility company model because you pay for what you use.

- **Platform as a Service (PaaS):** PaaS provides everything except applications. Services provided by this model include all phases of the system development life cycle (SDLC) and can use application programming interfaces (APIs), website portals, or gateway software. These solutions tend to be proprietary, which can cause problems if the customer moves away from the provider's platform.

- **Software as a Service (SaaS):** SaaS is designed to provide a complete packaged solution. The software is rented out to the user. The service is usually provided through some type of front end or web portal. While the end user is free to use the service from anywhere, the company pays a per-use fee.

NOTE   NIST Special Publication 500-292, "NIST Cloud Computing Reference Architecture," is another resource to learn more about cloud architecture.

DevOps, Continuous Integration (CI), Continuous Delivery (CD), and DevSecOps

DevOps (including the underlying technical, architectural, and cultural practices) characterizes a convergence of many technical, project management, and management movements. Before we start to define what is DevOps, let's take a look at the history of development methodologies. There are decades of lessons learned from software development, high reliability organizations, manufacturing, high-trust management models, and others that have evolved to the DevOps practices we know today.

The Waterfall Development Methodology

The waterfall model is a software and hardware development and project management methodology that has at least five to seven phases that follow in strict linear order. Each phase cannot start until the previous phase has been completed.

Figure 9-1 illustrates the typical phases of the waterfall development methodology.

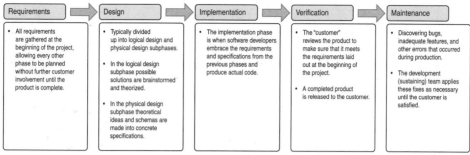

Figure 9-1   *The Typical Phases of the Waterfall Development Methodology*

There a few reasons why organizations use the waterfall methodology. One of the main reasons is because project requirements are agreed upon from the beginning; consequently, planning and scheduling is simple and clear. With a fully laid-out project schedule, an accurate estimate can be given, including development project cost, resources, and deadlines. Another reason is because measuring progress is easy as you move through the phases and hit the different milestones. Your end customer is not perpetually adding new requirements to the project, thus delaying production.

There are several disadvantages in the waterfall methodology, as well. One of the disadvantages is that it can be difficult for customers to enumerate and communicate all of their needs at the beginning of the project. If your end customer is dissatisfied with the product in the verification phase, it can be very costly to go back and design the code again. In the waterfall methodology, a linear project plan is rigid and lacks flexibility for adapting to unexpected events.

The Agile Methodology

Agile is a software development and project management process where a project is managed by breaking it up into several stages and involving constant collaboration with stakeholders and continuous improvement and iteration at every stage. The Agile methodology begins with end customers describing how the final product will be used and clearly articulating what problem it will solve. Once the coding begins, the respective teams cycle through a process of planning, executing, and evaluating. This process may allow the final deliverable to change in order to better fit the customer's needs. In an Agile environment, continuous collaboration is key. Clear and ongoing communication among team members and project stakeholders allows for fully informed decisions to be made.

The Agile methodology was originally developed by 17 people in 2001 in written form, and it is documented at "The Manifesto for Agile Software Development" (https://agilemanifesto.org).

Figure 9-2 illustrates Agile's four main values, as documented in "The Manifesto for Agile Software Development."

In Agile, the input to the development process is the creation of a business objective, concept, idea, or hypothesis. Then the work is added to a committed "backlog." From there, software development teams that follow the standard Agile or iterative process will transform that idea into "user stories" and some sort of feature specification. This specification is then implemented in code. The code is then checked in to a version control repository (for

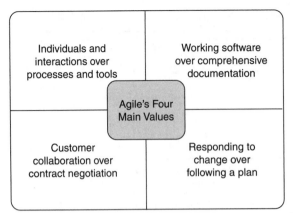

Figure 9-2 *The Agile Methodology's Four Main Values*

example, GitLab or GitHub), where each change is integrated and tested with the rest of the software system.

In Agile, value is created only when services are running in production; consequently, you must ensure that you are not only delivering fast flow, but that your deployments can also be performed without causing chaos and disruptions, such as service outages, service impairments, or security or compliance failures.

Figure 9-3 illustrates the general steps of the Agile methodology.

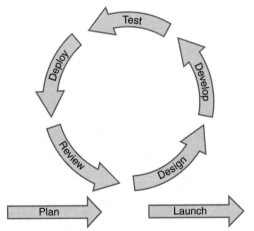

Figure 9-3 *The Agile Methodology's General Steps*

There is a concept adopted by many organizations related to Agile called "Scrum." Scrum is a framework that helps organizations work together because it encourages teams to learn through experiences, self-organize while working on a solution, and reflect on their wins and losses to continuously improve. Scrum is used by software development teams; however, its principles and lessons can be applied to all kinds of teamwork. Scrum describes a set of meetings, tools, and roles that work in concert to help teams structure and manage their work.

TIP   Scrum.org has a set of resources, certification, and training materials related to Scrum.

Figure 9-4 illustrates the high-level concepts of the Scrum framework. The Scrum framework uses the concept of "sprints" (a short, time-boxed period when a Scrum team works to complete a predefined amount of work). Sprints are one of the key concepts of the Scrum and Agile methodologies.

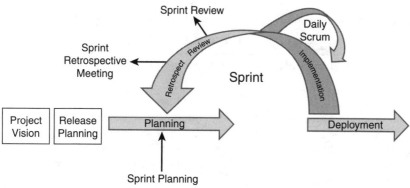

Figure 9-4   *The Scrum Framework and Sprints*

Agile is an implementation of the Lean management philosophy created to eliminate waste of time and resources across all aspects of business. The Lean management philosophy was derived from the "Toyota Production System" from the 1980s.

TIP   The following video provides a good overview of the Agile methodology: https://www.youtube.com/watch?v=Z9QbYZh1YXY. The following GitHub repository includes a very detailed list of resources related to the Agile methodology: https://github.com/lorabv/awesome-agile.

Agile also uses the Kanban process. Kanban is a scheduling system for lean development and just-in-time (JIT) manufacturing originally developed by Taiichi Ohno from Toyota.

There is yet another concept called Extreme Programming (EP). EP is a software development methodology designed to improve quality and for teams to adapt to the changing needs of the end customer. EP was originally developed by Ken Beck, who used it in the Chrysler Comprehensive Compensation System (C3) to help manage the company's payroll software. EP is similar to Agile, as its main goal is to provide iterative and frequent small releases throughout the development process. This enables both team members and customers to assess and review the development progress throughout the entire software development life cycle (SDLC).

NOTE   SDLC is also often used as an acronym for *secure development life cycle*. You will learn more about the secure development life cycle later in this chapter.

Figure 9-5 provides a good high-level overview of the Lean, Agile, Scrum, Kanban, and Extreme Programming concepts and associations.

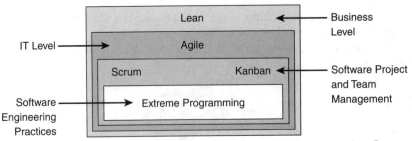

Figure 9-5 *Lean, Agile, Scrum, Kanban, and Extreme Programming Concepts*

DevOps

DevOps is the outcome of many trusted principles—from software development, manufacturing, and leadership to the information technology value stream. DevOps relies on bodies of knowledge from Lean, Theory of Constraints, resilience engineering, learning organizations, safety culture, human factors, and many others. Today's technology DevOps value stream includes the following areas:

- Product management

- Software (or hardware) development

- Quality assurance (QA)

- IT operations

- Infosec and cybersecurity practices

Figure 9-6 illustrates the steps to embrace DevOps within an organization.

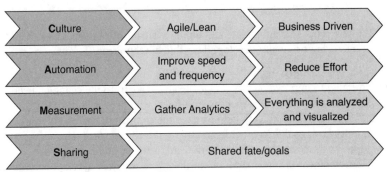

Figure 9-6 *Embracing DevOps*

There are "three general ways" to DevOps. The first "way" (illustrated in Figure 9-7) includes systems and flow. In this way (or method), you make work visible by reducing the work "batch" sizes, reducing intervals of work, and preventing defects from being introduced by building in quality and control.

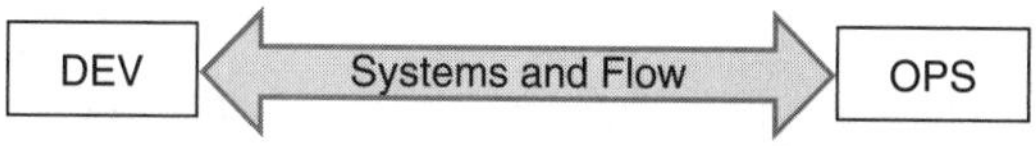

| DEV | ⟵ Systems and Flow ⟶ | OPS |

Figure 9-7   *DevOps Systems and Flow*

The second way is illustrated in Figure 9-8. This way includes a feedback loop to prevent problems from happening again (enabling faster detection and recovery by seeing problems as they occur and maximizing opportunities to learn and improve).

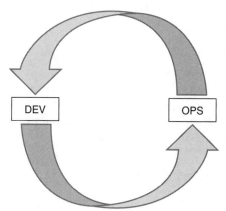

Figure 9-8   *DevOps Feedback Loop*

Figure 9-9 illustrates the third way (continuous experimentation and learning). In a true DevOps environment, you conduct dynamic, disciplined experimentation and take risks. You also define the time to fix issues and make systems better. The creation of shared code repositories help tremendously to achieve this continuous experimentation and learning process.

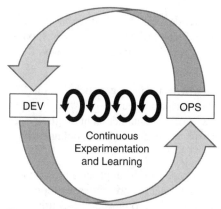

Figure 9-9   *DevOps Feedback Loop*

CI/CD Pipelines

Continuous Integration (CI) is a software development practice where programmers merge code changes in a central repository multiple times a day. Continuous Delivery (CD) sits on top of CI and provides a way for automating the entire software release process. When you adopt CI/CD methodologies, each change in code should trigger an automated build-and-test sequence. This automation should also provide feedback to the programmers who made the change.

CI/CD has been adopted by many organizations that provide cloud services (that is, SaaS, PaaS, and so on). For instance, CD can include cloud infrastructure provisioning and deployment, which traditionally have been done manually and consist of multiple stages. The main goal of the CI/CD processes is to be fully automated, with each run fully logged and visible to the entire team.

With CI/CD, most software releases go through the set of stages illustrated in Figure 9-10. A failure in any stage typically triggers a notification. For example, you can use Cisco WebEx Teams or Slack to let the responsible developers know about the cause of a given failure or to send notifications to the whole team after each successful deployment to production.

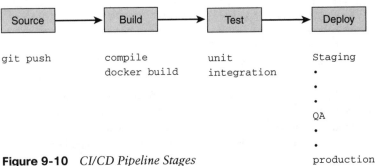

Figure 9-10 *CI/CD Pipeline Stages*

In Figure 9-10, the pipeline run is triggered by a source code repository (Git in this example). The code change typically sends a notification to a CI/CD tool, which runs the corresponding pipeline. Other notifications include automatically scheduled or user-initiated workflows, as well as results of other pipelines.

The Build stage includes the compilation of programs written in languages such as Java, C/C++, and Go. On the contrary, Ruby, Python and JavaScript programs work without this step; however, they could be deployed using Docker and other container technologies. Regardless of the language, cloud-native software is typically deployed with containers (in a microservice environment).

In the Test stage, automated tests are run to validate the code and the application behavior. The Test stage is an important stage, since it acts as a "safety net" to prevent easily reproducible bugs from being introduced. This concept can be applied to preventing security vulnerabilities, since at the end of the day, a security vulnerability is typically a software (or hardware) bug. The responsibility of writing test scripts can fall to a developer or a dedicated QA engineer. However, it is best done while new code is being written.

TIP   Depending on the size and complexity of the project, the Test phase can last from seconds to hours. Many organizations with large-scale projects run tests in multiple stages, starting with tests (typically called "smoke tests") that perform quick sanity checks from the user's point of view. Large-scale tests are typically parallelized to reduce runtime. It's very important that the test stage produce feedback to developers quickly, while the code is still fresh in their minds and they can maintain the state of flow.

Once you have a built your code and passed all predefined tests, you are ready to deploy it (the Deploy stage). Traditionally, there have been multiple deploy environments used by engineers (for example, a "beta" or "staging" environment used internally by the product team and a "production" environment).

NOTE   Organizations that have adopted the Agile methodology usually deploy work-in-progress manually to a staging environment for additional manual testing and review, and automatically deploy approved changes from the master branch to production.

The Serverless Buzzword

First, *serverless* does not mean that you do not need a "server" somewhere. Instead, it means that you will be using cloud platforms to host and/or to develop your code. For example, you might have a "serverless" app that is distributed in a cloud provider like AWS, Azure, or Google Cloud Platform.

Serverless is a cloud computing execution model where the cloud provider (AWS, Azure, Google Cloud, and so on) dynamically manages the allocation and provisioning of servers. Serverless applications run in stateless containers that are ephemeral and event-triggered (fully managed by the cloud provider).

AWS Lambda is one of the most popular serverless architectures in the industry. Figure 9-11 shows an example of a "function" or application in AWS Lamda.

NOTE   In AWS Lambda, you run code without provisioning or managing servers, and you only pay for the compute time you consume. When you upload your code, Lambda takes care of everything required to run and scale your application (offering high availability and redundancy).

As demonstrated in Figure 9-12, computing has evolved from traditional physical (bare-metal) servers to virtual machines (VMs), containers, and serverless architectures.

Container Orchestration

There have been multiple technologies and solutions to manage, deploy, and orchestrate containers in the industry. The following are the most popular:

- **Kubernetes:** One of the most popular container orchestration and management frameworks, originally developed by Google, Kubernetes is a platform for creating, deploying, and managing distributed applications. You can download Kubernetes and access its documentation at https://kubernetes.io.

9

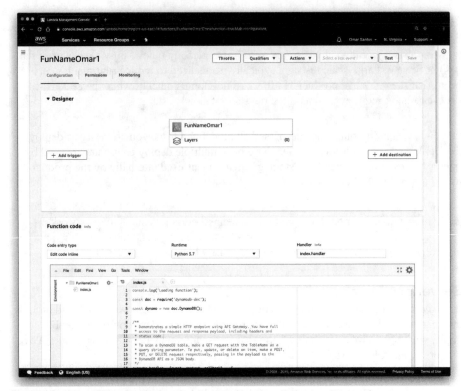

Figure 9-11 *AWS Lamda Serverless Application Function Example*

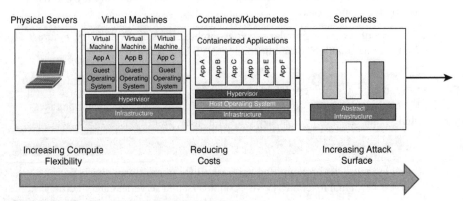

Figure 9-12 *The Evolution of Computing*

- **Nomad:** A container management and orchestration platform by HashCorp. You can download and obtain detailed information about Nomad at https://www.nomadproject.io.

- **Apache Mesos:** A distributed Linux kernel that provides native support for launching containers with Docker and AppC images. You can download Apache Mesos and access its documentation at https://mesos.apache.org.

■ **Docker Swarm:** A container cluster management and orchestration system integrated with the Docker Engine. You can access the Docker Swarm documentation at https://docs.docker.com/engine/swarm.

A Quick Introduction to Containers and Docker

Before you can even think of building a distributed system, you must first understand how the container images that contain your applications make up all the "underlying pieces" of such distributed system. Applications are normally composed of a language runtime, libraries, and source code. For instance, your application may use third-party or open source shared libraries such as libc and OpenSSL. These shared libraries are typically shipped as shared components in the operating system that you installed on a system. The dependency on these libraries introduces difficulties when an application developed on your desktop, laptop, or any other development machine (dev system) has a dependency on a shared library that isn't available when the program is deployed out to the production system. Even when the dev and production systems share the exact same version of the operating system, bugs can occur when programmers forget to include dependent asset files inside a package that they deploy to production.

The good news is that you can package applications in a way that makes it easy to share them with others. This is an example where containers become very useful. Docker, one of the most popular container runtime engines, makes it easy to package an executable and push it to a remote registry where it can later be pulled by others.

> **NOTE** Container registries are available in all of the major public cloud providers (for example, AWS, Google Cloud Platform, and Microsoft Azure) as well as services to build images. You can also run your own registry using open source or commercial systems. These registries make it easy for developers to manage and deploy private images, while image-builder services provide easy integration with continuous delivery systems.

Container images bundle a program and its dependencies into a single artifact under a root file system. Containers are made up of a series of file system layers. Each layer adds, removes, or modifies files from the preceding layer in the file system. The overlay system is used both when packaging up the image and when the image is actually being used. During runtime, there are a variety of different concrete implementations of such file systems, including *aufs*, *overlay*, and *overlay2*.

> **TIP** The most popular container image format is the Docker image format, which has been standardized by the Open Container Initiative (OCI) to the OCI image format. Kubernetes supports both Docker and OCI images. Docker images also include additional metadata used by a container runtime to start a running application instance based on the contents of the container image.

Let's take a look at an example of how container images work. In Figure 9-13 are three container images: A, B, and C. Container Image B is "forked" from Container Image A. Then, in Container Image B, Python version 3 is added. Furthermore, Container Image C is built upon Container Image B and the programmer adds OpenSSL and ngnix to develop a web server and enable TLS.

9

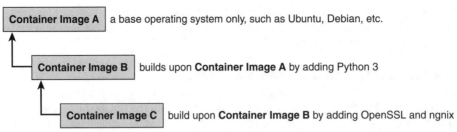

Container Image A a base operating system only, such as Ubuntu, Debian, etc.

Container Image B builds upon **Container Image A** by adding Python 3

Container Image C build upon **Container Image B** by adding OpenSSL and ngnix

Figure 9-13 *How Container Images Work*

Abstractly, each container image layer builds upon the previous one. Each parent reference is a pointer. The example in Figure 9-13 includes a simple set of containers; in many environments, you will encounter a much larger directed acyclic graph.

Even though the SCOR exam does not cover Docker in detail, it is still good to see a few examples of Docker containers, images, and related commands. Figure 9-14 shows the output of the **docker images** command.

```
root@kali:~
File  Edit  View  Search  Terminal  Help
root@kali:~# docker images
REPOSITORY              TAG          IMAGE ID        CREATED         SIZE
santosomar/hackme-rtov  latest       2e9fe51489a0    3 months ago    265MB
santosomar/bwapp        latest       35d0ed4da0ae    4 months ago    441MB
santosomar/hackazon     latest       ac1ccdb947ab    7 months ago    767MB
santosomar/dvwa         latest       105011d6abeb    7 months ago    393MB
santosomar/webgoat      latest       8711f9bf8156    7 months ago    665MB
santosomar/mutillidae_2 latest       7c041505981b    7 months ago    800MB
santosomar/juice-shop   latest       4ec87f6cdf3b    7 months ago    367MB
santosomar/dvna         latest       4d74d852a9aa    13 months ago   270MB
root@kali:~#
```

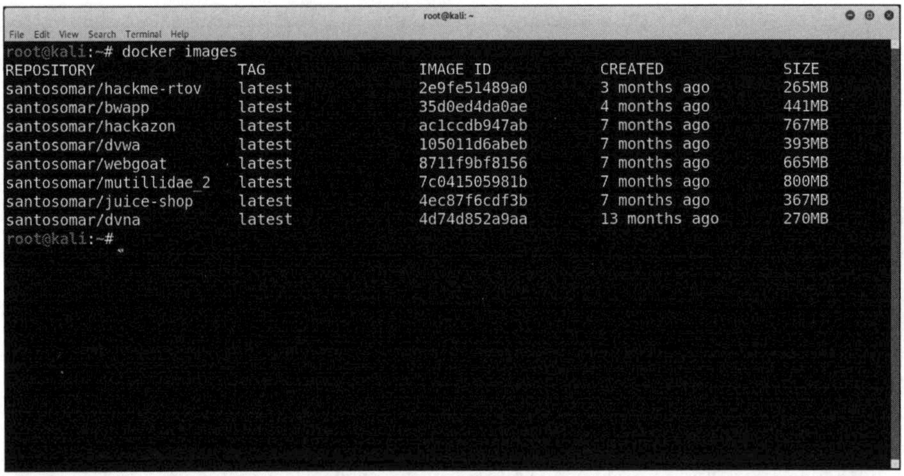

Figure 9-14 *The docker images Command*

TIP The Docker images shown in Figure 9-14 are images of intentionally vulnerable applications that you can also use to practice your skills. These Docker images and containers are included in a VM created by Omar Santos called WebSploit (websploit.org). The VM is built on top of Kali Linux and includes several additional tools, along with the aforementioned Docker containers. This can be a good tool, not only to get familiarized with Docker, but also to learn and practice offensive and defensive security skills.

Figure 9-15 shows the output of the **docker ps** command used to see all the running Docker containers in a system.

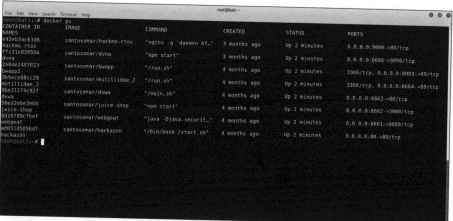

Figure 9-15 *The docker ps Command*

As you learned earlier in this chapter, you can use a public, cloud provider, or private Docker image repository. Docker's public image repository is called Docker Hub (https://hub.docker. com). You can find images by going to the Docker Hub website or by using the **docker search** command, as demonstrated in Figure 9-16.

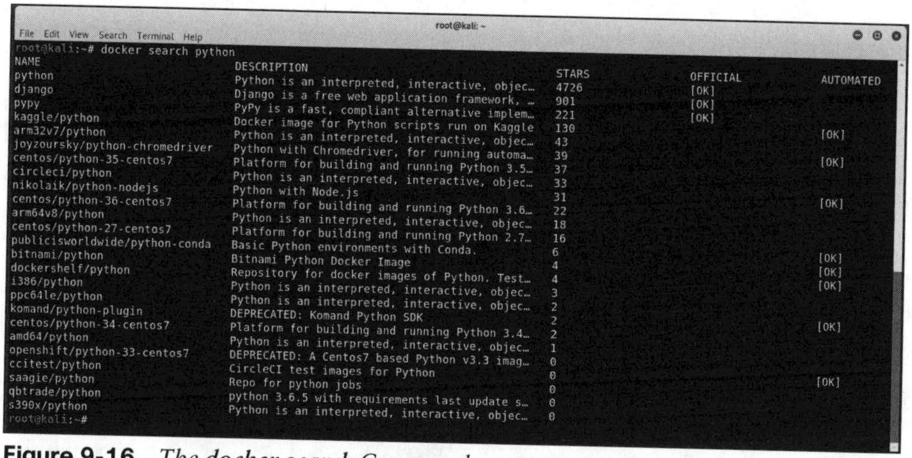

Figure 9-16 *The docker search Command*

In Figure 9-16, the user searches for a container image that matches the "python" keyword.

TIP You can practice and deploy your first container by using Katacoda (an interactive system that allows you to learn many different technologies, including Docker, Kubernetes, Git, Tensorflow, and many others). You can access Katacoda at https://www.katacoda.com. There are numerous interactive scenarios provided by Katacoda. For instance, you can use the "Deploying your first container" scenario to learn (hands-on) Docker: https://www.katacoda. com/courses/docker/deploying-first-container.

9

A Dockerfile can be used to automate the creation of a Docker container image. Example 9-1 shows an example of a Dockerfile. The Dockerfile shown in Example 9-1 is the official Python Docker image from Docker Hub.

Example 9-1   *A Dockerfile Example*

```
FROM alpine:3.10
# ensure local python is preferred over distribution python
ENV PATH /usr/local/bin:$PATH
# http://bugs.python.org/issue19846
# > At the moment, setting "LANG=C" on a Linux system *fundamentally
# breaks Python 3*, and that's not OK.
ENV LANG C.UTF-8
# install ca-certificates so that HTTPS works consistently
# other runtime dependencies for Python are installed later
RUN apk add --no-cache ca-certificates
ENV GPG_KEY E3FF2839C048B25C084DEBE9B26995E310250568
ENV PYTHON_VERSION 3.8.0
RUN set -ex \
        && apk add --no-cache --virtual .fetch-deps \
                gnupg \
                tar \
                xz \
        \
        && wget -O python.tar.xz "https://www.python.org/ftp/python/${PYTHON_
VERSION%%[a-z]*}/Python-$PYTHON_VERSION.tar.xz" \
        && wget -O python.tar.xz.asc "https://www.python.org/ftp/python/${PYTHON_
VERSION%%[a-z]*}/Python-$PYTHON_VERSION.tar.xz.asc" \
        && export GNUPGHOME="$(mktemp -d)" \
        && gpg --batch --keyserver ha.pool.sks-keyservers.net --recv-keys
<output omitted for brevity>
CMD ["python3"]
```

The full code in Example 9-1 can be obtained from https://github.com/The-Art-of-Hacking/h4cker/blob/master/SCOR/Dockerfile_example.

Let's create a simple Docker container based on the Dockerfile in Example 9-1. First, put the Dockerfile in a new directory/folder and then execute the command shown in Example 9-2 to create a Docker image called **mypython**.

Example 9-2   *Building the Docker Image*

```
┌─[omar@us-dev1]─[~/mypython]
└─── $ docker build -t mypython
Sending build context to Docker daemon  5.632kB
Step 1/13 : FROM alpine:3.10
 ---> 965ea09ff2eb
Step 2/13 : ENV PATH /usr/local/bin:$PATH
 ---> Using cache
 ---> 3801354cb4a4
```

```
Step 3/13 : ENV LANG C.UTF-8
 ---> Using cache
 ---> f5ee976b0ef2
Step 4/13 : RUN apk add --no-cache ca-certificates
<output omitted for brevity>
Step 12/13 : RUN set -ex;
wget -O get-pip.py "$PYTHON_GET_PIP_URL";
echo "$PYTHON_GET_PIP_SHA256 *get-pip.py" | sha256sum -c -;
python get-pip.py  --disable-pip-version-check  --no-cache-dir
pip==$PYTHON_PIP_VERSION"; pip --version; find /usr/local -depth
\(\( -type d -a \( -name test -o -name tests -o -name idle_test \) \) -o \( -type f
-a \( -name '*.pyc' -o -name '*.pyo' \) \)\) -exec rm -rf '{}' +; rm -f get-pip.py
 ---> Using cache
 ---> 646992bb197a
Step 13/13 : CMD ["python3"]
 ---> Using cache
 ---> 4790dbb6b084
Successfully built 4790dbb6b084
Successfully tagged mypython:latest
```

Example 9-3 shows the newly created image using the **docker images** command.

Example 9-3   *The Newly Created Docker Image*

```
┌─[omar@us-dev1]─[~/mypython]
└── $ docker images
REPOSITORY          TAG              IMAGE ID          CREATED            SIZE
mypython            latest           4790dbb6b084      2 minutes ago      110MB
```

You can now execute the **docker run mypython** command to run a new container.

TIP   You can access the Docker documentation at https://docs.docker.com. You can also complete a free and quick hands-on tutorial to learn more about Docker containers at https://www.katacoda.com/courses/container-runtimes/what-is-a-container-image.

Kubernetes

In larger environments, you will not deploy and orchestrate Docker containers in a manual way. You want to automate as much as possible. This is where Kubernetes comes into play. Kubernetes (often referred to as k8s) automates the distribution, scheduling, and orchestration of application containers across a cluster. Figure 9-17 illustrates the Kubernetes cluster concept.

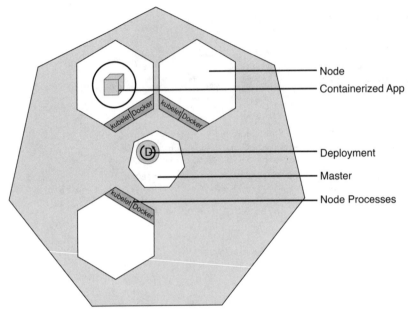

Figure 9-17   *The Kubernetes Cluster*

The following are the Kubernetes components:

- **Master:** Coordinates all the activities in your cluster (scheduling, scaling, and deploying applications).

- **Node:** A VM or physical server that acts as a worker machine in a Kubernetes cluster.

- **Pod:** A group of one or more containers with shared storage and networking, including a specification for how to run the containers. Each pod has an IP address, and it is expected to be able to reach all other pods within the environment.

Table 9-2 lists the differences between the legacy "rules" of standalone Docker containers and Kubernetes deployments.

Table 9-2   Differences Between the Legacy "Rules" of Standalone Docker Containers and Kubernetes Deployments

Standalone Docker Legacy "Rules"	Kubernetes "Rules"
No native container-to-container networking unless on the same VM.	All containers can communicate with each other without NAT.
Proxies or port forwarding needed.	All nodes can communicate with all containers (and vice versa) without NAT.
Built-in segmentation.	The IP that a container sees itself as is the same IP that others will see it as.

Figure 9-18 shows a high-level overview of a Kubernetes deployment.

One of the easiest way to learn Kubernetes is to use minikube (a lightweight Kubernetes implementation). Example 9-4 demonstrates how to start minikube with the **minikube start** command.

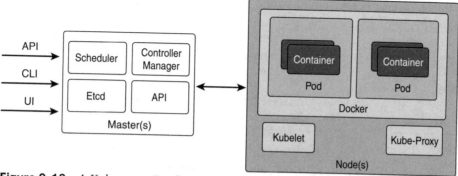

Figure 9-18   *A Kubernetes Deployment*

Example 9-4   *Starting Kubernetes (minikube)*

```
$ minikube start
* minikube v1.3.0 on Ubuntu 18.04
* Running on localhost (CPUs=2, Memory=2461MB, Disk=47990MB) ...
* OS release is Ubuntu 18.04.2 LTS
* Preparing Kubernetes v1.15.0 on Docker 18.09.5 ...
  - kubelet.resolv-conf=/run/systemd/resolve/resolv.conf
* Pulling images ...
* Launching Kubernetes ...
* Waiting for: apiserver proxy etcd scheduler controller dns
* Done! kubectl is now configured to use "minikube"
$
```

Once Kubernetes is deployed, you can check the version running the **kubctl** (the official Kubernetes client) **version** command, as demonstrated in Example 9-5.

Example 9-5   *Verifying the Version of Kubernetes*

```
$ kubectl version
Client Version: version.Info{Major:"1", Minor:"15", GitVersion:"v1.15.2", GitCommit:
"f6278300bebbb750328ac16ee6dd3aa7d3549568", GitTreeState:"clean", BuildDate:
"2019-08-05T09:23:26Z", GoVersion:"go1.12.5", Compiler:"gc", Platform:"linux/amd64"}
Server Version: version.Info{Major:"1", Minor:"15", GitVersion:"v1.15.0", GitCommit:
"e8462b5b5dc2584fdcd18e6bcfe9f1e4d970a529", GitTreeState:"clean", BuildDate:
"2019-06-19T16:32:14Z", GoVersion:"go1.12.5", Compiler:"gc", Platform:"linux/amd64"}
$
```

You can view the nodes in a cluster by running the **kubectl get nodes** command, as demonstrated in Example 9-6. In Example 9-6, only one node is deployed (minikube).

Example 9-6   *Displaying the Kubernetes Nodes*

```
$ kubectl get nodes
NAME        STATUS    ROLES     AGE       VERSION
minikube    Ready     master    2m12s     v1.15.0
```

9

Example 9-7 shows a deployment of a new app (omar-k8s-example).

Example 9-7   *Deploying a New App*

```
$ kubectl create deployment omar-k8s-example --image=g omar-k8s-example-image:v1
deployment.apps/omar-k8s-example created
$ kubectl get deployments
NAME                DESIRED   CURRENT   UP-TO-DATE   AVAILABLE   AGE
omar-k8s.example    1         1         1            1           10s
```

TIP   In a multi-node Kubernetes cluster, you use *kubeadm* to manage the nodes. Kubeadm is a Kubernetes component that provides a way to get a minimum viable Kubernetes cluster up and running. It is also worth noting that kubeadm only performs bootstrapping services and it does provision machines.

The Kubernetes official website includes different free hands-on tutorials that will help get you started with Kubernetes at https://kubernetes.io/docs/tutorials. The Kadacoda website also has several hands-on Kubernetes tutorials at https://www.katacoda.com/courses/kubernetes.

The Google Cloud Platform (GCP) offers a hosted Kubernetes-as-a-Service called Google Kubernetes Engine (GKE). Azure and AWS offer similar solutions. Figure 9-19 shows a Kubernetes cluster in GCP called **omar-k8s-cluster-1**.

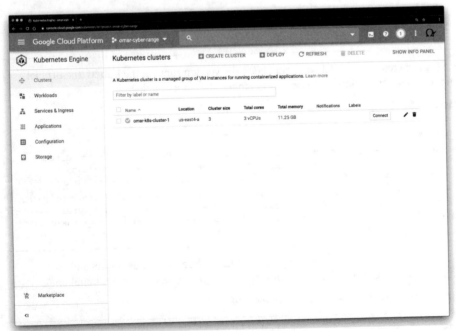

Figure 9-19   *The omar-k8s-cluster-1 Kubernetes Cluster in GCP*

Figure 9-20 shows the details of the Kubernetes cluster.

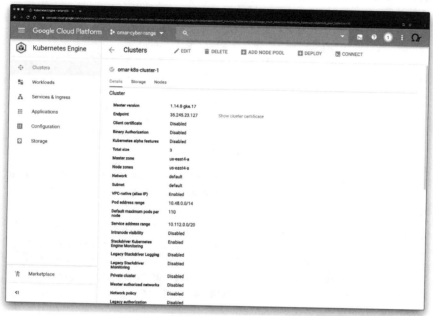

Figure 9-20   *The omar-k8s-cluster-1 Kubernetes Cluster Details*

Figure 9-21 shows the nodes within the Kubernetes cluster.

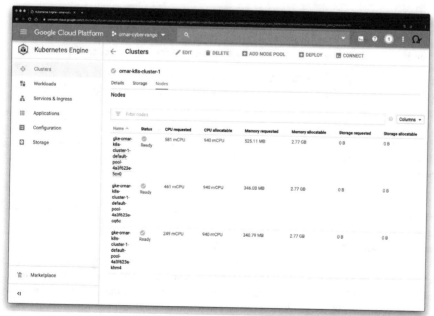

Figure 9-21   *The Kubernetes Cluster Nodes*

Example 9-8 shows the output of the **kubectl get nodes** command in Google's Cloud Shell (the Google Cloud Platform interactive shell).

Example 9-8 *The Output of the **kubectl get nodes** Command in Google's Cloud Shell*

```
santosomar@cloudshell:~ (omar-cyber-range)$ kubectl get nodes
NAME                                    STATUS   ROLES    AGE     VERSION
gke-omark8s-cluster1-4a3f623e-5cv0      Ready    <none>   5m28s   v1.14.8-gke.17
gke-omark8s-cluster1-4a3f623e-cq6c      Ready    <none>   5m27s   v1.14.8-gke.17
gke-omark8s-cluster1-4a3f623e-khm4      Ready    <none>   5m28s   v1.14.8-gke.17
santosomar@cloudshell:~ (omar-cyber-range)$
```

The GCP project in Example 9-8 is called **omar-cyber-range**.

Kubernetes supports a proxy that is responsible for routing network traffic to load-balanced services in a cluster. When deployed, the proxy must be present on every node in the Kubernetes cluster.

Kubernetes also runs a DNS server that provides naming and discovery for the services that are defined in the cluster. The Kubernetes DNS server also runs as a replicated service on the cluster. In other words, depending on how large your cluster is, you might see one or more DNS servers running at all times. Example 9-9 shows **kube-dns** running in the previously created Kubernetes cluster in the **omar-cyber-range** GCP project.

Example 9-9 *The Kubernetes DNS Server*

```
santosomar@cloudshell:~ (omar-cyber-range)$ kubectl get deployments
--namespace=kube-system kube-dns
NAME       READY   UP-TO-DATE   AVAILABLE   AGE
kube-dns   2/2     2            2           36m
```

In Kubernetes 1.12, Kubernetes transitioned from the **kube-dns** DNS server to the **core-dns** DNS server. If you are running a newer Kubernetes cluster, you may see **core-dns** instead.

Kubernetes also has a GUI. The Kubernetes GUI is run as a single replica, but it is still managed by a Kubernetes deployment for reliability and upgrades. You can see this UI server using the command shown in Example 9-10.

Example 9-10 *Deploying the Kubernetes System GUI (Dashboard)*

```
santosomar@cloudshell:~ (omar-cyber-range)$ kubectl get deployments
--namespace=kube-system kubernetes-dashboard
NAME                    DESIRED   CURRENT   UP-TO-DATE   AVAILABLE   AGE
kubernetes-dashboard    1         1         1            1           1d
```

Microservices and Micro-Segmentation

The ability to enforce network segmentation in container and VM environments is what people call *micro-segmentation*. Micro-segmentation is at the VM level or between containers regardless of a VLAN or a subnet. Micro-segmentation solutions need to be "application aware." This means that the segmentation process starts and ends with the application itself.

Most micro-segmentation environments apply a *zero-trust model*. This model dictates that users cannot talk to applications and that applications cannot talk to other applications unless a defined set of policies permits them to do so.

In Chapter 3, "Software-Defined Networking Security and Network Programmability," you learned about Contiv. As a refresher, Contiv is an open source project that allows you to deploy micro-segmentation policy-based services in container environments. It offers a higher level of networking abstraction for microservices by providing a policy framework. Contiv has built-in service discovery and service routing functions to allow you to scale out services.

NOTE   You can download Contiv and access its documentation at https://contiv.io.

With Contiv, you can assign an IP address to each container. This feature eliminates the need for host-based port NAT. Contiv can operate in different network environments, such as traditional Layer 2 and Layer 3 networks, as well as overlay networks. Contiv can be deployed with all major container orchestration platforms (or schedulers), such as Kubernetes and Docker Swarm. For instance, Kubernetes can provide compute resources to containers, and then Contiv provides networking capabilities.

TIP   The Contiv website includes several tutorials and step-by-step integration documentation at https://contiv.io/documents/tutorials/index.html.

DevSecOps

DevSecOps is a concept used in recent years to describe how to move security activities to the start of the development life cycle and have built-in security practices in the CI/CD pipeline. The business environment, culture, law compliance, and external market drive relate to how a secure development life cycle (also referred to as SDLC) and a DevSecOps program is implemented in an organization.

TIP   The DevSecOps project (https://devsecops.github.io) includes a set of tools and tutorials about DevSecOps and underlying practices. The Open DevSecOps GitHub organization (https://github.com/opendevsecops) includes a series of open source tools and resources, too.

9

The OWASP Proactive Controls (https://www.owasp.org/index.php/OWASP_Proactive_Controls) is a collection of secure development practices and guidelines that any software developer should follow to build secure applications. These practices will help you to shift security earlier into design, coding, and testing. Here are the OWASP Top 10 Proactive Controls:

1. Define Security Requirements
2. Leverage Security Frameworks and Libraries
3. Secure Database Access
4. Encode and Escape Data

5. Validate All Inputs

6. Implement Digital Identity

7. Enforce Access Controls

8. Protect Data Everywhere

9. Implement Security Logging and Monitoring

10. Handle All Errors and Exceptions

Additional best practices include security functions and tools in automated testing in CI/CD pipelines, parameterize queries to prevent SQL injection, and encoding data to protect against cross-site scripting (XSS) and other injection attacks. You should safely encode data before passing it on to an interpreter. Validate all inputs and treat all data as untrusted! Validate parameters and data elements using whitelisting and other techniques. Implement proper identity and authentication controls, as well as access controls. You should always have the "secure by default" and "shifting security to the left" mentality. This means "making it easy" for developers to write secure code and difficult for them to make dangerous mistakes, wiring secure defaults into their templates and frameworks, and building in the proactive controls you learned earlier.

Before you can begin adding security checks and controls, you need to understand the workflows and tools that the engineering teams are using in a CI/CD pipeline. You should ask the questions listed in Figure 9-22.

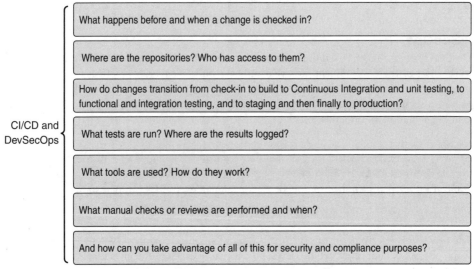

Figure 9-22 *CI/CD and DevSecOps*

You should also use software assurance tools and methods, including fuzzing, static application security testing (SAST), and dynamic application security testing (DAST). For example, you can use tools like Findsecbugs and SonarQube. Findsecbugs is a tool designed to find bugs in applications created in the Java programming language. It can be used with continuous integration systems such as Jenkins and SonarQube. Findsecbugs provides support for popular Java frameworks, including Spring-MCV, Apache Struts, Tapestry, and others. You can download and obtain more information about Findsecbugs at https://find-sec-bugs.github.io.

SonarQube is a tool that can be used to find vulnerabilities in code, and it provides support for continuous integration and DevOps environments. You can obtain additional information about SonarQube at https://www.sonarqube.org.

Fuzz testing, or *fuzzing*, is a technique that can be used to find software errors (or bugs) and security vulnerabilities in applications, operating systems, infrastructure devices, IoT devices, and other computing devices. Fuzzing involves sending random data to the unit being tested in order to find input validation issues, program failures, buffer overflows, and other flaws. Tools that are used to perform fuzzing are referred to as "fuzzers." Examples of popular fuzzers are Peach, Munity, American Fuzzy Lop, and Synopsys Defensics.

The Mutiny Fuzzing Framework is an open source fuzzer created by Cisco. It works by replaying packet capture files (pcaps) through a mutational fuzzer. You can download and obtain more information about Mutiny Fuzzing Framework at https://github.com/Cisco-Talos/mutiny-fuzzer. The Mutiny Fuzzing Framework uses Radamsa to perform mutations. Radamsa is a tool that can be used to generate test cases for fuzzers. You can download and obtain additional information about Radamsa at https://gitlab.com/akihe/radamsa.

American Fuzzy Lop (AFL) is a tool that provides features of compile-time instrumentation and genetic algorithms to automatically improve the functional coverage of fuzzing test cases. You can obtain additional information about AFL at http://lcamtuf.coredump.cx/afl/.

Peach is one of the most popular fuzzers in the industry. There is a free (open source) version, the Peach Fuzzer Community Edition, and a commercial version. You can download the Peach Fuzzer Community Edition and obtain additional information about the commercial version at https://www.peach.tech.

TIP   I have additional examples of fuzzers and fuzzing at my GitHub repository at https://h4cker.org/github.

Describing the Customer vs. Provider Security Responsibility for the Different Cloud Service Models

Cloud service providers (CSPs) such as Azure, AWS, and GCP have no choice but to take their security and compliance responsibilities very seriously. For instance, Amazon created a Shared Responsibility Model to describe what are the responsibilities of the AWS customers and Amazon's responsibilities in detail. The Amazon Shared Responsibility Model can be accessed at https://aws.amazon.com/compliance/shared-responsibility-model.

The "shared responsibility" depends on the type of cloud model (SaaS, PaaS, or IaaS). Figure 9-23 shows the responsibilities of a CSP and their customers in a SaaS environment.

9

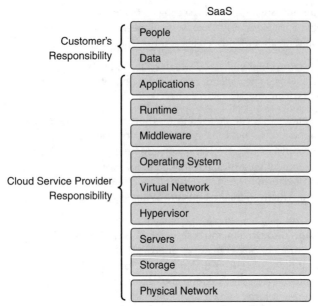

Figure 9-23 *SaaS Shared Security Responsibility*

Figure 9-24 shows the responsibilities of a CSP and their customers in a PaaS environment.

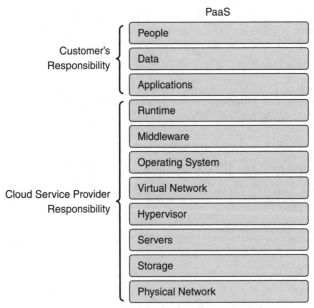

Figure 9-24 *PaaS Shared Security Responsibility*

Figure 9-25 shows the responsibilities of a CSP and their customers in an IaaS environment.

IaaS

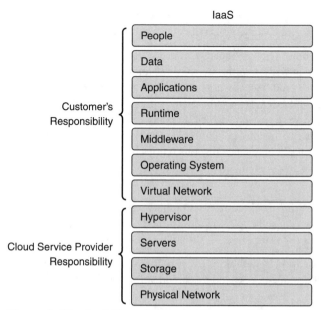

Figure 9-25 *IaaS Shared Security Responsibility*

Patch Management in the Cloud

Patch management in the cloud is also a shared responsibility in IaaS and PaaS environments, but not in a SaaS environment. For example, in a SaaS environment, the CSP is the one responsible for patching all software and hardware vulnerabilities. However, in an IaaS environment, the CSP is responsible only for patching the hypervisors, physical compute and storage servers, and the physical network. You are responsible for patching the applications, operating systems (VMs), and any virtual networks you deploy.

Security Assessment in the Cloud and Questions to Ask Your Cloud Service Provider

When performing penetration testing in the cloud, you must first understand what you can do and what you cannot do. Most CSPs have detailed guidelines on how to perform security assessments and penetration testing in the cloud. Regardless, there are many potential threats when organizations move to a cloud model. For example, although your data is in the cloud, it must reside in a physical location somewhere. Your cloud provider should agree in writing to provide the level of security required for your customers. As discussed in Chapter 1 the following are questions to ask a cloud provider before signing a contract for its services:

- **Who has access?** Access control is a key concern because insider attacks are a huge risk. Anyone who has been approved to access the cloud is a potential hacker, so you want to know who has access and how they were screened. Even if it was not done with malice, an employee can leave, and then you find out that you don't have the password, or the cloud service gets canceled because maybe the bill didn't get paid.

- **What are the provider's regulatory requirements?** Organizations operating in the United States, Canada, or the European Union have many regulatory requirements that they must abide by (for example, ISO/IEC 27002, EU-U.S. Privacy Shield Framework,

9

ITIL, FedRAMP, and COBIT). You must ensure that your cloud provider can meet these requirements and is willing to undergo certification, accreditation, and review.

■ **Do you have the right to audit?** This particular item is no small matter in that the cloud provider should agree in writing to the terms of the audit. With cloud computing, maintaining compliance could become more difficult to achieve and even harder to demonstrate to auditors and assessors. Of the many regulations touching upon information technology, few were written with cloud computing in mind. Auditors and assessors may not be familiar with cloud computing generally or with a given cloud service in particular. Division of compliance responsibilities between cloud provider and cloud customer must be determined before any contracts are signed or service is started.

■ **What type of training does the provider offer its employees?** This is a rather important item to consider because people will always be the weakest link in security. Knowing how your provider trains its employees is an important item to review.

■ **What type of data classification system does the provider use?** Questions you should be concerned with here include what data classification standard is being used and whether the provider even uses data classification.

■ **How is your data separated from other users' data? Is the data on a shared server or a dedicated system?** A dedicated server means that your information is the only thing on the server. With a shared server, the amount of disk space, processing power, bandwidth, and so on, is limited because others are sharing this device. If it is shared, the data could potentially become comingled in some way.

■ **Is encryption being used?** Encryption should be discussed. Is it being used while the data is at rest and in transit? You will also want to know what type of encryption is being used. For example, there are big technical differences between DES and AES; however, for both of these algorithms, the basic questions are the same: Who maintains control of the encryption keys? Is the data encrypted at rest in the cloud? Is the data encrypted in transit, or is it encrypted at rest and in transit?

■ **What are the service level agreement (SLA) terms?** The SLA serves as a contracted level of guaranteed service between the cloud provider and the customer that specifies what level of services will be provided.

■ **What is the long-term viability of the provider?** How long has the cloud provider been in business, and what is its track record? If it goes out of business, what happens to your data? Will your data be returned, and, if so, in what format?

■ **Will they assume liability in the case of a breach?** If a security incident occurs, what support will you receive from the cloud provider? While many providers promote their services as being "unhackable," cloud-based services are an attractive target to hackers.

■ **What is the disaster recovery/business continuity plan (DR/BCP)?** Although you may not know the physical location of your services, it is physically located somewhere. All physical locations face threats such as fire, storms, natural disasters, and loss of power. In case of any of these events, how will the cloud provider respond, and what guarantee of continued services is it promising?

> **TIP**   Even when you end a contract, you must ask what happens to the information after your contract with the cloud service provider ends. Insufficient due diligence is one of the biggest issues when moving to the cloud. Security professionals must verify that issues such as encryption, compliance, incident response, and so forth are all worked out before a contract is signed.

Cisco has several security solutions that can help protect the cloud and/or that are delivered from the cloud. The following sections cover these solutions.

Cisco Umbrella

Cisco Umbrella is a solution that evolved from the OpenDNS acquisition. The Cisco Umbrella is a cloud-delivered solution that blocks malicious destinations using DNS.

> **TIP**   Cisco Umbrella can be used on any device, including IoT devices, on any network, at any time. This is because its implementation is very straightforward and can be accomplished by forwarding DNS queries to the Umbrella cloud on existing DNS servers, running the Umbrella virtual appliances, and/or using the Microsoft Windows or macOS roaming client or the Cisco Security Connector for iOS.

OpenDNS is a suite of consumer products aimed at making your Internet faster, safer, and more reliable. The following website provides information on how to use OpenDNS to protect your system and your home: https://www.opendns.com/home-internet-security/.

Cisco Umbrella has the ability to see attacks before the application connection occurs. This limits the load on a firewall or any other network security infrastructure device. In addition, it helps to reduce alerts and improve security operations and incident response.

Umbrella looks at the patterns of DNS requests from devices and uses them to detect the following:

- Compromised systems
- Command-and-control callbacks
- Malware and phishing attempts
- Algorithm-generated domains
- Domain co-occurrences
- Newly registered domains
- Malicious traffic and payloads that never reach the target

The Cisco Umbrella Architecture

The Cisco Umbrella global infrastructure includes dozens of data centers around the world that resolve more than 100 billion DNS requests from millions of users every day. Umbrella data centers are peered with more than 500 of the top ISPs and content delivery networks to exchange BGP routes and ensure that requests are routed efficiently, without adding any latency over regional DNS providers.

9

Cisco Umbrella uses Anycast IP routing in order to provide reliability of the recursive DNS service. All data centers announce the same IP address, and all requests are transparently sent to the fastest and lowest-latency data center available.

> **TIP**   You can use Cisco Umbrella (OpenDNS) by just pointing your DNS configuration to the Anycast IP addresses 208.67.222.222 and 208.67.220.220. The website https://use.opendns.com provides additional instructions on how to configure your system to use OpenDNS/Umbrella.

Its scale and speed give Umbrella a massive amount of data and, perhaps more importantly, a very diverse data set that is not just from one geography or one protocol. This diversity enables Umbrella to offer unprecedented insight into staged and launched attacks. The data and threat analytics engines learn where threats are coming from, who is launching them, where they are going, and the width of the net of the attack—even before the first victim is hit. Umbrella uses authoritative DNS logs to find the following:

- Newly staged infrastructures

- Malicious domains, IP addresses, and ASNs

- DNS hijacking

- Fast flux domains

- Related domains

Fast flux is a DNS technique used by botnets to hide phishing and malware delivery sites behind an ever-changing network of compromised hosts acting as proxies. Umbrella is able to find these types of threats by using modeling inside the data analytics. Machine learning and advanced algorithms are used heavily to find and automatically block malicious domains.

The following are a few examples of available models:

- **Co-occurrence model:** This model identifies domains queried right before or after a given domain. This model helps uncover domains linked to the same attack, even if they're hosted on separate networks.

- **Traffic spike model:** This model recognizes when spikes in traffic to a domain match patterns seen with other attacks. For example, if the traffic for one domain matches the request patterns seen with exploit kits, you might want to block the domain before the full attack launches.

- **Predictive IP space monitoring model:** This model starts with domains identified by the spike rank model and scores the steps attackers take to set up infrastructure (for example, hosting provider, name server, and IP address) to predict whether the domain is malicious. This identifies other destinations that can be proactively blocked before an attack launches.

Secure Internet Gateway

When Cisco Umbrella servers receive a DNS request, they first identify which end customer the request came from and which policy to apply. Next, Cisco Umbrella determines whether the request is safe or whitelisted, malicious or blacklisted, or "risky." Safe requests are allowed to be routed as usual, and malicious requests are routed to a block page. Risky requests can be routed to the cloud-based proxy for deeper inspection.

This concept of a cloud-based proxy is the basis for the secure Internet gateway (SIG). Before looking at the functionality of the proxy, it is helpful to understand what traffic is typically sent to the proxy. Most phishing, malware, ransomware, and other threats are hosted at domains that are classified as malicious. Yet some domains host both malicious and safe content; these are the domains that are classified as risky. These sites (such as facebook. com, reddit.com, pastebin.com, and so on) allow users to upload and share content, making them difficult to police.

Traditional web proxies or gateways examine all Internet requests, which adds latency and complexity. The Cisco Umbrella secure Internet gateway proxy intercepts and inspects only requests for risky domains. When the secure Internet gateway identifies a risky domain and begins to proxy that traffic, it uses the URL inspection engine to first classify the URL. The Cisco Umbrella secure Internet gateway uses Cisco Talos threat intelligence, the Cisco web reputation system, and third-party feeds to determine if a URL is malicious.

If the disposition of a web resource is still unknown after the URL inspection, if a file is present, the secure Internet gateway can also look at the file's reputation. The file is inspected by both antivirus (AV) engines and Cisco Advanced Malware Protection (AMP) to block malicious files based on known signatures before they are downloaded.

Figure 9-26 shows the Cisco Umbrella Overview dashboard, which breaks down the different network requests, including the total number of DNS requests, proxy requests, total blocks, and security blocks.

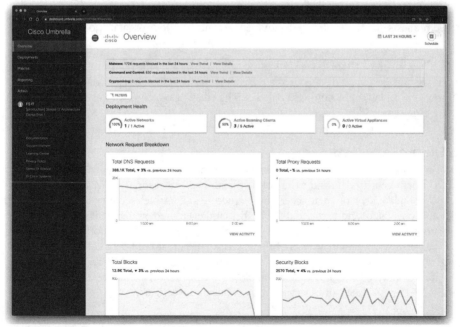

Figure 9-26   *The Cisco Umbrella Overview Dashboard*

Cisco Umbrella provides different dashboards and detailed reports. Figure 9-27 shows the Security Overview report dashboard.

Figure 9-27   *The Cisco Umbrella Security Overview Report*

The report shown in Figure 9-27 can be used to obtain a high-level overview of all the blocked requests and the Umbrella deployment activity for your organization.

> **TIP**   You can schedule Umbrella reports to be emailed to a specific recipient at regular intervals. The report will be displayed as a table showing an HTML version of the report and an attached .csv file with the complete data set. The email also includes a link to the "live version" of the same report.

Cisco Umbrella Investigate

Cisco Umbrella Investigate provides organizations access to global intelligence that can be used to enrich security data and events or help with incident response. Investigate provides the most complete view of an attacker's infrastructure and enables security teams to discover malicious domains, IP addresses, and file hashes, and even predict emergent threats. Investigate provides access to this intelligence via a web console or an application programming interface (API). With the integration of the Cisco AMP Threat Grid data in Investigate, intelligence about an attacker's infrastructure can be complemented by AMP Threat Grid's intelligence about malware files, providing a complete view of the infrastructure used in an attack. Figure 9-28 shows a query of binarycousins.com using Investigate.

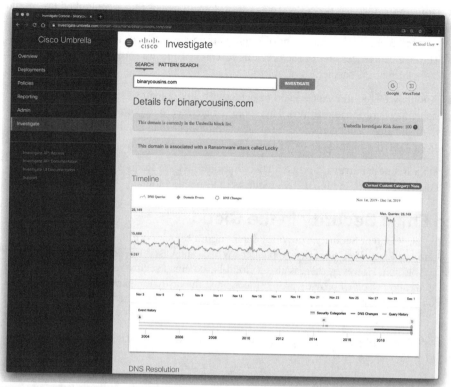

Figure 9-28 *The Cisco Umbrella Investigate*

NOTE   You can use Investigate, if your Umbrella license allows, by going to https://investigate.umbrella.com.

Cisco Umbrella Investigate provides a single, correlated source of intelligence and adds the security context needed to help organizations uncover and predict attacks. Investigate provides the following features:

- **Passive DNS Database:** Provides historical DNS data.

- **WHOIS Record Data:** Allows you to see domain ownership and uncover malicious domains registered with the same contact information.

- **Malware File Analysis:** Provides behavioral indicators and network connections of malware samples with data from Cisco AMP Threat Grid.

- **Autonomous System Number (ASN) Attribution:** Provides IP-to-ASN mappings.

- **IP Geolocation:** Allows you to see in which country an IP address is located.

- **Domain and IP Reputation Scores:** Allows you to leverage Investigate's risk scoring across a number of domain attributes to assess suspicious domains.

- **Domain Co-Occurrences:** Returns a list of domain names that were looked up around the same time as the domain being checked. The score next to the domain name in a co-occurrence is a measurement of requests from client IPs for these related domains. The co-occurrences are for the previous seven days and are shown whether the co-occurrence is suspicious or not.

- **Anomaly Detection:** Allows you to detect fast flux domains and domains created by domain generation algorithms. This score is generated based on the likeliness of the domain name being generated by an algorithm rather than a human. This score ranges from −100 (suspicious) to 0 (benign).

- **DNS Request Patterns and Geo Distribution of Those Requests:** Allows you to see suspicious spikes in global DNS requests to a specific domain.

Cisco Email Security in the Cloud

Chapter 10, "Content Security," provides details about the Cisco Email Security Appliance (ESA) and the Cisco Web Security Appliance (WSA). The Cisco ESA is an on-premises email security solution. However, there is also a cloud-based email security solution provided by Cisco. This allows you to provide protection against threats like ransomware, business email compromise (BEC), phishing, spear phishing, whaling, and many other email-driven attacks.

> **TIP** The Cisco ESA, Cisco WSA, and the cloud-based email security solution use AsyncOS as the main operating system.

Cisco cloud email security supports several techniques to create the multiple layers of security needed to defend against the aforementioned attack types. These techniques include the following:

- **Geolocation-based filtering:** To protect against sophisticated spear phishing by quickly controlling email content based on the location of the sender.

- **The Cisco Context Adaptive Scanning Engine (CASE):** CASE leverages hundreds of thousands of adaptive attributes that are tuned automatically based on real-time analysis of cyber threats. CASE combines adaptive rules and the real-time outbreak rules published by Cisco Talos to evaluate every message and assign a unique threat level. Based on the threat level, CASE recommends a period of time to quarantine the message to prevent an outbreak (as well as rescan intervals to reevaluate the message based on updated outbreak rules from Talos). The higher the threat level, the more often CASE rescans the message while it is quarantined.

- **Automated threat data drawn from Cisco Talos:** Threat intelligence information from Cisco's security research organization (Talos) to provide a deeper understanding of underlying cybersecurity threats.

- **Advanced Malware Protection (AMP):** To provide global visibility and continuous analytics across all components of the AMP architecture for endpoints and mobile devices. The Cisco AMP integration with Cisco Email Security also provides persistent protection against URL-based threats via real-time analysis of potentially malicious links.

Forged Email Detection

Cisco Email Security also provides a feature called Forged Email Detection (FED). FED is used to detect spear phishing attacks by examining one or more parts of the SMTP message for manipulation, including the "Envelope-From," "Reply To," and "From" headers.

Sender Policy Framework

Cisco Email Security also has the Sender Policy Framework (SPF) for sender authentication and DomainKeys Identified Mail (DKIM) and Domain-based Message Authentication, Reporting, and Conformance (DMARC) for domain authentication.

Email Encryption

Cisco Email Security supports advanced encryption key services to manage email recipient registration, authentication, and per-message/per-recipient encryption keys. The email security gateway also gives compliance and security officers the control of and visibility into how sensitive data is delivered. The cloud-based Cisco Email Security solution also provides a customizable reporting dashboard to access information about encrypted email traffic, including the delivery method used and the top senders and receivers.

The Cisco Email Security cloud service also supports S/MIME. Secure/Multipurpose Internet Mail Extensions (S/MIME) is a standards-based method for sending and receiving secure, verified email messages. S/MIME uses a public/private key pair to encrypt or sign messages.

NOTE   The Cisco Email Security administrator guide includes details the features and deployment of the cloud email security solution at https://www.cisco.com/c/en/us/td/docs/security/ces/user_guide/sma_user_guide_11-4/b_SMA_Admin_Guide_11_4.html.

Cisco Email Security for Office 365

Cisco Email Security can provide protection for Office 365 deployments. Figure 9-29 shows how a traditional Office 365 email exchange is done without Cisco Email Security.

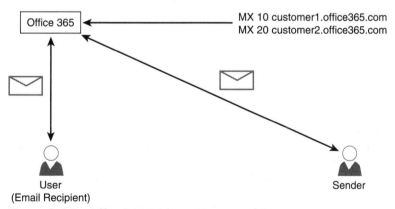

Figure 9-29   *Office 365 Without Cisco Email Security*

Figure 9-30 shows how an Office 365 email exchange is done with Cisco Email Security. You can see that the MX records are changed to the Cisco ESAs in the cisco-ces.com domain in this example. The interaction between the Cisco Email Security solution and Office 365 relays all emails for inspection. The Cisco Email Security cloud service provides a "public" email listener to protect all incoming and outgoing emails.

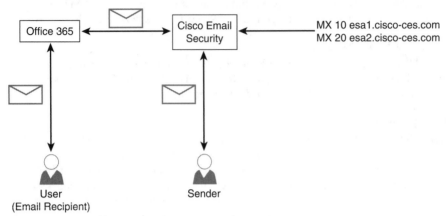

Figure 9-30   *Office 365 with Cisco Email Security*

Figure 9-31 shows the integration of the cloud-based Cisco Email Security solution and the AMP and Thread Grid clouds.

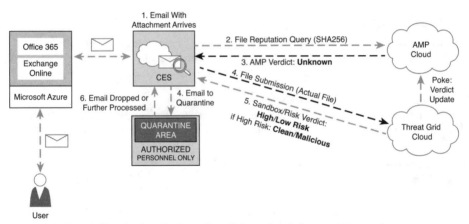

Figure 9-31   *Office 365 with Cisco Email Security Advanced Protection*

Cisco Cloudlock

Cloudlock was a company that Cisco acquired a few years ago. Now called Cisco Cloudlock, the solution is a cloud access security broker (CASB). A CASB provides visibility and compliance checks, protects data against misuse and exfiltration, and provides threat protections against malware like ransomware.

Cisco Cloudlock integrates with cloud services such as the following:

- Box
- Dropbox

- G Suite

- Office 365

- Okta

- OneLogin

- Salesforce

- ServiceNow

- Slack

- Webex Teams

Figure 9-32 shows the Cisco Cloudlock main dashboard. There you can see the different anomalies, top users with admin activity, and an overview of overall security risk based on location.

Figure 9-32 *Cisco Cloudlock Main Dashboard*

Figure 9-33 shows the Cisco Cloudlock Incidents dashboard. An incident in Cisco Cloudlock is a record of an instance of a document, object, event, or app triggering a Cloudlock policy. In the case of incidents related to data loss prevention (DLP), the incident includes information about the asset in violation, the platform account that owns the asset, how widely the asset is shared, and the history of the asset. Depending on the generating policy, the incident may contain a link to the asset itself and its content. Any access to content by security administrators is recorded in the Cloudlock Audit Log.

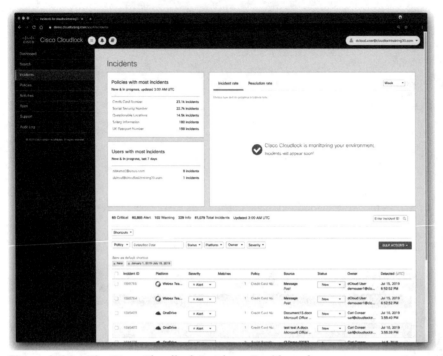

Figure 9-33   *The Cisco Cloudlock Incidents Dashboard*

A Cloudlock incident is typically created when an object or asset (document, object, field, post, and so on) in a platform monitored by Cisco Cloudlock meets three criteria:

1. The object/asset has been created or changed.

2. The object/asset is monitored by an active policy.

3. The object/asset is in violation of the criteria (content and/or context criteria) in the policy.

Figure 9-34 shows an example of a Cisco Cloudlock incident.

Figure 9-35 shows the Cisco Cloudlock Policies dashboard.

Policies are the automated rules you create in Cisco Cloudlock to customize information protection to match your organization's needs. Policies generate incidents that enable you to monitor and correct security issues. Cisco Cloudlock policies operate independently of one another; they do not interact directly, and there is no explicit order of execution. The response actions associated with one policy do not interact with the response actions of another policy. You can add a new policy by selecting a predefined policy or by creating your own policy. Figure 9-36 shows how to add a predefined policy. In this example, a policy is added to alert and block and transactions that may include United States Social Security numbers.

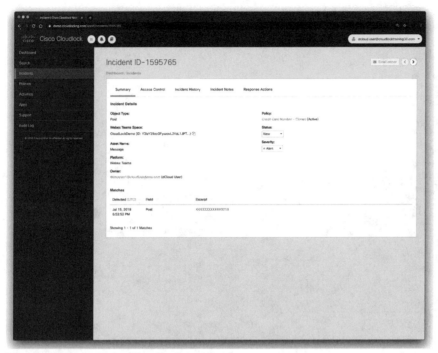

Figure 9-34   *A Cisco Cloudlock Incident*

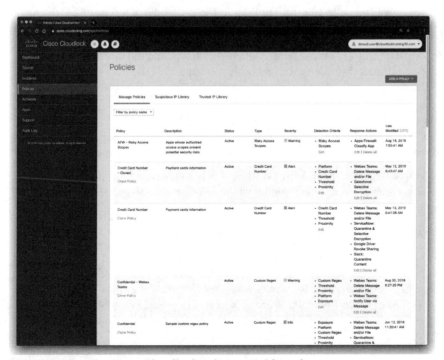

Figure 9-35   *The Cisco Cloudlock Policies Dashboard*

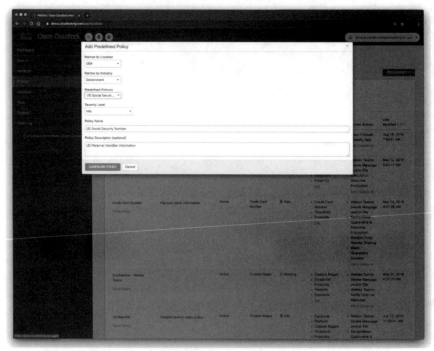

Figure 9-36   *Adding a Predefined Cisco Cloudlock Policy*

You can also design and build your own policies in Cisco Cloudlock by starting with one of these categories:

- **Context-Only:** Ignores the content contained in assets, and monitors only how widely assets are shared, their file types, and other metadata.

- **Custom Regex:** Monitors content matching a regular expression you create.

- **Event Analysis:** Monitors the platform events you select. Events are specific to each monitored platform. You can select individual raw events to monitor and/or events in normalized categories established by Cloudlock (for example, login events).

- **Salesforce Report Export Activity:** Applies only to the Salesforce platform. You can use it to monitor when, where, and by whom Salesforce reports are exported from the platform.

Figure 9-37 shows the Cisco Cloudlock App Discovery dashboard.

In order to use the Cisco Cloudlock App Discovery feature to review and investigate usage of cloud apps on your network, at least one log source must be connected to Cloudlock. A log source is a network device logging network activity, such as a Cisco Web Security Appliance (WSA) or Cisco Umbrella. When at least one log source is integrated with App Discovery, data will become available in the App Discovery dashboard and the Discovered Apps list page.

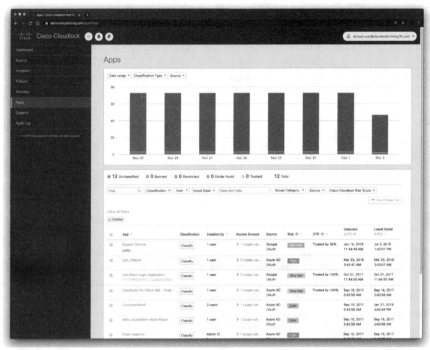

Figure 9-37 *The Cisco Cloudlock App Discovery Dashboard*

TIP   Cisco Cloudlock provides a Composite Risk Score (CRS) in order to assess the relative risk of cloud-connected apps and services according to the following factors:

- **Business Risk**, which incorporates the following:
    - The type of app usage—indirect use (such as a content delivery network), personal use (such as a game), or corporate use (such as a productivity tool).
    - The web reputation of the app, based on Cisco Talos research.
    - Financial viability risk to the provider of the app or service, based on Dun & Bradstreet's Dynamic Risk Score.
    - Data storage risk—an assessment of the nature of the data stored by the app or service. This ranges from no storage (lowest risk) to unstructured data (highest risk). Unstructured data consists of individual files such as emails, photos, documents, and the like.
- **Usage Risk**, which incorporates the following:
    - Traffic volume—the higher the volume of data flowing to and from an app, the higher the potential risk.
    - User count, as measured by number of unique IP addresses within your network that are used to access the app or service. The number of IP addresses is positively correlated with risk.
- **Vendor Compliance**, which includes the security controls put in place by the vendor of the app or service as well as any security-related compliance certifications earned by the vendor.

9

Stealthwatch Cloud

In Chapter 5, "Network Visibility and Segmentation," you learned about the Cisco Stealthwatch solution. The Cisco Stealthwatch solution uses NetFlow telemetry and contextual information from the Cisco network infrastructure. This solution allows network administrators and cybersecurity professionals to analyze network telemetry in a timely manner to defend against advanced cyber threats, including the following:

- Network reconnaissance

- Malware proliferation across the network for the purpose of stealing sensitive data or creating back doors to the network

- Communications between the attacker (or command-and-control servers) and the compromised internal hosts

- Data exfiltration

Cisco Stealthwatch aggregates and normalizes considerable amounts of NetFlow data to apply security analytics to detect malicious and suspicious activity. You can also monitor on-premises networks in your organizations using Cisco Stealthwatch Cloud. In order to do so, you need to deploy at least one Cisco Stealthwatch Cloud Sensor appliance (virtual or physical appliance). The Cisco Stealthwatch Cloud Sensor appliance can be deployed in two different modes (not mutually exclusive):

- By processing network metadata from a SPAN or a network TAP

- By processing metadata out of NetFlow or IPFIX flow records

Cisco Stealthwatch Cloud can also be integrated with Meraki and Cisco Umbrella. The following document details the integration with Meraki: https://www.cisco.com/c/dam/en/us/td/docs/security/stealthwatch/cloud/portal/SWC_Meraki_DV_1_1.pdf.

The following document outlines how to integrate Stealthwatch Cloud with the Cisco Umbrella Investigate REST API in order to provide additional information in the Stealthwatch Cloud environment for external entity domain information: https://www.cisco.com/c/dam/en/us/td/docs/security/stealthwatch/cloud/portal/SWC_Umbrella_DV_1_0.pdf.

> **TIP**   The Cisco Stealthwatch Cloud integrates with Kubernetes (in GCP/GKE, other public clouds, and on-premises Kubernetes deployments). The solution deploys as a pod on a Kubernetes node and shims into the node-level network communication abstraction layer. This provides visibility, baselining, and anomaly detection into container-to-container and pod-to-pod communications.

AppDynamics Cloud Monitoring

AppDynamics was another company acquired by Cisco. AppDynamics (or AppD for short) provides end-to-end visibility of applications and can provide insights about application performance. AppD is able to automatically discover the flow of all traffic requests in your environment by creating a dynamic topology map of all your applications.

AppD also provides cloud monitoring and supports the following platforms:

- AWS Monitoring

- Microsoft Azure

- Pivotal Cloud Foundry Monitoring

- Cloud Foundry Foundation

- Rackspace Monitoring

- Kubernetes Monitoring

- OpenShift Monitoring

- HP Cloud Monitoring

- Citrix Monitoring

- OpenStack Monitoring

- IBM Monitoring

- Docker Monitoring

- AWS Lambda Monitoring

AppDynamics can also be integrated with the Workload Optimization Manager, which is a server application running on a VM that you install on your network. You then assign Virtual Management services running on your network to be Workload Optimization Manager targets. Workload Optimization Manager discovers the devices each target manages, and then performs analysis, anticipates risks to performance or efficiency, and recommends actions you can take to avoid problems before they occur.

Figure 9-38 shows the Cloud Executive Dashboard of the Workload Optimization Manager.

Figure 9-39 shows the Workload Optimization Manager Cloud dashboard.

The Workload Optimization Manager is a Cisco agentless technology that detects relationships and dependencies between applications and the infrastructure layers. It provides a global topological mapping of your environment (local and remote, and across private and public clouds) and the interdependent relationships within the environment, mapping each layer of the full infrastructure stack to application demand. This allows you to get real-time actions that ensure workloads get the resources they need when they need them, enabling continuous placement, resizing, and capacity decisions that can be automated, driving continuous health in the environment.

Figure 9-40 shows the cloud applications used in the organization's cloud deployment. In this example, all applications are hosted in AWS.

Figure 9-40 shows the list of cloud applications within your environment. The integration with the Cisco Workload Optimization Manager helps you to monitor and analyze application performance across your data centers and into public clouds (AWS in this example). AppDynamics and the Cisco Workload Optimization Manager quickly model what-if scenarios based on the real-time environment to accurately forecast capacity needs. In addition, you can track compute, storage, and database consumption (CPU, memory, latency, and Database Transaction Unit [DTU]) across cloud regions and availability zones.

9

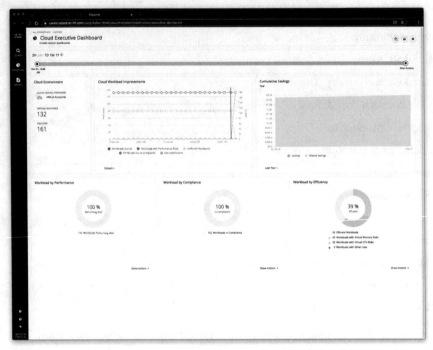

Figure 9-38   *The Cloud Executive Dashboard of the Workload Optimization Manager*

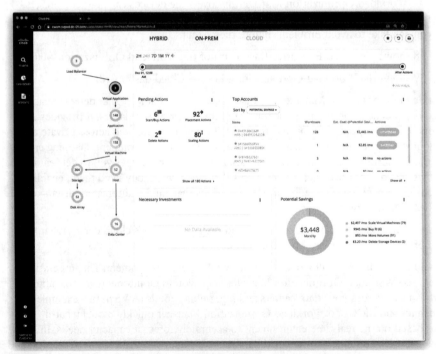

Figure 9-39   *The Workload Optimization Manager Cloud Dashboard*

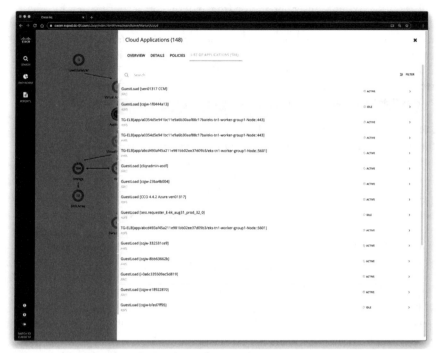

Figure 9-40 *The List of Cloud Applications*

> **NOTE**   The Cisco Workload Optimization Manager and AppDynamics support AWS, Google Cloud Platform, and Microsoft Azure.

Cisco Tetration

Cisco Tetration is a solution created by Cisco that utilizes rich traffic flow telemetry to address critical data center operationality use cases. It uses both hardware and software agents as telemetry sources and performs advanced analytics on the collected data. To access the information, Cisco Tetration provides a scalable point-and-click web UI to search information using visual queries and visualizes statistics using variety of charts and tables. In addition, all the administrative functions and cluster monitoring can be done through the same web UI. Cisco Tetration supports both on-premises and public cloud workloads.

Tetration Agents

Tetration uses software agents and can also obtain telemetry information from Cisco network infrastructure devices. The Tetration software agent is a piece of software running within a host operation system (such as Linux or Windows). Its core functionality is to monitor and collect network flow information. It also collects other host information such as network interfaces and active processes running in the system. The information collected by the agent is exported to a set of collectors running within Tetration cluster for further analytical processing. In addition, software agents also have capability to set firewall rules on the installed hosts.

In order for Tetration to import user annotation from external orchestrators, Tetration needs to establish outgoing connections to the orchestrator API servers (Vcenter, Kubernetes, F5 BIG-IP, and so on). Sometimes it is not possible to allow direct incoming connections to the orchestrators from the Tetration cluster. Secure Connector solves this issue by establishing an outgoing connection from the same network as the orchestrator to the Tetration cluster. This connection is used as a reverse tunnel to pass requests from the cluster back to the orchestrator API server.

Application Dependency Mapping

Application Dependency Mapping (ADM) is a functionality in Cisco Tetration that helps provide insight into the kind of complex applications that run in a data center or in the cloud.

ADM enables network admins to build tight network security policies based on various signals such as network flows, processes, and other side information like load balancer configs and route tags. Not only can these policies be exported in various formats for consumption by different enforcement engines, but Tetration can also verify policies against ongoing traffic in near real time.

Tetration Forensics Feature

The Forensics feature set enables monitoring and alerting for possible security incidents by capturing real-time forensic events and applying user-defined rules. The Forensics feature enables the following features:

- Defining of rules to specify forensic events of interest

- Defining trigger actions for matching forensic events

- Searching for specific forensic events

- Visualizing event-generating processes and their full lineages

TIP For each Workload we compute a Forensics Score. A Workload's Forensics Score is derived from the Forensic Events observed on that Workload based on the profiles enabled for this scope. A score of 100 means no Forensic Events were observed via configured rules in enabled profiles, and a score of 0 means there is a Forensic Event detected that requires immediate action. The Forensics Score for a Scope is the average Workload score within that Scope. The Forensics Score for a given hour is a minimum of all scores within that hour.

Tetration Security Dashboard

The Tetration Security Dashboard, shown in Figure 9-41, presents actionable security scores by bringing together multiple signals available in Tetration. It helps in understanding the current security position and improving it. Security Score is a number between 0 and 100. It indicates the security position in the category. A score of 100 is the best score, and a score of 0 is the worst. Scores closer to 100 are better.

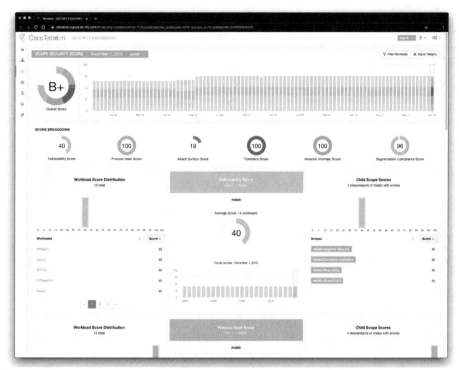

Figure 9-41   *The Tetration Security Dashboard*

The Security Score computation takes into account vulnerabilities in installed software packages, consistency of process hashes, open ports on different interfaces, forensic and network anomaly events, and compliance/non-compliance to policies.

The Vulnerability Dashboard, shown in Figure 9-42, enables administrators to focus their effort on critical vulnerabilities and workloads that need the most attention. Administrators can select relevant scope at the top of this page as well as select the Common Vulnerability Scoring System (CVSS) score. The new page highlights the distribution of vulnerabilities in the chosen scope. This new page also displays vulnerabilities by different attributes, such as the complexity of exploits, whether the vulnerabilities can be exploited over the network, and whether the attacker needs local access to the workload. Furthermore, there are statistics to quickly filter out vulnerabilities that are remotely exploitable and have the lowest complexity to exploit.

Tetration provides a REST API for interacting with all features in a programmatic way. Tetration also has the concept of *connectors*. Connectors are integrations that Tetration supports for a variety of use cases, including flow ingestion, inventory enrichment, and alert notifications.

9

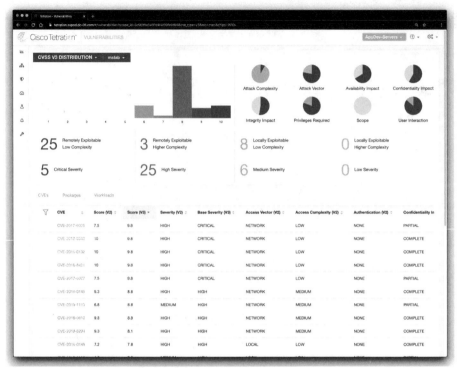

Figure 9-42   *The Tetration Vulnerability Dashboard*

Exam Preparation Tasks

As mentioned in the section "How to Use This Book" in the Introduction, you have a couple of choices for exam preparation: the exercises here, Chapter 12, "Final Preparation," and the exam simulation questions in the Pearson Test Prep Software Online.

Review All Key Topics

Review the most important topics in this chapter, noted with the Key Topic icon in the outer margin of the page. Table 9-3 lists these key topics and the page numbers on which each is found.

Table 9-3   Key Topics for Chapter 9

Key Topic Element	Description	Page Number
List	Identifying the essential characteristics of cloud computing	551
List	Understanding the different cloud deployment models	552
List	Identifying the different cloud service models	552
Paragraph	Understanding what is DevOps	552
Section	The Agile Methodology	553
Section	DevOps	556

Key Topic Element	Description	Page Number
Section	CI/CD Pipelines	558
Section	The Serverless Buzzword	559
Section	Container Orchestration	559
Section	A Quick Introduction to Containers and Docker	561
Section	Kubernetes	565
Section	Microservices and Micro-Segmentation	570
Section	DevSecOps	571
Section	Describing the Customer vs. Provider Security Responsibility for the Different Cloud Service Models	573
Section	Patch Management in the Cloud	575
Section	Security Assessment in the Cloud and Questions to Ask Your Cloud Service Provider	575
Section	Cisco Umbrella	577
Section	The Cisco Umbrella Architecture	577
Section	Secure Internet Gateway	578
Section	Cisco Umbrella Investigate	580
Section	Cisco Email Security in the Cloud	582
Section	Forged Email Detection	583
Section	Sender Policy Framework	583
Section	Email Encryption	583
Section	Cisco Email Security for Office 365	583
Section	Cisco Cloudlock	584
Section	Stealthwatch Cloud	590
Section	AppDynamics Cloud Monitoring	590
Section	Cisco Tetration	593

Define Key Terms

Define the following key terms from this chapter and check your answers in the glossary:

cloud access security broker (CASB), Continuous Integration (CI), Continuous Delivery (CD), DevOps, DevSecOps, Kubernetes (k8s), Nomad, Apache Mesos, Docker Swarm

9

Review Questions

1. What is Extreme Programming (EP)?

   a. A software development methodology designed to improve quality and for teams to adapt to the changing needs of the end customer

   b. A DevSecOps concept to provide better SAST and DAST solutions in a DevOps environment

   c. A software development methodology designed to provide cloud providers with the ability to scale and deploy more applications per workload

   d. None of these answers is correct.

2. Which of the following is a framework that helps organizations work together because it encourages teams to learn through experiences, self-organize while working on a solution, and reflect on their wins and losses to continuously improve?

   a. DevSecOps

   b. Scrum

   c. Waterfall

   d. None of these answers is correct.

3. Which of the following is the CI/CD pipeline stage that includes the compilation of programs written in languages such as Java, C/C++, and Go?

   a. Develop

   b. Build

   c. Deploy

   d. Package and Compile

4. Which of the following is a Kubernetes component that is a group of one or more containers with shared storage and networking, including a specification for how to run the containers?

   a. Pod

   b. k8s node

   c. kubectl

   d. kubeadm

5. Which of the following is a technique that can be used to find software errors (or bugs) and security vulnerabilities in applications, operating systems, infrastructure devices, IoT devices, and other computing devices? This technique involves sending random data to the unit being tested in order to find input validation issues, program failures, buffer overflows, and other flaws.

   a. Scanning

   b. DAST

   c. Fuzzing

   d. SAST

6. Which of the following is a Cisco Umbrella component that provides organizations access to global intelligence that can be used to enrich security data and events or help with incident response? It also provides the most complete view of an attacker's infrastructure and enables security teams to discover malicious domains, IP addresses, and file hashes and even predict emergent threats.

   a. Investigate

   b. Internet Security Gateway

   c. Cloudlock

   d. CASB

7. Cisco Cloud Email Security supports which of the following techniques to create the multiple layers of security needed to defend against?

   a. Geolocation-based filtering

   b. The Cisco Context Adaptive Scanning Engine (CASE)

   c. Advanced Malware Protection (AMP)

   d. All of these answers are correct.

8. Which of the following statements are true about the Cisco Email Security solution?

   a. The Sender Policy Framework (SPF) is used for sender authentication.

   b. DomainKeys Identified Mail (DKIM) is used for domain authentication.

   c. Domain-based Message Authentication, Reporting, and Conformance (DMARC) is used for domain authentication.

   d. All of these answers are correct.

9. You can design and build your own policies in Cisco Cloudlock by starting with which of the following categories?

   a. Custom Regex

   b. Event Analysis

   c. Salesforce Report Export Activity

   d. All of these answers are correct.

10. Cisco Cloudlock provides a _____ in order to assess the relative risk of cloud-connected apps and services according to business risk, usage risk, and vendor compliance.

   a. Composite Risk Score (CRS)

   b. Composite Risk Rating (CRR)

   c. Common Vulnerability Scoring System (CVSS)

   d. None of these answers is correct.

9

CHAPTER 10

Content Security

This chapter covers the following topics:

Content Security Fundamentals

Cisco Web Security Appliance (WSA)

Cisco Email Security Appliance (ESA)

Cisco Content Security Management Appliance (SMA)

The following SCOR 350-701 exam objectives are covered in this chapter:

- **Domain 4.0 Content Security**

    - 4.1 Implement traffic redirection and capture methods

    - 4.2 Describe web proxy identity and authentication, including transparent user identification

    - 4.3 Compare the components, capabilities, and benefits of local and cloud-based email and web solutions (ESA, CES, WSA)

    - 4.4 Configure and verify web and email security deployment methods to protect on-premises and remote users (inbound and outbound controls and policy management)

    - 4.5 Configure and verify email security features such as SPAM filtering, antimalware filtering, DLP, blacklisting, and email encryption

    - 4.6 Configure and verify secure Internet gateway and web security features such as blacklisting, URL filtering, malware scanning, URL categorization, web application filtering, and TLS decryption

"Do I Know This Already?" Quiz

The "Do I Know This Already?" quiz allows you to assess whether you should read this entire chapter thoroughly or jump to the "Exam Preparation Tasks" section. If you are in doubt about your answers to these questions or your own assessment of your knowledge of the topics, read the entire chapter. Table 10-1 lists the major headings in this chapter and their corresponding "Do I Know This Already?" quiz questions. You can find the answers in Appendix A, "Answers to the 'Do I Know This Already?' Quizzes and Q&A Sections."

Table 10-1 "Do I Know This Already?" Section-to-Question Mapping

Foundation Topics Section	Questions
Content Security Fundamentals	1
Cisco Web Security Appliance (WSA)	2–5
Cisco Email Security Appliance (ESA)	5–8
Cisco Content Security Management Appliance (SMA)	9–10

CAUTION The goal of self-assessment is to gauge your mastery of the topics in this chapter. If you do not know the answer to a question or are only partially sure of the answer, you should mark that question as wrong for purposes of the self-assessment. Giving yourself credit for an answer you incorrectly guess skews your self-assessment results and might provide you with a false sense of security.

1. Which of the following statements is not true about AsyncOS?

    a. AyncOS is the underlying operating system for Cisco WSA.

    b. AyncOS is the underlying operating system for Cisco ESA.

    c. AyncOS is the underlying operating system for Cisco SMA.

    d. AyncOS provides a user UNIX shell, and administrators can configure the system using a web admin portal (or a web-based GUI).

2. Which of the following is the Cisco WSA engine that analyzes and categorizes unknown URLs and blocks websites that fall below a defined security policy or threshold? The same engine analyzes more than 200 different factors related to web traffic and the network to determine the level of risk associated with a site.

    a. AVC engine

    b. Web reputation engine

    c. CASB engine

    d. File reputation engine

3. In which type of Cisco WSA deployment mode is the client configured to use the web proxy?

    a. Transparent mode

    b. Explicit forward mode

    c. WCCP mode

    d. None of these answers is correct.

4. Which of the following statements is not true?

   a. Because the client knows there is a proxy and sends all traffic to the proxy in explicit forward mode, the client does not perform a DNS lookup of the domain before requesting the URL. The Cisco WSA is responsible for DNS resolution, as well.

   b. When you configure the Cisco WSA in explicit mode, you do not need to configure any other network infrastructure devices to redirect client requests to the Cisco WSA. However, you must configure each client to send traffic to the Cisco WSA.

   c. In transparent mode, you can also configure the client's proxy settings using DHCP or DNS, using proxy auto-configuration (PAC) files, or with Microsoft Group Policy Objects (GPOs).

   d. You can advertise and configure clients with PAC settings by using the Web Proxy Auto-Discovery (WPAD) protocol. WPAD uses the auto-detect proxy settings found in every modern web browser.

5. You are hired to deploy a web security solution using Cisco WSA. Your boss asks for you to select the best deployment option where web clients do not require an agent or a special configuration in the web browser or operating system. Which of the following is the best approach to accomplish this task?

   a. Enabling WCCP in your infrastructure to redirect web traffic to the Cisco WSA, requiring a review of routing configurations and firewall policies

   b. Configuring the Cisco WSA in transparent mode using hardware load balancers and PAC files

   c. Configuring policy-based routing along with hardware load balancers in explicit web traffic mode

   d. Configuring the Cisco WSA in explicit mode using PAC files and policy-based routing in Cisco routers

6. Which of the following is the entity responsible for forwarding emails from a sender to the recipient, which most people refer to as the "mail server"?

   a. Mail transfer agent (MTA)

   b. Mail delivery agent (MDA)

   c. Mail submission agent (MSA)

   d. Mail user agent (MUA)

7. The Cisco ESA acts as a mail transfer agent. The Cisco ESA is the destination of which public records?

   a. AA

   b. MX

   c. C-NAME

   d. All of these answers are correct.

8. Which of the following is used by the Cisco ESA to handle incoming SMTP connection requests? These entities demarcate the email processing service configured on a Cisco ESA interface.

   a. WCCP redirects

   b. MX records

   c. SMTP MSAs

   d. Listeners

9. Which of the following provides a means for gateway-based cryptographic signing of outgoing messages? This technology allows you to embed verification data in an email header and for email recipients to verify the integrity of the email messages, and it uses DNS TXT records to publish public keys.

   a. SPF

   b. DKIM

   c. SenderBase

   d. Cisco SMA

10. You are hired to deploy an email and web security solution that can be managed from a centralized location. In addition, this solution must allow you to integrate with third-party solutions to monitor outgoing emails to make sure that no sensitive information is being transferred out of your company. Which of the following is the best approach to accomplish this task?

   a. Deploy Cisco FMC to manage and monitor Cisco ESA, Cisco WSA, and Cisco FTD with DLP services.

   b. Deploy Cisco SMA to manage and monitor Cisco ESA and Cisco WSA, and make sure that the Cisco ESA DLP email policies are enabled in the Outgoing Mail Policies table.

   c. Deploy Cisco SMA to manage and monitor Cisco ESA and Cisco WSA, and make sure that the Cisco FMC DLP email policies are enabled in the Outgoing Mail Policies table.

   d. Deploy Cisco FMC to manage and monitor Cisco ESA and Cisco WSA, and make sure that the Cisco ESA DLP email policies are enabled in the Outgoing Mail Policies table.

Foundation Topics

Content Security Fundamentals

Cyber actors (attackers) use email and the web as the two top threat vectors to carry out many of their attacks. Why? It is because email and web protocols are the most popular protocols used by individuals and organizations. In Chapter 1, "Cybersecurity Fundamentals," you learned the many different social engineering attacks that can be carried over email (phishing, spear phishing, whaling, and so on). You also learned how attackers can fool users to follow malicious links, impersonate websites, and attack different web-based applications.

Cisco acquired a company called Ironport that created what we know today as the Cisco Web Security Appliance (WSA) and the Cisco Email Security Appliance (ESA) to address this problem. The Cisco WSA and Cisco ESA are solutions designed to provide strong

10

protection, complete control, and operational visibility into threats to an organization. The Cisco WSA and Cisco ESA have been integrated with other Cisco solutions such as AMP, and they also can digest threat intelligence from Cisco Talos.

The Cisco WSA and Cisco ESA can be managed by the Cisco Content Security Management Appliance (SMA). The Cisco SMA provides a solution for centralizing the management and reporting functions of multiple Cisco ESA and Cisco WSA devices. When you deploy the Cisco SMA, it provides simplification of administration and planning, and it improves compliance monitoring. Another benefit of the Cisco SMA is that it allows administrators to enable consistent policy enforcement and enhances threat protection.

The underlying operating system of the Cisco ESA, Cisco WSA, and Cisco SMA is the Async Operating System (AsyncOS). You will learn more about AsyncOS in the following section.

Cisco Async Operating System (AsyncOS)

AsyncOS powers the Cisco WSA, Cisco ESA, and Cisco SMA, and it is based on a FreeBSD-based kernel. However, Cisco enhanced AsyncOS to address some of the limitations of traditional Linux and UNIX operating systems. One focus was scalability in order to support thousands of connections per minute. Cisco WSA, Cisco ESA, and Cisco SMA running AsyncOS take advantage of a high-performance file system and optimized asynchronous communication of email and web transactions (thus the name AsyncOS). AsyncOS does not have a user UNIX shell. Administrators can configure the system using a web admin portal (or a web-based GUI) or a fully scriptable command-line interface (CLI).

Cisco WSA

Under the hood, the Cisco Web Security Appliance (WSA) includes a web proxy, a threat analytics engine, antimalware engine, policy management, and reporting in a single physical or virtual appliance. The main use of the Cisco WSA is to protect users from accessing malicious websites and being infected by malware.

Organizations can also configure the Cisco WSA to give users access to the sites they need to do their work and deny other sites, including gaming sites, social media, and so forth.

The following are the different Cisco WSA feature engines:

- **Web Reputation engine:** Analyzes and categorizes unknown URLs and blocks websites that fall below a defined security policy or threshold. The Web Reputation engine analyzes more than 200 different factors related to web traffic and the network to determine the level of risk associated with a site. The Cisco WSA Web Reputation engine is very different in comparison to legacy URL blacklisting and whitelisting capabilities of traditional web proxies. The Cisco WSA engine analyzes a large data set and produces a granular reputation score of –10 to +10. This reputation score allows security professionals to make a better risk assessment.

- **Web filtering:** Syndicates traditional URL filtering with real-time dynamic content analysis. This, in turn, allows for granular acceptable use policy (AUP) creation and warns the user on certain quota and bandwidth conditions.

- **Application Visibility and Control (AVC):** Enables the Cisco WSA to inspect and/or block applications that are not allowed by the organization's security policy. You can allow users to use social media sites like Twitter and Facebook and then block micro-applications within those social media sites (like Facebook games).

- **Cloud access security:** The Cisco WSA can detect and stop hidden threats in cloud apps by leveraging built-in AVC along with integrations with cloud access security brokers (CASBs) such as Cisco Cloudlock.

- **Antivirus scanning:** The Cisco WSA supports different antivirus programs such as McAfee, Sophos, and Webroot.

- **File reputation:** Based on Cisco Talos threat intelligence, which is updated every three to five minutes.

- **Data-loss prevention (DLP):** The Cisco WSA can redirect all outbound traffic to a third-party DLP system, allowing deep content inspection for regulatory compliance and data exfiltration protection. This allows you to inspect web content by title, metadata, and size, and even to prevent users from storing files to cloud services, such as Box, Dropbox, iCloud, and Google Drive.

- **File sandboxing:** The Cisco WSA has been integrated with the Cisco AMP and Cisco Threat Grid sandboxing capabilities. This allows for putting an unknown file in a sandbox to inspect its behavior. Cisco AMP and Threat Grid use machine learning to analyze the file and determine the threat level. You will learn more about Cisco AMP and Threat Grid in Chapter 11, "Endpoint Protection and Detection."

- **File retrospection:** The Cisco WSA examines files that are downloaded and continues to cross-examine files over an extended period of time. The file disposition can be Unknown, Clean, Malware, and so on. A changed file disposition is referred to as a retrospective disposition.

- **Cognitive threat analytics:** The Cisco WSA supports anomaly detection of HTTP and HTTPS traffic. The state and results of the cognitive threat analytics metrics are fine-tuned based on new threat information discovered by the system and Cisco Talos. This allows the Cisco WSA to discover confirmed threats in an environment even when HTTPS traffic inspection has been disabled.

The Cisco WSA Proxy

The Cisco WSA virtual and physical appliances are typically placed either on the inside of the Internet edge firewall or in a demilitarized zone (DMZ). The reason you deploy the Cisco WSA behind the firewall or in a DMX is to be able to centralize proxying and to reduce the number of Cisco WSA appliances.

NOTE The Cisco WSA can be deployed as a physical appliance or as a virtual machine running on VMware's ESX, KVM, or Microsoft's Hyper-V. A proxy sits between HTTP clients (web browsers or APIs [in the case of machine-to-machine communication]) and HTTP servers. This specifically means that the WSA as a web proxy has two sets of TCP sockets per client request: one connection from the client to the WSA and another connection from the WSA to the web server.

10

Cisco WSA physical and virtual appliances have one or more of the following interface types:

- **M1:** Typically used for management. The M1 interface can be used for data traffic (otherwise known as a one-armed interface configuration).

- **P1/P2:** These are typically the interfaces used for web proxy traffic (that is, data interfaces). If you enable the P1 and P2 interfaces, each interface must be connected to different subnets. You can also combine M1 and P1. If doing so, M1 can be configured to proxy requests and P1 is used to send traffic to the Internet. If you use multiple interfaces for proxying, you need to configure static routes to direct the traffic to the correct interface.

- **T1/T2:** Typically used for Layer 4 traffic monitoring to listen to all TCP ports. When you enable the T1/T2 ports, they are not configured with an IP address because they are promiscuous monitoring ports. T1 can be configured alone for duplex communication, or T1 and T2 can be configured together in simplex mode. For instance, T1 can be configured to receive all outgoing traffic to the Internet, and the T2 interface can be configured to receive all incoming traffic from the Internet.

You can deploy the Cisco WSA in two different modes:

- Explicit forward mode

- Transparent mode

Cisco WSA in Explicit Forward Mode

In explicit forward mode, the client is configured to explicitly use the proxy, consequently sending all web traffic to the proxy, as demonstrated in Figure 10-1.

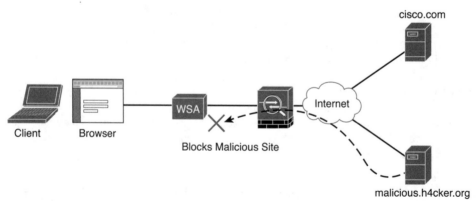

Figure 10-1   *Cisco WSA in Explicit Forward Mode*

TIP   Because the client knows there is a proxy and sends all traffic to the proxy in explicit forward mode, the client does not perform a DNS lookup of the domain before requesting the URL. The Cisco WSA is responsible for DNS resolution, as well.

When you configure the Cisco WSA in explicit mode, you do not need to configure any other network infrastructure devices to redirect client requests to the Cisco WSA. However, you must configure each client to send traffic to the Cisco WSA. In large environments, this could be problematic. However, you can also configure the client's proxy settings using DHCP or DNS, using proxy auto-configuration (PAC) files, or with Microsoft Group Policy Objects (GPOs). You can also lock browser proxy settings with solutions like Microsoft GPOs.

TIP   You can advertise and configure clients with PAC settings by using the Web Proxy Auto-Discovery (WPAD) protocol. WPAD uses the auto-detect proxy settings found in every modern web browser. Proxy server configurations can be provisioned to clients through DHCP option 252 with the URL as a string in the option (for example, https://secretcorp.org/wpad.dat) or with DNS by creating an A host record for wpad.secretcorp.org.

Figure 10-2 shows the proxy configuration of a macOS device.

Figure 10-2   *Proxy Configuration in a Mac OS X Device*

The Cisco WSA also supports SOCKS proxy configurations When it is configured as a SOCKS proxy, the client exchanges SOCKS protocol messages to negotiate a proxy connection. When a connection is established, the client communicates with the Cisco WSA by using the SOCKS protocol.

NOTE You need to configure a SOCKS policy in order to use the Cisco WSA SOCKS proxy. The SOCKS protocol (and consequently the Cisco WSA) only supports direct forward connections. The Cisco WSA does not forward traffic to any upstream proxies when configured as a SOCKS proxy. In addition, the Cisco WSA SOCKS proxy does not support scanning services, which are used by AVC, DLP, and malware detection. The Cisco WSA SOCKS proxy is not able to decrypt SSL traffic because it tunnels traffic from the client to the server.

Cisco WSA in Transparent Mode

When the Cisco WSA is in transparent mode, clients do not know there is a proxy deployed. Network infrastructure devices are configured to forward traffic to the Cisco WSA. In transparent mode deployments, network infrastructure devices redirect web traffic to the proxy. Web traffic redirection can be done using policy-based routing (PBR)—available on many routers—or using Cisco's Web Cache Communication Protocol (WCCP) on Cisco ASA, Cisco routers, or switches.

Figure 10-3 shows a Cisco WSA in transparent mode.

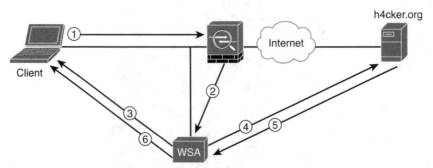

Figure 10-3 *Cisco WSA in Transparent Mode*

TIP WCCP is a Cisco-developed content-routing protocol that provides a mechanism to redirect traffic flows in real time. It has built-in load balancing, scaling, fault tolerance, and service-assurance (failsafe) mechanisms.

The following are the steps illustrated in Figure 10-3.

Step 1. The client initiates a connection to h4cker.org.

Step 2. The Cisco ASA redirects the request to the Cisco WSA using WCCP.

Step 3. The Cisco WSA verifies the request and replies to the client if the web request violates a policy or the security engine flags it.

Step 4. The Cisco WSA initiates a new connection to h4cker.org.

Step 5. The h4cker.org web server replies to the Cisco WSA. The Cisco WSA checks for malicious or inappropriate content and blocks it, if needed.

Step 6. If the content is acceptable, the Cisco WSA forwards the content to the client.

TIP   In Figure 10-3, the client is unaware its traffic is being sent to a proxy (Cisco WSA) and, as a result, the client uses DNS to resolve the domain name in the URL and send the web request destined for the web server (not the proxy). When you configure the Cisco WSA in transparent mode, you need to identify a network choke point with a redirection device (in this example, a Cisco ASA) to redirect traffic to the proxy.

When transparent mode is configured, you are able to force all traffic to the proxy if desired (without end-user interaction). Load balancing is inherent without the use of hardware load balancers or PAC files. Many organizations deploy transparent mode Cisco WSAs in phases by using access control lists (ACLs) with policy-based routing or WCCP.

NOTE   When you enable WCCP in your infrastructure, it requires review of routing configurations, firewall policies, and so on. For instance, when you configure WCCP in the Cisco ASA, the Cisco WSA and clients need to be within the same security zone.

Configuring WCCP in a Cisco ASA to Redirect Web Traffic to a Cisco WSA

The following are the steps to configure WCCP in the Cisco ASA:

Step 1.   Create an access control list (ACL) to define (match) the HTTP and HTTPS traffic from the 10.1.1.0/24 and 10.1.2.0/24 subnets, as shown in Example 10-1.

Example 10-1   *Matching the HTTP and HTTP Traffic*

```
access-list HTTP-TRAFFIC permit tcp 10.1.1.0 255.255.255.0 any eq www
access-list HTTP-TRAFFIC permit tcp 10.1.2.0 255.255.255.0 any eq www
access-list HTTPS-TRAFFIC permit tcp 10.1.1.0 255.255.255.0 any eq https
access-list HTTPS-TRAFFIC permit tcp 10.1.2.0 255.255.255.0 any eq https
```

Step 2.   You can also inspect FTP traffic in the Cisco WSA. In order to do so, create an ACL to match FTP traffic, as demonstrated in Example 10-2.

Example 10-2   *Matching FTP Traffic*

```
access-list FTP-TRAFFIC permit tcp 10.1.1.0 255.255.255.0 any eq ftp
access-list FTP-TRAFFIC permit tcp 10.1.1.0 255.255.255.0 any range 11000 11006
access-list FTP-TRAFFIC permit tcp 10.1.2.0 255.255.255.0 any eq ftp
access-list FTP-TRAFFIC permit tcp 10.1.2.0 255.255.255.0 any range 11000 11006
```

Step 3.   Create another ACL to include the IP address of the Cisco WSA (10.1.2.3) and create the WCCP redirect lists, as demonstrated in Example 10-3. You can configure WCCP redirection of HTTP traffic (TCP port 80 traffic) and also non-HTTP TCP traffic, as well as UDP packets. For instance, you can redirect packets used for proxy-web cache handling, File Transfer Protocol (FTP) caching, FTP proxy handling, audio and video applications, and so on. To achieve this task, you can configure multiple WCCP service groups. Service information

10

is specified in the WCCP configuration commands using dynamic services identification numbers (such as "10" or "20", as shown in Example 10-4) or a predefined service keywords (such as "web-cache"). The networking device uses that information to validate that service group members are all providing or using the same service.

Example 10-3 *Creating an ACL to Define Where to Send the Traffic and Creating the WCCP Redirect Lists*

```
access-list WSA extended permit ip host 10.1.2.3 any
wccp web-cache redirect-list HTTP-TRAFFIC group-list WSA
wccp 10 redirect-list FTP-TRAFFIC group-list WSA
wccp 20 redirect-list HTTPS-TRAFFIC group-list WSA
```

Step 4.   Finally, configure the WCCP redirection of traffic on the source interface (the inside interface in this example).

Example 10-4 *Configuring Redirection of Traffic on Source Interface*

```
wccp interface inside web-cache redirect in
wccp interface inside 10 redirect in
wccp interface inside 20 redirect in
```

You can also configure WCCP on a Cisco Firepower Threat Defense (FTD) device by using the Cisco Firepower Management Console (FMC) FlexConfig policies. A FlexConfig policy is a container of an ordered list of FlexConfig objects. Each object includes a series of Apache Velocity scripting language commands, Cisco ASA software configuration commands, and variables that you define. The contents of each FlexConfig object are essentially a program that generates a sequence of the Cisco ASA commands that will then be deployed to the assigned devices. This command sequence then configures the related feature on the Cisco FTD device.

The Cisco FTD devices use Cisco ASA configuration commands to implement some features, but not all features. There is no unique set of Cisco FTD configuration commands. Instead, the point of FlexConfig is to allow you to configure features that are not yet directly supported through the Cisco FMC policies and settings. Figure 10-4 shows the use of FlexConfig to configure WCCP on a Cisco FTD device via the Cisco FMC.

NOTE   Cisco strongly recommends using FlexConfig policies only if you are an advanced user with a strong Cisco ASA background and at your own risk. Enabling features through FlexConfig policies may cause unintended results with other configured features.

Configuring WCCP on a Cisco Switch

Let's take a look on how to configure WCCP on a Cisco switch to redirect traffic to the Cisco WSA. Refer to the topology shown in Figure 10-5.

The following are the steps to configure WCCP on a Cisco switch to send traffic to the Cisco WSA.

Figure 10-4 *Configuring WCCP on a Cisco FTD via FMC's FlexConfig*

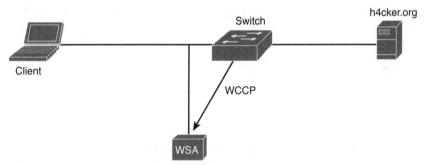

Figure 10-5 *Configuring WCCP on a Cisco Switch to Send Traffic to a Cisco WSA*

Step 1. Configure an access control list (ACL) to match the web traffic, as demonstrated in Example 10-5.

Example 10-5 *Matching HTTP and HTTPS Traffic*

```
ip access-list extended WEB-TRAFFIC
 permit tcp 10.1.1.0 0.0.0.255 any eq www
 permit tcp 10.1.2.0 0.0.0.255 any eq www
 permit tcp 10.1.1.0 0.0.0.255 any eq 443
 permit tcp 10.1.2.0 0.0.0.255 any eq 443
```

Step 2. You can also redirect FTP traffic to the Cisco WSA. In Example 10-6, an ACL called FTP-TRAFFIC is configured to redirect FTP traffic via WCCP. This ACL, along with the one configured in Example 10-5, will be associated to the WCCP configuration at a later step.

Example 10-6   *Matching FTP Traffic*

```
ip access-list extended FTP-TRAFFIC
 permit tcp 10.1.1.00.0.0.255 any eq ftp
 permit tcp 10.1.1.00.0.0.255 any range 11000 11006
 permit tcp 10.1.2.00.0.0.255 any eq ftp
 permit tcp 10.1.2.00.0.0.255 any range 11000 11006
```

Step 3.   Configure another ACL to define where to send the traffic (that is, the Cisco WSA's IP address), as shown in Example 10-7.

Example 10-7   *Defining Where to Send the HTTP, HTTPS, and FTP Traffic*

```
ip access-list standard WSA
 permit 10.1.3.3
```

Step 4.   Create the WCCP lists, as demonstrated in Example 10-8.

Example 10-8   *Creating the WCCP Lists*

```
ip wccp web-cache redirect-list HTTP-TRAFFIC group-list WSA
ip wccp 10 redirect-list FTP-TRAFFIC group-list WSA
ip wccp 20 redirect-list HTTPS-TRAFFIC group-list WSA
```

Step 5.   Configure the WCCP redirection of traffic on the source interface, as shown in Example 10-9.

Example 10-9   *Configuring the WCCP Redirection of Traffic on the Source Interface*

```
interface vlan88
 ip wccp web-cache redirect in
 ip wccp 10 redirect in
 ip wccp 20 redirect in
```

Configuring the Cisco WSA to Accept WCCP Redirection

Figure 10-6 shows how to configure WCCP on the Cisco WSA.

Navigate to **Network > Transparent Redirection** and click **Edit Device**. Select **WCCP v2 Router** from the drop-down and click **Submit**. Click **Add Service** to add a new WCCP redirection service, and the screen shown in Figure 10-6 is displayed.

NOTE   The WCCP configuration can be customized to use different service IDs for different traffic. Each service ID needs a separate entry on the Cisco WSA.

Traffic Redirection with Policy-Based Routing

You can also configure PBR on a Cisco router to redirect web traffic to the Cisco WSA.

NOTE   Configuring PBR can affect the router's performance if enabled in software (without hardware acceleration). You should review the respective router documentation to determine any impact.

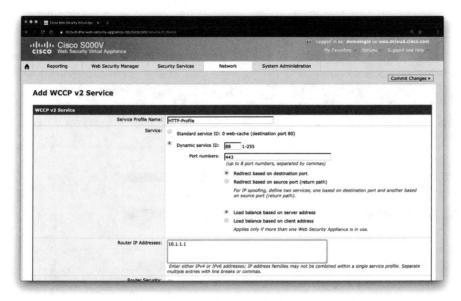

Figure 10-6 *Configuring WCCP on the Cisco WSA*

When you configure the Cisco WSA in transparent mode using a "Layer 4 switch," no additional configuration is needed on the Cisco WSA. You just navigate to **Network > Transparent Redirection** and select **Layer 4 Switch**. In this case, the redirection is controlled by the Layer 4 switch (or router). Why Layer 4? Because redirection is being done based on Layer 4 ports.

In Example 10-10, a PBR policy is configured in a Cisco router that matches traffic from two source subnets (10.1.1.0/24 and 10.1.1.2.0/24). The web traffic is received on interface VLAN 88. The traffic is sent to the Cisco WSA configured with IP address 10.1.2.3.

Example 10-10 *PBR Configuration in a Cisco Router*

```
access-list 101 permit tcp 10.1.1.0 0.0.0.255 any eq 80
access-list 101 permit tcp 10.1.2.0 0.0.0.255 any eq 80
access-list 101 permit tcp 10.1.1.0 0.0.0.255 any eq 443
access-list 101 permit tcp 10.1.2.0 0.0.0.255 any eq 443
!
route-map WebRedirect permit 10
  match ip address 101
  set ip next-hop 10.1.3.3
interface vlan88
  ip policy route-map WebRedirect
```

Cisco WSA Security Services

The Cisco WSA uses security components to protect end users from a range of malware threats. You can configure antimalware and web reputation settings for each policy group. When you configure Access Policies, you can also have AsyncOS for Web choose a combination of antimalware scanning and web reputation scoring to use when determining what content to block.

Figure 10-7 shows the Security Services options in the Cisco WSA.

Figure 10-7 *Configuring WCCP on the Cisco WSA*

> **NOTE** The CCNP Security 300-725 SWSA exam, "Securing the Web with Cisco Web Security Appliance (SWSA)," and the CCIE lab cover configuration and troubleshooting of the Cisco WSA.

Deploying Web Proxy IP Spoofing

When the Cisco WSA (as a web proxy) forwards a request, by default it changes the request source IP address to match its own address. However, you can change this behavior by enabling web proxy IP spoofing so that requests appear to come from the client rather than from the Cisco WSA.

IP spoofing is supported in transparent and explicitly forwarded proxy configurations. When the Cisco WSA is deployed in transparent mode, you can enable IP spoofing either for only transparently redirected connections or for all connections (transparently redirected *and* explicitly forwarded).

When you configure explicit proxy with IP spoofing, you must ensure that HTTP reply packets are routed back to the Cisco WSA.

NOTE   When you configure IP spoofing and the Cisco WSA is connected to a WCCP router, two WCCP services must be configured (one based on source ports and one based on destination ports) in order to track the underlying HTTP transactions.

Configuring Policies in the Cisco WSA

The Cisco WSA identifies and controls web requests using different policies. When a client initiates a web request to a web server, the Cisco WSA inspects the transaction and determines to which policy it belongs. The defined policy actions are applied to the request.

TIP   The Cisco WSA evaluates policies from the top down (similar to router and firewall ACLs). A best practice is to place the most accessed or used policies at the top to increase performance.

One of the policy types you can enable in the Cisco WSA is called identification policies. Identification policies are configured to identify the users behind the web requests, instead of just reporting based on the IP address of the system or device making the web request. You can configure the Cisco WSA to interact with Lightweight Directory Access Protocol (LDAP) or Active Directory (AD) authentication servers.

NOTE   LDAP supports only basic authentication, whereas AD supports NTLM, Kerberos, and basic authentication.

Traditionally, users can be identified by username and password and then their credentials are validated with an authentication. Subsequently, policies are applied based on the username. However, the WSA can be configured to authenticate users without prompting the end user for credentials (transparent identification). When you enable transparent identification, the user is authenticated using the authentication "state" obtained from another trusted source. Consequently, the Cisco WSA assumes that the user has already been authenticated by that trusted source and applies the configured policies. Transparent authentication is considered a single sign-on (SSO) environment, and the users are not aware that a proxy has been deployed. This is also useful when client devices are not capable of displaying an authentication prompt (such as a printer or an IP phone).

The Cisco WSA provides different options for the AD or LDAP realm (authentication). The following are the available schemes when using AD authentication (AD realm):

- **Basic authentication:** Done via a web browser. Basic authentication is not transparent.

- **NTLMSSP:** This is a type of transparent authentication. The web browser must be compatible and provide support for NTLMSSP. NTLMSSP uses AD domain credentials for login and is typically used in Windows AD environments (although it can also work with Mac, with additional configuration on the client side).

- **Kerberos:** Primarily used with Windows clients, Kerberos is considered the more secure option.

10

The Cisco WSA supports different authentication schemes for a wide range of client support. The Authentication Surrogates options enable you to configure how web transactions will be associated with a user after the user has been successfully authenticated. The following options are provided by the Cisco WSA:

- **IP Address:** The user's identity is used until the surrogate times out.

- **Persistent Cookie:** The user's identity is used until the surrogate times out.

- **Session Cookie:** The user's identity is used until the browser is closed or the session times out.

There are also access policies. Access policies configured in the Cisco WSA map the identification profiles and users. They also map time-based restrictions, to make sure that the necessary controls align with your business policies.

You can add a new policy by navigating to **Web Security Manager > Access Policies > Add Policy.** There you can assign a unique name for the policy and map the identification profile settings and optionally additional advanced settings. After submitting the new policy, you can do additional customization to adjust how the access policy behaves compared to the global policy settings.

TIP   You can use protocols and user agents to control policy access to protocols and configure blocking for specific client applications (including social media or instant messaging clients). You can also configure the Cisco WSA to tunnel HTTP CONNECT requests on specific ports.

You can also customize URL filtering using different policies to specify how a transaction based on the URL category of a particular HTTP or HTTPS request is handled by the Cisco WSA. When you configure URL filtering, you can also define custom URL categories. Once the custom URL category is created, you can specify whether to block, redirect, allow, monitor, warn, or apply quota-based or time-based filters for websites in the custom categories.

The following are some additional settings and customizations you can configure in the Cisco WSA:

- Earlier in this chapter you learned about the AVC engine. You can use the AVC engine to enforce acceptable-use policy components to block or allow applications by application type and by individual applications. In addition, you can control different application behaviors (for example, file transfers).

- You can also configure the Cisco WSA web proxy to block file downloads based on file characteristics, including the file size, file type, and MIME type.

- You can also define an access policy to apply antimalware and URL reputation.

- By default, the Cisco WSA only redirects and decodes port 80 HTTP traffic. However, you can configure the Cisco WSA to decrypt and evaluate SSL traffic. You can do this by navigating to **Security Services > HTTPS Proxy.** Furthermore, a root certificate used to sign web traffic must be created or uploaded to the Cisco WSA. You can create a certificate on the Cisco WSA and then install the certificate to all clients that will be connecting through the Cisco WSA. You can also use the HTTPS proxy to change the decryption options, invalid certificate handling, and Online Certificate Status Protocol (OCSP) options. Then you add a decryption policy or edit the global policy once the HTTPS proxy is configured.

■ You can also create an outbound malware policy on the Cisco WSA to block malware uploads.

■ The Cisco WSA supports DLP servers. To integrate the Cisco WSA with an external DLP server, you need to configure a data security policy to manage data uploads to the web. Then enable external DLP policies to redirect outbound traffic to external servers for scanning. You can define an external DLP by navigating to **Network > External DLP Servers** and configuring the server by selecting the communication protocol (Internet Content Adaptation Protocol [ICAP] or Secure ICAP) and setting the service address, service URL, and load-balancing method. ICAP is a lightweight HTTP-like protocol that is used to forward web requests to external DLP servers or content scanners.

Cisco WSA Reports

The Cisco WSA provides detailed reporting of all the web transactions, malware threats, URL categories, and many other web proxy transactions.

Figure 10-8 shows the Web Sites report (dashboard), which includes statistics about the top domains requested, top domains blocked, and several other statistics.

Figure 10-9 shows the Users report (dashboard), including the transactions blocked for the top users and the top users based on bandwidth usage.

Figure 10-8 *The Cisco WSA Web Sites report*

Figure 10-9 *The Cisco WSA Users Report*

The Cisco WSA can also provide reports about the top malware threats (files) monitored or blocked, as well as the trend of malware threat files detected, as shown in Figure 10-10. The report illustrated in Figure 10-10 also includes the top malware threat files by file type and the malware files by category.

Figure 10-11 shows the URL categories report displaying the top URL categories (total transactions) and the total URL blocked and warned transactions sorted by category.

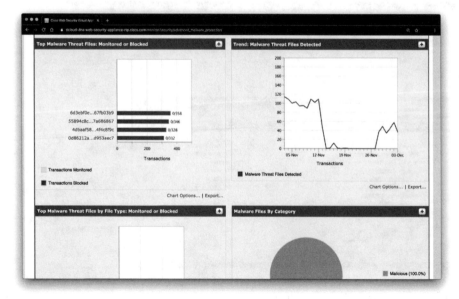

Figure 10-10 *The Cisco WSA Malware Threats Report*

Figure 10-11   *The Cisco WSA URL Categories Report*

Cisco ESA

In Chapter 9, "Securing the Cloud," you learned that the Cisco ESA can be deployed as a physical appliance, virtual appliance, or as a cloud service. In this section you will learn the details about the Cisco ESA acting as the email gateway to an organization, controlling the transfer of all email connections, accepting messages, and relaying messages to the appropriate email servers. As you probably already know, email transactions on the Internet use SMTP. The Cisco ESA can handle all SMTP connections for an organization acting as the SMTP gateway.

Reviewing a Few Email Concepts

You may already be familiar with email protocols and concepts. However, as a refresher, the following are some of the most important email concepts that you must be familiar with to understand how the Cisco ESA works:

- **Mail transfer agent (MTA):** Most people refer to the MTA as the "mail server" (the entity responsible for transferring emails from a sender to the recipient).

- **Mail delivery agent (MDA):** A component of an MTA responsible for the final delivery of an email message to a person's inbox (mailbox).

- **Mail user agent (MUA):** An email client or email reader installed on the user's system (or mobile device).

- **Mail submission agent (MSA):** A component of an MTA that accepts new mail messages from an MUA (using SMTP).

10

- **Internet Message Access Protocol (IMAP):** An email client communication protocol that allows users to keep messages on the server. An IMAP-enabled MUA displays messages directly from the server. However, you can also download messages using IMAP for archiving purposes.

- **Post Office Protocol (POP):** An application-layer protocol used by an email client to retrieve (download) email from a remote server.

People have called "mail servers" so many different things, such as an MTA, a mail router, a mail transport agent, and a mail exchanger (MX). DNS MX records are used to route the mail traffic on the Internet. An MX record is a type of verified resource record in DNS that specifies a mail server responsible for accepting email messages on behalf of a recipient's domain, and a preference value is used to prioritize mail delivery if multiple mail servers are available. The set of MX records of a domain name specifies how email should be routed with SMTP. Example 10-11 shows the output of the Linux **dig** command displaying the DNS resolution of the domain h4cker.org. The highlighted lines are the MX records for the domain.

Example 10-11   *An Example of DNS MX Records*

```
$dig h4cker.org MX
; <<>> DiG 9.10.6 <<>> h4cker.org MX
;; global options: +cmd
;; Got answer:
;; ->>HEADER<<- opcode: QUERY, status: NOERROR, id: 13242
;; flags: qr rd ra ad; QUERY: 1, ANSWER: 5, AUTHORITY: 0, ADDITIONAL: 1

;; OPT PSEUDOSECTION:
; EDNS: version: 0, flags:; udp: 4096
;; QUESTION SECTION:
;h4cker.org.            IN      MX

;; ANSWER SECTION:
h4cker.org.     3600    IN      MX      5 gmr-smtp-in.1.google.com.
h4cker.org.     3600    IN      MX      10 alt1.gmr-smtp-in.1.google.com.
h4cker.org.     3600    IN      MX      20 alt2.gmr-smtp-in.1.google.com.
h4cker.org.     3600    IN      MX      30 alt3.gmr-smtp-in.1.google.com.
h4cker.org.     3600    IN      MX      40 alt4.gmr-smtp-in.1.google.com.

;; Query time: 291 msec
;; SERVER: 192.168.88.1#53(192.168.88.1)
;; MSG SIZE  rcvd: 163
```

Cisco ESA Deployment

The Cisco ESA can be deployed in different ways. Similar to the Cisco WSA, the Cisco ESA can be deployed with a single physical interface to filter email to and from your mail servers or in a two-interface configuration. When you configure the Cisco ESA with two interfaces, one interface is used for email transfers to and from the Internet and the other interface is used for email transfers to and from the internal servers.

Cisco ESA deployments are fairly straightforward. The Cisco ESA acts as a mail transfer agent. The Cisco ESA is the destination of the public MX records. In other words, the MX records of the underlying domain should point to the Cisco ESA's public IP address. The Cisco ESA needs to be accessible through the public Internet and should be the first hop in the organization's email infrastructure. Let's take a look at the topology in Figure 10-12.

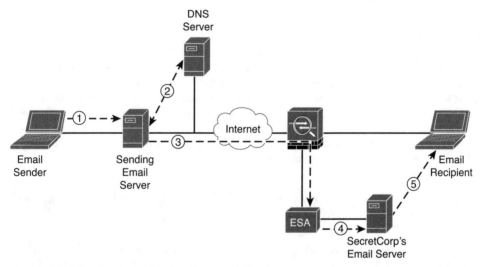

Figure 10-12 *An Example of a Cisco ESA Deployment*

The following are the steps illustrated in Figure 10-12:

1. The email sender attempts to send an email to omar@secretcorp.org. The email client sends the email to the "sending email server."

2. The sending mail server looks up the secretcorp.org MX record and receives the hostname of the Cisco ESA (mail.secretcorp.org). The sending email server also queries the DNS server for the IP address of mail.secretcorp.org.

3. The sending mail server opens an SMTP connection with the Cisco ESA.

4. The Cisco ESA inspects the email transaction and sends the mail to the internal mail server, if it conforms to the configured security policies and it is determined that the email is not malicious or spam.

5. The email recipient's client retrieves the email from the internal mail server by using IMAP or POP.

Cisco ESA Listeners

The Cisco ESA uses listeners to handle incoming SMTP connection requests. Such listener delimits the email processing service configured on a Cisco ESA interface.

Cisco ESA listeners apply to email entering the appliance from either the Internet (public listeners) or internal systems (private listeners), as demonstrated in Figure 10-13.

10

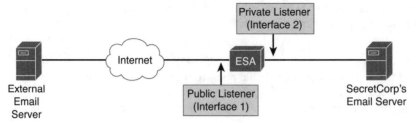

Figure 10-13   *Cisco ESA Listeners*

SenderBase

The Cisco ESA and Cisco WSA are products that evolved from a company that Cisco acquired called Ironport. Ironport's SenderBase (now Cisco SenderBase) is a reputation service that enables you to control the messages that come through the Cisco ESA email gateway based on the senders' trustworthiness (reputation). When the Cisco ESA receives messages from known or highly reputable senders, it delivers them directly to the end user without any content scanning. However, when the Cisco ESA receives email messages from unknown or less reputable senders, it performs antispam and antivirus scanning. The Cisco ESA uses a reputation score that ranges from –10 to +10.

> **TIP**   Cisco partnered with antivirus companies such as McAfee and Sophos to provide network antivirus scanning capabilities on the Cisco ESA.

The Cisco ESA has the concept of outbreak filters. Outbreak filters are enabled by default and provide a dynamic quarantine (also called a DELAY quarantine). The Cisco ESA can continue to hold or release back though antivirus and Advanced Malware Protection (AMP) for additional scans.

> **TIP**   These outbreak filters offer a significant catch rate for outbreaks over traditional solutions, since they provide the "human" element after signature, heuristics, and hash-based scanning. It has been proven that outbreak filters deliver more than nine hours of lead time over antivirus engines for zero-day outbreaks.

The Recipient Access Table (RAT)

The recipient access table (RAT), not to be confused with remote-access Trojan (also RAT), is a Cisco ESA term that defines which recipients are accepted by a public listener.

> **TIP**   At a minimum, RAT stipulates the listener address and whether to accept or reject it. For instance, a Cisco ESA might accept mail from secretcorp.com or secretcorp.org.

Cisco ESA Data Loss Prevention

The Cisco ESA has a DLP feature that allows you to secure your sensitive, proprietary information and intellectual property, preventing this data from leaving your network (maliciously or unintentionally).

TIP   You can specify the types of data your users are not allowed to send via email by creating DLP policies to scan outgoing messages.

The Cisco ESA's mail policy is a set of rules that specify the types of suspect, sensitive, or malicious content you might not want entering or leaving your network, such as the following:

- Marketing messages

- Spam

- Graymail

- Malware

- Phishing, spear phishing, whaling, and other targeted email-based attacks

- Confidential data

- Personally identifiable information (PII)

SMTP Authentication and Encryption

Sender Policy Framework (SPF) enables recipients to verify the sender's IP addresses by looking up DNS records that list authorized mail gateways for a particular domain. SPF is an industry standard defined in RFC 4408. SPF uses DNS TXT resource records. The Cisco ESA supports SPF to verify HELO/EHLO and MAIL FROM identity (FQDN). When you enable SPF, the Cisco ESA adds headers in the message, allowing you to obtain additional intelligence on the email sender. One challenge is that the effectiveness of SPF implementations is based on participation. Also, you need to invest time to make sure SPF records are up to date. Many organizations implement SPF because some of the most prevalent email providers nowadays do not accept mail without SPF records.

TIP   Some organizations go to the extent of not allowing any emails that do not have an SPF record. Doing so will block more spam emails; however, some legitimate mail might also be dropped if the sending entity hasn't configured SPF correctly.

Domain Keys Identified Mail (DKIM)

DKIM is an industry standard defined in RFC 5585. DKIM provides a means for gateway-based cryptographic signing of outgoing messages. This allows you to embed verification data in an email header and for email recipients to verify the integrity of the email messages.

DKIM uses DNS TXT records to publish public keys.

10

NOTE   A few additional specifications related to DKIM exist. RFC 6376 introduces DKIM signatures, RFC 5863 provides information on DKIM development, deployment, and operation, and RFC 5617 addresses Author Domain Signing Practices (ADSP).

You configure SPF and DKIM verification in mail flow policies (by navigating to **Mail Policies > Mail Flow Policy**).

Cisco Content Security Management Appliance (SMA)

The Cisco Content SMA provides centralized management and monitoring (reporting) of Cisco WSAs and Cisco ESAs. The Cisco SMA simplifies the planning and administration of Cisco ESA and Cisco WSA deployments. In Figure 10-14, a Cisco SMA is deployed to manage and monitor two Cisco ESAs and three Cisco WSAs.

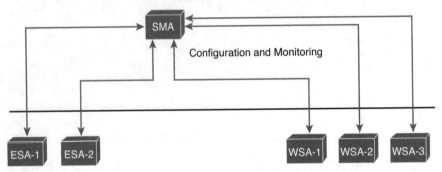

Figure 10-14 *Cisco SMA Managing Cisco ESAs and Cisco WSAs*

The centralized configuration and management provided by the Cisco SMA in Figure 10-14 helps with consistent enforcement of acceptable-use policies and to enhance threat protection. In other words, this centralized reporting and management helps determine which users are in violation of acceptable use policies.

TIP In the Cisco SMA, data is aggregated from multiple Cisco ESAs, including data categorized by sender, recipient, message subject, and other parameters. Scanning results, such as spam and virus verdicts, are also displayed, as are policy violations.

Figure 10-15 shows the Cisco SMA Monitoring Mail Flow Summary dashboard for Email Security. There you can see the number of attempted incoming email messages along with the email messages that were categorized as security threats.

Similarly, Figure 10-16 shows the number of outgoing email messages processed along with the email messages that were categorized as "clean" and other outgoing email statistics.

Figure 10-17 shows the Advanced Malware Protection (AMP) summary dashboard for incoming files within email messages. Statistics about the disposition of each file is displayed, including clean, unknown, unscannable, low risk, and malicious. The malicious category is further decomposed into malware, custom detection, and custom threshold.

Figure 10-18 shows the AMP Reputation dashboard in the Cisco SMA. Keep in mind that these are statistics summarized from all managed Cisco ESA appliances (physical or virtual).

Figure 10-19 shows the Cisco SMA AMP File Analysis dashboard. The dashboard shows the time and verdict (or interim verdict) for each file sent for analysis. Each managed Cisco ESA checks for analysis results every 30 minutes. You can also export the data as a .csv file to view more than 1000 File Analysis results.

Figure 10-15 *Cisco SMA Mail Flow Summary*

Figure 10-16 *Cisco SMA Outgoing Email Statistics*

10

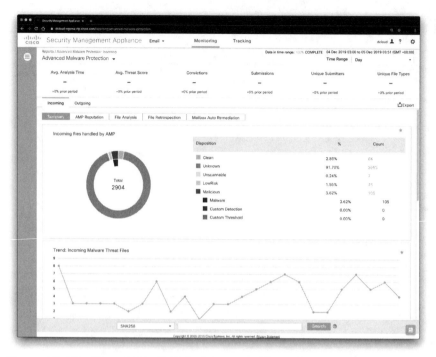

Figure 10-17   *Cisco SMA AMP Summary Dashboard for Incoming Email Messages*

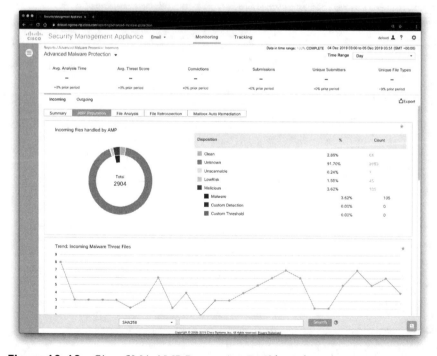

Figure 10-18   *Cisco SMA AMP Reputation Dashboard*

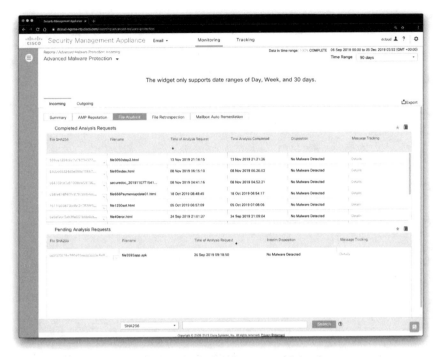

Figure 10-19 *Cisco SMA AMP File Analysis Dashboard*

As you learned earlier in this chapter, the Cisco ESA and Cisco WSA can integrate with the Cisco AMP Threat Grid cloud and on-premises appliances. Files that are whitelisted on the AMP Threat Grid appliance show as "clean." You can drill down to view detailed analysis results, including the threat characteristics for each file.

TIP You can also search for additional information about an SHA value, or you can click the link at the bottom of the File Analysis details page to view additional details on the server that analyzed the file. If a file extracted from a compressed or archived file is sent for analysis, only the SHA value of the extracted file is included in the File Analysis report.

You can use the File Analysis view of the AMP dashboard to view the following:

- The number of incoming and outgoing files that are uploaded for file analysis by the File Analysis service of the Advanced Malware Protection engine

- A list of incoming and outgoing files that have completed File Analysis requests

- A list of incoming and outgoing files that have pending File Analysis requests

The File Retrospection dashboard (shown in Figure 10-20) lists the files processed by the managed Cisco ESAs for which the verdict has changed since the message was received. Because Advanced Malware Protection is focused on targeted and zero-day threats, threat verdicts can change as aggregated data might reveal more information.

10

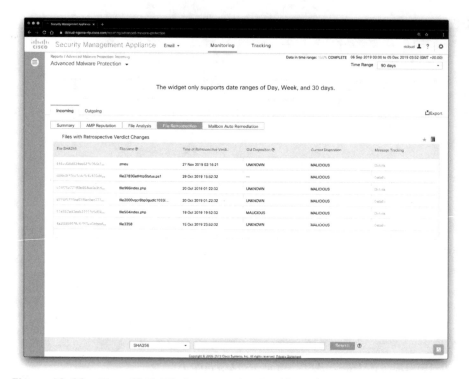

Figure 10-20 *Cisco SMA File Retrospection Dashboard*

Figure 10-21 shows the DLP Incident Summary dashboard, which includes the incidents of DLP policy violations occurring in outgoing mail. As discussed earlier in this chapter, the Cisco ESA uses DLP email policies enabled in the Outgoing Mail Policies table to detect sensitive data sent by your users. Every occurrence of an outgoing message violating a DLP policy is reported as an incident.

You can leverage the DLP Incident Summary report to see what type of sensitive data is being sent by your users and how severe such DLP incidents are. Additionally, you can see how many of these messages are being delivered, dropped, and who is sending these messages.

The DLP Incident Summary page contains two main sections:

- The DLP Incident Trend graphs summarizing the top DLP incidents by severity (Low, Medium, High, Critical) and policy matches

- The DLP Incident Details listing

Figure 10-21 *DLP Incident Summary Dashboard*

Exam Preparation Tasks

As mentioned in the section "How to Use This Book" in the Introduction, you have a couple of choices for exam preparation: the exercises here, Chapter 12, "Final Preparation," and the exam simulation questions in the Pearson Test Prep Software Online.

Review All Key Topics

Review the most important topics in this chapter, noted with the Key Topic icon in the outer margin of the page. Table 10-2 lists these key topics and the page numbers on which each is found.

Table 10-2 Key Topics for Chapter 10

Key Topic Element	Description	Page Number
Section	Cisco Async Operating System (AsyncOS)	604
List	Defining the different Cisco WSA feature engines	604
Section	The Cisco WSA Proxy	605
List	Listing the Cisco WSA deployment modes	606
Section	Cisco WSA in Explicit Forward Mode	606
Section	Cisco WSA in Transparent Mode	608

10

Key Topic Element	Description	Page Number
Section	Deploying Web Proxy IP Spoofing	614
Section	Configuring Policies in the Cisco WSA	615
List	Listing the available schemes when using AD authentication (AD realm) in the Cisco WSA	615
List	Recognizing the different customization and configurations supported in the Cisco WSA	616
List	Reviewing email fundamentals	619
Section	Cisco ESA Deployment	620
Section	Cisco ESA Listeners	621
Section	SenderBase	622
Section	The Recipient Access Table (RAT)	622
Section	Cisco ESA Data Loss Prevention	622
Section	SMTP Authentication and Encryption	623
Section	Domain Keys Identified Mail (DKIM)	623
Section	Cisco Content Security Management Appliance (SMA)	624

Define Key Terms

Define the following key terms from this chapter and check your answers in the glossary:

Domain Keys Identified Mail (DKIM), Sender Policy Framework (SPF), SenderBase, mail transfer agent (MTA), mail delivery agent (MDA), mail user agent (MUA), Internet Message Access Protocol (IMAP), Post Office Protocol (POP), Mail Exchanger (MX) record, Web Proxy Auto-Discovery (WPAD) protocol, proxy auto-configuration (PAC) files, Web Cache Communication Protocol (WCCP)

Review Questions

1. You have been asked to configure the company's network in a way that web traffic is redirected to a Cisco WSA in real time. You must pick a solution that has built-in load balancing, scaling, fault tolerance, and service-assurance (failsafe) mechanisms. Which of the following technologies and deployment modes can accomplish this task?

    a. Cisco WSA in explicit client mode with policy-based routing in Cisco routers

    b. Cisco WSA in explicit client mode with WCCP Cisco switches and firewalls

    c. Cisco WSA in transparent mode with WCCP enabled in Cisco switches and firewalls

    d. Cisco SMA using WCCP and policy-based routing to redirect traffic to the Cisco WSA

2. The Cisco WSA supports SOCKS proxy configurations. Which of the following is not true about Cisco WSA SOCKS proxy configurations?

    a. WCCP can be used to redirect traffic in explicit SOCKS proxy configuration mode.

    b. The SOCKS protocol (and consequently the Cisco WSA) only supports direct forward connections.

    c. The Cisco WSA does not forward traffic to any upstream proxies when configured as a SOCKS proxy.

    d. The Cisco WSA SOCKS proxy does not support scanning services, which are used by AVC, DLP, and malware detection.

3. You can advertise and configure clients with PAC settings by using Web Proxy Auto-Discovery (WPAD) protocol. WPAD uses the auto-detect proxy settings found in every modern web browser. Proxy server configurations can be provisioned to clients through which of the following options?

    a. DHCP option 252 with the URL as a string in the option

    b. DHCP option 252 with the IP address of the Cisco WSA as a string in the option

    c. DHCP option 110 with the IP address of the Cisco WSA as a string in the option

    d. DHCP option 110 with the URL as a string in the option

4. Which of the following can be used in transparent mode Cisco WSA deployments?

    a. Policy-based routing in Cisco routers

    b. WCCP in Cisco ASA and Cisco FTD devices

    c. WCCP in Cisco switches and routers

    d. All of these answers are correct.

5. The Cisco WSA provides different options for AD or LDAP realm (authentication). Which of the following is an available scheme when using AD authentication (AD realm)?

    a. Basic authentication via a web browser

    b. NTLMSSP

    c. Kerberos

    d. All of these answers are correct.

6. The Authentication Surrogates option in the Cisco WSA enables you to configure how web transactions will be associated with a user after the user has been successfully authenticated. Which of the following is not an option provided by the Cisco WSA?

    a. The IP address of the user

    b. Persistent cookies

    c. Session cookies

    d. The WCCP ID

7. Which of the following is an email client communication protocol that allows users to keep messages on the server?

    a. IMAP

    b. POP3

    c. SMTP

    d. None of these answers is correct.

10

8. Which of the following is a component of a mail transfer agent (MTA) responsible for the final delivery of an email message to a person's inbox (mailbox)?

    a. Mail Submission Agent (MSA)

    b. Mail User Agent (MUA)

    c. Mail Delivery Agent (MDA)

    d. DKIM

9. Which of the following statements is not true?

    a. The Cisco ESA can be deployed with a single physical interface to filter email to and from your mail servers or in a two-interface configuration.

    b. The Cisco WSA can be deployed with a single physical interface to inspect web traffic or in a two-interface configuration.

    c. When you configure the Cisco ESA with two interfaces, one interface is used for email transfers to and from the Internet and the other interface is used for email transfers to and from the internal servers.

    d. The Cisco ESA can be deployed with a single physical interface to filter email to and from your mail servers or in a two-interface configuration. However, the Cisco WSA can only be deployed in a two-interface configuration.

10. Which of the following protocols is used by the Cisco ESA to verify HELO/EHLO and MAIL FROM identity (FQDN)? When you enable this protocol, the Cisco ESA adds headers in the message, allowing you to obtain additional intelligence on the email sender.

    a. Domain Keys Identified Mail (DKIM)

    b. SenderBase

    c. Sender Policy Framework (SPF)

    d. WCCP

CHAPTER 11

Endpoint Protection and Detection

This chapter covers the following topics:

> Introduction to Endpoint Protection and Detection
>
> Cisco AMP for Endpoints
>
> Cisco Threat Response

The following SCOR 350-701 exam objectives are covered in this chapter:

- **Domain 5.0 Endpoint Protection and Detection**

  - 5.1 Compare Endpoint Protection Platforms (EPP) and Endpoint Detection & Response (EDR) solutions

  - 5.2 Explain antimalware, retrospective security, indicator of compromise (IOC), antivirus, dynamic file analysis, and endpoint-sourced telemetry

  - 5.3 Configure and verify outbreak control and quarantines to limit infection

  - 5.4 Describe justifications for endpoint-based security

  - 5.5 Describe the value of endpoint device management and asset inventory such as MDM

  - 5.7 Describe endpoint posture assessment solutions to ensure endpoint security

  - 5.8 Explain the importance of an endpoint patching strategy

"Do I Know This Already?" Quiz

The "Do I Know This Already?" quiz allows you to assess whether you should read this entire chapter thoroughly or jump to the "Exam Preparation Tasks" section. If you are in doubt about your answers to these questions or your own assessment of your knowledge of the topics, read the entire chapter. Table 11-1 lists the major headings in this chapter and their corresponding "Do I Know This Already?" quiz questions. You can find the answers in Appendix A, "Answers to the 'Do I Know This Already?' Quizzes and Q&A Sections."

Table 11-1 "Do I Know This Already?" Section-to-Question Mapping

Foundation Topics Section	Questions
Introduction to Endpoint Protection and Detection	1
Cisco AMP for Endpoints	2–9
Cisco Threat Response	10

1. Which of the following is not a feature of the AMP solution?

    a. File reputation

    b. File sandboxing

    c. File retrospection

    d. Web content filtering and redirect

2. You are hired to deploy AMP for Endpoints. In order to allow a connector to communicate with Cisco cloud servers for file and network disposition lookups, a firewall must allow the clients to connect to the Cisco servers over which of the following protocols and ports?

    a. TCP port 443 and TCP port 80

    b. TCP port 443 or TCP port 32137

    c. UDP port 32137 and TCP port 443

    d. TCP port 443, UDP port 53, and UDP port 500

3. Which of the following AMP for Endpoints features allow you to create lists for Custom Detections, Application Control, Network, and Endpoint indicators of compromise (IOC)?

    a. Inbox feature

    b. Group Policies

    c. Outbreak Control

    d. None of these answers is correct.

4. Advanced custom detections offer many more signature types to the detection, including which of the following?

    a. File body–based signatures

    b. MD5 signatures

    c. Logical signatures

    d. All of these answers are correct.

5. You can use outbreak control IP lists in conjunction with _____ detections, which allows you to flag or even block suspicious network activity.

    a. device flow correlation (DFC)

    b. PAC files

    c. group policies

    d. AVC

6. You are hired to deploy AMP for Endpoints, and one of the requirements is that you must use an exclusion set to resolve conflicts with other security products or mitigate performance issues by excluding directories that contain large files that are frequently written to, like databases. Which of the following is an exclusion type available in AMP for Endpoints that can help you accomplish this task?

   a. Threat-based exclusion

   b. Extension-based exclusion

   c. Wildcards

   d. All of these answers are correct.

7. Cisco AMP for Endpoints has connectors for which of the following operating systems?

   a. Windows

   b. macOS

   c. Android

   d. All of these answers are correct.

8. Which of the following is used by the Cisco ESA to handle incoming SMTP connection requests? These entities demarcate the email-processing service configured on a Cisco ESA interface.

   a. WCCP redirects

   b. MX records

   c. SMTP MSAs

   d. Listeners

9. Which of the following clients allow you to aid the distribution of the AMP for Endpoints connector and can be used for remote access VPN, secure network access, and posture assessments with Cisco's Identity Services Engine?

   a. DUO

   b. AnyConnect

   c. Tetration

   d. Cisco SMA

10. Which of the following is a "one-pane-of-glass" console that automates integrations across Cisco security products (including AMP for Endpoints) and threat intelligence sources?

    a. Cisco SMA

    b. Cisco Threat Response (CTR)

    c. Tetration

    d. Firepower Management Console

Foundation Topics

Introduction to Endpoint Protection and Detection

Throughout this book, you have been learning about the various technologies that can also help detect threats in endpoint devices. You have learned that security technologies and processes should not just focus on detection but should also provide the capability to mitigate the impact of an attack. Organizations must maintain visibility and control across the extended network during the full attack continuum:

- Before an attack takes place

- During an active attack

- After an attacker starts to damage systems or steal information

In Chapter 4, "Authentication, Authorization, Accounting (AAA) and Identity Management," you learned about the Cisco ISE, 802.1X, Network Access Control (NAC), endpoint posture assessment, and how clients like AnyConnect are used to interact with network devices and Cisco solutions to protect not only the endpoint, but also the underlying network. You also learned the uses and importance of a multifactor authentication (MFA) strategy.

In Chapter 7, "Cisco Next-Generation Firewalls and Cisco Next-Generation Intrusion Prevention Systems," you learned all about the components that make up the AMP architecture and the AMP cloud.

In Chapter 7, you also learned that the AMP solution enables malware detection, blocking, continuous analysis, and retrospective views with the following features:

- **File reputation:** AMP allows you to analyze files inline and block or apply policies.

- **File sandboxing:** AMP allows you to analyze unknown files to understand true file behavior.

- **File retrospection:** AMP allows you to continue to analyze files for changing threat levels.

TIP Remember that the architecture of AMP can be broken down into three main components: the AMP cloud, AMP client connectors, and intelligence sources. This chapter focuses on the AMP for Endpoints client connector.

This chapter will go over the Cisco Advanced Malware Protection (AMP) for Endpoints in more detail. This chapter looks at where AMP for Endpoints fits into the AMP architecture. You'll also learn about the types of AMP for Endpoints connectors, how to create policies for them, and how to install them. The chapter describes how to use the AMP cloud console, and you will even get a look at AMP detecting and remediating malware.

NOTE After Cisco acquired SourceFire, the solution previously known as FireAMP was renamed AMP for Endpoints. You still see the term FireAMP in legacy documentation.

Endpoint Threat Detection and Response (ETDR) and Endpoint Detection and Response (EDR)

Before diving deep into AMP for Endpoints, let's define what the industry refers to as Endpoint Threat Detection and Response (ETDR) and Endpoint Detection and Response (EDR). Gartner defines EDR as the "tools primarily focused on detecting and investigating suspicious activities (and traces of such) other problems on hosts/endpoints."

11

NOTE Cisco also provides a great definition and an overview of EDR at the following website: https://www.cisco.com/c/en/us/products/security/endpoint-security/what-is-endpoint-detection-response-edr.html.

EDR solutions monitor endpoint and network events and record the information in a central database so that you can perform further analysis, detection, investigation, and reporting. Typically, software (an agent) is installed on the endpoint that allows ongoing monitoring and detection of potential security threats.

TIP Not all EDR solutions work the same way or offer the same range of capabilities. Some EDR solutions perform more analysis on the agent and others focus on the backend (using a management console). Modern EDR solutions integrate with threat intelligence delivered from the cloud.

The minimum capabilities of a good EDR solution are as follows:

- **Filtering:** The ability to filter out false positives (to help reduce "alert fatigue," which increases the potential for real threats to go undetected).

- **Threat blocking:** At the end of the day, the EDR solution must be able to contain the threats, not just detect them.

- **Help with digital forensics and incident response (DFIR):** The ability to allow an organization to effectively perform DFIR tasks, as well as threat hunting to prevent data loss (data breaches).

TIP Another term in the industry is Endpoint Protection Platform (EPP). An EPP provides not only detection, but also protection (threat blocking). In many cases, people refer to EPP and EDR as the same thing. The following blog post provides a good overview of EPP, EDR, and AMP for Endpoints: https://blogs.cisco.com/security/epp-edr-cisco-amp-for-endpoints-is-next-generation-endpoint-security.

Cisco AMP for Endpoints

AMP for Endpoints provides more than just endpoint-level visibility into files. It also provides cloud-based detection of malware, in which the cloud constantly updates itself. This enables very rapid detection of known malware because the cloud resources are used instead of endpoint resources. This architecture has a number of benefits. With the majority of the processing power being performed in the cloud, the endpoint software remains very lightweight.

The AMP cloud is able to provide a historical view of malware activity, segmented into two activity types:

- **File trajectory:** What endpoints have seen the files

- **Device trajectory:** Actions the files performed on given endpoints

With the data storage and processing in the cloud, the AMP solution is able to provide powerful and detailed reporting, as well as provide very robust management.

The AMP for Endpoints agent is also able to take action. For example, it can block malicious network connections based on custom IP blacklists or intelligent dynamic lists of malicious IP addresses.

AMP for Endpoints is the connector that resides on—you guessed it—endpoints. It resides on Windows, Mac, Linux, and Android endpoints. Unlike traditional endpoint protection software that uses a local database of signatures to match a known bad piece of software or a bad file, AMP for Endpoints remains lightweight, sending a hash to the cloud and allowing the cloud to make intelligent decisions and return the verdicts Clean, Malware, and Unknown.

Figure 11-1 shows the AMP for Endpoints high-level architecture.

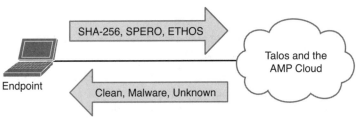

Figure 11-1 *AMP for Endpoints High-Level Architecture*

AMP for Endpoints connectors must be able to reach the AMP cloud. That means the agents may have to be able to go through firewalls and proxy servers to reach the Internet. If traversing a firewall and/or web proxy to reach the Internet, those products must allow connectivity from the AMP connector to the Cisco servers over HTTPS (TCP 443).

To allow a connector to communicate with Cisco cloud servers for file and network disposition lookups, a firewall must allow the clients to connect to the Cisco servers over TCP 443 by default or TCP 32137.

Outbreak Control

With a solution as powerful and extensive as AMP for Endpoints, it is difficult to determine where to start describing how to configure and use the system; however, it makes logical sense to begin with Outbreak Control because the objects you create within Outbreak Control are key aspects of endpoint policies.

Outbreak Control allows you to create lists that customize AMP for Endpoints to your organization's needs. You can view the main lists from the AMP cloud console by clicking the **Outbreak Control** menu, which offers options in the following categories: Custom Detections, Application Control, Network, and Endpoint IOC (indicators of compromise), as shown in Figure 11-2.

You can think of custom detections as a blacklist. You use them to identify files that you want to detect and quarantine. When a custom detection is defined, not only do endpoints quarantine matching files when they see them, but any AMP for Endpoints agents that have seen the files before the custom detection was created can also quarantine the files through retrospection, also known as cloud recall.

11

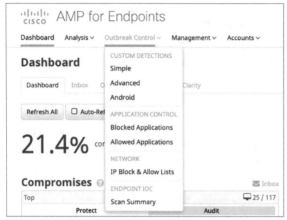

Figure 11-2 *AMP for Endpoints Outbreak Control*

Simple custom detection allows you to add file signatures, while the advanced custom detections are more like traditional antivirus signatures.

Creating a simple custom detection is similar to adding new entries to a blacklist. You define one or more files that you are trying to quarantine by building a list of SHA-256 hashes. If you already have the SHA-256 hash of a file, you can paste that hash directly into the UI, or you can upload files directly and allow the cloud to create the SHA-256 hash for you.

To create a simple custom detection, navigate to **Outbreak Control > Custom Detections > Simple** and the list of all existing simple custom detections appears, as shown in Figure 11-3. To add a new one, you must type it in the Name box and click **Save**, as shown in Figure 11-3.

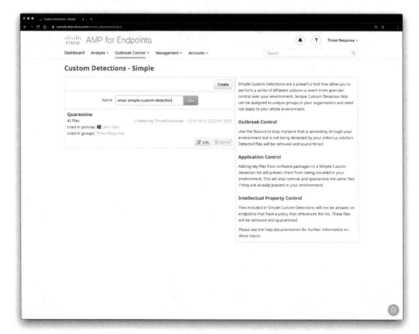

Figure 11-3 *AMP for Endpoints Outbreak Control Options*

The detection is then added to the list, as shown in Figure 11-4, and automatically edited—with the contents displayed on the right side.

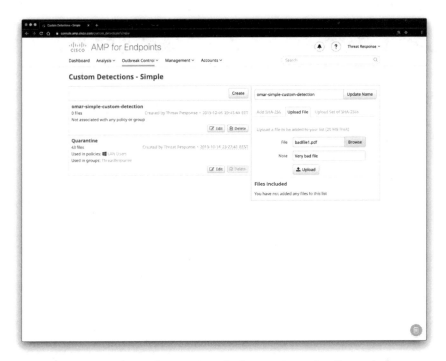

Figure 11-4 *AMP for Endpoints Simple Custom Detection Example*

If you already have the SHA-256 hash of a file, simply paste it in, add a note, and click **Save**; otherwise, you can upload a file, add a note, and click **Upload**. Once the file is uploaded, the hash is created and shown on the bottom-right side. You must click **Save**, or the hash will not be stored as part of your simple custom detection.

Simple custom detections just look for the SHA-256 hash of a file. Advanced custom detections offer many more signature types to the detection, based on ClamAV signatures, including the following:

- File body–based signatures

- MD5 signatures

- MD5, PE section–based signatures

- An extended signature format (with wildcards, regular expressions, and offsets)

- Logical signatures

- Icon signatures

11

To create an advanced custom detection, navigate to **Outbreak Control > Custom Detections > Advanced**, and the list of all existing advanced custom detections appears, as shown in Figure 11-5.

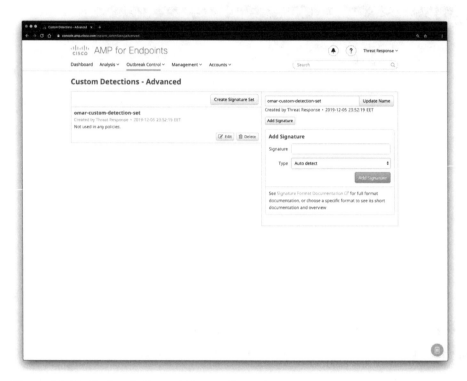

Figure 11-5   *Outbreak Control Custom Detections*

To add a new custom detection, you must type it in the Name box and click **Save** to add it to the list. Click **Edit** to display the contents of the new advanced detection object on the right side, as shown in Figure 11-6. The signature types can be auto-detected, or you can manually select them from the drop-down list. Figure 11-6 shows the available signature types.

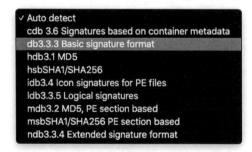

Figure 11-6   *Outbreak Control Custom Detections Available Signature Types*

Next, you click the **Build Database From Signature Set** button, and a success message is displayed, showing the successful creation of the advanced custom detection signature set. A View Changes link is visible with every custom detection, both simple and advanced. The AMP cloud maintains an audit log for each of the detection lists, and you can view it by clicking that link.

Android detections are defined separately from the ones used by Windows or Mac. These detections provide granular control over Android devices in an environment. The detections look for specific applications, and you build them by either uploading the app's .apk file or selecting that file from the AMP console's inventory list.

You can choose to use Android custom detections for two main functions: outbreak control and application control.

When using an Android custom detection for outbreak control, you are using the detection to stop malware that is spreading through mobile devices in the organization. When a malicious app is detected, the user of the device is notified and prompted to uninstall it.

You don't have to use these detections just for malware, but you can also use them to stop applications that you don't want installed on devices in your organization. This is what AMP refers to as "application control." Simply add apps to an Android custom detection list that you don't want installed, and AMP notifies the user of the unwanted application and prompts the user to uninstall it, just as if it were a malicious app.

IP Blacklists and Whitelists

You can use outbreak control IP lists in conjunction with device flow correlation (DFC) detections. DFC allows you to flag or even block suspicious network activity. You can use policies to specify the behavior of AMP for Endpoints when a suspicious connection is detected and also to specify whether the connector should use addresses in the Cisco intelligence feed, the custom IP lists you create yourself, or a combination of both.

You use an IP whitelist to define IPv4 addresses that should not be blocked or flagged by DFC. AMP bypasses or ignores the intelligence feeds as they relate to the IPv4 addresses in the whitelist (also often called the "allow list").

You use IP blacklists to create DFC detections. Traffic that matches entries in the blacklist are flagged or blocked, as the DFC rule dictates.

To create an IP list, navigate to **Outbreak Control > Network > IP Block and Allow Lists**, as shown in Figure 11-7.

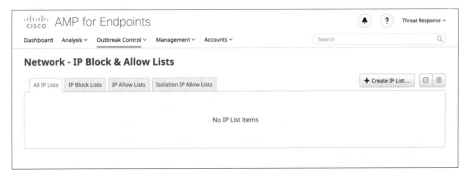

Figure 11-7 *AMP for Endpoints IP Block & Allow Lists*

In the screen shown in Figure 11-7, you click **Create IP List** to start a new IP list, and you're brought to the **New IP List** configuration screen, where you can create an IP list either by typing the IPv4 addresses in classless interdomain routing (CIDR) notation or by uploading a file that contains a list of IPs. You can also specify port numbers to block or allow. After the list is created, you can edit it only by downloading the resulting file and uploading it back to the AMP console. Figure 11-8 shows the New IP List screen, with a mixture of entries entered as text. You name the list, choose whether it is a whitelist (IP Allow List) or a black-list (IP Block List), and enter a series of IPv4 addresses, one line at a time. Each line must contain a single IP or CIDR.

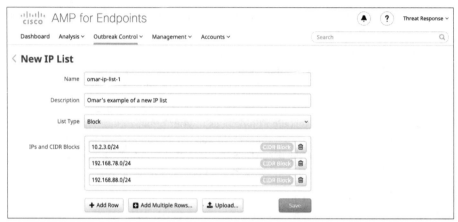

Figure 11-8 *Creating a New IP List*

You click **Create IP List** to create the text file in the cloud console, and your new IP list is shown on the screen. If you click **Edit**, you can change the name of the IP list. To update the contents of the list, you must click **Download** and then delete the list. Then you create a new list with the same name and upload the modified file.

As for custom detections, the AMP console maintains an audit trail for IP lists that you can view by clicking **View Changes**.

AMP for Endpoints Application Control

Like files, applications can be detected, blocked, and whitelisted. As with the other files, AMP does not look for the name of the application but the SHA-256 hash.

To create a new application control list for blocking an application, navigate to **Outbreak Control > Application Control > Blocked Applications**. Figure 11-9 demonstrates this process. As you would expect, to create a new list, you click **Create**. You must name the list and click **Save** before you can add any applications to the blocking list.

Once the list has been created and saved, click **Edit** to add any applications. If you already have the SHA-256 hash, add it. Otherwise, you can upload one application at a time and have the AMP cloud console calculate the hash for you, as long as the file is not larger than the 20MB limit. You can also upload an existing list. Figure 11-10 shows a blocking list with an existing application hash shown at the bottom of the right-hand column, while another file is being uploaded for hash calculation.

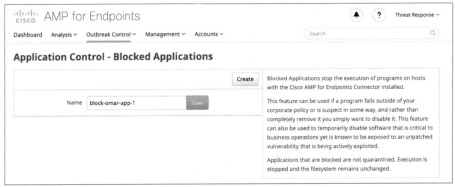

Figure 11-9   *Application Control—Blocked Applications*

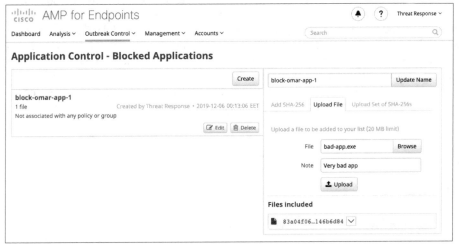

Figure 11-10   *Blocking a New Application*

Exclusion Sets

There is one more object that you should try to create before you build your policies, and that is an exclusion set. An exclusion set is a list of directories, file extensions, or even threat names that you do not want the AMP agent to scan and subsequently not convict as malware.

You can use an exclusion set to resolve conflicts with other security products or mitigate performance issues by excluding directories that contain large files that are frequently written to, like databases. If you are running an antivirus product on computers with the AMP for Endpoints connector, you should exclude the location where that product is installed.

It's important to remember that any files stored in a location that has been added to an exclusion set will not be subjected to application blocking, simple custom detections, or advanced custom detections.

11

These are the available exclusion types:

- **Threat:** This type excludes specific detections by threat name.

- **Extension:** This type excludes files with a specific extension.

- **Wildcard:** This type excludes files or paths using wildcards for filenames, extensions, or paths.

- **Path:** This type excludes files in a given path.

For Windows, path exclusions may use constant special ID lists (CSIDL), which are Microsoft given names for common file paths. For more on CSIDL, see https://docs.microsoft.com/en-us/windows/win32/shell/csidl?redirectedfrom=MSDN.

Cisco-maintained exclusions are created and maintained by Cisco to provide better compatibility between the AMP for Endpoints Connector and antivirus, security, or other software. Click the **Cisco-Maintained Exclusions** button to view the list of exclusions. These cannot be deleted or modified and are presented so you can see which files and directories are being excluded for each application. These exclusions may also be updated over time with improvements, and new exclusions may be added for new versions of an application. When one of these exclusions is updated, any policies using the exclusion will also be updated so the new exclusions are pushed to your connectors. Figure 11-11 shows Cisco-maintained exclusions.

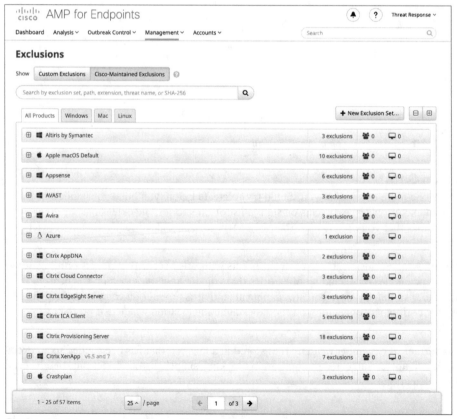

Figure 11-11 *Cisco-Maintained Exclusions*

Each row displays the operating system, exclusion set name, the number of exclusions, the number of groups using the exclusion set, and the number of computers using the exclusion set. You can use the search bar to find exclusion sets by name, path, extension, threat name, or SHA-256. You can also filter the list by operating system by clicking the respective tabs.

To create a new exclusion set, navigate to **Management > Exclusions**, as shown in Figure 11-12.

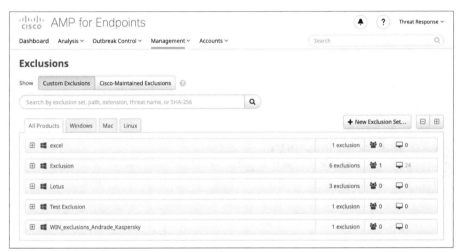

Figure 11-12   *AMP for Endpoints Exclusions*

In Figure 11-12, you see a list of any existing exclusions and can create new ones. Click **New Exclusion Set** and select the operating system for the new exclusion set (Windows, Mac, or Linux), as shown in Figure 11-13.

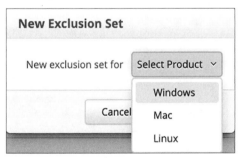

Figure 11-13   *Selecting the Operating System for the New AMP for Endpoints Exclusion Set*

After selecting the operating system for the new AMP for Endpoints exclusion set, the screen in Figure 11-14 is displayed.

In the example in Figure 11-14, Mac OS was selected. You can use the search bar to find exclusion sets by name, path, extension, threat name, or SHA-256. You can also filter the list by operating system by clicking the respective tabs. Click **View All Changes** to see a filtered list of the Audit Log showing all exclusion set changes. Click any exclusion set to expand its details. You can click **View Changes** in this view to see changes made to just that particular set.

11

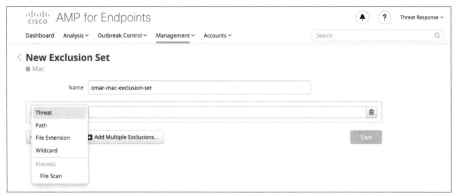

Figure 11-14 *Adding an AMP for Endpoints Exclusion Set*

AMP for Endpoints Connectors

AMP for Endpoints is available for multiple platforms: Windows, Android, Mac, and Linux. You can see the available connectors from the cloud console by navigating to **Management > Download Connector**. In Figure 11-15, you see the types of endpoints.

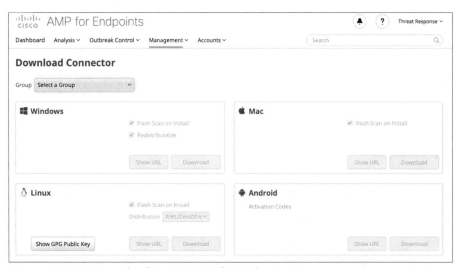

Figure 11-15 *Downloading the AMP for Endpoint Connector*

AMP for Endpoints Policies

You can configure different policies for each of the supported platforms, respectively. To create a new policy, navigate to **Management > Policies**, as shown in Figure 11-16.

A policy is applied to an endpoint via groups. Groups allow the computers in an organization to be managed according to their function, location, or other criteria as determined by the administrator.

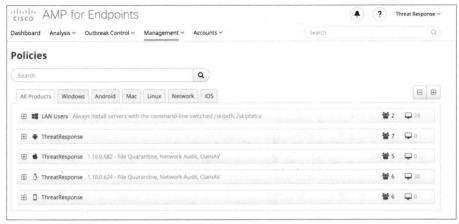

Figure 11-16 *AMP for Endpoints Policies*

TIP   Outbreak Control and Exclusion sets are combined with other settings into a policy. The policy affects the behavior and certain settings of the connector.

You can create a new group by navigating to **Management > Groups**. The screen shown in Figure 11-17 will be displayed.

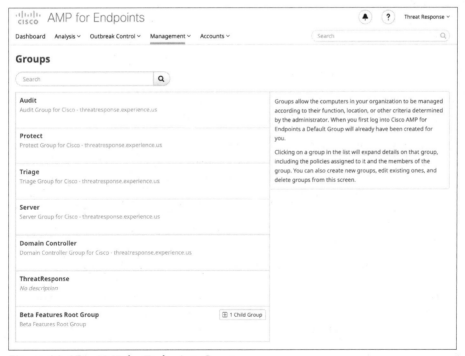

Figure 11-17 *AMP for Endpoints Groups*

AnyConnect AMP Enabler

You can use the AMP Enabler add-on to AnyConnect to aid in the distribution of the AMP connector to clients who use AnyConnect for remote access VPN, secure network access, posture assessments with Cisco's Identity Services Engine, and more. Figure 11-18 shows the AnyConnect Secure Mobility Client with the AMP Enabler tile.

Figure 11-18 *AnyConnect AMP Enabler*

AMP for Endpoints Engines

There are three detection and protection "engines" in AMP for Endpoints:

- **TETRA:** A full client-side antivirus solution. Do not enable the use of TETRA if there is an existing antivirus product in place. The default AMP setting is to leave TETRA disabled, as it changes the nature of the AMP connector from being a very lightweight agent to being a "thicker" software client that consumes more disk space for signature storage and more bandwidth for signature updates. When you enable TETRA, another configuration subsection is displayed, allowing you to choose what file scanning options you wish to enable.

- **Spero:** A machine learning–based technology that proactively identifies threats that were previously unknown. It uses active heuristics to gather execution attributes, and because the underlying algorithms come up with generic models, they can identify malicious software based on its general appearance rather than basing identity on specific patterns or signatures.

- **Ethos:** A "fuzzy fingerprinting" engine that uses static or passive heuristics.

AMP for Endpoints Reporting

AMP for Endpoints includes a series of reporting dashboards that can be very useful to understand what's happening in your endpoints. The main dashboard, illustrated in Figure 11-19, provides a view of threat activity in your organization over the past 30 days, as well as the percentage of compromised computers. You can filter by platform, date ranges, and other attributes.

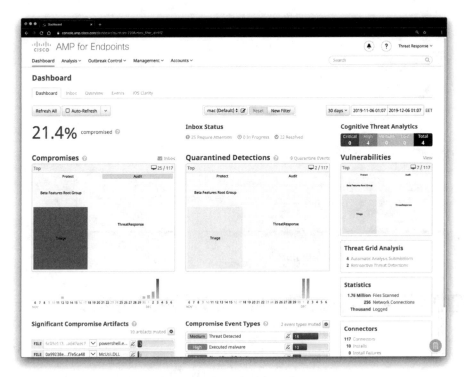

Figure 11-19 *AMP for Endpoints Main Dashboard*

Figure 11-20 shows the AMP for Endpoint Inbox report. The Inbox is a tool that allows you to see compromised computers in your business and track the status of compromises that require manual intervention to resolve. You can filter computers to work on by selecting Groups in the heat map, selecting a day with compromises in the bar chart, selecting an SHA-256 hash checksum from the Significant Compromise Artifacts list, or selecting from the Compromise Event Types list. These filters can be saved and set as your default view. You can also filter the computer list by those that require attention, those that are in progress, and those that have been resolved by clicking the matching tabs. You can order the list by date or severity by selecting from the Sort drop-down menu. When a computer is marked as resolved, it is no longer reflected in data on the Dashboard or Inbox.

Figure 11-21 shows the AMP for Endpoints Overview dashboard, which displays the status of your environment and highlights recent threats and malicious activity in your AMP for Endpoints deployment. You can click the headings of each section to navigate directly to relevant pages in the console to investigate and remedy situations.

11

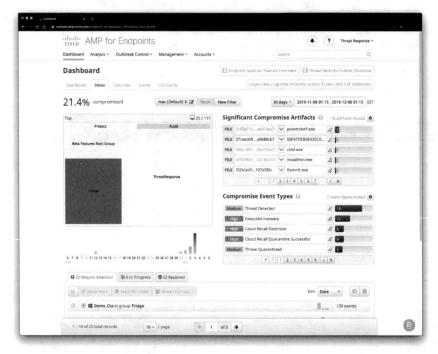

Figure 11-20 *AMP for Endpoints Inbox*

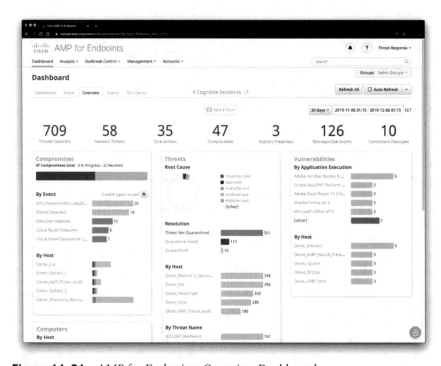

Figure 11-21 *AMP for Endpoints Overview Dashboard*

Figure 11-22 shows the AMP for Endpoints Events dashboard, displaying the most recent events in your AMP for Endpoints deployment.

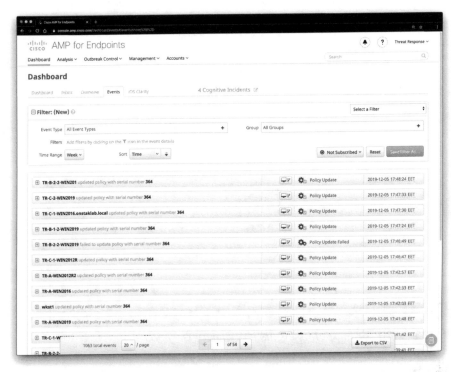

Figure 11-22 *AMP for Endpoints Events Dashboard*

Figure 11-23 shows the iOS Clarity dashboard in AMP for Endpoints. If you have already linked your Meraki SM or other mobile device manager (MDM), this tab displays a summary of activity, a list of the most recently observed applications on your managed iOS devices, and a list of devices that have not reported back in more than seven days.

TIP   The Cisco Meraki SM provides MDM functionality that ensures that diverse user equipment (mobile phones, tablets, laptops, and so on) is configured to a consistent standard and a supported set of applications, functions, or corporate policies. MDMs (such as the Cisco Meraki SM) also provide capabilities to update equipment, applications, functions, or policies in a scalable manner. This automated update or patch provisioning not only ensures that the mobile device performs consistently, but also that security vulnerabilities are patched in a very effective manner. You can also use MDMs to monitor and track mobile devices (including location, status, activity, jailbreak detection, and so on). MDMs also allow you to efficiently diagnose and troubleshoot mobile devices remotely.

11

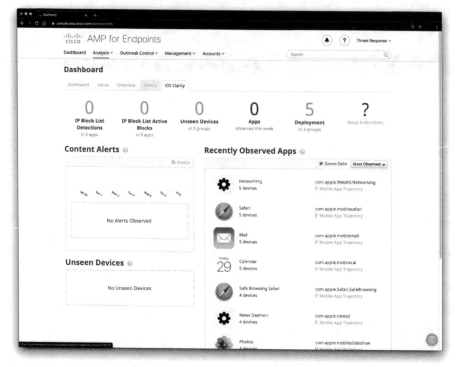

Figure 11-23 *AMP for Endpoints Events iOS Clarity Dashboard*

Cisco Threat Response

Cisco Threat Response is a "one-pane-of-glass" console that automates integrations across Cisco security products and threat intelligence sources. This is an ongoing effort from Cisco to provide a single console for the management of most of its security products.

Cisco Threat Response integrates with the following Cisco security solutions:

- Cisco Advanced Malware Protection

- AMP for Endpoints

- Cisco Threat Grid

- Cisco Umbrella

- Cisco Email Security

- Cisco Next-Generation Firewalls (NGFW)

- Next-Generation Intrusion Prevention System (NGIPS)

The screenshots in the previous sections were collected using a Cisco Threat Response console.

TIP The following video provides an overview of the Cisco Threat Response solution: https://www.youtube.com/watch?v=ycwkY53ve1U. The following video provides a detailed demo and walkthrough of Cisco Threat Response: https://www.youtube.com/watch?v=sHEbKivwTJM.

Exam Preparation Tasks

As mentioned in the section "How to Use This Book" in the Introduction, you have a couple of choices for exam preparation: the exercises here, Chapter 12, "Final Preparation," and the exam simulation questions in the Pearson Test Prep Software Online.

Review All Key Topics

Review the most important topics in this chapter, noted with the Key Topic icon in the outer margin of the page. Table 11-2 lists these key topics and the page numbers on which each is found.

Table 11-2 Key Topics for Chapter 11

Key Topic Element	Description	Page Number
List	Surveying the AMP for endpoints core features	637
Section	Endpoint Threat Detection and Response (ETDR) and Endpoint Detection and Response (EDR)	637
List	Reviewing file trajectory and device trajectory	638
Figure 11-1	Understanding the AMP for Endpoints architecture	639
Section	Outbreak Control	639
Section	IP Blacklists and Whitelists	643
Section	AMP for Endpoints Application Control	644
Section	Exclusion Sets	645
Section	AMP for Endpoints Connectors	648
Section	AMP for Endpoints Policies	648
Section	AMP for Endpoints Engines	650
Tip	Describing the value of endpoint device management and asset inventory such as MDM	653

Define Key Terms

Define the following key terms from this chapter and check your answers in the glossary.

TETRA, Spero, Ethos, mobile device management (MDM)

11

Review Questions

1. Which of the following is an AMP for Endpoints engine that uses machine learning to proactively identify threats that were previously unknown? This solution uses active heuristics to gather execution attributes, and because the underlying algorithms come up with generic models, they can identify malicious software based on its general appearance rather than basing identity on specific patterns or signatures.

   a. TETRA

   b. Ethos

   c. Spero

   d. All of these answers are correct.

2. Which of the following is a list of directories, file extensions, or even threat names that you do not want the AMP agent to scan and definitely not to convict as malware?

   a. Exclusion set

   b. Application blacklist

   c. TETRA blacklist

   d. None of these answers is correct.

3. Like files, applications can be detected, blocked, and whitelisted with AMP for Endpoints. AMP for Endpoints does not look for the name of the application but which of the following elements?

   a. An SHA hash (checksum)

   b. An outbreak signature

   c. A custom signature

   d. A Spero signature

4. Device flow correlation (DFC) can be used for which of the following scenarios?

   a. To correlate AMP for Endpoint logs with Cisco ISE logs

   b. To correlate AMP for Endpoint logs with Cisco FMC logs

   c. To correlate AMP for Endpoint logs with Tetration logs

   d. To flag or block suspicious network activity

5. You are investigating a potential threat outbreak in your organization. Which of the following can be used to see what endpoints have seen a potential malware file?

   a. Outbreak groups

   b. Outbreak filters

   c. Device trajectory

   d. File trajectory

CHAPTER 12

Final Preparation

The first 11 chapters of this book cover the technologies, protocols, design concepts, and considerations required for your preparation in passing the Cisco Implementing and Operating Cisco Security Core Technologies (SCOR 350-701) exam. This is the exam required for the CCNP Security, CCIE Security, and Cisco Certified Specialist – Security Core Certifications. If you are pursuing the CCNP Security certification, you still need to select and pass a concentration exam to achieve that certification. If you are preparing for the CCIE certification, you must also pass the hands-on 8-hour lab.

Chapter 1 through 11 cover the information necessary to pass the exam. However, most people need more preparation than simply reading the first 11 chapters of this book. This chapter, along with the Introduction of the book, suggests hands-on activities and a study plan that will help you complete your preparation for the exam.

Hands-on Activities

As mentioned, you should not expect to pass the SCOR 350-701 exam by just reading this book. The SCOR 350-701 exam requires hands-on experience with many of the Cisco technologies, tools, and techniques discussed in this book. These include Cisco routers, switches, firewalls, Cisco AMP, Cisco Threat Response, Cisco ISE, Cisco Stealthwatch, Cisco pxGrid, Cisco Umbrella Investigate, Cisco Cognitive Threat Analytics, Cisco Encrypted Traffic Analytics, Cisco AnyConnect Secure Mobility Client, Cisco ESA, Cisco WSA, Software Defined Networking (SDN) solutions, and the different APIs supported by those products. Building your own lab, breaking it, and fixing it is the most effective way to learn the skills necessary to pass the exam.

Suggested Plan for Final Review and Study

This section provides a suggested study plan from the point at which you finish reading Chapter 11 until you take the Implementing and Operating Cisco Security Core Technologies (SCOR 350-701) exam. You can ignore this five-step plan, use it as is, or modify it to better meet your needs:

Step 1. **Review key topics and DIKTA questions:** You can use the table at the end of each chapter that lists the key topics in each chapter or just flip through the pages looking for key topics. Also, reviewing the "Do I Know This Already" (DIKTA) questions from the beginning of the chapter can be helpful.

Step 2. **Review the exam Blueprint:** Cisco maintains a list of testable content known as the Implementing and Operating Cisco Security Core Technologies (SCOR 350-701) Exam Blueprint. Review it and make sure you are familiar with every item that is listed. You can download a copy at https://www.cisco.com/c/dam/en_us/training-events/le31/le46/cln/marketing/exam-topics/350-701-SCOR.pdf.

Step 3. Study "Review Questions" sections: Go through the review questions at the end of each chapter to identify areas in which you need more study.

Step 4. Use the Pearson Cert Practice Test engine to practice: The Pearson Test Prep software provides a bank of unique exam-realistic questions available only with this book.

Summary

The tools and suggestions listed in this chapter have been designed with one goal in mind: to help you develop the skills required to pass the Implementing and Operating Cisco Security Core Technologies (SCOR 350-701) exam to achieve the CCNP Security, CCIE Security, or the Cisco Certified Specialist – Security Core Certifications.

This book has been developed from the beginning both to present you with a collection of facts and to help you learn how to apply those facts. Regardless of your experience level before reading this book, it is our hope that the broad range of preparation tools, and even the structure of the book, will help you pass the exam with ease. Please remember to take advantage of the resources of my GitHub repository at https://h4cker.org/github. I wish you success in your exam and hope that our paths cross again as you continue to grow in your IT and cybersecurity career.

GLOSSARY OF KEY TERMS

NUMERICS

802.1Q 802.1Q is an IEEE standard protocol used for VLAN tagging of Ethernet frames. 802.1Q defines the procedures to be used by switches, wireless access points, and other network devices when handling such frames. The most critical piece of information in an 802.1Q VLAN tag is the VLAN ID.

802.1X 802.1X is an IEEE standard that is used to implement port-based access control. In simple terms, an 802.1X access device will allow traffic on the port only after the device has been authenticated and authorized.

A

AAA Authentication, authorization, and accounting.

Access control list (ACL) This is the simplest way to implement a DAC-based system. ACLs can apply to different objects (like files) or they can also be configured statements (policies) in network infrastructure devices (routers, firewalls, and so on). For instance, an ACL, when applied to an object, will include all the subjects that can access the object and their specific permissions. There is also the concept of ACLs in routers and firewalls. In those implementations, the ACL provides packet filtering to protect "internal" networks from the "outside" systems and to filter traffic leaving the inside network. ACL criteria could be the source address of the traffic, the destination address of the traffic, destination port, source port, and the upper-layer protocol (otherwise known as the five-tuple).

Access control matrix (ACM) This is an access control mechanism that is usually associated with a DAC-based system. An ACM includes three elements: the subject, the object, and the set of permissions. Each row of an ACM is assigned to a subject, while each column represents an object. The cell that identifies a subject/object pair includes the permission that subject has on the object. An ACM could be seen as a collection of access control lists or a collection of capability tables, depending on how you want to read it.

Accounting Accounting is the process of auditing and monitoring what a user does once a specific resource is accessed. This process is sometimes overlooked; however, as a security professional, it is important to be aware of accounting and to advocate that it be implemented because of the great help it provides during detection and investigation of cybersecurity breaches.

Apache Mesos A distributed Linux kernel that provides native support for launching containers with Docker and AppC images. You can download Apache Mesos and access its documentation at https://mesos.apache.org.

Asymmetric algorithm An encryption algorithm that uses two different keys—a public key and a private key. Together they make a key pair.

Attribute-based access control (ABAC)   Attribute-based access control (ABAC) is a logical access control model that controls access to objects by evaluating rules against the attributes of entities (both subject and object), operations, and the environment relevant to a request.

Authentication server   An entity that provides an authentication service to an authenticator. The authentication server determines whether the supplicant is authorized to access the service. This is sometimes referred to as the Policy Decision Point (PdP). The Cisco Identity Services Engine (ISE) is an example of an authentication server.

Authenticator   An entity that facilitates authentication of other entities attached to the same LAN. This is sometimes referred to as the Policy Enforcement Point (PeP). Cisco switches, wireless routers, and access points are examples of authenticators.

Authorization   Authorization is the process of assigning an authenticated subject's permission to carry out a specific operation. The authorization model defines how access rights and permission are granted. The three primary authorization models are object capability, security labels, and ACLs.

B

BeyondCorp   Google's implementation of zero trust. This model shifts access control from the network perimeter firewalls and other security devices to individual devices and users.

Black hat hackers   These individuals perform illegal activities, such as organized crime.

Block cipher   A block cipher is a symmetric key (same key to encrypt and decrypt) cipher that operates on a group of bits called a block. A block cipher encryption algorithm may take a 64-bit block of plain text and generate a 64-bit block of cipher text. With this type of encryption, the same key to encrypt is also used to decrypt.

BPDU Guard   A Cisco switch feature that allows a switch to protect itself if bridge protocol data units (BPDUs) show up where they should not.

C

Capability table   A collection of objects that a subject can access, together with the granted permissions. The key characteristic of a capability table is that it is subject centric instead of being object centric, like in the case of an access control list.

CDP   Cisco Systems introduced the Cisco Discovery Protocol (CDP) in 1994 to provide a mechanism for the management system to automatically learn about devices connected to the network. CDP runs on Cisco devices (routers, switches, phones, and so on) and is also licensed to run on some network devices from other vendors. Using CDP, network devices periodically advertise their own information to a multicast address on the network, making it available to any device or application that wishes to listen and collect it.

Certificate authority   A system that generates and issues digital certificates to users and systems.

Cisco FDM   The Cisco Firepower Device Manager (FDM) is used to configure small Cisco FTD deployments. To access the Cisco FDM, you just need to point your browser at the firewall in order to configure and manage the device.

Cisco FMC   Cisco FTD devices, Cisco Firepower devices, and the Cisco ASA FirePOWER modules can be managed by the Firepower Management Center (FMC), formerly known as the FireSIGHT Management Center.

Cloud access security broker (CASB)   CASB provides visibility and compliance checks, protects data against misuse and exfiltration, and provides threat protections against malware such as ransomware.

Continuous Delivery (CD)   Continuous Delivery (CD) sits on top of Continuous Integration (CI) and provides a way for automating the entire software release process. When you adopt CI/CD methodologies, each change in code should trigger an automated build-and-test sequence. This automation should also provide feedback to the programmers who made the change.

Continuous Integration (CI)   A software development practice where programmers merge code changes in a central repository multiple times a day.

Contiv   Contiv is an open source project that allows you to deploy micro-segmentation policy-based services in container environments.

Control plane   The control plane includes protocols and traffic that the network devices use on their own without direct interaction from an administrator. An example is a routing protocol. A routing protocol can dynamically learn and share routing information that the router can then use to maintain an updated routing table. If a failure occurs in the control plane, a router might lose the capability to share or correctly learn dynamic routing information, and as a result might not have the routing intelligence to be able to route for the network.

Crypter   A crypter functions to encrypt or obscure the code. Some crypters obscure the contents of the Trojan by applying an encryption algorithm. Crypters can use anything from AES, RSA, to even Blowfish, or might use more basic obfuscation techniques such as XOR, Base64 encoding, or even ROT13. Again, these techniques are used to conceal the contents of the executable program, making it undetectable by antivirus and resistant to reverse-engineering efforts.

D

Data plane   The data plane includes traffic that is being forwarded through the network (sometimes called transit traffic). An example is a user sending traffic from one part of the network to access a server in another part of the network; the data plane represents the traffic that is either being switched or forwarded by the network devices between clients and servers. A failure of some component in the data plane results in the customer's traffic not being able to be forwarded. Other times, based on policy, you might want to deny specific types of traffic that is traversing the data plane.

Designated port   The switch port that can send the best bridge protocol data unit (BPDU) for a particular VLAN on a switch is considered the designated port.

DevOps   DevOps is the outcome of many trusted principles from software development, manufacturing, and leadership to the information technology value stream. DevOps relies on bodies of knowledge from Lean, Theory of Constraints, resilience engineering, learning organizations, safety culture, human factors, and many others.

DevSecOps   DevSecOps is a concept used in recent years to describe how to move security activities to the start of the development lifecycle and have built-in security practices in the CI/CD pipeline. The business environment, culture, law compliance, and external market drive relate to how a secure development life cycle (also referred to as SDLC) and a DevSecOps program are implemented in an organization.

DHCP Snooping   A Cisco switch feature that prevents rogue DHCP servers from impacting the network.

Diameter   An authentication protocol. RADIUS and TACACS+ were created with the aim of providing AAA services to network access via dial-up protocols or terminal access. Due to their success and flexibility, they have been used in several other scopes. To respond to newer access requirements and protocols, the IETF has proposed a new protocol called Diameter, which is described in RFC 6733.

Diffie-Hellman   Diffie-Hellman is a key agreement protocol that enables two users or devices to authenticate each other's pre-shared keys without actually sending the keys over the unsecured medium. R1 sends the Key Exchange (KE) payload and a randomly generated value called a nonce.

Digital certificate   A digital entity used to verify that a user is who he or she claims to be, and to provide the receiver with the means to encode a reply. Digital certificates also apply to systems, not only to individuals.

Discretionary access control (DAC)   A discretionary access control (DAC) is defined by the owner of the object. DACs are used in commercial operating systems. The object owner builds an ACL that allows or denies access to the object based on the user's unique identity. The ACL can reference a user ID or a group (or groups) that the user is a member of. Permissions can be cumulative.

DMVPN   DMVPN is a technology created by Cisco that aims to reduce the hub router configuration. In a legacy hub-and-spoke IPsec configuration, each spoke router has a separate block of configuration lines on the hub router that defines the crypto map characteristics, the crypto ACLs, and the GRE tunnel interface. When deploying DMVPN, you configure a single mGRE tunnel interface, a single IPsec profile, and no crypto access lists on the hub router. The main benefit is that the size of the configuration on the hub router remains the same even if spoke routers are added at a later point.

Docker Swarm   A container cluster management integrated with the Docker Engine. You can access the Docker Swarm documentation at https://docs.docker.com/engine/swarm.

Domain Keys Identified Mail (DKIM)   DKIM is an industry standard defined in RFC 5585. DKIM provides a means for gateway-based cryptographic signing of outgoing messages. This allows you to embed verification data in an email header and for email recipients to verify the integrity of the email messages. DKIM uses DNS TXT records to publish public keys.

Downloadable ACL (dACL)   A downloadable ACL (dACL), also called a *per-user ACL*, is an ACL that can be applied dynamically to a port. The term *downloadable* stems from the fact that these ACLs are pushed from the authenticator server (for example, from a Cisco ISE) during the authorization phase. When a client authenticates to the port (for example, by using 802.1X), the authentication server can send a dACL that will be applied to the port and that will limit the resources the client can access over the network.

Dropper   A dropper is software designed to install a malware payload on the victim's system. Droppers try to avoid detection and evade security controls by using several methods to spread and install the malware payload.

DTLS   Datagram Transport Layer Security (DTLS), defined in RFC 6347, provides security and privacy for UDP packets. This allows UDP-based applications to send and receive traffic in a secure fashion without concern about packet tampering and message forgery. Thus, applications can avoid the delays associated with TCP but still communicate securely by using DTLS.

Dynamic ARP Inspection   A Cisco switch feature that prevents spoofing of Layer 2 information by hosts.

E

EAP over LAN (EAPoL)   An encapsulation defined in 802.1X that's used to encapsulate EAP packets to be transmitted from the supplicant to the authentication server.

Endpoint group (EPG)   Cisco ACI allows organizations to automatically assign endpoints to logical security zones called endpoint groups (EPGs). EPGs are used to group VMs within a tenant and apply filtering and forwarding policies to them. These EPGs are based on various network-based or VM-based attributes.

Ethos   Ethos is a "fuzzy fingerprinting" engine that uses static or passive heuristics. The engine creates generic file signatures that can match polymorphic variants of a threat. This is useful because when a threat morphs or a file is changed, the structural properties of that file often remain the same, even though the content has changed. Unlike most other signature tools, Ethos uses distributed data mining to identify suitable files. It uses in-field data for sources, which provides a highly relevant collection from which to generate the signatures.

Exploit   An exploit refers to a piece of software, a tool, a technique, or a process that takes advantage of a vulnerability that leads to access, privilege escalation, loss of integrity, or denial of service on a computer system. Exploits are dangerous because all software has vulnerabilities; hackers and perpetrators know that there are vulnerabilities and seek to take advantage of them.

Extensible Authentication Protocol (EAP)   An authentication protocol used between the supplicant and the authentication server to transmit authentication information.

F

Federated Identity Management   A collection of shared protocols that allows user identities to be managed across organizations.

Federation Provider   An identity provider that offers single sign-on, consistency in authorization practices, user management, and attributes-exchange practices between identity providers (issuers) and relying parties (applications).

"Five-tuple"   The source and destination IP addresses, source and destination ports, and IP protocol.

FlexVPN   FlexVPN is an IKEv2-based VPN technology that provides several benefits beyond traditional site-to-site VPN implementations. FlexVPN is a standards-based solution that can interoperate with non-Cisco IKEv2 implementations. It supports different VPN topologies, including point-to-point, remote-access, hub-spoke, and dynamic mesh (including per-user or per-peer policies). FlexVPN combines all these different VPN technologies using one command-line interface (CLI) set of configurations. FlexVPN supports unified configuration and **show** commands, underlying interface infrastructure, and features across different VPN topologies.

FlowCollector   A physical or virtual appliance that collects NetFlow data from infrastructure devices.

FlowReplicator   A physical appliance used to forward NetFlow data as a single data stream to other devices. The FlowReplicator is also known as the UDP Director.

FlowSensor   A physical or virtual appliance that can generate NetFlow data when legacy Cisco network infrastructure components are not capable of producing line-rate, unsampled NetFlow data.

Forest   A collection of domains managed by a centralized system.

G

GDOI   GDOI is defined as the ISAKMP Domain of Interpretation (DOI) for group key management. The GDOI protocol operates between a group member and a group controller or key server (GCKS), which establishes SAs among authorized group members.

GETVPN   Cisco's Group Encrypted Transport VPN (GETVPN) provides a collection of features and capabilities to protect IP multicast group traffic or unicast traffic over a private WAN. GETVPN combines the keying protocol Group Domain of Interpretation (GDOI) and IPsec. GETVPN enables the router to apply encryption to "native" (non-tunneled) IP multicast and unicast packets and removes the requirement to configure tunnels to protect multicast and unicast traffic. DMVPN allows Multiprotocol Label Switching (MPLS) networks to maintain full-mesh connectivity, natural routing path, and Quality of Service (QoS).

Gray hat hackers   These individuals usually follow the law but sometimes venture over to the darker side of black hat hacking. It would be unethical to employ these individuals to perform security duties for your organization because you are never quite clear where they stand.

GRE   Generic Routing Encapsulation (GRE) Protocol is defined by RFC 2784 and extended by RFC 2890. GRE provides a simple mechanism to encapsulate packets of any protocol (the payload packets) over any other protocol (the delivery protocol) between two endpoints. In a GRE tunnel implementation, the GRE protocol adds its own header (4 bytes plus options) between the payload (data) and the delivery header.

H

Hashed Message Authentication Code (HMAC)   HMAC uses the mechanism of hashing, but instead of using a hash that anyone can calculate, it includes in its calculation a secret key of some type. Thus, only the other party who also knows the secret key and can calculate the resulting hash can correctly verify the hash. When this mechanism is used, an attacker who is eavesdropping and intercepting packets cannot inject or remove data from those packets without being noticed because he cannot recalculate the correct hash for the modified packet because he does not have the key or keys used for the calculation.

Hashing   Hashing is a method used to verify data integrity. An example of using a hash to verify integrity is the sender running a hash algorithm on a packet and attaching that hash to it. The receiver runs the same hash against the packet and compares his results against the results the sender had (which are attached to the packet as well). If the hash generated matches the hash that was sent, they know that the entire packet is intact. If a single bit of the hashed portion of the packet is modified, the hash calculated by the receiver will not match, and the receiver will know that the packet had a problem (specifically with the integrity of the packet). Examples of hashing algorithms are MD5 and SHA.

Hashing algorithms   Algorithms used to verify data integrity.

I

IaaS   IaaS describes a cloud solution in which you rent infrastructure. You purchase virtual power to execute your software as needed. This is much like running a virtual server on your own equipment, except you are now running a virtual server on a virtual disk. This model is similar to a utility company model because you pay for what you use.

Identification   Identification is the process of providing the identity of a subject or user. This is the first step in the authentication, authorization, and accounting process. Providing a username, a passport, an IP address, or even pronouncing your name is a form of identification.

Identity certificate   An identity certificate is similar to a root certificate, but it describes the client and contains the public key of an individual host (the client). An example of a client is a web server that wants to support Secure Sockets Layer (SSL) or a router that wants to use digital signatures for authentication of a VPN tunnel.

Identity Provider (IdP)   An application, website, or service responsible for coordinating identities between users and clients. IdPs can provide a user with identifying information and provide that information to services when the user requests access.

Implicit deny   If no rule is specified for the transaction of the subject/object, the authorization policy should deny the transaction.

Indicator of compromise (IOC)   An indicator of compromise (IOC) is any observed artifact on a system or a network that could indicate an intrusion. There may be artifacts left on a system after an intrusion or a breach, and they can be expressed in a language that describes the threat information, known as an IOC. The sets of information describe how and where to detect the signs of the intrusion or breach. IOCs can be host-based and/or network-based artifacts, but the scan actions are carried out on the host only.

Internet Message Access Protocol (IMAP)   An email client communication protocol that allows users to keep messages on the server. An IMAP-enabled mail user agent (MUA) displays messages directly from the server. However, you can also download messages using IMAP for archiving purposes.

IP Source Guard   A Cisco switch feature that prevents spoofing or Layer 2 information by hosts.

IPFIX   The Internet Protocol Flow Information Export (IPFIX) is a network flow standard led by the Internet Engineering Task Force (IETF). IPFIX was created to provide a common, universal standard of export for flow information from routers, switches, firewalls, and other infrastructure devices.

K–M

Kerberos   A ticket-based protocol for authentication built on symmetric-key cryptography.

Kubernetes (k8s)   A container and application orchestration platform. Kubernetes automates the distribution, scheduling, and orchestration of application containers across a cluster.

L2F   Layer 2 Forwarding (L2F) Protocol is a legacy VPN protocol created by Cisco for Layer 2 VPN implementations.

LLDP   802.1AB (Station and Media Access Control Connectivity Discovery, or Link Layer Discovery Protocol [LLDP]). LLDP, which defines basic discovery capabilities, was enhanced to specifically address the voice application; this extension to LLDP is called LLDP-MED or LLDP for Media Endpoint Devices.

MAC Authentication Bypass (MAB)   MAB is feature that relies on a MAC address for authentication. For instance, you can "whitelist" a MAC address to bypass 802.1X authentication. This is done for devices that do not have an 802.1X supplicant (such as printers, IP phones, and so on). A MAC address is a globally unique identifier that is assigned to all network-attached devices. Subsequently, it can be used in authentication. However, since you can spoof a MAC address, MAB is not a strong form of authentication and can be abused by attackers.

Mail delivery agent (MDA)   A component of a mail transfer agent (MTA) responsible for the final delivery of an email message to a person's inbox (mailbox).

Mail Exchanger (MX) record   DNS MX records are used to route the mail traffic on the Internet. An MX record is a type of verified resource record in DNS that specifies a mail server responsible for accepting email messages on behalf of a recipient's domain, and a preference value is used to prioritize mail delivery if multiple mail servers are available. The set of MX records of a domain name specifies how email should be routed with Simple Mail Transfer Protocol (SMTP).

Mail transfer agent (MTA)   The entity responsible for transferring emails from a sender to the recipient.

Mail user agent (MUA)   A component of a mail transfer agent (MTA) that accepts new mail messages from a mail server. The MUA is also known as the "email client".

Management plane   The management plane includes the protocols and traffic that an administrator uses between his workstation and the router or switch itself. An example is using a remote management protocol such as Secure Shell (SSH) to monitor or configure the router or switch.

Mandatory access control (MAC)   A mandatory access control (MAC) is defined by policy and cannot be modified by the information owner. MACs are primarily used in secure military and government systems that require a high degree of confidentiality. In a MAC environment, objects are assigned a security label that indicates the classification and category of the resource. Subjects are assigned a security label that indicates a clearance level and assigned categories (based on the need to know).

Mobile device management (MDM)   Software that is used for the administration of mobile devices, including smartphones, tablets, and laptops.

Multifactor authentication   Multifactor authentication is when two or more factors are presented.

Multilayer authentication   Multilayer authentication is when two or more of the same type of factors are presented.

Multitenancy   A term in computing architecture referring to the serving of many users (tenants) from a single instance of an application. Software as a Service (SaaS) offerings are examples of multitenancy. They exist as a single instance but have dedicated shares served to many companies and teams.

N

NAT Traversal (NAT-T)   NAT Traversal (NAT-T) is a technology to encapsulate IPsec packets in UDP. Traditionally, the IPsec tunnels fail to pass traffic if there is a PAT device between the peers. By default, IPsec devices use the Encapsulated Security Payload (ESP) protocol, which does not have any Layer 4 information, and therefore the PAT device ends up dropping the IPsec packet. NAT Traversal (NAT-T) is used to encapsulate the ESP packets into a UDP port connection on port 4500 so that any intermediate PAT device would have no trouble translating the encrypted packets. NAT-T is dynamically negotiated if both VPN peers are NAT-T capable or if there is a NAT or PAT device between the peers. If both conditions are met, VPN peers start their communication using ISAKMP (UDP port 500), and as soon as a NAT or PAT device is detected, they switch to UDP port 4500 to complete the rest of their negotiations. NAT-T is globally enabled on the Cisco ASA by default. In many cases, the NAT/PAT devices time out the NAT-T encrypted connection on UDP port 4500 entries if there is no active traffic passing through them.

Need to know   A subject should be granted access to an object only if the access is needed to carry out the job of the subject.

NETCONF   Defined in RFCs 6241 and 6242, NETCONF is a network management protocol created to overcome the challenges in legacy Simple Network Management Protocol (SNMP) implementations.

NetFlow   NetFlow is a technology originally created by Cisco that provides comprehensive visibility into all network traffic that traverses a Cisco-supported device.

Network Functions Virtualization (NFV)   NFV is a technology that addresses the virtualization of Layer 4 through Layer 7 services. These include things like load balancing and security capabilities such as firewall-related features. In short, with NFV, you convert certain types of network appliances into VMs. NFV was created to address the inefficiencies that were introduced by virtualization.

Neutron   Neutron is the networking component in OpenStack. Neutron is designed to provide "networking as a service" in private, public, and hybrid cloud environments. Other OpenStack components, such as Horizon (Web UI) and Nova (compute service), interact with Neutron using a set of APIs to configure the networking services. Neutron uses plug-ins to deliver advanced networking capabilities and allow third-party vendor integration. Neutron has two main components: the neutron server and a database that handles persistent storage and plug-ins to provide additional services.

Nomad   A container management and orchestration platform by HashCorp. You can download and obtain detailed information about Nomad at https://www.nomadproject.io.

Nondesignated port   A switch port that does not forward packets so as to prevent the existence of loops within networks.

O

OAuth   An open standard for authorization used by many APIs and modern applications. You can access OAuth and OAuth 2.x specifications and documentation at https://oauth.net/2.

One-time passcode (OTP)   A one-time passcode (OTP) is a set of characteristics that can be used to prove a subject's identity one time and one time only. Because the OTP is valid for only one access, if it's captured, additional access would be automatically denied. The OTP is generally delivered through a hardware or software token device. The token displays the code, which must then be typed in at the authentication screen. Alternatively, the OTP may be delivered via email, text message, or phone call to a predetermined address or phone number.

Open vSwitch   An open source implementation of a multilayer virtual switch inside the hypervisor.

OpenDaylight (ODL)   OpenDaylight is a popular open source project that is focused on the enhancement of software-defined networking (SDN) controllers to provide network services across multiple vendors. OpenDaylight interacts with Neutron via a northbound interface and manages multiple interfaces southbound, including the Open vSwitch Database Management Protocol (OVSDB) and OpenFlow.

OpenID (or OpenID Connect)   An open standard for authentication. OpenID Connect allows third-party services to authenticate users without clients needing to collect, store, and subsequently become liable for a user's login information. Detailed information about OpenID can be accessed at https://openid.net/connect/.

Out-of-band authentication   Out-of-band authentication requires communication over a channel that is distinct from the first factor. A cellular network is commonly used for out-of-band authentication. For example, a user enters her name and password at an application logon prompt (factor 1). The user then receives a call on her mobile phone; the user answers and provides a predetermined code (factor 2). For the authentication to be compromised, the attacker would have to have access to both the computer and the phone.

P

PaaS   PaaS provides everything except applications. Services provided by this model include all phases of the system development life cycle (SDLC) and can use application programming interfaces (API), website portals, or gateway software. These solutions tend to be proprietary, which can cause problems if the customer moves away from the provider's platform.

Packer   A packer is similar to a program such as WinZip, Rar, or Tar because it compresses files. However, whereas compression programs compress files to save space, packers do this to obfuscate the activity of the malware. The idea is to prevent anyone from viewing the malware's code until it is placed in memory.

PFS   Perfect Forward Secrecy (PFS) is a cryptographic technique where the newly generated keys are unrelated to any previously generated key.

Port security   A Cisco switch feature that limits the number of MAC addresses to be learned on an access switch port.

Post Office Protocol (POP)   An application-layer protocol used by an email client to retrieve (download) email from a remote server.

Posture assessment   Posture assessment includes a set of rules in a security policy that define a series of checks before an endpoint is granted access to the network. Posture assessment checks include the installation of operating system patches, host-based firewalls, antivirus and antimalware software, disk encryption, and more.

PPTP   Point-to-Point Tunneling Protocol (PPTP) is a legacy VPN protocol.

Proxy auto-configuration (PAC) files   Files used to configure an end-user client's web proxy settings.

pxGrid   Cisco pxGrid provides a cross-platform integration capability between security monitoring applications, threat detection systems, asset management platforms, network policy systems, and practically any other IT operations platform. Cisco ISE supports Cisco pxGrid to provide a unified ecosystem to integrate multivendor tools to exchange information either unidirectionally or bidirectionally.

R

RADIUS   The Remote Authentication Dial-In User Service (RADIUS) is an AAA protocol mainly used to provide network access services. Due to its flexibility, it has been adopted in other scenarios as well. The authentication and authorization parts are specified in RFC 2865, while the accounting part is specified in RFC 2866.

Ransomware   A piece of malware that is designed to encrypt personal files on the victim's system until a ransom is paid to the attacker.

Representational State Transfer (REST)   REST is an API standard. REST is easier to use than SOAP. It uses JSON instead of XML, and it uses standards like Swagger and the OpenAPI specification (https://www.openapis.org) for ease of documentation and to help with adoption.

RESTCONF   A REST-based variant of NETCONF used to manage networking devices.

Role-based access control (RBAC)   A role-based access control (also called a "nondiscretionary control") is an access permission based on a specific role or function. Administrators grant access rights and permissions to roles. Users are then associated with a single role. There is no provision for assigning rights to a user or group account.

Root certificate   A root certificate contains the public key of the CA server and the other details about the CA server.

Root Guard   A Cisco switch feature that controls which ports are not allowed to become root ports to remote root switches.

Root port   The switch port that is closest to the root bridge in terms of STP path cost (that is, it receives the best BPDU on a switch) is considered the root port. All switches, other than the root bridge, contain one root port.

S

SaaS   SaaS is designed to provide a complete packaged solution. The software is rented out to the user. The service is usually provided through some type of front end or web portal. While the end user is free to use the service from anywhere, the company pays a per-use fee.

Scalable Group Tag Exchange Protocol (SXP)   The Scalable Group Tag (SGT) Exchange Protocol (SXP) is a control plane protocol used to convey IP-to-SGT mappings to network devices when you cannot perform inline tagging. SXP provides capabilities to identify and classify IP packets to corresponding SGTs tracked in the mapping table within network devices. SPX uses peer-to-peer TCP connections over TCP port 64999.

Security Assertion Markup Language (SAML)   SAML is an open standard for exchanging authentication and authorization data between identity providers. SAML is used in many single sign-on (SSO) implementations.

Security group–based ACL (SGACL)   An ACL that implements access control based on the security group assigned to a user (for example, based on his role within the organization) and the destination resources. SGACLs are implemented as part of Cisco TrustSec policy enforcement. Cisco TrustSec is described in a bit more detail in the sections that follow. The enforced ACL may include both Layer 3 and Layer 4 access control entries (ACEs).

Security zone   A *security zone* is a collection of one or more inline, passive, switched, or routed interfaces (or ASA interfaces) that you can use to manage and classify traffic in different policies. Interfaces in a single zone could span multiple devices. Furthermore, you can also enable multiple zones on a single device to segment your network and apply different policies.

Sender Policy Framework (SPF)   SPF enables recipients to verify the sender's IP addresses by looking up DNS records that list authorized mail gateways for a particular domain. SPF is an industry standard defined in RFC 4408. SPF uses DNS TXT resource records. The Cisco ESA supports SPF to verify HELO/EHLO and MAIL FROM identity (FQDN).

SenderBase   A reputation service that enables you to control the messages that come through the Cisco ESA email gateway based on the senders' trustworthiness (reputation).

Simple Object Access Protocol (SOAP)   SOAP is a standards-based web services access protocol that was originally developed by Microsoft and has been used by numerous legacy applications for many years. SOAP exclusively uses XML to provide API services. XML-based specifications are governed by XML Schema Definition (XSD) documents. SOAP was originally created to replace older solutions such as the Distributed Component Object Model (DCOM) and Common Object Request Broker Architecture (CORBA).

SKEYID   The SKEYID is a string derived from secret material that is known only to the active participants in the IKE exchange.

Social Identity Provider (Social IdP)   A type of identity provider originating in social services like Google, Facebook, Twitter, and so on.

Spero   A machine learning–based technology that proactively identifies threats that were previously unknown. It uses active heuristics to gather execution attributes, and because the underlying algorithms come up with generic models, they can identify malicious software based on its general appearance rather than basing identity on specific patterns or signatures.

Split tunneling   After the tunnel is established, typically the default behavior of a VPN client is to encrypt traffic to all the destination IP addresses. This means that if an SSL VPN user wants to browse to a given site over the Internet, the packets are encrypted and sent to the VPN headend. After decrypting them, the VPN headend searches its routing table and forwards the packets to the appropriate next-hop IP address in clear text. These steps are reversed when traffic returns from the web server and is destined to the SSL VPN client. With split tunneling, the VPN headend notifies the client about the secured subnets. The VPN client, using the secured routes, encrypts only those packets that are destined for the networks behind the security appliance. With split tunneling, the remote computer is susceptible to threat actors, who can potentially take control over the computer and direct traffic over the tunnel.

Stealthwatch Management Console (SMC)   The main management application in the Cisco Stealthwatch solution that provides detailed dashboards and the ability to correlate network flow and events.

Stream cipher   A stream cipher is a symmetric key cipher (meaning the same key is used to encrypt and decrypt), where the plaintext data to be encrypted is done a bit at a time against the bits of the key stream, also called a cipher digit stream. The resulting output is a ciphertext stream. Because a cipher stream does not have to fit in a given block size, there may be slightly less overhead than with a block cipher that requires padding to complete a block size.

Stream Control Transmission Protocol (SCTP)   IPFIX uses SCTP, which provides a packet transport service designed to support several features beyond TCP or UDP capabilities. Many refer to SCTP as a simpler state machine than features provided by TCP with an "a la carte" selection of features.

Structured Threat Information eXpression (STIX)   STIX is a language to represent threat intelligence information. STIX details can contain data such as the IP addresses or domain names of command-and-control servers (often referred to C2 or CnC), malware hashes, and so on. STIX was originally developed by MITRE and is now maintained by OASIS. You can obtain more information at https://oasis-open.github.io/cti-documentation. You can review numerous examples of STIX content, objects, and properties at https://oasis-open.github.io/cti-documentation/stix/examples.

Supplicant   An entity that seeks to be authenticated by an authenticator. For example, this could be a client laptop connected to a switch port. An example of a supplicant software is the Cisco AnyConnect Secure Mobility Client.

Symmetric algorithm   An encryption algorithm that uses the same key to encrypt the data and decrypt the data.

T

TACACS+   Terminal Access Controller Access Control System Plus (TACACS+) is a proprietary protocol developed by Cisco. It also uses a client-server model, where the TACACS+ client is the access server and the TACACS+ server is the machine providing TACACS+ services (that is, authentication, authorization, and accounting).

TETRA   An AMP for Endpoints engine that provides a full client-side antivirus solution.

Threat   A threat is any potential danger to an asset. If a vulnerability exists but has not yet been exploited—or, more importantly, it is not yet publicly known—the threat is latent and not yet realized. Fire, someone stealing an asset, a cybersecurity attack are all examples of threats.

Threat intelligence   Threat intelligence is referred to as the knowledge about an existing or emerging threat to assets, including networks and systems. Threat intelligence includes context, mechanisms, indicators of compromise (IoCs), implications, and actionable advice. Threat intelligence is referred to as the information about the observables, the intent of the IoCs, and capabilities of internal and external threat actors and their attacks. Threat intelligence includes specifics on the tactics, techniques, and procedures of these adversaries.

Trusted Automated eXchange of Indicator Information (TAXII)   An open transport mechanism that standardizes the automated exchange of cyber-threat information. TAXII was originally developed by MITRE and is now maintained by OASIS. You can also obtain detailed information about TAXII at https://oasis-open.github.io/cti-documentation.

TrustSec Security Group Tag (SGT)   Cisco TrustSec is a solution for identity and policy enforcement. ISE can use Security Group Tags for authentication and authorization. SGTs are values that are inserted into the client's data frames by a network device (for example, a switch, firewall, or wireless AP). This tag is then processed by another network device receiving the data frame and used to apply a security policy.

U–W

uSeg EPG   A micro-segment in ACI is also often referred to as a uSeg EPG. You can group endpoints in existing application EPGs into new micro-segment (uSeg) EPGs and configure network or VM-based attributes for those uSeg EPGs. With these uSeg EPGs, you can apply dynamic policies. You can also apply policies to any endpoints within the tenant.

VLAN   One way to identify a local area network is to say that all the devices in the same LAN have a common Layer 3 IP network address and that they also are all located in the same Layer 2 broadcast domain. A virtual LAN (VLAN) is another name for a Layer 2 broadcast domain. VLANs are controlled by the switch. The switch also controls which ports are associated with which VLANs.

VLAN ACL   A VLAN access control list (VLAN ACL), also called a VLAN map, is not specifically a Layer 2 ACL; however, it is used to limit the traffic within a specific VLAN. A VLAN map can apply a MAC access list, a Layer 3 ACL, and a Layer 4 ACL to the inbound direction of a VLAN to provide access control.

Vulnerability   A vulnerability is a weakness in the system design, implementation, software, or code, or the lack of a mechanism. A specific vulnerability might manifest as anything from a weakness in system design to the implementation of an operational procedure. Vulnerabilities might be eliminated or reduced by the correct implementation of safeguards and security countermeasures.

Web Cache Communication Protocol (WCCP)   WCCP is a Cisco-developed content-routing protocol that provides a mechanism to redirect traffic flows in real time. It has built-in load balancing, scaling, fault tolerance, and service-assurance (failsafe) mechanisms.

Web Identity   Identifying characters obtained from an HTTP request (often these are retrieved from an authenticated email address).

Web Proxy Auto-Discovery (WPAD) protocol   You can advertise and configure clients with PAC settings by using the Web Proxy Auto-Discovery (WPAD) protocol. WPAD uses the auto-detect proxy settings found in every modern web browser.

White hat hackers   These individuals perform ethical hacking to help secure companies and organizations. Their belief is that you must examine your network in the same manner as a criminal hacker to better understand its vulnerabilities.

Windows Identity   This is how Active Directory in Microsoft Windows environments organizes user information.

Wrapper   A wrapper is a program used to combine two or more executables into a single packaged program. Wrappers are also referred to as binders, packagers, and EXE binders because they are the functional equivalent of binders for Windows Portable Executable files. Some wrappers only allow programs to be joined; others allow the binding together of three, four, five, or more programs. Basically, these programs perform like installation builders and setup programs. Besides allowing you to bind a program, wrappers add additional layers of obfuscation and encryption around the target file, essentially creating a new executable file.

WS-Federation   A common infrastructure (federated standard) for identity, used by web services and browsers on Windows Identity Foundation. Windows Identity Foundation is a framework created by Microsoft for building identity-aware applications.

Y–Z

YANG   YANG is an API contract language used in many networking devices. In other words, you can use YANG to write a specification for what the interface between a client and networking device (server) should be on a particular topic. A YANG model typically concentrates on the data that a client processes using standardized operations.

Zero trust   This concept assumes that no system or user will be "trusted" when requesting access to the corporate network, systems, and applications hosted on the premises or in the cloud. You must first verify their trustworthiness before granting access.

Zone-Based Firewall (ZBFW)   The Cisco IOS Zone-Based Firewall is a stateful firewall used in Cisco IOS devices. ZBFW is the successor of the legacy IOS firewall or the context-based access control (CBAC).

Answers to the "Do I Know This Already?" Quizzes and Q&A Sections

Do I Know This Already? Quiz Answers

Chapter 1

1. b
2. b
3. a
4. a
5. d
6. d
7. a
8. e
9. d
10. e
11. b
12. d
13. a
14. d
15. b
16. d
17. e
18. a

Chapter 2

1. a
2. a
3. c
4. d
5. a
6. b

7. d

8. c

9. a

10. b

Chapter 3

1. a

2. d

3. b

4. a

5. a

6. b

7. b

8. c

9. c

10. d

Chapter 4

1. a

2. b

3. a

4. a

5. b

6. a and b

7. c

8. d

9. e

10. c

11. a

12. f

13. a

14. a

15. d

Chapter 5

1. d

2. a

3. a

4. b

5. d

6. d

7. a

8. d

9. d

10. a

11. e

12. b

13. a and c

14. b

15. a

Chapter 6

1. d

2. e

3. c

4. e

5. d

6. a and b

7. d

8. c

9. d

10. a

11. a

12. d

Chapter 7

1. b

2. b

3. c

4. c

5. d

6. d

7. d

8. c

9. d

10. a

Appendix A: Answers to the "Do I Know This Already?" Quizzes and Q&A Sections   681

A

Chapter 8

1. e
2. a
3. a
4. c
5. a
6. b
7. d
8. a and c
9. d
10. d

Chapter 9

1. b
2. b
3. a
4. a
5. d
6. a
7. c
8. a and b
9. d
10. b

Chapter 10

1. d
2. b
3. b
4. c
5. a
6. a
7. b
8. d
9. b
10. b

Chapter 11

1. d
2. b
3. c
4. d
5. a
6. d
7. d
8. d
9. b
10. b

Review Question Answers

Chapter 1

1. d
2. a
3. e
4. a
5. c
6. a
7. e
8. a
9. a
10. b

Chapter 2

1. a, b, and c
2. a, b, and d
3. b, c, and d
4. a and b
5. a and c
6. a and d
7. c
8. b and c
9. d
10. a

Appendix A: Answers to the "Do I Know This Already?" Quizzes and Q&A Sections   683

A

Chapter 3

1. d
2. b
3. c
4. a
5. a
6. a
7. b
8. d
9. d
10. d

Chapter 4

1. a
2. d
3. a and b
4. d
5. b
6. d
7. a
8. d
9. c
10. d
11. a
12. a

Chapter 5

1. b
2. c
3. d
4. a
5. e
6. d
7. d
8. a and c
9. b
10. b

Chapter 6

1. a
2. b
3. b
4. d
5. e
6. b
7. a
8. c
9. c
10. a

Chapter 7

1. d
2. b
3. d
4. a
5. b
6. a
7. d
8. d
9. d
10. a

Chapter 8

1. a
2. b
3. a
4. d
5. b and d
6. a
7. d
8. a
9. c
10. d

A

Chapter 9

1. a
2. b
3. b
4. a
5. c
6. a
7. d
8. d
9. d
10. a

Chapter 10

1. c
2. a
3. a
4. d
5. d
6. d
7. a
8. c
9. d
10. c

Chapter 11

1. c
2. a
3. a
4. d
5. d

CCNP Security Core SCOR (350-701) Exam Updates

Over time, reader feedback allows Pearson to gauge which topics give our readers the most problems when taking the exams. To assist readers with those topics, the authors create new materials clarifying and expanding on those troublesome exam topics. As mentioned in the Introduction, the additional content about the exam is contained in a PDF on this book's companion website, at http://www.ciscopress.com/title/9780135971970.

This appendix is intended to provide you with updated information if Cisco makes minor modifications to the exam upon which this book is based. When Cisco releases an entirely new exam, the changes are usually too extensive to provide in a simple update appendix. In those cases, you might need to consult the new edition of the book for the updated content. This appendix attempts to fill the void that occurs with any print book. In particular, this appendix does the following:

- Mentions technical items that might not have been mentioned elsewhere in the book

- Covers new topics if Cisco adds new content to the exam over time

- Provides a way to get up-to-the-minute current information about content for the exam

Always Get the Latest at the Book's Product Page

You are reading the version of this appendix that was available when your book was printed. However, given that the main purpose of this appendix is to be a living, changing document, it is important that you look for the latest version online at the book's companion website. To do so, follow these steps:

Step 1. Browse to www.ciscopress.com/title/9780135971970.

Step 2. Click the Updates tab.

Step 3. If there is a new Appendix B document on the page, download this latest version.

> **NOTE** The downloaded document has a version number. Comparing the version of the print Appendix B (Version 1.0) with the latest online version of this appendix, you should do the following:
>
> - **Same version:** Ignore the PDF that you downloaded from the companion website.
> - **Website has a later version:** Ignore this Appendix B in your book and read only the latest version that you downloaded from the companion website.

Technical Content

The current Version 1.0 of this appendix does not contain additional technical coverage.

Index

B

F

J - K

L

P

T

W

X

Y - Z

REGISTER YOUR PRODUCT at CiscoPress.com/register
Access Additional Benefits and SAVE 35% on Your Next Purchase

- Download available product updates.
- Access bonus material when applicable.
- Receive exclusive offers on new editions and related products.
  (Just check the box to hear from us when setting up your account.)
- Get a coupon for 35% for your next purchase, valid for 30 days.
  Your code will be available in your Cisco Press cart. (You will also find
  it in the Manage Codes section of your account page.)

Registration benefits vary by product. Benefits will be listed on your account page
under Registered Products.

CiscoPress.com – Learning Solutions for Self-Paced Study, Enterprise, and the Classroom
Cisco Press is the Cisco Systems authorized book publisher of Cisco networking technology,
Cisco certification self-study, and Cisco Networking Academy Program materials.

At **CiscoPress.com** you can
- Shop our books, eBooks, software, and video training.
- Take advantage of our special offers and promotions (ciscopress.com/promotions).
- Sign up for special offers and content newsletters (ciscopress.com/newsletters).
- Read free articles, exam profiles, and blogs by information technology experts.
- Access thousands of free chapters and video lessons.

Connect with Cisco Press – Visit CiscoPress.com/community
Learn about Cisco Press community events and programs.

Cisco Press

ALWAYS LEARNING

PEARSON